W9-CKH-071

Organic Reactions

Organic Reactions

VOLUME 21

JOHN WILEY & SONS
NEW YORK · LONDON · SYDNEY · TORONTO

WH
547
O68A
16561

Published by John Wiley & Sons, Inc.

Library of Congress Catalogue Card Number: 42-20265

ISBN 0-471-19622-3

Printed in the United States of America.

10 9 8 7 6 5 4 3 2 1

PREFACE TO THE SERIES

In the course of nearly every program of research in organic chemistry the investigator finds it necessary to use several of the better-known synthetic reactions. To discover the optimum conditions for the application of even the most familiar one to a compound not previously subjected to the reaction often requires an extensive search of the literature; even then a series of experiments may be necessary. When the results of the investigation are published, the synthesis, which may have required months of work, is usually described without comment. The background of knowledge and experience gained in the literature search and experimentation is thus lost to those who subsequently have occasion to apply the general method. The student of preparative organic chemistry faces similar difficulties. The textbooks and laboratory manuals furnish numerous examples of the application of various syntheses, but only rarely do they convey an accurate conception of the scope and usefulness of the processes.

For many years American organic chemists have discussed these problems. The plan of compiling critical discussions of the more important reactions thus was evolved. The volumes of *Organic Reactions* are collections of chapters each devoted to a single reaction, or a definite phase of a reaction, of wide applicability. The authors have had experience with the processes surveyed. The subjects are presented from the preparative viewpoint, and particular attention is given to limitations, interfering influences, effects of structure, and the selection of experimental techniques. Each chapter includes several detailed procedures illustrating the significant modifications of the method. Most of these procedures have been found satisfactory by the author or one of the editors, but unlike those in *Organic Syntheses* they have not been subjected to careful testing in two or more laboratories.

Each chapter contains tables that include all the examples of the reaction under consideration that the author has been able to find. It is inevitable, however, that in the search of the literature some examples will be missed, especially when the reaction is used as one step in an extended synthesis. Nevertheless, the investigator will be able to use the tables and their accompanying bibliographies in place of most or all of the literature search so often required.

Because of the systematic arrangement of the material in the chapters and the entries in the tables, users of the books will be able to find information desired by reference to the table of contents of the appropriate chapter. In the interest of economy the entries in the indices have been kept to a minimum, and, in particular, the compounds listed in the tables are not repeated in the indices.

The success of this publication, which will appear periodically, depends upon the cooperation of organic chemists and their willingness to devote time and effort to the preparation of the chapters. They have manifested their interest already by the almost unanimous acceptance of invitations to contribute to the work. The editors will welcome their continued interest and their suggestions for improvements in *Organic Reactions*.

Chemists who are considering the preparation of a manuscript for submission to Organic Reactions are urged to write either secretary before they begin work.

CONTENTS

Organic Reactions

CHAPTER 1

FLUORINATION BY SULFUR TETRAFLUORIDE

G. A. Boswell, Jr., W. C. Ripka, R. M. Scribner, and C. W. Tullock

Central Research Department, Experimental Station, E. I. du Pont de Nemours and Company, Inc., Wilmington, Delaware

CONTENTS

* We thank Mrs. Aster Wu and W. T. Boyce, du Pont Experimental Station, for assistance in searching the chemical literature.

INTRODUCTION

Although sulfur tetrafluoride is a remarkably versatile reagent for selectively introducing fluorine atoms into organic compounds, its existence was doubted as late as 1950. A colorless gas boiling at −38°, sulfur tetrafluoride was probably first made by treatment of sulfur with uranium hexafluoride,[1] and subsequently by reaction of sulfur with cobalt(III) fluoride;[2] both products, however, were of questionable purity. In 1950 sulfur tetrafluoride was unequivocally synthesized by electrical decomposition of trifluoromethylsulfur pentafluoride,[3] but it was not until 1955 that gram amounts were prepared by passing elemental fluorine over sulfur at −75°, thus making sulfur tetrafluoride a viable compound.[4] A more practical laboratory synthesis involves heating sulfur dichloride at atmospheric pressure with sodium fluoride in acetonitrile at 50–70° in conventional glass equipment.[5, 6] Volatile

$$3\,SCl_2 + 4\,NaF \xrightarrow{CH_3CN} SF_4 + S_2Cl_2 + 4\,NaCl$$

sulfur tetrafluoride escapes as it is formed and is conveniently collected in a cooled trap.

With a practical synthesis of sulfur tetrafluoride in hand, the way was clear for the discovery of its remarkable fluorinating ability.

NATURE OF THE REACTION

The ability of sulfur tetrafluoride to replace carbonyl oxygen selectively with a *gem*-difluoro group is probably the most useful synthetic application of this reagent.[7] The reaction has broad scope and is effective with virtually all carbonyl compounds, including acid halides, aldehydes, ketones, and quinones. This is one of the very few reactions generally useful for introducing fluorine at a specific site in a molecule under mild conditions. Because

$$>C{=}O + SF_4 \rightarrow\ >CF_2 + SOF_2$$

carbonyl reactivity varies widely depending on the type of functional group of which it is a part (*e.g.*, aldehydes react more readily than do

[1] O. Ruff and A. Heinzelmann, *Z. Anorg. Chem.*, **72**, 63 (1911).

[2] J. Fischer and W. Jaenckner, *Z. Angew. Chem.*, **42**, 810 (1929).

[3] G. A. Silvey and G. H. Cady, *J. Amer. Chem. Soc.*, **72**, 3624 (1950).

[4] F. Brown and P. L. Robinson, *J. Chem. Soc.*, **1955**, 3147.

[5] W. C. Smith, C. W. Tullock, E. L. Muetterties, W. R. Hasek, F. S. Fawcett, V. A. Engelhardt, and D. D. Coffman, *J. Amer. Chem. Soc.*, **81**, 3165 (1959).

[6] C. W. Tullock, F. C. Fawcett, W. C. Smith, and D. D. Coffman, *J. Amer. Chem. Soc.*, **82**, 539 (1960).

[7] W. R. Hasek, W. C. Smith, and V. A. Engelhardt, *J. Amer. Chem. Soc.*, **82**, 543 (1960).

acid fluorides or esters), fluorination of a specific carbonyl group can be easily achieved in multifunctional molecules. The reaction is not restricted to carbonyl oxygen; the oxygen atom doubly bonded to arsenic, iodine, or phosphorus is similarly replaced by two fluorine atoms. The thiocarbonyl group is fluorinated in an analogous manner.

Replacement of hydroxyl groups of primary, secondary, and tertiary polyhalogenated alcohols, polynitroalcohols, carboxylic acids, and sulfonic acids with a fluorine atom is another very useful synthetic application of sulfur tetrafluoride. Certain enolic hydroxyl groups such as those found in tropolone are also replaced with fluorine by reaction with sulfur tetrafluoride at moderate temperatures. Linear polyfluoroalcohols react smoothly at room temperature to give the corresponding fluorides in high yield. Carboxylic acids react stepwise to give first acyl fluorides which may then undergo fluorination of the carbonyl group to yield trifluoromethyl derivatives. The first step occurs below room temperature, and conditions have now been found for carrying out the second step on sensitive molecules at temperatures as low as room temperature.

Although less success has been achieved with conventional alcohols as the result of side reactions and decomposition, it may be possible to find conditions under which the reaction will become useful for these substrates too.

Carboxylic acid derivatives such as anhydrides, esters, and salts give products like those from the acids or corresponding acid fluorides but require more vigorous conditions. Some anhydrides react without loss of the ring oxygen atom, *e.g.*, dichloromaleic anhydride gives 3,4-dichloro-2,2,5,5-tetrafluoro-2,5-dihydrofuran. Esters require temperatures up to 300° to give trifluoromethyl compounds. α,α-Difluoroethers and acyl fluorides appear to be intermediates. Esters will react with sulfur tetrafluoride at 130° when catalyzed by boron trifluoride or titanium tetrafluoride, but no reaction takes place with sulfur tetrafluoride and hydrogen fluoride at temperatures to 170°.

The amide group, however, is quite sensitive to sulfur tetrafluoride. Two types of amides may be distinguished: those with at least one nitrogen-hydrogen bond and those without such bonds. In the first group the carbon-nitrogen bond breaks to give an acyl fluoride which may react further. With amides having no nitrogen-hydrogen bond, the carbonyl-nitrogen bond may also be cleaved to give the acyl fluoride or retained to give the α,α-difluoroamine.

Compounds having carbon-nitrogen multiple bonds are converted by sulfur tetrafluoride to iminosulfur difluorides, unusual thermally stable substances of the general formula $RN{=}SF_2$. A variety of polyhalogenated compounds undergo exchange with sulfur tetrafluoride, as exemplified by the fluorination of chloropyrimidines. Although sulfur tetrafluoride adds to olefins with difficulty, and then only to fluoroolefins or fluorovinyl ethers (*e.g.*, $CH_3OCF{=}CF_2$), halogenated olefins are readily fluorinated by sulfur tetrafluoride in the presence of lead dioxide, presumably via lead tetrafluoride generated *in situ*. Primary amines and sulfur tetrafluoride react to form alkyl or aryl iminosulfur difluorides.

Several excellent surveys of sulfur tetrafluoride and its use in the fluorination of organic materials have appeared in the literature since 1960.[7–13]

MECHANISTIC CONSIDERATIONS

No systematic mechanistic studies of fluorination by sulfur tetrafluoride have yet been described. Hence the following discussion necessarily represents current tentative views, some of which appear contradictory.

The reaction of sulfur tetrafluoride with the hydroxyl group has been postulated to occur by elimination of hydrogen fluoride to give an alkoxysulfur trifluoride intermediate,[8, 14, 15] which then undergoes either an internal (S_Ni) or an external S_N2 displacement of OSF_3 by fluoride ion to form the carbon-fluorine bond.[8, 14, 16] There is most likely a continuum of mechanisms operative from S_Ni to S_N2.

The proposal that the mechanism involves an S_Ni decomposition of an alkoxysulfur trifluoride intermediate stems from the reaction's analogy to the conversion of alcohols to alkyl chlorides by thionyl chloride; here intermediate alkyl chlorosulfites decompose on heating to give alkyl

$$\gt C{-}OH + SF_4 \underset{+HF}{\overset{-HF}{\rightleftharpoons}} \gt C(O)(F)SF_2 \longrightarrow \gt C{-}F + SOF_2$$

[8] W. C. Smith, *Angew. Chem.*, **74**, 742 (1962).

[9] D. G. Martin, *Ann. N.Y. Acad. Sci.*, **145**, 161 (1967).

[10] P. Boissin and M. Carles, Commis. Energ. At. (Fr.), Serv. Doc., Ser. Bibliogr., No. **98**, (1967) (Fr.) [*C.A.*, **68**, 110866 (1968)].

[11] H. L. Roberts, *Quart. Rev.* (London), **15**, 30 (1961).

[12] H. L. Roberts in *Inorganic Sulfur Chemistry*, G. Nickless, Ed., p. 419, Elsevier, New York, 1968.

[13] S. M. Williamson in *Progress in Inorganic Chemistry*, G. A. Cotton, Ed., Vol. 7, pp. 39–77, Wiley–Interscience, New York, 1966.

[14] R. E. A. Dear and E. E. Gilbert, *J. Org. Chem.*, **33**, 819 (1968).

[15] K. Baum, *J. Amer. Chem. Soc.*, **91**, 4594 (1969).

[16] W. A. Sheppard and C. M. Sharts, *Organic Fluorine Chemistry*, pp. 164–166, Benjamin, New York, 1969.

chlorides and sulfur dioxide.[17, 18] The related decompositions of chloro-[17] and fluoro-formates[16] with elimination of carbon dioxide and formation of alkyl chlorides and alkyl fluorides, respectively, are also considered to involve collapse of cyclic intermediates by an S_Ni mechanism. Such cyclic intermediates are now considered to be ion pairs with a partial positive charge on the α-carbon atom.[17–19] These reactions, therefore, exhibit certain characteristics of S_N1 reactions.[17]

The ease with which sulfur tetrafluoride replaces a hydroxyl group by a fluorine atom would be expected to depend on the facility with which the reversibly formed alkoxysulfur trifluoride undergoes decomposition, and this factor in turn depends on the ability of the α-carbon atom to support a positive charge.[20] Experimental support for this view is provided by

$$\rangle C{-}OSF_3 \longrightarrow \left[\overset{\pm}{C}\begin{smallmatrix} \cdot\cdot\bar{O} \\ \cdot\cdot F \end{smallmatrix} \rangle SF_2 \right] \xrightarrow{F^-} \begin{cases} \rangle C{-}F + SOF_2 \\ F{-}C\langle + SOF_3^- \end{cases}$$

reports showing that, with increasing substitution of the α-carbon atom of polyhalogenated alcohols by strongly electron-withdrawing groups, correspondingly higher reaction temperatures are required. Thus 1H,1H,3H-tetrafluoro-1-propanol[21] and α-trifluoromethylbenzyl alcohol[14, 22, 23] are converted to the corresponding fluorides by treatment at 45–90°, whereas α,α-bis(trifluoromethyl)benzyl alcohol apparently requires 150° for a reasonable rate of reaction.[24] 2-Trichloromethylhexafluoro-2-propanol was recovered unchanged after overnight treatment with sulfur tetrafluoride at 200°; reaction finally occurred at 300° to give rearranged fluorinated products (pp. 17–18).[25]

An alkoxysulfur trifluoride intermediate with three strongly electron-withdrawing groups attached to the α-carbon atom is expected to form an ion pair with difficulty. Alcohols with trifluoromethyl groups attached

[17] E. S. Gould, *Mechanism and Structure in Organic Chemistry*, pp. 294–296, Holt, Rinehart and Winston, New York, 1959.

[18] J. Hine, *Physical Organic Chemistry*, pp. 114–116, McGraw-Hill Book Company, Inc., New York, 1956.

[19] J. D. Roberts and M. C. Caserio, *Basic Principles of Organic Chemistry*, p. 393, Benjamin, New York, 1964.

[20] D. J. Cram, *J. Amer. Chem. Soc.*, **75**, 332 (1953).

[21] W. R. Hasek and A. C. Haven, Jr., U.S. Pat. 2,980,740 (1961) [*C.A.*, **55**, 23342 (1961)].

[22] L. A. Wall and J. M. Antonucci, U.S. Pat. 3,265,746 (1966) [*C.A.*, **65**, 13602 *b* (1966)].

[23] J. M. Antonucci and L. A. Wall, *SPE* (*Soc. Plast. Engrs.*) *Trans.*, **3**, 225 (1963).

[24] W. A. Sheppard, *J. Amer. Chem. Soc.*, **87**, 2410 (1965).

[25] R. E. A. Dear, E. E. Gilbert, and J. J. Murray, *Tetrahedron*, **27**, 3345 (1971).

to the α-carbon atom fail to form carbonium ions; protonation takes place instead.[26]

$$CF_3-\underset{H}{\overset{H_2\overset{+}{O}}{C}}-CF_3 \qquad CF_3-\underset{H}{\overset{H_2\overset{+}{O}}{C}}-CH_3 \qquad C_6H_5-\underset{CF_3}{\overset{H_2\overset{+}{O}}{C}}-CF_3$$

In marked contrast, polyfluorinated tertiary alcohols having vinyl or ethynyl groups attached to the α-carbon atom react with remarkable ease, being converted at room temperature or lower to fluoroolefins[14, 27] or fluoroallenes,[14] respectively; the latter are accompanied by varying amounts of the unrearranged fluorinated alkyne derivatives (see, pp. 14–17).[14, 25]

$$R-\underset{R'}{\overset{OH}{C}}-C\equiv CH \xrightarrow{SF_4} R-\underset{R'}{\overset{F}{C}}-C\equiv CH \text{ and/or } R-\underset{R'}{C}=C=C\begin{matrix}H\\F\end{matrix}$$

where R and R′ contain halogen.

This markedly increased reactivity of polyfluorinated tertiary alcohols having unsaturated groups attached to the α-carbon atom is consistent with an S_Ni mechanism involving ion pairs. Resonance stabilization of the carbonium ion of the ion-pair intermediate facilitates ionization of the carbon-oxygen bond by compensating for the inductive destabilizing effects of the fluoroalkyl groups.

It has been stated that there is a rough correlation between the acidity of the hydroxyl group and its reactivity toward sulfur tetrafluoride[28] and the yield of the corresponding fluoride.[7] This is based on the facile conversion at room temperature, or lower, of carboxylic and sulfonic acids, which possess very acidic hydroxyl groups, to acyl and sulfonyl fluorides in high yield; only moderate yields of fluorinated products are obtained with less acidic compounds such as tropolone, which is converted to α-fluorotropone. Polyfluoroalcohols and polynitroalcohols, both being relatively acidic, are readily converted to the corresponding alkyl fluorides, but higher temperatures are often necessary. The less acidic conventional aliphatic alcohols give some of the corresponding alkyl fluorides, but alkyl ethers are major by-products.[7, 8]

Subsequent experience with a variety of polyfluorinated alcohols appears to be at variance with the contention that the reactivity of the

[26] Ref. 16, p. 323.

[27] B. M. Lichstein and C. Woolf, U.S. Pat. 3,472,905 (1969) [*C.A.* **71**, 123491 (1969)].

[28] K. Baum, *J. Org. Chem.*, **33**, 1293 (1968).

hydroxyl group toward sulfur tetrafluoride is greater for strongly acidic alcohols. Increasing the number of strongly electron-withdrawing substituents on the α-carbon atom of polyfluorinated alcohols actually necessitates a corresponding increase in reaction temperature notwithstanding the greater acidity of the hydroxyl group. Polyfluoroalcohols and perfluoroalcohols of the structure R_fCH_2OH, R_fCHOHR_f, and $(R_f)_3COH$ are highly acidic. The acidity of the hydroxyl group increases with increasing perfluoroalkyl substitution, so that perfluoro-*t*-butyl alcohol is more acidic than phenol.[29] Yet it is likely that this highly acidic alcohol would require a reaction temperature greater than 300° in view of the extreme reluctance with which 2-trichloromethylhexafluoro-2-propanol reacts. On the other hand, the nonhalogenated alcohol 2-hydroxy-2-methyl-3-butyne undergoes an explosively vigorous reaction (presumably at −78°) to give unrearranged 2-fluoro-2-methyl-3-butyne along with 2-methyl-1-buten-3-yne formed by dehydration.[14]

Very acidic hydroxyl groups are cleanly replaced with a fluorine atom by sulfur tetrafluoride, even though relatively high reaction temperatures are often required, indicating that side reactions such as ether formation are minimized with alcohols of this character. The clean reactions and good yields obtained with acidic alcohols may be the result of fortuitously selecting substrates which are thermally stable and unable to undergo side reactions and may not be related to acidity *per se*.* Since structural features which increase acidity, such as fluoroalkyl substituents on the α-carbon atom, also destabilize the carbonium ion of the ion pair intermediates, high acidity of alcohols might be expected to decrease rather than increase reactivity.

Alkoxysulfur trifluorides have been proposed as intermediates in the replacement of carbonyl groups by fluorine (Scheme 1).[7, 8] Since carbonyl

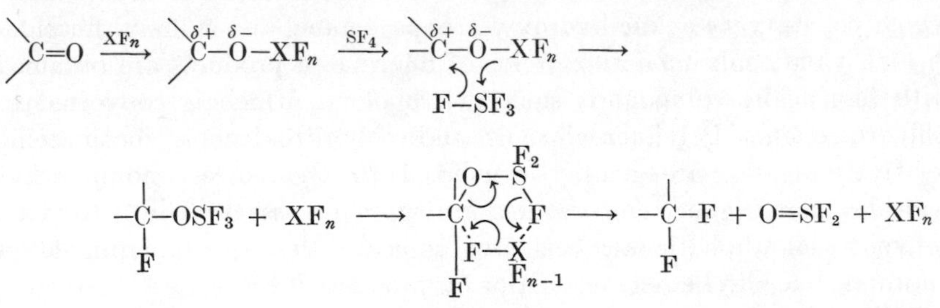

$$>C{=}O \xrightarrow{XF_n} >\overset{\delta+}{C}{-}\overset{\delta-}{O}{-}XF_n \xrightarrow{SF_4} -\overset{\delta+}{C}{-}\overset{\delta-}{O}{-}XF_n \quad (F{-}SF_3) \longrightarrow$$

$$-\underset{F}{\overset{|}{C}}{-}OSF_3 + XF_n \longrightarrow \left[-C(F)\cdots O{-}SF_2{-}F\cdots F{-}XF_{n-1}\right] \longrightarrow -\underset{F}{\overset{|}{C}}{-}F + O{=}SF_2 + XF_n$$

SCHEME 1

* Although published reports suggest that the reaction is not satisfactory with conventional aliphatic alcohols,[7, 8, 15, 28] there is a paucity of published experimental data to support this view.

[29] Ref. 120, p. 436.

reactions are catalyzed by Lewis acids, several variations of this pathway have been suggested to account for the acceleration of the reaction by substantial amounts of hydrogen fluoride.[30] In each, α-fluoroalkoxysulfur trifluoride intermediates have been invoked as direct precursors of carbon-fluorine bond formation (Scheme 2).

$$SF_4 + HF \longrightarrow \underset{F}{\overset{\delta-}{H}}\cdots F \cdots \overset{\delta+}{S}F_3 \xrightarrow{\;\rangle C=O\;} \text{[cyclic transition state: } \rangle C(\cdots F^{\delta-}\!-\!H \cdots F)(=O \cdots \overset{\delta+}{S}F_3)\text{]}$$

$$SF_4 + HF \longrightarrow {}^{-}HF_2 + \overset{+}{S}F_3 \xrightarrow{\;\rangle C=O\;} \rangle \overset{+}{C}\!-\!O\!-\!SF_3 \;(\leftarrow {}^{-}H\!-\!F_2) \longrightarrow \left[-\underset{F}{\overset{|}{C}}-OSF_3 \right]$$

$$\rangle C\!=\!O + HF \rightleftharpoons -\underset{F}{\overset{|}{C}}-OH \xrightarrow[HF]{SF_4} \left[-\underset{F}{\overset{|}{C}}-OSF_3 \right]$$

Scheme 2

The initial step is suggested to be polarization of the carbonyl group by coordination with a Lewis acid XF_n, followed by reaction with sulfur tetrafluoride to give the alkoxysulfur trifluoride intermediate.[7, 8] The role of XF_n may be assumed by sulfur tetrafluoride at higher temperature, but reactions were observed to proceed under much milder conditions when XF_n was hydrogen fluoride, boron trifluoride, antimony trifluoride, titanium tetrafluoride, or phosphorus pentafluoride. This led to the proposal that the ability of a substance to act as a catalyst could be correlated with its strength as a Lewis acid with respect to the carbonyl group. Additional evidence that the reaction may be initiated by coordination of a Lewis acid was believed to be provided by the behavior of compounds having two strongly electron-withdrawing groups attached to the carbonyl group. Such compounds, whose coordination to Lewis acids is presumably greatly decreased as a result of reduced Lewis base character of the carbonyl group, require more vigorous conditions than compounds with only one electron-withdrawing substituent. Thus 250° and boron

[30] (a) D. G. Martin and F. Kagan, *J. Org. Chem.*, **27**, 3164 (1962); (b) Upjohn Co., Brit. Pat. 930,888 (1963) [*C.A.*, **60**, 641 (1964)]; (c) F. Kagan and D. G. Martin, U.S. Pat. 3,211,723 (1965).

trifluoride as a catalyst were required to convert 2,2,3,3-tetrafluoropropionyl fluoride to 1(H)-perfluoropropane;[8] whereas propionic acid, which is rapidly converted to propionyl fluoride and hydrogen fluoride, gave 1,1,1-trifluoropropane in 89% yield after 8 hours at 150°.[7,*]

A study of the reaction of steroid ketones and carboxylic acids revealed that sulfur tetrafluoride containing 20% or more hydrogen fluoride effected transformations of these compounds to fluorinated derivatives at room temperature in moderate to high yields, thus markedly extending the synthetic utility of the reaction.[30] This observation suggested that the reaction might involve the prior formation of a complex of hydrogen fluoride and sulfur tetrafluoride[30] or, alternatively,[31] that hydrogen fluoride adds across the carbonyl group to give an α-fluoroalcohol[32] which is, in turn, converted to an α-fluoroalkoxysulfur trifluoride intermediate (Scheme 2). Sulfur tetrafluoride-hydrogen fluoride is generally the preferred reagent for converting aldehydes, ketones, and acyl fluorides to the corresponding *gem*-difluoro derivatives.

The replacement of carbonyl oxygen with two fluorine atoms by sulfur tetrafluoride may then be regarded as simply the reaction of an α-fluoroalcohol, or an α,α-difluoroalcohol in the case of an acyl fluoride. An equilibrium between the carbonyl compound, hydrogen fluoride, and the α-fluoroalcohol explains the need for an increasing concentration of hydrogen fluoride, which so dramatically improves the reaction in terms of very mild reaction conditions, good to high yields of desired product even with sensitive compounds, and the absence of side reactions or troublesome by-products. Moreover, it now becomes apparent that the factors which control the reactivity of alcohols similarly determine the reactivity of the carbonyl group.

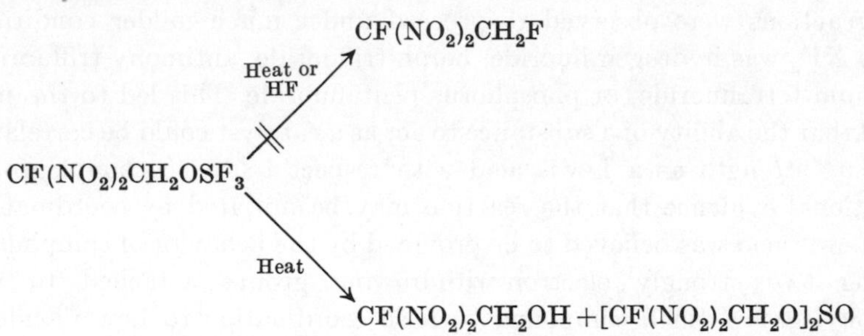

Any proposed mechanism which invokes alkoxysulfur trifluoride intermediates in the replacement of hydroxyl and carbonyl groups with

* More recent work (*cf.* ref. 30) indicates that this reaction could be carried out at room temperature simply by increasing the amount of hydrogen fluoride.

[31] Ref. 30a, footnote 11.

[32] S. Andreades and D. C. England, *J. Amer. Chem. Soc.*, **83**, 4670 (1961).

sulfur tetrafluoride must take into account the report that 2-fluoro-2,2-dinitroethoxysulfur trifluoride did not form 1,2-difluoro-1,1-dinitroethane on heating in an inert solvent or by treating with hydrogen fluoride.[15, *] Although alkoxysulfur trifluoride intermediates have been used to explain the replacement of carbonyl and hydroxyl groups by fluorine with sulfur tetrafluoride, none was known until 2-fluoro-2,2-dinitroethoxysulfur trifluoride was isolated. This colorless liquid, which is readily hydrolyzed by atmospheric moisture, was formed along with the expected 1,2-difluoro-1,1-dinitroethane and bis(2-fluoro-2,2-dinitroethyl)sulfite from the reaction of sulfur tetrafluoride with 2-fluoro-2,2-dinitroethanol.[15]

$$CF(NO_2)_2CH_2OH \xrightarrow{SF_4} CF(NO_2)_2CH_2OSF_3 + CF(NO_2)_2CH_2F + [CF(NO_2)_2CH_2O]_2SO$$

The finding that similar yields of 1,2-difluoro-1,1-dinitroethane and the alkoxysulfur trifluoride were obtained at 25 and 100° under otherwise identical conditions is difficult to explain on the basis of mechanisms involving 2-fluoro-2,2-dinitroethoxysulfur trifluoride as a direct intermediate. However, these results were preliminary and additional control experiments were needed. A mechanism was proposed in which the alkoxysulfur trifluoride is not a direct intermediate in the formation of a carbonfluorine bond.[15]

$$ROH \xrightarrow{SF_4} R\!-\!\overset{H}{\underset{+}{O}}\!-\!\underset{-}{S}F_4 \xrightarrow{-H^+} R\!<\!\overset{O}{\underset{F}{}}\!>\!\bar{S}F_3 \longrightarrow RF + SOF_3^-$$

$$R\!-\!\overset{H}{\underset{+}{O}}\!-\!\underset{-}{S}F_4 \xrightarrow{-HF} ROSF_3 \qquad SOF_3^- \longrightarrow SOF_2 + F^-$$

The failure to detect alkoxysulfur trifluoride intermediates in previous sulfur tetrafluoride reactions with alcohols (or carbonyl compounds) might be due to reversion to the starting alcohol (or carbonyl compound) on hydrolysis during workup.

Additional aspects of the reaction of sulfur tetrafluoride with alcohols and carbonyl compounds are discussed under "Scope and Limitations."

Fluorination of thiocarbonyl groups by sulfur tetrafluoride yields sulfur as a by-product. The reaction has been explained by a mechanism which parallels that proposed for a carbonyl group, the sulfur arising by an oxidation-reduction step.[33]

* Treatment of an alkoxysulfur trifluoride with hydrogen fluoride might be expected to reverse the reaction, *i.e.*, $ROH + SF_4 \rightleftharpoons ROSF_3 + HF$.

[33] R. J. Harder and W. C. Smith, *J. Amer. Chem. Soc.*, **83**, 3422 (1961).

$$>C{=}S + SF_4 \longrightarrow >C(F){-}SSF_3$$

$$2\ >C(F){-}S{-}SF_2(F) \longrightarrow 2\ >CF_2 + 2\,S_2F_2 \longrightarrow 3\,S + SF_4$$

The following speculative mechanism rationalizes the formation of iminosulfur difluorides from the reaction of sulfur tetrafluoride with compounds having carbon-nitrogen multiple bonds.[34]

$$RN{=}C{=}O + SF_4 \longrightarrow \left[\begin{array}{c} R{-}N{-}COF \\ | \\ F_2S{-}F \end{array}\right] \longrightarrow RN{=}SF_2 + COF_2$$

$$RC{\equiv}N + SF_4 \longrightarrow \left[\begin{array}{c} RCF{=}N \\ | \\ F{-}SF_2 \end{array}\right] \longrightarrow RCF_2N{=}SF_2$$

The sulfur tetrafluoride fluorination of organo-phosphorus and organo-arsenic compounds containing doubly bonded oxygen and hydroxyl groups may be viewed as analogous to fluorinations of carbonyl and organic hydroxyl groups.

SCOPE AND LIMITATIONS

Alcohols

Reaction with alcohols to give compounds having a fluorine atom at the former site of the hydroxyl group is an important synthetic use of sulfur tetrafluoride.[35] Highly acidic alcohols are converted in high yield to the corresponding fluorides. Conventional alcohols, however, to date have given mixtures of products, a fact that limits the applicability of this reaction.

Carboxylic and sulfonic acids, representing the extreme in highly acidic hydroxyl groups, are easily converted to acyl and sulfonyl fluorides (see pp. 30–31), whereas alcohols such as methanol, ethanol, and 2-propanol give mixtures of alkyl fluorides and alkyl ethers.[7, 8]

Interestingly, there are no published reports on the reaction with phenols or oximes as general classes.

In general, the reaction has been synthetically useful only with alcohols that are roughly equal to or greater than tropolone in acidity ($pK_a = 6.42$).

[34] W. C. Smith, C. W. Tullock, R. D. Smith, and V. A. Engelhardt, *J. Amer. Chem. Soc.*, **82**, 551 (1960); W. C. Smith, U.S. Pat. 2,862,029 (1958) [*C.A.* **53**, 9152 (1959)].

[35] C. M. Sharts and W. A. Sheppard, *Org. Reactions*, **21**, 125 (1973) survey the methods for introducing one fluorine atom into organic molecules.

Haloalcohols. Polyfluorinated alcohols with the exception of tertiary alcohols substituted by three strongly electron-withdrawing groups on the α-carbon atom react readily, and the reaction has proved to be generally useful for converting these alcohols to polyfluoroalkanes in good to excellent yields. Linear polyfluoroalcohols of 3 to 9 carbon atoms have been converted to the corresponding fluoroalkanes without rearrangement.[21]

Polyfluorinated hydrocarbons are technically important because they possess an unusual combination of properties, including excellent thermal stability and a high degree of resistance to oxidative degradation. Before the discovery that they could be made from polyfluorinated alcohols and sulfur tetrafluoride, they were prepared by processes which employed elemental fluorine or required relatively inaccessible reactants. α,α,ω-Trihydroperfluoroalkanols of the general formula $H(CF_2CF_2)_nCH_2OH$, prepared by telomerization of tetrafluoroethylene with methanol, are useful for conversion to polyfluoroalkanes. In general, alcohols of low molecular weight undergo reaction at lower temperatures than longer-chain alcohols. The usual range lies between 30 and 150°, but reaction can occur as low as −20° with alcohols of low molecular weight. Autogenous pressures are generally used for batch reactions, although pressure is not a critical factor. In a typical experiment, 1H,1H,7H-dodecafluoro-1-heptanol **(1)** and sulfur tetrafluoride in a molar ratio of about 1:2 were allowed to react in a pressure vessel at temperatures up to 80° for 10 hours to give a quantitative yield of 1H,1H,7H-tridecafluoroheptane.[21]

$$\underset{\mathbf{1}}{CHF_2(CF_2)_5CH_2OH} \xrightarrow{SF_4} CHF_2(CF_2)_5CH_2F + SOF_2 + HF$$

A preponderance of fluorine atoms (or other electronegative substituents) in the alcohol is apparently not required to obtain satisfactory yields of the corresponding fluorides; it appears necessary only to have electron-withdrawing atoms or groups in reasonable proximity to the hydroxyl group. The limits of separation remain to be determined. Thus α,α-bis-(trifluoromethyl)benzyl alcohols, prepared by the Friedel-Crafts reaction of hexafluoroacetone with benzene derivatives, are converted to heptafluoroisopropylbenzene derivatives in high yield.[24, 36] Treating α,α-bis-(trifluoromethyl)benzyl alcohol **(2)** with sulfur tetrafluoride at 150° for 8 hours afforded heptafluoroisopropylbenzene in 91% yield. Similarly,

[36] E. I. du Pont de Nemours and Co., Fr. Pat. 1,325,204 (1963) [*C.A.*, **59**, 11339 (1963)]; D. C. England, U.S. Pat. 3,236,894 (1966).

p-methyl-, *p*-isopropyl-, *p*-fluoro-, *p*-bromo-, *m*-bromo-, and *p*-nitro-α,α-bis(trifluoromethyl)benzyl alcohols were converted to the corresponding substituted heptafluoroisopropylbenzene derivatives in 49–70% yields.[24]

$$C_6H_6 + (CF_3)_2CO \xrightarrow{AlCl_3} \underset{\mathbf{2}}{C_6H_5C(CF_3)_2OH} \xrightarrow{SF_4} C_6H_5C(CF_3)_2F$$

Owing to reaction of sulfur tetrafluoride with the amino group. amino substitution on the phenyl ring considerably lowers the yield, *o*-Amino-α,α-bis(trifluoromethyl)benzyl alcohol gave *o*-heptafluoroisopropylaniline in only 8% yield. Reaction of the less hindered *para* isomer gave a slightly better yield (19%) of fluorinated product. Alkyl-substituted amines fare better, as shown by the conversion of *p*-dimethylamino-α,α-bis(trifluoromethyl)benzyl alcohol to *p*-heptafluoroisopropyl-N,N-dimethylaniline in 62% yield. Anhydrous hydrogen fluoride has been used as a combination catalyst and solvent for amino-substituted benzyl alcohols.[24] It protects the amino group by protonation.

α-(Pentafluorophenyl)ethyl alcohol gave *trans*-1,3-bis(pentafluorophenyl)-1-butene **(3)** rather than α-(pentafluorophenyl)ethyl fluoride owing to facile carbonium ion formation followed by elimination and

$$C_6F_5CHOHCH_3 \xrightarrow{SF_4} C_6F_5\overset{+}{C}HCH_3 \longrightarrow$$

$$C_6F_5CH{=}CH_2 \xrightarrow{C_6F_5\overset{+}{C}HCH_3} \underset{\mathbf{3}}{C_6F_5CH{=}CHCH(CH_3)C_6F_5}$$

subsequent attack by a second pentafluorophenyl cation.[37] However, α-(pentafluorophenyl)-α-hydroxyacetic acid afforded the polyfluorinated ethylbenzene derivative **4** in modest yield (35%) because carbonium ion formation and elimination are not possible. Sulfur tetrafluoride treatment

$$C_6F_5CHOHCO_2H + 3\,SF_4 \rightarrow \underset{\mathbf{4}}{C_6F_5CHFCF_3} + 3\,SOF_2 + 2\,HF$$

converted (pentafluorophenyl)methanol to the corresponding pentafluorobenzyl fluoride in 80% yield.[37]

Polyhalo-acetylenic and-allylic Alcohols. Polyfluorinated tertiary alcohols having alkenyl or alkynyl groups attached to the hydroxyl-bearing carbon atom are prone to undergo fluorination with rearrangement. Thus the acetylenic alcohols derived from hexafluoroacetone and other haloacetones give mainly fluorinated allenes along with small amounts

[37] B. G. Oksenenko, V. A. Sokolenko, V. M. Vlasov, and G. G. Yakobson, *Izv. Sib. Otd. Akad. Nauk SSR, Ser. Khim. Nauk*, **1970**, 102 [*C.A.*, **73**, 3558 (1970)].

of the corresponding fluorinated acetylenes formed by direct replacement of the hydroxyl group by fluorine. The fluorinated acetylenes were formed in minor yield from the acetylenic alcohols **5d**, **e**, and **f** and not at all from alcohols **5a**, **b**, and **c**. The major (or sole) product in each of

$$\underset{\mathbf{5}}{(CF_2Y)(CF_2Z)C(OH)C{\equiv}CH} \rightarrow (CF_2Y)(CF_2Z)C(F)C{\equiv}CH \text{ and/or } (CF_2Y)C(CF_2Z){=}C{=}CHF$$

a, Y = Z = F
b, Y = F, Z = Cl
c, Y = Z = Cl
d, Y = F, Z = H
e, Y = Z = H
f, Y = Cl, Z = H

these reactions is the corresponding allene.[14] Although these acetylenic alcohols are unusually inert toward acidic reagents, the fluorination reaction proceeds readily at room temperature and below.

The mechanism of allene formation has not been determined, but a reasonable pathway has been formulated.[14, 25] The first stage is formation of intermediate **6**, which can rearrange by an intramolecular S_Ni' pathway to the allene. The strongly electron-withdrawing influence of the trifluoromethyl groups would be expected to facilitate nucleophilic attack by

$$\underset{\mathbf{5a}}{(CF_3)_2C(OH)C{\equiv}CH} \xrightarrow[-HF]{SF_4} \underset{\mathbf{6}}{(CF_3)_2C(O{-}SF_2{-}F){-}C{\equiv}C{-}H} \longrightarrow (CF_3)_2C{=}C{=}CHF + SOF_2$$

fluorine on the terminal carbon of the triple bond by polarizing the triple bond, leaving the terminal carbon with a partial positive charge. The fact that the acetylenic alcohols **5a–f** can be distilled from hot concentrated sulfuric acid without structural change indicates that the carbonium ion is not readily formed and would certainly not be expected to form under the mild conditions used in the reaction with sulfur tetrafluoride.[14]

This mechanism is analogous to that proposed for the reaction of allylic alcohols with thionyl chloride to give rearranged allyl chlorides.[17] The S_Ni' reaction is reported to proceed more readily with thionyl chloride and allylic alcohols than with saturated secondary and tertiary alcohols.

$$RCH{=}CHCH_2OH \xrightarrow{SOCl_2} RCH{=}CH{-}CH_2{-}O{-}S(O){-}Cl \longrightarrow RCHClCH{=}CH_2 + SO_2$$

The formation of acetylenic products from acetylenic carbinols in which the fluorine atoms of the trifluoromethyl groups are replaced by chlorine and hydrogen atoms has been explained on the basis of two competing reactions. Replacing the fluorine atoms by atoms of reduced electronegativity considerably lowers the positive character of the terminal acetylenic carbon atom, retarding the intramolecular production of the allene and allowing an intermolecular displacement to compete.[14] A

$$F^- + RR'C(C{\equiv}CH){-}O{-}SF_2{-}F \longrightarrow RR'CF{-}C{\equiv}CH + SOF_2 + F^-$$

$$(R', R \neq CF_3)$$

reasonable alternative would be a competing S_Ni intramolecular process becoming significant as the electrophilic character of the terminal acetylenic carbon atom decreases and the ability of the α-carbon atom to support a positive charge increases.

$$RR'C(C{\equiv}CH){-}O{-}SF_2F \longrightarrow RR'CF{-}C{\equiv}CH + SOF_2$$

Similar results are observed when the ethynyl hydrogen atom is replaced by a chlorine atom. Thus in the chloroethynylated alcohol **7** the influence of the two trifluoromethyl groups is sufficiently offset by the chlorine atom that the alkyne is the predominant product. The partially fluorinated alcohol yields only the acetylene derivative (Eq. 1).[14] Thus it appears that a high degree of positive charge on the terminal

$$\underset{\mathbf{7}}{(CF_3)_2C(OH)C{\equiv}CCl} \xrightarrow{SF_4} (CF_3)_2CFC{\equiv}CCl + (CF_3)_2C{=}C{=}CClF$$

acetylenic carbon atom is required for an S_Ni' reaction leading to the isomeric fluorinated allene derivative.

$$(CHF_2)_2C(OH)C{\equiv}CCl \xrightarrow{SF_4} (CHF_2)_2CFC{\equiv}CCl \qquad \text{(Eq. 1)}$$

Significantly, reaction of the nonhalogenated acetylenic alcohol, dimethylethynylcarbinol, gave no allene but did give a fluoroacetylene and a vinylacetylene derivative in an explosively vigorous reaction.[14]

$$(CH_3)_2C(OH)C{\equiv}CH \xrightarrow{SF_4} (CH_3)_2CFC{\equiv}CH + CH_2{=}C(CH_3)C{\equiv}CH$$

As might be expected, reaction of sulfur tetrafluoride with allylic tertiary alcohols (prepared from vinylmagnesium halides and perhalogenated acetones) results in an analogous rearrangement. The sole products are perhaloalkyl-substituted fluoroolefins with the fluorine atoms attached to primary carbon atoms (Eq. 2).[14, 27] Reaction of sulfur tetrafluoride

$$(CF_3)_2C(OH)CH{=}CH_2 + SF_4 \rightarrow (CF_3)_2C{=}CHCH_2F + SOF_2 + HF \qquad \text{(Eq. 2)}$$

with 1,1,1-trifluoro-2-trifluoromethyl-3-buten-2-ol at room temperature for 17 hours gave 1,1,1-trifluoro-2-trifluoromethyl-4-fluoro-2-butene in 59% yield. This reaction may involve initial replacement of the hydroxyl group by a fluorine atom followed by rearrangement to give a stable internal olefin with the fluorine attached to the primary or terminal carbon atom.[27] However, in light of the results of the reaction of sulfur tetrafluoride with halogenated dialkylethynylcarbinols, an intramolecular S_Ni' mechanism would appear more reasonable.

$$(CF_3)_2C(CH{=}CH_2)O{-}SF_2{-}F \longrightarrow (CF_3)_2C{=}CHCH_2F + SOF_2$$

Saturated Polyhalotertiary Alcohols. In contrast to unsaturated tertiary alcohols, saturated halomethyl-substituted tertiary alcohols react very slowly with sulfur tetrafluoride at ambient temperature. Several hours at 90° are required in most instances for complete conversion. In the case of *t*-butyl alcohol with two of the methyl groups substituted by two or more fluorine atoms, fluorination does not occur, and olefinic products are formed. Under these conditions, 1,1,1,3,3,3-hexafluoro-2-methyl-2-propanol was converted to 1,1-bis(trifluoromethyl)ethylene.[25] No addition of hydrogen fluoride to the double bond was observed. The less fluorinated alcohols, 1,1,1,3,3-pentafluoro-2-methyl-2-propanol and

$$(CF_3)_2C(OH)CH_3 \xrightarrow[-HF]{SF_4} F_3C{-}C(CF_3)(H_2C{-}H){-}O{-}SF_2{-}F \longrightarrow (CF_3)_2C{=}CH_2 + HF + SOF_2$$

1,1,3,3-tetrafluoro-2-methyl-2-propanol, reacted similarly. In each case, however, a small amount of by-product resulting from hydrogen fluoride addition to the double bond was formed.

Interestingly, replacement of a trifluoromethyl group with a chlorodifluoromethyl group in such alcohols completely changes the course of the reaction. Rearranged fluorinated products are formed in which the incoming fluorine has become attached to the carbon atom originally bearing the chlorine atom and the chlorine has migrated to the former site of the departed hydroxyl group. In this way tertiary alcohol **8** was converted to a rearranged polyhaloalkane at 90°.[38]

$$\underset{\mathbf{8}}{(CClF_2)_2C(OH)CH_3} \xrightarrow{SF_4,\ 90°} \underset{(65\%)}{(CClF_2)(CF_3)C(Cl)CH_3}$$

[38] Fr. Pat. 1,370,349 (1964) [*C.A.* **62** 3935 (1965)]; E. E. Frisch, U.S. Pat. 3,322,840 (1967).

Similarly, 1-chloro-1,1,3,3,3-pentafluoro-2-methyl-2-propanol **(9)** was converted to 1,1,1,3,3,3-hexafluoro-2-methyl-2-chloropropane.[25] Thus, when tertiary alcohol **9** and sulfur tetrafluoride were allowed to react at 50° for 16 hours, an acyl fluoride believed to be 1-chloro-1-trifluoromethylpropionyl fluoride **(10)** was isolated as the major fraction. Its characterization was based on spectral properties and its conversion to 1,1,1,3,3,3-hexafluoro-2-methyl-2-chloropropane by prolonged treatment (about 72 hours) with sulfur tetrafluoride at 50°. A mechanism has been proposed to account for this surprising observation.[25]

$$\underset{\mathbf{9}}{CClF_2C(OH)(CH_3)CF_3} \xrightarrow{SF_4, 50°} \underset{\mathbf{10}}{F_3C-C(Cl)(CH_3)-COF} \xrightarrow{SF_4, 50°} (CF_3)_2C(CH_3)Cl$$

2-Trichloromethylhexafluoro-2-propanol **(11)** was recovered unchanged after overnight treatment with sulfur tetrafluoride at 200°. At 300°, reaction occurred giving 2-chloro-2-(dichlorofluoromethyl)hexafluoropropane and 2-chloro-2-(chlorodifluoromethyl)hexafluoropropane in 53 and 10% yield, respectively. A mechanism has been proposed for the

$$\underset{\mathbf{11}}{(CF_3)_2C(OH)CCl_3} \xrightarrow{SF_4, 300°} \underset{\text{(Major)}}{(CF_3)_2C(CCl_2F)Cl} + \underset{\text{(Minor)}}{(CF_3)_2C(CClF_2)Cl}$$

formation of the major product. The minor product is probably formed from the major product by simple halogen exchange of chlorine with fluorine (see p. 44).[25]

Nitroalcohols. The conversion of 2,2,2-trinitroethanol to 2,2,2-trinitrofluoroethane in 63% yield after 4 days at ambient temperature is in keeping with the observation that the hydroxyl group of an alcohol bearing strongly electron-withdrawing substituents undergoes a facile replacement by fluorine with sulfur tetrafluoride.[28] 2,2-Dinitropropanol and 2,2-dinitro-1,3-propanediol are reported to be less reactive. The former gave 2,2-dinitro-1-fluoropropane in 46% yield after 20 hours at 85–90°. The latter gave 2,2-dinitro-1,3-difluoropropane in 62% yield after 8 hours at 85–90°. The reported failure of 2-methyl-2-nitropropanol to react after an 8-hour reaction period at 110° is difficult to understand.[28]

2-Fluoro-2,2-dinitroethanol was observed to give 2-fluoro-2,2-dinitroethoxysulfur trifluoride as well as bis(2-fluoro-2,2-dinitroethyl) sulfite and the expected 1,2-difluoro-1,1-dinitroethane (Eq. 3).[15] The isolation of the alkoxysulfur trifluoride is particularly pertinent because such

$$CF(NO_2)_2CH_2OH \xrightarrow{SF_4} CF(NO_2)_2CH_2OSF_3 + [CF(NO_2)_2CH_2O]_2SO + CF(NO_2)_2CH_2F \quad \text{(Eq. 3)}$$

compounds have been invoked as intermediates in mechanisms proposed to explain hydroxyl and carbonyl reactions with sulfur tetrafluoride (see pp. 5–12).

Miscellaneous Alcohols. Tropolone reacted with sulfur tetrafluoride to give 2-fluorocycloheptatrienone in modest yield. However, 3,5,7-tribromotropolone gave 3,5,7-tribromo-2-fluorocycloheptatrienone in 57% yield under comparable conditions.

$$\text{Tropolone} \xrightarrow[60^\circ/10\ \text{hr}]{SF_4} \text{2-fluorocycloheptatrienone}\ (28\%)$$

$$\text{3,5,7-Tribromotropolone} \xrightarrow[60^\circ/8\ \text{hr}]{SF_4} \text{3,5,7-tribromo-2-fluorocycloheptatrienone}\ (57\%)$$

The reaction of sulfur tetrafluoride with glycolic acid at 160° gave 1,1,1,2-tetrafluoroethane in 48% yield.[7]

An Alternative Method for Direct Replacement of Hydroxyl Groups by Fluorine. Diethyl-(2-chloro-1,1,2-trifluoroethyl)amine (p. 125), prepared from diethylamine and 1-chloro-1,2,2-trifluoroethylene, is a useful reagent for converting primary or secondary alcohols to the corresponding fluoro compounds.[35] The reagent reacts with primary and secondary hydroxysteroids in methylene chloride at 25° to give fluorine

$$\text{3-hydroxy-}\Delta^5\text{-steroid-17-one} + (C_2H_5)_2NCF_2CHFCl \longrightarrow \text{3-fluoro-}\Delta^5\text{-steroid} \qquad \text{(Eq. 4)}$$

derivatives with essentially complete inversion of configuration except with Δ^5-3β-hydroxysteroids, where retention of configuration is observed. Thus 3β-hydroxy-5-androsten-17-one gave the corresponding 3β-fluoro derivative in 78% yield (Eq. 4).[39] Other reports indicate that the reaction is sensitive to solvent and temperature effects which can bring about elimination and rearrangement.[40]

The reaction of this reagent and polyfluorinated alcohols apparently has not been investigated.

Aldehydes and Ketones

Much of the interest in *geminal* difluoro compounds stems from their potential value as pharmaceuticals,[41, 42] particularly among the many fluoro steroids that have been prepared.[43–46]

Measurement of the conformational properties of cyclic compounds has been greatly facilitated by examination of the ^{19}F resonance spectra of their *gem*-difluoro derivatives.[47,48] One advantage of this "fluorine labeling" technique is that it is usually possible to work with the relatively large chemical shift differences between axial and equatorial fluorine atoms.

In pioneering work on the use of sulfur tetrafluoride in organic synthesis, *gem*-difluorides were prepared from aldehydes and ketones lacking α-hydrogens by reaction with sulfur tetrafluoride at high temperatures (generally 150–200°).[5, 7, 49] However, aldehydes possessing α-hydrogen

$$OHC\text{-}C_6H_4\text{-}CHO \xrightarrow[150°,\ 8\ hr]{SF_4} F_2CH\text{-}C_6H_4\text{-}CHF_2 \quad (88\%)$$

$$CH_3(CH_2)_nCHO \xrightarrow[50\text{–}60°,\ 8\ hr]{SF_4} CH_3(CH_2)_nCHF_2 \quad n=0\ (35\%),\ n=5\ (43\%)$$

atoms had to be treated at much lower temperatures to prevent charring, with the consequence that yields of difluorides were relatively low.[7]

[39] D. E. Ayer, *Tetrahedron Lett.*, **1962**, 1065.

[40] L. H. Knox, E. Velarde, S. Berger, D. Caudriello, and A. D. Cross, *Tetrahedron Lett.*, **1962**, 1249; *J. Org. Chem.*, **29**, 2187 (1964).

[41] W. R. Sherman, M. Freifelder, and G. R. Stone, *J. Org. Chem.* **25**, 2048 (1960).

[42] M. P. Mertes and S. E. Saheb, *J. Med. Chem.*, **6**, 619 (1963).

[43] J. Tadanier and W. Cole, *J. Org. Chem.*, **26**, 2436 (1961).

[44] G. A. Boswell, Jr., A. L. Johnson, and J. P. McDevitt, *J. Org. Chem.*, **36**, 575 (1971).

[45] E. I. du Pont de Nemours and Co., Belg. Pat. 765,554 (1971).

[46] A. L. Johnson, *J. Med. Chem.*, **15**, 360, 784, 854 (1972); *Steroids*, **20**, 263 (1972).

[47] J. D. Roberts, *Chem. Brit.*, **2**, 529 (1966), and later papers.

[48] K. Grychtol, H. Musso, and J. F. M. Oth, *Chem. Ber.* **105**, 1798 (1972).

[49] S. A. Fuqua, R. M. Parkhurst, and R. M. Silverstein, *Tetrahedron*, **20**, 1625 (1964).

An even more severe limitation was that ketones with α-hydrogen atoms, although nearly as susceptible to charring, required higher temperatures for reaction with sulfur tetrafluoride. For example, at 39°, cyclohexanone was converted by 1 molar equivalent of sulfur tetrafluoride to 1,1-difluorocyclohexane in 31% yield; at 50°, extensive charring occurred.[7]

Benzophenone was also found to be quite resistant to reaction with sulfur tetrafluoride, and temperatures considerably higher than those employed for other ketones were required. However, the yield of diphenyldifluoromethane was increased from 10 to 97% by addition of hydrogen

$$C_6H_5COC_6H_5 \xrightarrow[180°]{SF_4,\ HF} C_6H_5CF_2C_6H_5$$

fluoride (20 mole% relative to ketone).[7] Lewis acids such as boron trifluoride, arsenic trifluoride, and titanium tetrafluoride were even more potent catalysts.[7]

This catalytic effect was exploited in the preparation of *gem*-difluoro steroids from steroid ketones in yields ranging from 2 to 48% under very mild conditions.[43] Reactions were carried out at 40° in chloroform containing hydrogen fluoride or boron trifluoride. Hydrogen fluoride was generated *in situ* from sulfur tetrafluoride and ethanol by doping the chloroform with 0.75–3% ethanol prior to reaction, the amount depending on the reactivity of the particular ketone involved. Although catalysis with boron trifluoride in one instance gave a higher yield of difluorosteroid, in general its use was accompanied by considerable resinification and low recovery of starting material.

Further improvements in the reaction of steroid aldehydes and ketones with sulfur tetrafluoride were devised[30] and are the best available today for nonsteroid as well as steroid aldehydes and ketones. Substantial amounts of hydrogen fluoride conveniently generated *in situ* by hydrolysis ($SF_4 + H_2O \rightarrow SOF_2 + 2\ HF$) are used and the reaction is carried out

C_8H_{17} O H $\xrightarrow{SF_4,\ HF}$ F_2 H

at 10–30° in an inert solvent, usually methylene chloride. Thus cholestan-3-one at 10° overnight with excess sulfur tetrafluoride containing 22 mole% hydrogen fluoride gave 3,3-difluorocholestane in 78% yield. In contrast to the uncatalyzed reaction of cyclohexanone mentioned earlier, conditions similar to the above gave pure 1,1-difluorocyclohexane in 70% yield.[50]

Many steroid ketones and a few steroid aldehydes have been treated with the sulfur tetrafluoride-hydrogen fluoride reagent. These examples provide an excellent means of ascertaining the effects of small differences in molecular environment on carbonyl reactivity, especially when the behavior of two or more carbonyl groups in the same molecule can be compared. The order of decreasing reactivity of steroid carbonyl functions toward sulfur tetrafluoride-hydrogen fluoride is: 3-ketone $\cong$ 6-formyl > 17-ketone > 20-ketone of an 11-keto steroid > 20-ketone of an 11-deoxy steroid > conjugated 3-ketone > 11-ketone.[30] This order approximates the relative reactivity of steroid carbonyl groups in addition reactions, *e.g.*, in cyanohydrin formation.[51]

The low reactivity of conjugated steroid 3-ketones is illustrated by androst-4-ene-3,17-dione, which on treatment with excess sulfur tetrafluoride containing 20 mole% hydrogen fluoride at room temperature overnight in methylene chloride afforded the corresponding 17,17-difluoro steroid in 63% yield and recovered starting material in 4% yield.[30]

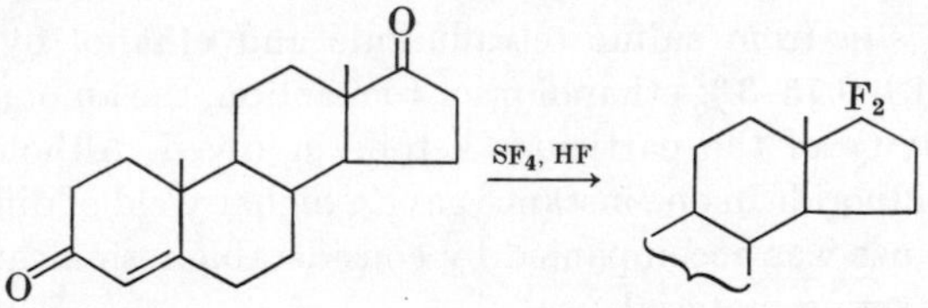

SF_4, HF

Under similar conditions 5α-androst-1-ene-3,11,17-trione gave mainly the corresponding 17,17-difluoro steroid and a small amount of what probably is the 3,3,17,17-tetrafluoro steroid.[30]

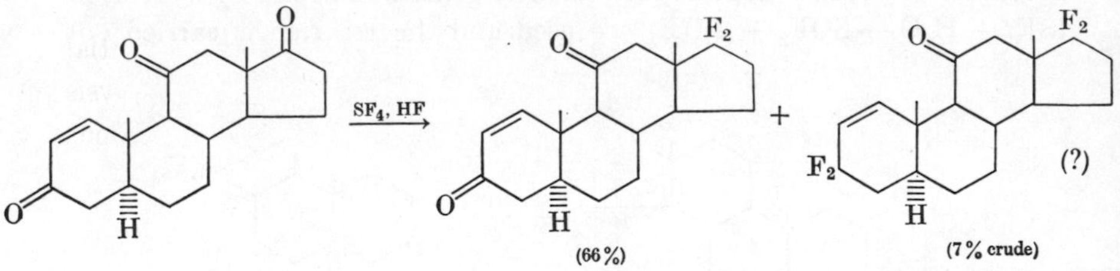

SF_4, HF

(66%)

(?) (7% crude)

Pregnane 20-ketones are more susceptible to fluorination with sulfur tetrafluoride than are conjugated 3-ketones, but they are more resistant

[50] D. R. Strobach and G. A. Boswell, Jr. *J. Org. Chem.*, **36**, 818 (1971).

[51] J. F. W. Keana in *Steroid Reactions*, C. Djerassi, Ed., pp. 53–56, Holden-Day, Inc., San Francisco, 1963; O. H. Wheeler and J. L. Mateos, *Can. J. Chem.*, **36**, 712 (1958).

than 17-ketones. Thus progesterone and 11-keto progesterone gave the corresponding 20,20-difluorosteroids in modest yields. Comparison of the

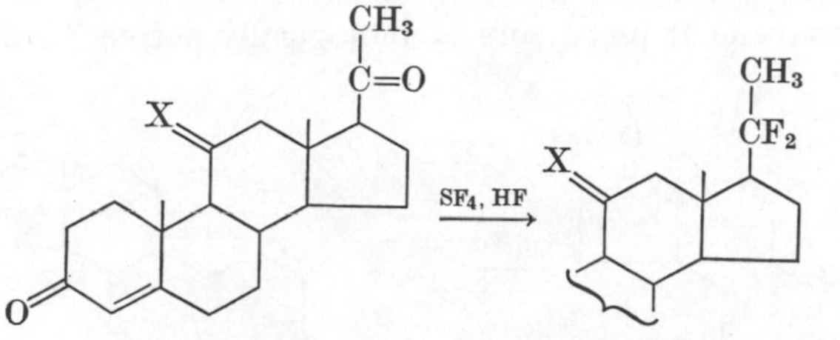

$\xrightarrow{SF_4,\ HF}$

X = H_2, X = O → X = H_2 (36%), X = O (37%)

yields of 20,20-difluorosteroids obtained from the corresponding 3α-acetoxy-5β-pregnan-20-ones under similar conditions suggests that in this series the 11-keto group exerts a favorable influence on the reaction of the C_{20} carbonyl group.[30]

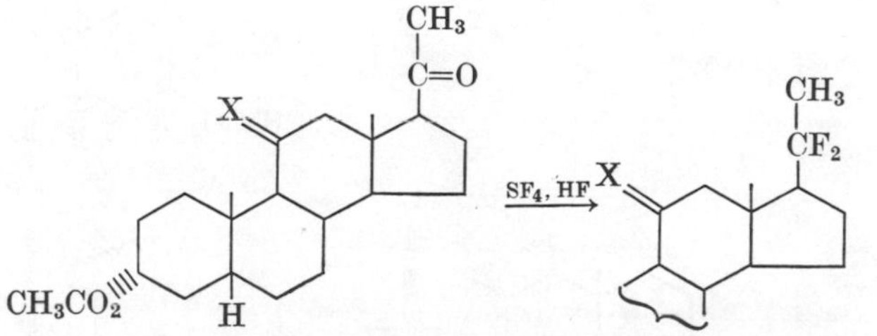

$\xrightarrow{SF_4,\ HF}$

X = H_2, X = O → X = H_2 (23%), X = O (58%)

Sulfur tetrafluoride reacts preferentially with the 6-keto group of 5α-fluoro-6-ketones[35] containing additional carbonyl groups at C_{17} or C_{20}.[52] However, by employing longer reaction times and larger amounts of hydrogen fluoride, carbonyl groups at C_{17} or C_{20} can also be converted to the corresponding *gem*-difluoro groups. Treatment of dione **12** with excess sulfur tetrafluoride containing about 11 mole % hydrogen fluoride in methylene chloride for 10 hours at 20° gave pure trifluoride by simple

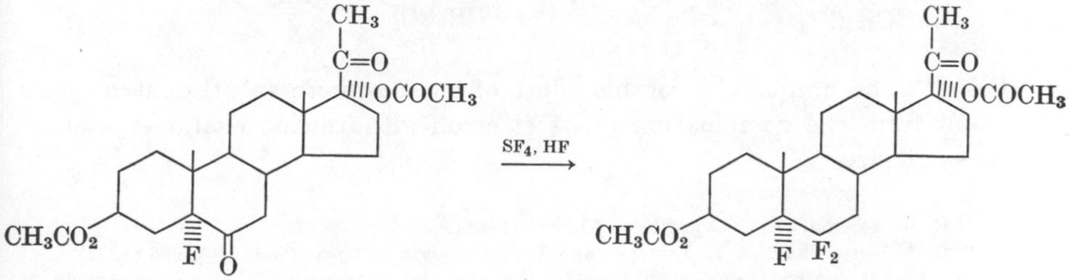

$\xrightarrow{SF_4,\ HF}$

12 → (67%)

crystallization. Dione **13** and excess sulfur tetrafluoride containing 47 mole% hydrogen fluoride for 20 hours at 20° gave mainly the pentafluoride and a small amount of the trifluoride.[52] 5α,6,6-Trifluoro-3β-acetates are convenient precursors to biologically active 6,6-difluoro-Δ^4-3-ketones.[44, 52, 53]

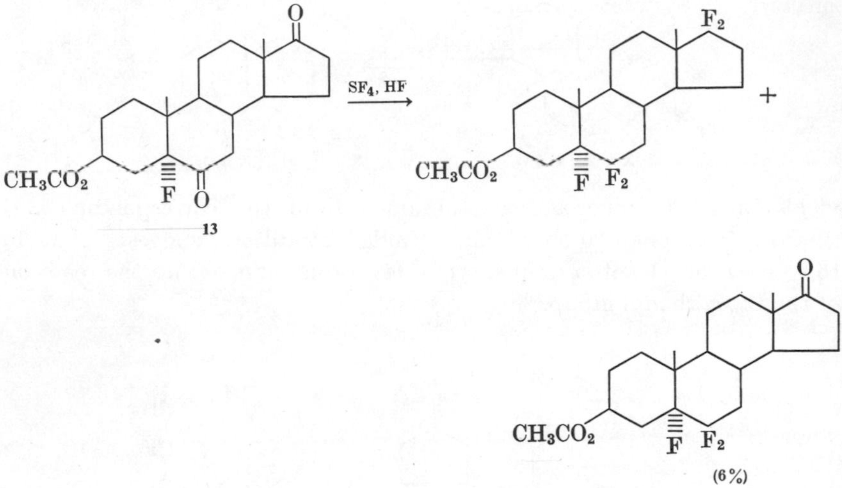

CH_3CO_2 ... **13** $\xrightarrow{SF_4,\ HF}$ CH_3CO_2 ... F F_2 + CH_3CO_2 ... F F_2 (6%)

CH_3CO_2 ... F F_2 → → O ... F_2

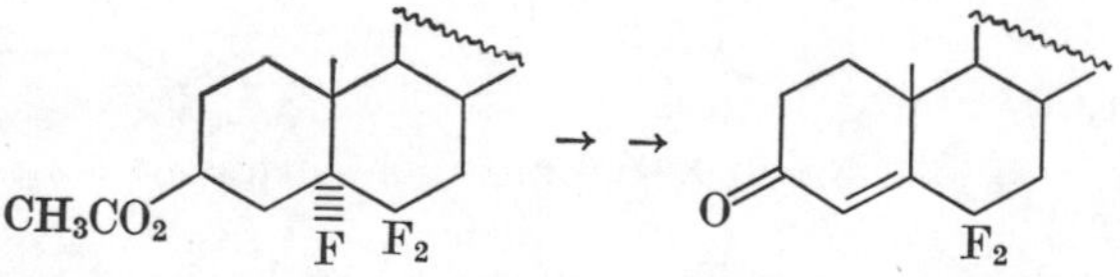

Steroid 5α-fluoro-6-ketones react more readily with sulfur tetrafluoride than do the corresponding 5α-hydro-6-ketones. For a series of 5α-X-pregnane-6,20-diones, the order of decreasing yields of the corresponding 5α-X-6,6-difluorosteroids was found to be X = F > H > Cl ≫ Br,

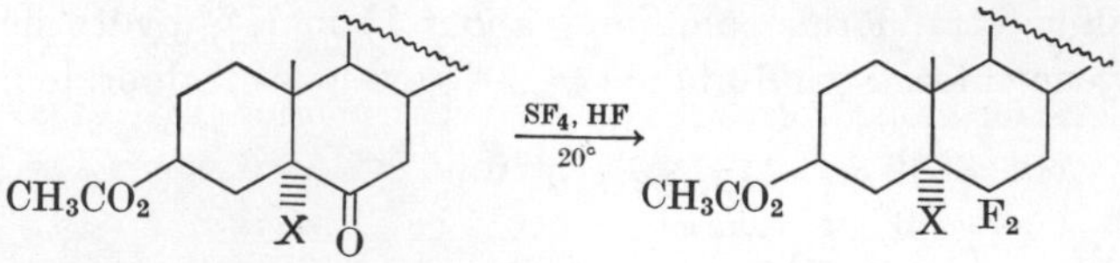

CH_3CO_2 ... X O $\xrightarrow[20°]{SF_4,\ HF}$ CH_3CO_2 ... X F_2

OAc.[54a] The uniquely favorable effect of the 5α-fluoro substituent may result from the combination of its electron-withdrawing character and its small size.

[52] G. A. Boswell, Jr., *J. Org. Chem.*, **31**, 991 (1966).

[53] G. A. Boswell, Jr., A. L. Johnson, and J. P. McDevitt, *Angew. Chem.*, **83**, 116 (1971).

[54] (a) R. M. Scribner, unpublished results from this laboratory; (b) R. M. Scribner and L. G. Wade, Jr., unpublished results from this laboratory.

Steroid aldehydes are quite reactive toward sulfur tetrafluoride. The 6α-formyl steroid **14** was converted by sulfur tetrafluoride containing only 0.6 mole % hydrogen fluoride in methylene chloride and tetrahydrofuran to the corresponding 6α-difluoromethyl derivative in better than 60 % yield.[55] A higher concentration of hydrogen fluoride gave a complex mixture which includes fluoroethers arising from addition of hydrogen

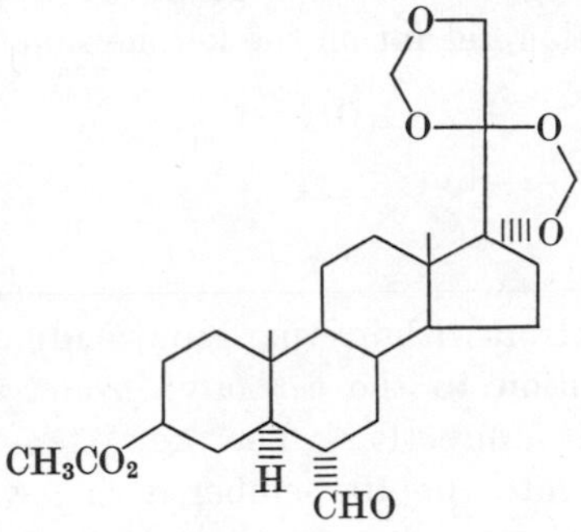

CH_3CO_2 H CHO OHC H

14 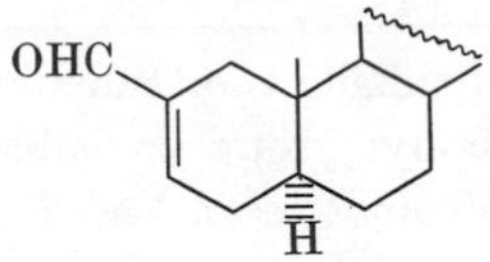 **15**

fluoride to the bismethylenedioxy group. In contrast to conjugated ketones, which are unreactive toward sulfur tetrafluoride, a conjugated aldehyde (**15**) has been converted to the Δ^2-difluoromethyl steroid in good yield.[56]

Unconjugated steroid 3-ketones (*e.g.*, 5α-cholestan-3-one) are among the most reactive of steroid carbonyl compounds.

Nonsteroidal alicyclic ketones which have been converted to the corresponding *gem*-difluorides by sulfur tetrafluoride-hydrogen fluoride include the homologous cycloalkanones ranging from cyclobutanone to cyclooctanone and cyclododecanone.[50, 57, 58] In contrast to cycloheptanone

$$(CH_2)_nC{=}O$$
$$n = 3\text{–}7, 11$$

and cyclododecanone, which gave the corresponding 1,1-difluorocycloalkanes in yields of 49 and 23 %, respectively, cyclooctanone gave a mixture from which 1,1-difluorocyclooctane could be isolated in only 0.5 % yield.[50]

Recently a similar situation has been encountered in the case of cyclodecanone. No 1,1-difluorocyclodecane could be isolated after treatment

[55] D. G. Martin and J. E. Pike, *J. Org. Chem.*, **27**, 4086 (1962).

[56] J. A. Edwards, P. G. Holten, J. C. Orr, L. C. Ibanez, E. Necoechea, A. de la Roz, E. Segovia, R. Urquiza, and A. Bowers, *J. Med. Chem.*, **6**, 174 (1963).

[57] C. A. Tolman and R. C. Lord, personal communication.

[58] C. A. Tolman, Ph.D. Thesis, University of California, Berkeley, June 1964 [*C.A.* **62**, 15607 (1965)].

with either sulfur tetrafluoride or phenylsulfur trifluoride.[59] These results may reflect the low carbonyl reactivities of cyclooctanone and cyclodecanone relative to cycloheptanone and cyclododecanone as measured, for example, by dissociation of their cyanohydrins.[60] In terms of the mechanistic Schemes 1 and 2 outlined earlier (pp. 8, 9), this could be interpreted to mean that for cyclooctanone and cyclodecanone the equilibrium for reversible addition to the carbonyl group by hydrogen fluoride in the first step of the reaction lies far on the ketone side.

$$>C{=}O + HF \rightleftharpoons >C(OH)F$$

By utilizing the tendency for electron-withdrawing groups adjacent to a carbonyl group to enhance addition to the carbonyl group, 1,1-difluorocyclododecane was prepared indirectly.[59] The key step which introduced the *gem*-difluoro group into the 10-membered ring was the fluorination of sebacoin tosylate with a mixture of sulfur tetrafluoride and hydrogen fluoride. Treatment of the difluorotosylate with potassium

TsO ... O $\xrightarrow[19-23°]{SF_4,\ HF}$ TsO ... F_2 (14%)

t-butoxide gave 3,3-difluoro-*trans*-cyclodecene in 42% yield; this was hydrogenated over palladium to afford 1,1-difluorocyclodecane in 33% yield from the difluoro tosylate.

Several cycloalkanediones, *e.g.*, cyclooctane-1,2-dione, have been converted to the corresponding tetrafluorocycloalkanes in modest yields.[61, 62] Diphenyltriketone, $C_6H_5COCOCOC_6H_5$, gives 1,3-diphenylhexafluoropropane in 50% yield.[7]

Highly fluorinated ketones can be remarkably sluggish toward sulfur tetrafluoride-hydrogen fluoride, *e.g.*, α,α,α-trifluoroacetophenone required 100° with sulfur tetrafluoride to give perfluoroethylbenzene in 65% yield.[7] $C_6H_5CO(CF_2)_3COC_6H_5$ (**16**) gave a significant amount of monoketone after

[59] E. A. Noe and J. D. Roberts, *J. Amer. Chem. Soc.*, **94**, 2020 (1972).

[60] M. S. Newman in *Steric Effects in Organic Chemistry*, p. 238, John Wiley and Sons, New York, 1956.

[61] S. L. Spasov, D. L. Griffith, E. S. Glazer, K. Nagarajan, and J. D. Roberts, *J. Amer. Chem. Soc.*, **89**, 88 (1967).

[62] J. E. Andersen, E. S. Glazer, D. L. Griffith, R. Knorr, and J. D. Roberts, *J. Amer. Chem. Soc.*, **91**, 1386 (1969).

only 5 hours at 220°; after 34 hours at this high temperature both carbonyl groups were converted to *gem* difluoro groups.[63] Although both reactions were carried out in the absence of added hydrogen fluoride, it is unlikely that adventitious hydrogen fluoride was completely absent.

Abnormal Reactions. A few quinones such as anthraquinone and chloranil react normally with sulfur tetrafluoride to give products in which each carbonyl group is replaced by two fluorine atoms,[7] but most quinones give at least some isomeric products. For example, sulfur tetrafluoride containing hydrogen fluoride reacts with perhalonaphthoquinones to give mixtures which include products corresponding to addition of fluorine atoms to the carbon atoms β to the carbonyl groups.[64] Hexafluoro-1,2-naphthoquinone gave perfluoro-1,2-dihydronaphthalene and the β-addition product perfluoro-1,4-dihydronaphthalene in approximately equal amounts. At 75–140°, hexafluoro-1,4-naphthoquinone gave the same perfluorodihydronaphthalenes, but in a different ratio.[64]

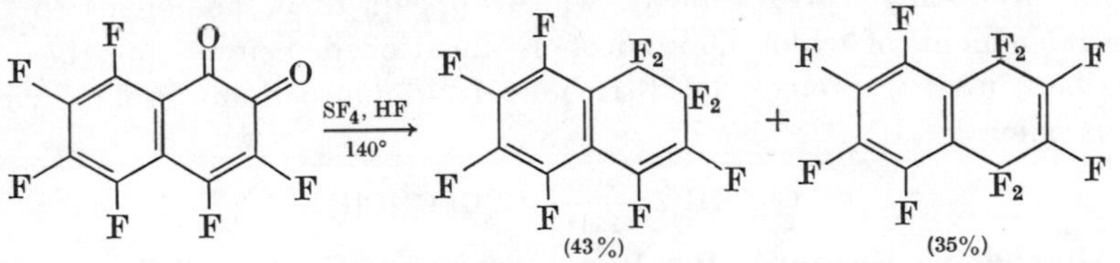

$\xrightarrow[140°]{SF_4,\ HF}$ (43%) + (35%)

Formation of isomeric products may result from delocalization of the positive charge generated by protonation of the quinone carbonyl groups.

Another quinone which is irregular in its behavior toward sulfur tetrafluoride is 2,5-dihydroxy-1,4-benzoquinone; it gave a heptafluorocyclohexene.[7]

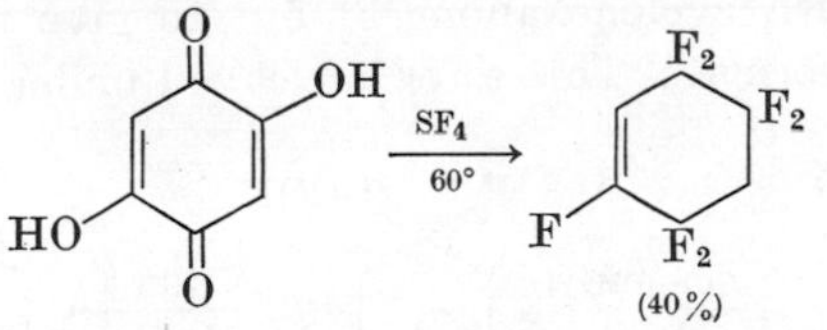

$\xrightarrow[60°]{SF_4}$ (40%)

In a few instance, products corresponding to replacement of hydrogen atoms by fluorine atoms have been isolated from reactions with ketones. From anthrone, 10,10-difluoroanthrone is obtained.[65] Addition of a radical scavenger and an antioxidant inhibited to some extent the formation of

[63] F. S. Fawcett and W. A. Sheppard, *J. Amer. Chem. Soc.*, **87**, 4341 (1965).

[64] V. D. Shteingarts and B. G. Oksenenko, *Zh. Org. Khim.*, **6**, 1611 (1970) [*C.A.*, **73**, 109548 (1970)].

[65] D. E. Applequist and R. Searle, *J. Org. Chem.*, **29**, 987 (1964).

the difluoroanthrone. This was suggested as evidence for a free-radical mechanism analogous to halogenations by molecular chlorine or bromine.[65]

$$\text{Anthrone} \xrightarrow[69^\circ]{SF_4,\ HF} \text{10,10-difluoroanthrone } (F_2)\ (85\%)$$

An attempt to prepare 2-methyl-1,1-difluorocyclohexane from 2-methylcyclohexanone gave instead what appears to be 2-methyl-1,1,2-trifluorocyclohexane in 10% yield.[61] Formally, at least, precedents for these reactions are found in the oxidative fluorination of triphenylphosphines and arsines by sulfur tetrafluoride.[66]

$$(C_6H_5)_3P \xrightarrow[150^\circ]{SF_4} \underset{(69\%)}{(C_6H_5)_3PF_2} + S^0$$

An interesting rearrangement, which appears to be analogous to the rearrangement of chlorofluoroalcohols mentioned before (pp. 17–18), has been used to prepare 1,1,2-trichloro-1,2-difluoroethane in 62% yield from chloral.[67]

$$Cl_3CCHO \xrightarrow[150^\circ]{SF_4} Cl_2CFCHClF$$

Alternative Reagents for Replacement of Carbonyl Oxygen by Fluorine. Carbonyl fluoride reacts with ketones and aldehydes to give *gem*-difluorides,[68] though in general it is less reactive than sulfur tetrafluoride. In contrast to the latter, the reactions of carbonyl fluoride are base-catalyzed and occasionally give isolable intermediates. For example, in the presence of a trace of N,N-dimethylformamide, carbonyl fluoride reacts with cyclohexanone at 50° to give a distillable fluorocyclohexyl fluoroformate. This ester is cleaved under mild conditions by

$$\text{Cyclohexanone} \xrightarrow{COF_2(DMF)} \underset{(64\%)}{\text{1-fluoro-1-(OCOF)cyclohexane}} \xrightarrow{BF_3} \underset{(54\%)}{\text{1,1-difluorocyclohexane } (F_2)}$$

boron trifluoride etherate to 1,1-difluorocyclohexane.[68] Benzaldehyde and carbonyl fluoride reacted at 150° in the presence of a trace of dimethyl formamide to give benzylidene difluoride directly in 80% conversion;

[66] W. C. Smith, *J. Amer. Chem. Soc.*, **82**, 6176 (1960).

[67] P. B. Sargeant, *J. Org. Chem.*, **35**, 678 (1970).

[68] F. S. Fawcett, C. W. Tullock, and D. D. Coffman, *J. Amer. Chem. Soc.*, **84**, 4275 (1962).

in the absence of dimethylformamide or pyridine catalyst the conversion dropped to 18%. Benzophenone, carbonyl fluoride, and pyridine catalyst at 250° gave difluorodiphenylmethane in 62% conversion. Use of cesium fluoride as catalyst in place of pyridine resulted in 40% conversion; but, with no added catalyst, difluorodiphenylmethane was obtained in only 3% conversion.[68]

Although 1,2-diones are converted to the corresponding tetrafluoro compounds by sulfur tetrafluoride, carbonyl fluoride reacts with several acyclic 1,2-diones to give cyclic carbonates. Thus 2,3-pentanedione reacts with carbonyl fluoride in pyridine at 100° to give 1-ethyl-2-methyl-1,2-difluoroethylene carbonate. Both *cis* and *trans* forms of the cyclic carbonate can be isolated.[68]

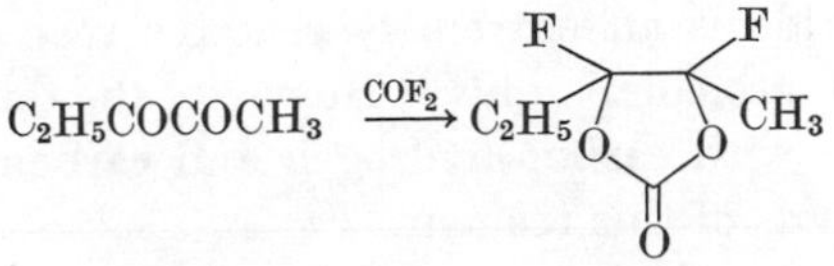

$$C_2H_5COCOCH_3 \xrightarrow{COF_2} C_2H_5CF{-}CFCH_3 \text{ (cyclic carbonate: } {-}O{-}C(=O){-}O{-}\text{ bridge)} \quad (46\%)$$

Phenylsulfur trifluoride converts carbonyl compounds to *gem*-difluoro derivatives.[69] Prepared by a procedure described in *Organic Syntheses*,[69]

$$C_6H_5SF_3 + C_6H_5CHO \xrightarrow[71\text{–}80\%]{} C_6H_5CHF_2 + C_6H_5SOF$$

this liquid reagent can be stored indefinitely in bottles of aluminum or polytetrafluoroethylene. It is not so toxic as sulfur tetrafluoride, and its use does not require pressure equipment. Reaction with benzaldehyde may be conducted at 100° in a glass flask; the benzal difluoride is removed by distillation under reduced pressure.[69] Cyclooctanone was converted at 100° to 1,1-difluorocyclooctane in 9% yield, which compares with a yield of only 0.5% from the corresponding reaction with sulfur tetrafluoride at 30°.[50]

Selenium tetrafluoride has been employed as a reagent for the fluorination of ketones.[70] Yields appear to be somewhat erratic. 1,1-Difluorocyclohexane was obtained in 7% yield from cyclohexanone, and 3,3-difluorocholestane was obtained in 89% yield from 3-cholestanone. Reactants were mixed at −70° and then warmed gradually to 24°.

Molybdenum hexafluoride* in methylene chloride with boron trifluoride

* Peninsular Chem Research, Inc. (P.O. Box 14318, Gainesville, Florida) sells an 0.8 *M* solution of activated molybdenum hexafluoride in methylene chloride under the name "Fluoreze-M."

[69] W. A. Sheppard, *J. Amer. Chem. Soc.*, **84**, 3058 (1962); *Org. Syn.*, **44**, 39, 82 (1964).

[70] P. W. Kent and K. R. Wood, Brit. Pat. 1,136,075 (1968) [*C.A.* **70**, 88124 (1969)].

as catalyst is a new reagent for conversion of aldehydes and ketones to difluoromethylene compounds.[71] Yields of products from simple aldehydes and ketones appear to be somewhat lower than from the corresponding reactions of sulfur tetrafluoride-hydrogen fluoride. However, reactions can be carried out at room temperature in glass apparatus. The fact that benzophenone is converted to difluorodiphenylmethane at room temperature, in contrast to the vigorous conditions (100–180°, 8 hours)

$$MoF_6 + C_6H_5COC_6H_5 \xrightarrow[CH_2Cl_2]{BF_3} MoOF_4 + \underset{(55\%)}{C_6H_5CF_2C_6H_5}$$

which have been employed for the same conversion effected by sulfur tetrafluoride-hydrogen fluoride,[7] is noteworthy.

Bromine trifluoride is an extremely reactive reagent which converts ketones to *gem*-difluoro compounds.[72] However, the tendency for bromine trifluoride to react with carbon-hydrogen and carbon-carbon bonds has limited the usefulness of this reagent.

gem-Difluorides have been prepared by the reaction of alkynes with liquid hydrogen fluoride at −23°.[73]

$$CH_3C{\equiv}CH \xrightarrow[-23^\circ]{HF} \underset{(64\%)}{CH_3CF_2CH_3}$$

Carboxylic Acids

A general and direct method for preparation of trifluoromethyl compounds is the reaction of sulfur tetrafluoride with aliphatic or aromatic carboxyl groups. The reaction takes place in two steps: the carboxyl group is converted to an acyl fluoride by one equivalent of sulfur tetrafluoride and then subsequently to the trifluoromethyl derivative by a second equivalent.[7]

$$RCO_2H + SF_4 \longrightarrow RCOF + HF + SOF_2$$

$$RCOF + SF_4 \xrightarrow{HF} RCF_3 + SOF_2$$

Since conversion of carboxylic acids to acyl fluorides occurs at significantly lower temperatures than replacement of the oxygen atom of acyl fluorides by two fluorine atoms, intermediate acyl fluorides may be isolated. Thus mild treatment of malonic acid with sulfur tetrafluoride at 40° for 16 hours yields malonyl fluoride. However, when the temperature

[71] F. Mathey and J. Bensoam, *Tetrahedron*, **27**, 3965 (1971).

[72] T. E. Stevens, *J. Org. Chem.*, **26**, 1627 (1961).

[73] A. L. Henne, *Org. Reactions*, **11**, 49 (1944).

is raised to 150° for 8 hours, only 1,1,1,3,3,3-hexafluoropropane is obtained.[7]

$$CH_2(CO_2H)_2 \xrightarrow{SF_4} \begin{cases} \xrightarrow{40°, 16\,hr} CH_2(COF)_2 \\ \xrightarrow[150°, 8\,hr]{} CH_2(CF_3)_2 \end{cases}$$

Intermediate acyl fluorides can also be prepared by treatment of the sodium salts of carboxylic acids with sulfur tetrafluoride. The sodium salt of benzoic acid gave benzoyl fluoride in 48% yield.[7] In contrast, under the same reaction conditions used with its sodium salt, benzoic acid gave a mixture of benzoyl fluoride and benzotrifluoride,[7] clearly illustrating the essential role of a Lewis acid catalyst in effecting replacement of the carbonyl oxygen of the acyl fluoride by two fluorine atoms.

$$C_6H_5CO_2H \xrightarrow{SF_4, 120°} \underset{(41\%)}{C_6H_5COF} + \underset{(22\%)}{C_6H_5CF_3}$$

The catalyst is hydrogen fluoride liberated by the initial reaction of the carboxylic acid with sulfur tetrafluoride. Most intermediate acyl fluorides need not be isolated because, by selecting the appropriate reaction conditions, carboxylic acids can be converted directly to the corresponding trifluoromethyl derivatives. Halogenated solvents such as methylene chloride are advantageous in these reactions when the substrates are high-melting. However, solvents which act as Lewis bases, such as diethyl ether, tetrahydrofuran, and ethylene glycol dimethyl ether, inhibit the fluorination reaction, presumably by complexing with the catalyst.[74] Since acid fluorides do not require Lewis acids for their formation, they are readily produced in the presence of these solvents.

Aliphatic Carboxylic Acids. Aliphatic carboxylic acids are readily converted by sulfur tetrafluoride to the corresponding trifluoromethyl derivatives. In a typical case, *n*-heptanoic acid reacted at 130° to give 1,1,1-trifluoroheptane in 80% yield.[7, 75, 76] Under sufficiently mild reaction conditions, the configuration is preserved at the carbon atom adjacent to the carboxyl group. For example, *cis*-4-*t*-butylcyclohexanecarboxylic acid by treatment with sulfur tetrafluoride at 70° for 10 days gave *cis*-4-*t*-butylcyclohexyltrifluoromethane in 57% yield.[77] Similarly,

[74] F. Kagan and D. G. Martin, U.S. Pat. 3,297,728 (1967) [*C.A.*, **67**, 117107 (1967)].
[75] W. R. Hasek, *Org. Syn.*, **41**, 104 (1961).
[76] W. C. Smith, U.S. Pat. 2,859,245 (1958) [*C.A.*, **53**, 12236a (1959)].
[77] E. W. Della, *J. Amer. Chem. Soc.*, **89**, 5221 (1967).

themselves cyclize thermally by a 1,5-hydrogen shift and electrocyclic ring closure sequence to chromenes.[7]

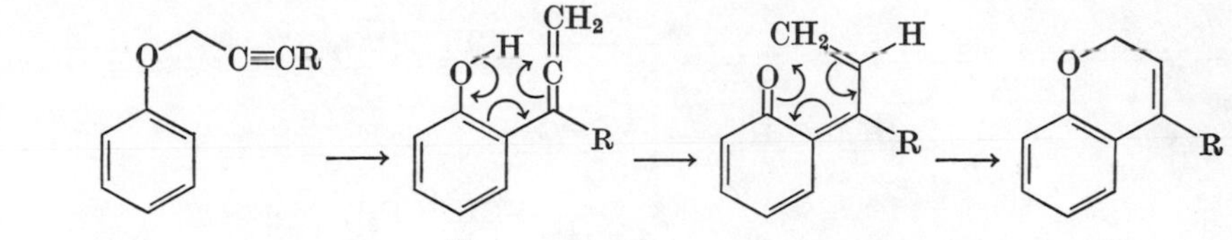

18

19

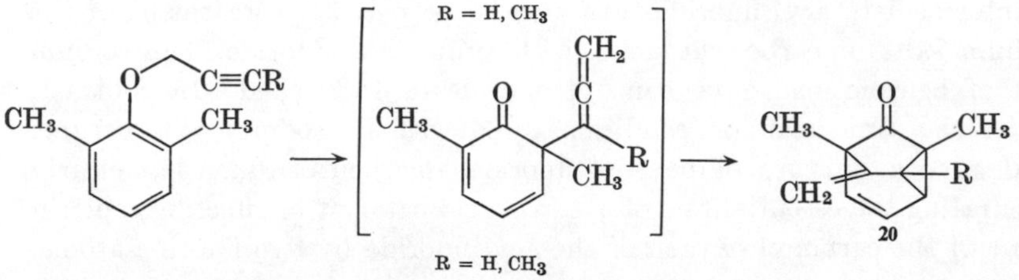

R = H, CH_3

20

R = H, CH_3

An interesting application of the propargyl aryl ether rearrangement is found in the thermal conversion of 1,4-bis(phenoxy)-2-butyne to the benzofurobenzopyran **21**.[218] This reaction proceeds through the formation of the Δ^3-chromene followed by a second rearrangement of an allyl aryl ether and finally coumaran ring closure of the *o*-allyl phenol.

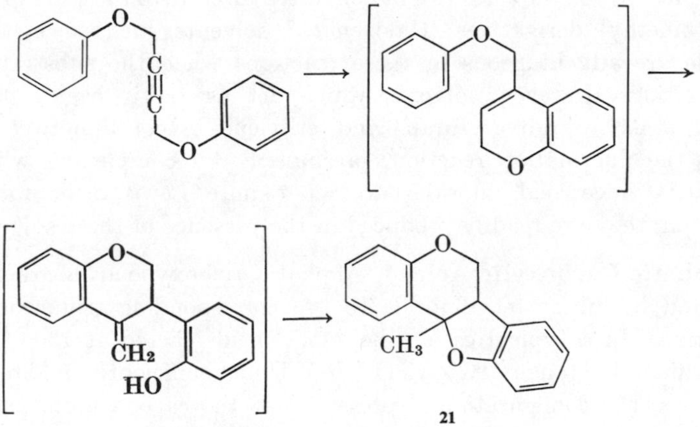

21

The only *para* migration of a propargyl side chain reported thus far is that of butynyl 2,6-dimethylphenyl ether. In addition to the internal Diels-Alder adduct **20** (R = CH_3), a small amount of the *p*-rearrangement product **22** (p. 33) was found.[110]

[218] B. S. Thyagarajan, K. K. Balasubramanian, and R. B. Rao, *Tetrahedron*, **23**, 1893 (1967).

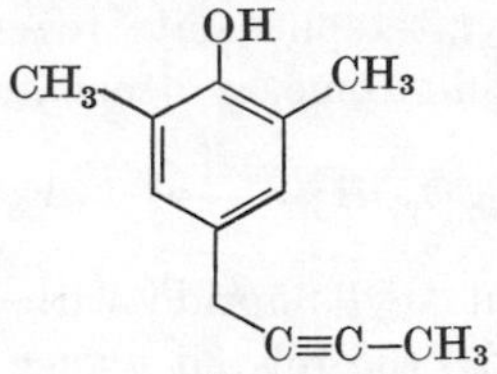

4-(2-butynyl)-2,6-dimethylphenol (OH; CH_3, CH_3; CH_2–C≡C–CH_3)

22

Experimental Conditions. Although many Claisen rearrangements have been accomplished simply by heating the aryl ethers in the temperature range 150–200°, better yields and more consistent results are obtained by using a solvent of the appropriate boiling point.[1] It has been pointed out in the foregoing discussion that the nature of the solvent can strongly affect the product distribution by its influence on the relative rates of rearrangement to available open positions and the extent of secondary reactions; these factors should be carefully considered in choosing the reaction medium. The traditional use of tertiary aromatic amines such as dimethylaniline and diethylaniline[1] appears well justified for most reactions, although dimethylformamide offers advantages in certain systems and other aprotic, polar solvents could well prove equally useful. Reaction temperature and duration also are important and should, of course, be minimized in order to achieve maximum yields of the normal product. The use of trapping agents such as butyric anhydride to capture unusually labile *ortho*-rearrangement products has been successful.[201,219] Isolation of the rearrangement products usually offers no complications; the value of Claisen's alkali[1] for separation of weakly acidic phenolic products from neutral material has already been emphasized.

Aliphatic Claisen Rearrangements

Aliphatic Claisen rearrangements have been successful not only with open-chain systems but also with structures in which the vinyl or the allyl group is part of a ring and with systems in which the allyl portion of the ether is replaced by propargyl or allenyl groups (refs. 116, 120, 126, 127, 133, 151, 220–222). The ether oxygen itself may be part of a ring as in derivatives of partially reduced furans and pyrans.[80, 121, 122, 223] Under certain circumstances the double bond of the allyl moiety may be part of

[219] R. D. H. Murray and M. M. Ballantyne, *Tetrahedron,* **26,** 4667 (1970).

[220] S. Julia, M. Julia, and P. Graffin, *Bull. Soc. Chim. Fr.*, **1964,** 3218.

[221] R. Gardi, R. Vitali, and P. P. Castelli, *Tetrahedron Lett.*, **1966,** 3203.

[222] R. Vitali and R. Gardi, *Gazz. Chim. Ital.*, **96,** 1125 (1966) [*C.A.*, **66,** 28976j (1967)].

[223] S. J. Rhoads and C. F. Brandenburg, *J. Amer. Chem. Soc.*, **93,** 5805 (1971).

For example, monomethyl terephthalate reacted only at the carboxyl group to give methyl *p*-trifluoromethylbenzoate (63% yield).[7]

$$p\text{-}HO_2CC_6H_4CO_2CH_3 \xrightarrow[130^\circ]{SF_4} p\text{-}CF_3C_6H_4CO_2CH_3$$

Whereas reaction of an acyl fluoride with sulfur tetrafluoride gives trifluoromethyl compounds, the related transformation of acyl chlorides is not always straightforward. For example, benzoyl chloride appears to undergo exchange of the chlorine for a fluorine atom first, followed by replacement of the carbonyl oxygen with two fluorine atoms and chlorination of the ring to give *m*-chlorobenzotrifluoride.[7] The nature of the chlorinating agent is unknown. Pentafluorobenzoyl chloride reacted in

$$C_6H_5COCl \xrightarrow{SF_4} C_6H_5COF \xrightarrow[\text{Chlorination}]{SF_4,\ HF} m\text{-}ClC_6H_4CF_3$$

the expected manner to give pentafluorobenzotrifluoride.[37]

$$C_6F_5COCl \xrightarrow{SF_4} C_6F_5COF \xrightarrow{SF_4,\ HF} \underset{(95\%)}{C_6F_5CF_3}$$

An alternative method for the preparation of aryl trifluoromethyl derivatives is the reaction of benzotrichloride and antimony trifluoride to give benzotrifluoride in 60–65% yield.[81]

$$C_6H_5CCl_3 \xrightarrow[130\text{–}140^\circ]{SbF_3} C_6H_5CF_3$$

Heterocyclic Carboxylic Acids. A variety of heterocyclic ring systems are stable under conditions necessary to convert a carboxyl group to the corresponding trifluoromethyl group. Thus 5-nitro-2-furoic acid gave 2-trifluoromethyl-5-nitrofuran in 37% yield.[41] 5-Carboxy-6-azauracil treated with a 40-molar excess of sulfur tetrafluoride at 50° for 24 hours gave 5-trifluoromethyl-6-azauracil.[82] Imidazolecarboxylic acids

[Structure: 5-carboxy-6-azauracil (O, HN, CO_2H, N, N, H, O) $\xrightarrow{SF_4}$ 5-trifluoromethyl-6-azauracil (O, HN, CF_3, N, N, H, O) (53%)]

are converted to the corresponding trifluoromethyl derivatives in high yield.[83]

[81] A. L. Henne, *Org. Reactions*, **11**, 62 (1944).

[82] M. P. Mertes and S. E. Saheb, *J. Heterocycl. Chem.*, **2**, 491 (1965).

[83] M. S. Raasch, *J. Org. Chem.*, **27**, 1406 (1962).

Amino Acids. Conversion of the carboxyl group of aliphatic amino acids to trifluoromethyl derivatives by sulfur tetrafluoride is accomplished by carrying out the reaction in hydrogen fluoride.[83] This serves not only to catalyze conversion of intermediate acyl fluorides to trifluoromethyl groups, but also simultaneously protects the amino group by protonation. When glycine was allowed to react at 120° for 8 hours, 2,2,2-trifluoroethylamine was obtained in 24% yield.[83] When an optically active amino acid is used, optical activity is retained in the resulting amine. For example, L-leucine reacted to give optically active 1,1,1-trifluoro-2-amino-4-methylpentane (22% yield).[83] Unfortunately in these experi-

$$\text{L-}(CH_3)_2CHCH_2CH(NH_2)CO_2H \xrightarrow{SF_4,\ HF} \text{L-}(CH_3)_2CHCH_2CH(NH_2)CF_3$$

ments there was no independent measure or proof of optical purity. Several structural features of the reactant amino acids affected the yields. Amino acids capable of forming lactams (*e.g.*, 4-aminobutyric acid) give low yields. If diketopiperazine formation is hindered, yields are better. Thus N-dodecylalanine, in which diketopiperazine formation is hindered by the N-substituent, gave 1-trifluoromethylethyldodecylamine in 61% yield, whereas alanine gave 1,1,1-trifluoro-2-aminopropane in only 29% yield.[83]

$$CH_3CH(CO_2H)NHC_{12}H_{25}\text{-}n \xrightarrow{SF_4,\ HF} CH_3CH(CF_3)NHC_{12}H_{25}\text{-}n$$

$$CH_3CH(NH_2)CO_2H \xrightarrow{SF_4,\ HF} CH_3CH(NH_2)CF_3$$

Low yields were obtained at 120° (8 hours) from amino acids with sensitive substituents. Methionine gave the corresponding trifluoromethyl derivative in only 1.5% yield, and serine, cystine, tyrosine, tryptophan, and histidine gave no isolable products.[83] The milder conditions used for steroid carboxylic acids[30, 74] and the optically active deuterated phenylbutanoic acids[78] might be more successful with these sensitive compounds.

$$CH_3SCH_2CH_2CH(NH_2)CO_2H \xrightarrow{SF_4} CH_3SCH_2CH_2CH(NH_2)CF_3$$

An efficient route to ω,ω,ω-trifluoroamino acids has been devised which has been used to synthesize DL-5,5,5-trifluoronorvaline from glutamic acid in three steps (in 50% overall yield).[84] The key step in the synthesis is conversion of the hydantoin **17** to the trifluoromethyl derivative **18**. While the hydantoin moiety of acid **18** was stable to the reaction conditions, the product **18** was largely racemized. In independent experiments, L-5-hydantoin-β-propionic acid **(17)** was shown to lose 97% of its optical

[84] R. M. Babb and F. W. Bollinger, *J. Org. Chem.*, **35**, 1438 (1970).

$$HO_2C(CH_2)_2CH(NH_2)CO_2H \xrightarrow[\text{2. HCl}]{\text{1. KNCO}} \underset{\mathbf{17}}{\text{(hydantoin with } HO_2CCH_2CH_2\text{)}} \xrightarrow{SF_4,\ HF}$$

$$\underset{\mathbf{18}}{\text{(hydantoin with } CF_3CH_2CH_2\text{)}} \xrightarrow[100^\circ]{Ba(OH)_2} CF_3(CH_2)_2CH(NH_2)CO_2H$$

activity on exposure to anhydrous hydrogen fluoride at 100° for 3 hours and all of its optical activity on exposure to constant-boiling hydrochloric acid at 108° for 3 hours.[84]

Polycarboxylic Acids. The degree of fluorination of polycarboxylic acids can be controlled by varying the amount of sulfur tetrafluoride. Thus from sebacic acid either 1,1,1,10,10,10-hexafluorodecane or 10,10,10-trifluorodecanoyl fluoride can be produced as the principal product depending on the molar ratio of sulfur tetrafluoride to substrate.[7] 1,1-Dicarboxylic acids such as malonic acid react to give bis(trifluoromethyl) derivatives.[7] Acetylenedicarboxylic acid gave 1,1,1,4,4,4-hexafluoropropyne in 80% yield when allowed to react at 170° for 8 hours.[7] 3-Methylcyclobut-2-ene-1,2-dicarboxylic acid afforded 3-methyl-1,2-bis(trifluoromethyl)cyclobut-2-ene in 31% yield on heating with sulfur tetrafluoride at 170° for 8 hours.[7]

$$HO_2C(CH_2)_8CO_2H \xrightarrow{6\ SF_4} \underset{(87\%)}{CF_3(CH_2)_8CF_3}$$

$$HO_2C(CH_2)_8CO_2H \xrightarrow{3\ SF_4} \underset{(45\%)}{CF_3(CH_2)_8COF} + \underset{(27\%)}{CF_3(CH_2)_8CF_3} + \underset{(21\%)}{FCO(CH_2)_8COF}$$

$$HO_2CC{\equiv}CCO_2H \xrightarrow{SF_4} CF_3C{\equiv}CCF_3$$

$$\text{3-methylcyclobutene-1,2-}(CO_2H)_2 \xrightarrow{SF_4} \text{3-methylcyclobutene-1,2-}(CF_3)_2$$

Steric hindrance in polycarboxylic acids can lead to selective reaction at the less hindered carboxyl groups. Thus 1,2,3-benzenetricarboxylic acid gave 3-trifluoromethylphthalic acid in 69% yield after hydrolysis of the acyl fluoride groups.[85] Similarly, treatment of 1,2,3,4-benzenetetra-

$$\text{C}_6\text{H}_3(\text{CO}_2\text{H})_3\text{-1,2,3} \xrightarrow{SF_4} \text{C}_6\text{H}_3(\text{CF}_3)(\text{COF})_2 \longrightarrow \text{C}_6\text{H}_3(\text{CF}_3)(\text{CO}_2\text{H})_2$$

carboxylic acid with sulfur tetrafluoride afforded 3,6-bis(trifluoromethyl)-phthalic acid in 76% yield.[85]

$$\text{C}_6\text{H}_2(\text{CO}_2\text{H})_4 \xrightarrow{SF_4} \text{C}_6\text{H}_2(\text{CF}_3)_2(\text{COF})_2 \longrightarrow \text{C}_6\text{H}_2(\text{CF}_3)_2(\text{CO}_2\text{H})_2$$

Certain polycarboxylic acids cyclize on reaction with sulfur tetrafluoride to give tetrafluoro ethers.[7] This presumably will occur in any 1,3-dicarboxylic acid which can form the corresponding anhydride by cyclization

$$HO_2CCH_2OCH_2CO_2H \xrightarrow{SF_4} \underset{(35\%)}{CF_3CH_2OCH_2CF_3} + \underset{(14\%)}{O(CH_2CF_2)_2O}$$

$$HO_2CCH_2CH(CO_2H)CH_2CO_2H \xrightarrow{SF_4} \underset{(20\%)}{\text{CF}_3\text{-}\overline{\text{C}_5\text{H}_5\text{F}_4\text{O}}}$$

of the intermediate monoacyl fluoride.

$$O(CH_2CO_2H)_2 \xrightarrow{SF_4} O(CH_2COF)(CH_2CO_2H) \xrightarrow{-HF} \text{cyclic anhydride} \xrightarrow{SF_4} \text{2,2,6,6-tetrafluoro-1,4-dioxane}$$

Esters

Carboxylic acid esters usually require vigorous reaction conditions to convert them to the corresponding trifluoromethyl compounds. For example, it was necessary to heat methyl benzoate to 360° for 6 hours to obtain benzotrifluoride.[7] α,α-Difluoro ethers and acyl fluorides appear

[85] A. I. Burmakov, L. A. Alekseeva, and L. M. Yagupol'skii, *J. Gen. Chem. USSR.*, **5**, 1892 (1969) [*C.A.*, **72**, 21463 (1970)].

$$C_6H_5CO_2CH_3 \xrightarrow{SF_4,\ HF} C_6H_5CF_3 \ (55\%)$$

to be intermediates in these conversions. Thus in the reaction of methyl formate with sulfur tetrafluoride to give mainly fluoroform and methyl fluoride, methyl difluoromethyl ether was detected in the crude product

$$HCO_2CH_3 \xrightarrow{SF_4} HCF_3 + CH_3F + HCF_2OCH_3$$

by mass spectroscopy.[7] On warming sulfur tetrafluoride with ethyl acetate at temperatures which permitted recovery of most of the ethyl acetate, ethyl 1,1-difluoroethyl ether was produced in small amounts.[7] Acyl fluorides are formed either directly from the ester or by elimination of

$$CH_3CO_2C_2H_5 \xrightarrow{SF_4} CH_3CF_2OCH_2CH_3$$

alkyl fluoride from an α,α-difluoro ether.[7] Acyl fluorides have been

$$RCO_2R' \xrightarrow{SF_4} RCOF + R'F + SOF_2$$

$$RCO_2R' \xrightarrow{SF_4} RCF_2OR' \rightarrow RCOF + R'F$$

isolated in some reactions of esters with sulfur tetrafluoride.[7] Certain esters will react with sulfur tetrafluoride at 130° when catalyzed by boron trifluoride or titanium tetrafluoride, whereas hydrogen fluoride is ineffective at temperatures up to 170°. As mentioned previously (p. 33), this permits selective fluorination of carboxyl, hydroxyl, keto, and aldehydic groups in the presence of esters.

Generally, there is no synthetic advantage in using esters instead of carboxylic acids to convert aliphatic or aryl derivatives to the corresponding trifluoromethyl compounds. If the corresponding carboxylic acid is unstable, the reaction of sulfur tetrafluoride with the ester provides a convenient route to the trifluoromethyl derivative. Thus 2,4-bis(dicarbethoxymethylene)-1,3-dithietane and sulfur tetrafluoride in the presence of hydrogen fluoride reacted at up to 200° to give 2,4-bis(hexafluoroisopropylidene)-1,3-dithietane.[86] This compound is a convenient source of bis(trifluoromethyl)thioketene.[86]

$$(C_2H_5CO_2)_2C{=}\langle S_2 \rangle{=}C(CO_2C_2H_5)_2 \xrightarrow{SF_4,\ HF} (CF_3)_2C{=}\langle S_2 \rangle{=}C(CF_3)_2 \ (69\%) \xrightarrow{650^\circ} (CF_3)C{=}C{=}S$$

[86] M. S. Raasch, U.S. Pat. 3,275,609 (1966) [*C.A.*, **66**, 3168 q (1967)].

Reaction of sulfur tetrafluoride with oxalate esters did not give the expected hexafluoroethane but instead gave a polymeric material of high molecular weight which is claimed to be a copolymer of the oxalate ester and sulfur tetrafluoride.[87]

Perfluoroacyl esters, in contrast to aliphatic esters, react with sulfur tetrafluoride to give the corresponding α,α-difluoro ethers in moderate to good yields. Thus aryl trifluoroacetates and heptafluorobutyrates, prepared from the corresponding anhydrides[88] or acid chlorides, give aryl pentafluoroethyl and heptafluorobutyl ethers. 4-Nitrophenyl trifluoroacetate gave 4-nitrophenyl pentafluoroethyl ether in 69% yield.[89] Fluorine substitution in the *beta* position of α,α-difluoro ethers apparently

$$4\text{-}O_2NC_6H_4OCOCF_3 \xrightarrow{SF_4} 4\text{-}O_2NC_6H_4OCF_2CF_3$$

results in substantial stabilization and, in contrast to the case with β-hydrogen atoms, permits isolation of the ethers in good yields.

Interestingly, the yields of ethers from perfluoroacyl esters of phenols decreased with increasing chain length of the perfluoroacyl group. Thus *p*-nitrophenyl nonafluoro-*n*-butyl ether was obtained in only 30% yield compared to 69 and 80% for the corresponding pentafluoroethyl and trifluoromethyl ethers, respectively.[89] Bis(trifluoroacetoxy)methane reacts in an unusual manner to give fluoromethyl pentafluoroethyl ether resulting from replacement of an acetoxyl group by fluoride ion.[90]

$$(CF_3CO_2)_2CH_2 \xrightarrow{SF_4} CF_3CF_2OCH_2F$$

1,1-Difluoroformals are produced by treating carbonates of fluoroalkylmethyl alcohols with sulfur tetrafluoride and hydrogen fluoride at 25°.[90]

$$(R_fCH_2O)_2C{=}O \xrightarrow{SF_4,\ HF} \underset{(20\text{–}30\%)}{(R_fCH_2O)_2CF_2}$$

Trifluoromethyl Ethers from Fluoroformates. Trifluoromethyl aryl ethers are obtained on heating sulfur tetrafluoride with aryl fluoroformates in hydrogen fluoride medium at 160–175°.[89, 91] It is more convient to treat the phenol with carbonyl fluoride[92] at 100° and then,

$$ArOCOF \xrightarrow{SF_4,\ HF} ArOCF_3$$

87 J. W. Dale and G. J. O'Neill, U.S. Pat. 3,309,344, (1967) [*C.A.*, **67**, 3350 (1967)].

88 R. F. Clark and J. H. Simons, *J. Amer. Chem. Soc.*, **75**, 6305 (1953); M. Green, *Chem. Ind.* (London), **1961**, 435.

89 W. A. Sheppard, *J. Org. Chem.*, **29**, 1 (1964); U.S. Pat. 3,265,741 (1966) [*C.A.*, **65**, 13610 (1966)].

90 P. E. Aldrich and W. A. Sheppard, *J. Org. Chem.*, **29**, 11 (1964).

91 W. A. Sheppard, *J. Amer. Chem. Soc.*, **83**, 4860 (1961).

92 M. W. Farlow, E. H. Mann, and C. W. Tullock, *Inorg. Syn.*, **6**, 155 (1960).

without isolation of the fluoroformate intermediate, allow it to react with sulfur tetrafluoride at 150–175°.[89] The hydrogen fluoride by-product from the carbonyl fluoride reaction serves as catalyst for the fluorination. By this two-step process, phenol is converted to phenyl trifluoromethyl ether in 62% overall yield. Careful selection of proper conditions is

$$C_6H_5OH + COF_2 \longrightarrow C_6H_5OCOF \xrightarrow{SF_4} C_6H_5OCF_3$$

critical for successful reaction.[89] For the reaction times investigated, temperatures lower than 150° gave small amounts of ether mixed with unreacted fluoroformate, whereas temperatures greater than 175° gave lower yields of ether with increased tar formation. Greater than catalytic amounts of hydrogen fluoride were necessary for successful reaction. Although hydrogen fluoride may serve as solvent, the use of inert solvents such as nitrobenzene was found necessary for successful reaction with higher-melting phenols. For example, 4,4′-dihydroxydiphenyl sulfone gave only a yield of 5% of the corresponding bis(trifluoromethyl) ether without solvent, but a yield of 56% was obtained with nitrobenzene

$$(4\text{-}HOC_6H_4)_2SO_2 \rightarrow (4\text{-}CF_3OC_6H_4)_2SO_2$$

as solvent.[89] Apparently, the reactants must either be liquids or in solution for reaction with carbonyl fluoride or sulfur tetrafluoride to proceed to completion at a reasonable rate.

When applied to hydroquinone, the two-step reaction sequence above afforded

$$4\text{-}(CF_3OC_6H_4O)_2CO$$

19

as the main by-product and, as expected, the ratio of **19** to 1,4-bis(trifluoromethoxy)benzene increased as the molar ratio of carbonyl fluoride to hydroquinone was decreased. When *m*-cresol was allowed to react sequentially with carbonyl fluoride and sulfur tetrafluoride, *m*-tolyl trifluoromethyl ether was isolated in 10% yield; the main product was a sulfide, tentatively assigned structure **20**.[89] Formation of sulfide **20**

$$m\text{-}HOC_6H_4CH_3 \xrightarrow[2.\ SF_4]{1.\ COF_2} m\text{-}F_3COC_6H_4CH_3 + [F_3CO(CH_3)C_6H_3]_2S$$

20 (40%)

was postulated to occur via a Friedel-Crafts reaction of thionyl fluoride with tolyl trifluoromethyl ether followed by oxidation-reduction with additional thionyl fluoride. The analogy of the reaction of anisole with thionyl chloride to give bis(4-methoxyphenyl) sulfide was cited.[89]

$$CF_3O-C_6H_4(CH_3) + SOF_2 \xrightarrow{HF} CF_3O-C_6H_3(CH_3)-SO-C_6H_3(CH_3)-OCF_3 \xrightarrow{SOF_2}$$

$$CF_3O-C_6H_3(CH_3)-S-C_6H_3(CH_3)-OCF_3 + SO_2F_2$$

$$CH_3OC_6H_5 + SOCl_2 \rightarrow (4\text{-}CH_3OC_6H_4)_2S$$

o-Nitrophenol, perhaps because of hydrogen bonding of the hydroxyl group with the nitro group, does not react with carbonyl fluoride unless sodium fluoride is used as base.

For extending the preparation of aryl trifluoromethyl ethers to alkyl trifluoromethyl ethers, the reaction was useful only if the β-carbon atom of the alcohol was substituted with one or more electron-withdrawing groups such as fluorine, chlorine, bromine, trifluoromethoxyl, or fluoroalkyl.[90] Methanol afforded methyl trifluoromethyl ether, but its homologs gave decomposition products. With one or two electronegative β-substituents, the yield of ether was 30–40%; the yield was raised to 50–60% when three electron-withdrawing substituents were present. Thus 2-haloethanols were converted to the 2-haloethyl trifluoromethyl ethers in yields of 15–42%, whereas 2,2,2-trichloroethanol gave the corresponding trifluoromethyl ether in 72% yield.[90]

$$XCH_2CH_2OH \xrightarrow[\substack{2.\ SF_4,\ HF,\\ X\,=\,Cl,\ Br}]{1.\ COF_2} XCH_2CH_2OCF_3$$

$$CCl_3CH_2OH \xrightarrow[2.\ SF_4,\ HF]{1.\ COF_2} CCl_3CH_2OCF_3$$

Selected steroid fluoroformates have been converted to the corresponding trifluoromethoxyl-substituted steroids under very mild conditions.[93] Thus 5α-fluoroandrostane-3β,6β,17β-triol 6-fluoroformate 3,17-diacetate, obtained by the sequence **21** to **24**, on treatment with sulfur tetrafluoride and hydrogen fluoride at 20° for 65 hours, afforded the 6β-trifluoromethyl ether **25** in 25% yield.[93] (Equations are on p. 42.)

[93] G. A. Boswell, Jr., and W. C. Ripka, U.S. Pat. 3,546,259 (1970) [*C.A.*, **74**, 100297 (1971)].

$OCOCH_3$

CH_3CO_2 **21** → 1. NOF, 2. $Al_2O_3 \cdot H_2O$ → CH_3CO_2, F, O **22**

→ $NaBH_4$ → CH_3CO_2, F, OH **23** → COF_2 → CH_3CO_2, F, OCOF **24**

→ SF_4, HF → CH_3CO_2, F, OCF_3 **25**

Using similar conditions, testosterone and 19-nortestosterone have been converted to the 17-trifluoromethoxyl derivatives, demonstrating that an electron-withdrawing group such as a fluorine atom on the β-carbon atom is not necessary.[94] The finding that the Δ^4-3-keto system was not attacked under these conditions again illustrates the high degree of selective fluorination possible with sulfur tetrafluoride and hydrogen fluoride.

OCOF, R, O → SF_4, HF, 20° → OCF_3, R, O

($R = CH_3$, H)

Amides

Amide groups react with sulfur tetrafluoride, and the course of reaction is dependent on the presence or absence of nitrogen-hydrogen bonds in the amide groupings. With amides containing nitrogen-hydrogen bonds, the carbonyl-nitrogen bond breaks to give an acyl fluoride, which may then undergo further reaction to give the trifluoromethyl derivative.

[94] G. A. Boswell, Jr., unpublished results from this laboratory.

Thus benzamide and N-methylbenzamide give benzotrifluoride and benzoyl fluoride. With amides lacking a nitrogen-hydrogen bond, the carbonyl-nitrogen bond may or may not be broken. For example, tetramethylurea gives dimethylcarbamyl fluoride, and N,N-dimethylbenzamide affords either benzoyl fluoride or N N-dimethyl-α,α-difluorobenzylamine.[7]

$$C_6H_5CON(CH_3)_2 \xrightarrow{SF_4} C_6H_5COF \text{ or } C_6H_5CF_2N(CH_3)_2 \ (\mathbf{26})$$

Preparation of the difluoro amine **26** could not be repeated consistently and failed completely if the starting material was contaminated with benzoic acid.[7] These results suggest that cleavage is caused by trace amounts of hydrogen fluoride.

N,N-Bis(trifluoromethyl)anilines have been prepared in low yield (2–12%) by the reaction of sulfur tetrafluoride with N-fluoroformyl-N-trifluoromethylanilines or N,N-bis(fluoroformyl)anilines.[95] Trichlorofluoromethane was added as a solvent in some of these experiments. The N-fluoroformyl-N-trifluoromethylanilines were prepared by heating mercuric fluoride, aryl isothiocyanate, carbonyl fluoride, and cesium fluoride in a pressure vessel at 200°.[96]

$$XC_6H_4N(COF)(CF_3) \text{ or } XC_6H_4N(COF)_2 \xrightarrow{SF_4,\ HF} XC_6H_4N(CF_3)_2$$

X = H; *m*-NO₂; *p*-NO₂; *m*,*p*-NH₂; *m*,*p*-CH₃, *m*,*p*-F; *p*-Br; *m*,*p*-CO₂H

$$ArNCS + HgF_2 \longrightarrow ArN{=}CF_2 \xrightarrow[CsF]{COF_2} ArN(CF_3)COF$$

Anhydrides

Phthalic anhydride reacted with sulfur tetrafluoride at 180° to give phthaloyl fluoride in 92% yield. At 350°, *o*-bis(trifluoromethyl)benzene was obtained in 45% yield.[7] Some anhydrides react without loss of the

$$C_6H_4(CO)_2O \xrightarrow[180°]{SF_4} o\text{-}C_6H_4(COF)_2 \xrightarrow[350°]{SF_4} o\text{-}C_6H_4(CF_3)_2$$

[95] F. S. Fawcett and W. A. Sheppard, *J. Amer. Chem. Soc.*, **87**, 4340 (1965).
[96] W. A. Sheppard, *J. Amer. Chem. Soc.*, **87**, 4338 (1965).

ring oxygen atom. Dichloromaleic anhydride, for example, is converted to 3,4-dichloro-2,2,5,5-tetrafluoro-2,5-dihydrofuran (**27**) in 46% yield.[7] The

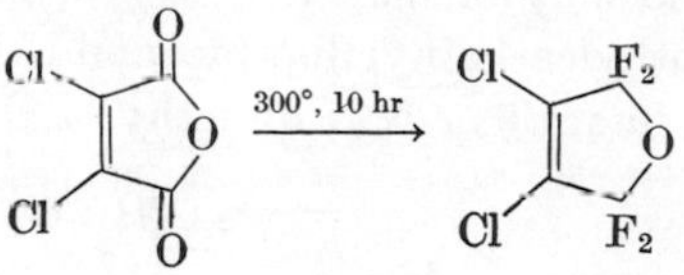

300°, 10 hr

27

reaction of dichloromaleic anhydride with sulfur tetrafluoride has been studied extensively. Dihydrofuran **27**, 2,3-dichloro-4,4-difluoro-2-butenolactone, and *cis*- and *trans*-2,3-dichlorohexafluoro-2-butene are produced.[97, 98] Since the butenolactone can be converted to dihydrofuran **27**, it is a probable intermediate in the reaction.[98] The highest yields (about 100%) of the difluorolactone were obtained by using a low ratio of sulfur tetrafluoride to anhydride (about 2.4:1) and heating at 300° for 15 hours. The tetrafluoro ether **27** was the major product (55% yield) when the ratio of sulfur tetrafluoride to anhydride was at least 6.4:1 and the reactants were heated to 300° for 71 hours.[98]

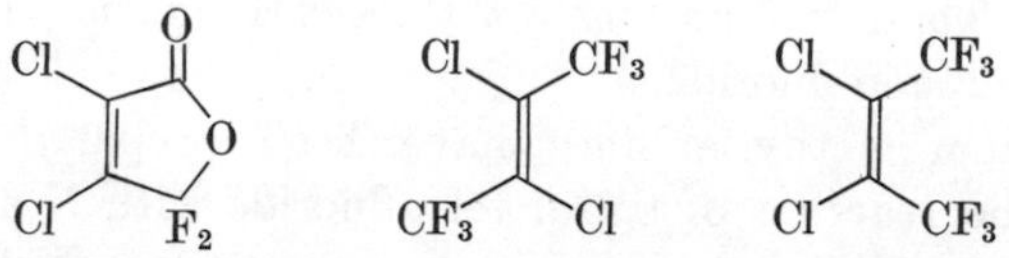

Organic Halides

One of the most widely used methods for introducing fluorine into organic compounds is replacement of chlorine, bromine, or iodine atoms by fluorine. Agents used include hydrogen fluoride, potassium fluoride, antimony trifluoride, silver fluoride, antimony pentafluoride or chlorofluoride, and, less frequently, sodium fluoride, manganese trifluoride, cobalt trifluoride, and chromium trifluoride.[99]

Similar halogen exchanges occur with sulfur tetrafluoride but often only at relatively high temperatures; this seriously limits its usefulness.[100] Metathesis of carbon tetrachloride with excess sulfur tetrafluoride at 225–235° under pressure gives a mixture of trichlorofluoromethane,

[97] E. S. Blake and J. L. Schaar, *Ind. Eng. Chem., Prod. Res. Develop.*, **8**, 212 (1969).

[98] W. J. Feast, W. K. R. Musgrave, and N. Reeves, *J. Chem. Soc., C*, **1970**, 2429.

[99] A. M. Lovelace, D. A. Rausch, and W. Postelnek, *Aliphatic Fluorine Compounds*, pp. 2 ff., Reinhold Publishing Co., New York, 1958.

[100] C. W. Tullock, R. A. Carboni, R. J. Harder, W. C. Smith, and D. D. Coffman, *J. Amer. Chem. Soc.*, **82**, 5107 (1960).

dichlorodifluoromethane, and chlorotrifluoromethane in combined yields up to 85%. The product ratio depends upon conditions. By operating

$$CCl_4 \xrightarrow[225-235^\circ]{SF_4} CCl_3F + CCl_2F_2 + CClF_3$$

at 225°, trichlorofluoromethane is the principal product, whereas chlorotrifluoromethane is the chief product at 325°.[100] Halogen exchange with carbon tetrachloride may be effected with sulfur tetrafluoride formed *in situ*. Thus trichlorofluoromethane and dichlorodifluoromethane were formed in 75 and 9% yields, respectively, on heating carbon tetrachloride with sodium fluoride, sulfur dichloride, and chlorine at 235° under pressure.[100]

Metathesis of carbon tetrabromide with sulfur tetrafluoride gives products which suggest that replacement of bromine by fluorine is easier than replacement of chlorine. Thus dibromodifluoromethane is the main fluorinated product formed along with trifluoromethane at 225°, and bromotrifluoromethane is the principal product formed at 325° in reaction of carbon tetrabromide with excess sulfur tetrafluoride. At higher temperatures, carbon tetrabromide gives some carbon tetrafluoride, whereas carbon tetrachloride does not.[100]

Higher haloalkanes are less prone to exchange with sulfur tetrafluoride than are halomethanes. Under extreme conditions, hexachloroethane was converted only to tetrachlorodifluoroethane, and 2,2,3,3-tetrachlorohexafluorobutane was converted to 2,2,3-trichloroheptafluorobutane.[100]

Reaction of sulfur tetrafluoride with chloroalkenes gives chlorofluoroalkanes. Thus tetrachloroethylene is converted to 1,2-dichlorotetrafluoroethane and chloropentafluoroethane. The same products were obtained by reaction of sulfur tetrafluoride with trichloroethylene.

$$CCl_2{=}CCl_2 \xrightarrow{SF_4} CClF_2CClF_2 + CF_3CClF_2$$

At the elevated temperatures used, disproportionation by dehalogenation and/or by dehydrohalogenation and subsequent addition of the fragments to yield saturated products must precede or accompany fluorination. Conversion of hexachlorocyclopentadiene to a mixture of pentachlorotrifluorocyclopentenes must involve similar processes. Hexachlorobenzene is converted by sulfur tetrafluoride at 200–400° to a mixture of dichlorooctafluorocyclohexene and trichlorononafluorocyclohexane.[100]

Reaction of sulfur tetrafluoride with monohalogenated organic compounds as a route to organic fluorides has not been reported.

The behavior of halogenated heterocycles in sulfur tetrafluoride-halogen exchanges contrasts markedly with that of chloroalkenes and hexachlorobenzene in that saturated or nearly saturated chlorofluorides are not obtained. 2,4-Dichloro- and 4,6-dichloro-pyrimidine undergo halogen exchange with formation of the corresponding difluoropyrimidines.

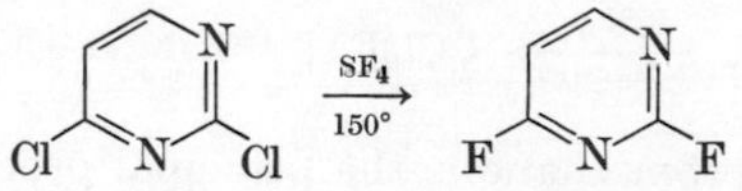

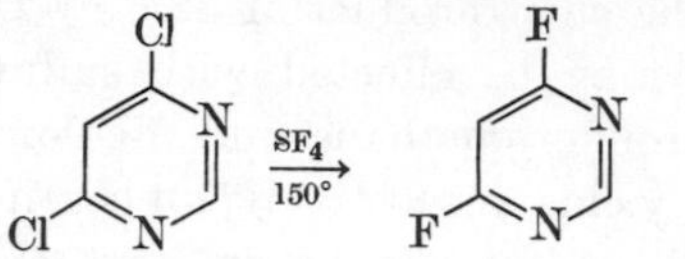

2,4,6-Trichloropyrimidine is similarly converted to a mixture of 4,6-dichloro-2-fluoro- and 2,6-dichloro-4-fluoropyrimidine. Reaction of cyanuric chloride with sulfur tetrafluoride in 1:6 molar ratio at 250° yields

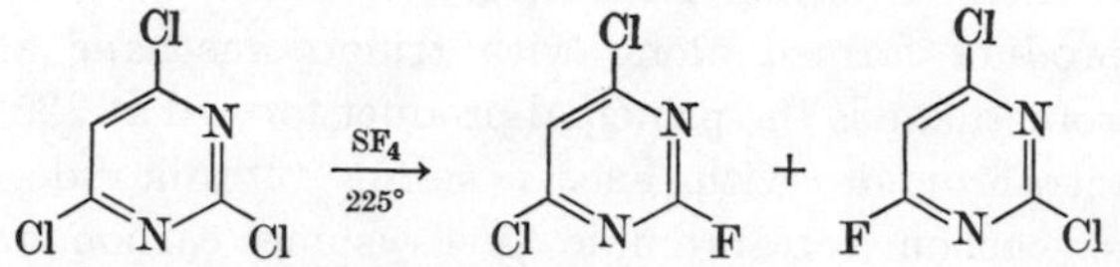

mainly cyanuric fluoride. With lower proportions of sulfur tetrafluoride, cyanuric chlorofluorides are formed.[100]

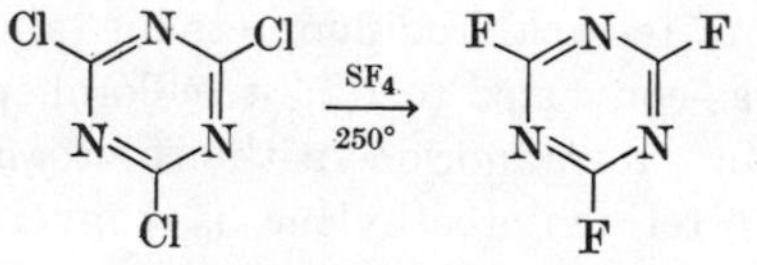

Amines

Ammonia reacts with sulfur tetrafluoride at room temperature to form the pungent, unstable gas, thiazyl fluoride, N≡SF.[101] Similarly methyl- or ethyl-amine reacts with sulfur tetrafluoride at −78° to afford the alkyliminosulfur difluoride.[102–104] Methyliminosulfur difluoride is a gas

$$3\ RNH_2 + SF_4 \rightarrow RN{=}SF_2 + 2\ RNH_2{\cdot}HF$$

(bp 16°) that slowly polymerizes at room temperature.[102] Excess methylamine reacts with sulfur tetrafluoride to give bis(methylimino) sulfur (**28**), which is also formed from the reaction of methylamine with methyliminosulfur difluoride.[102] Phenyliminosulfur difluoride, unlike

[101] O. Glemser, *Endeavour*, **28**, 86 (1969), and references therein.

[102] B. Cohen and A. G. MacDiarmid, *J. Chem. Soc., A*, **1966**, 1780.

[103] B. Cohen and A. G. MacDiarmid, *Angew. Chem.*, **75**, 207 (1963).

[104] B. G. Demitras, R. A. Kent, and A. G. MacDiarmid, *Chem. Ind.* (London), **1964**, 1712.

$CH_3N{=}SF_2$ and $C_6H_5N{=}SCl_2$, is thermally stable; it is best prepared by reaction of sulfur tetrafluoride with phenyl isocyanate[34] rather than

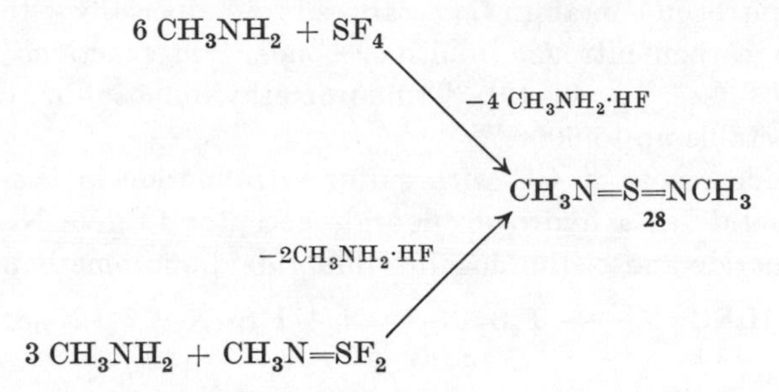

with aniline. Aniline reacts with phenyliminosulfur difluoride in the presence of trimethylamine to give the bromine-colored bis(phenylimino)-

$$C_6H_5N{=}SF_2 + C_6H_5NH_2 \rightarrow C_6H_5N{=}S{=}NC_6H_5$$

sulfur, also obtainable by reaction of aniline and sulfur tetrafluoride.[105] Pentafluoroaniline and *p*-chloroaniline are converted by sulfur tetrafluoride to the corresponding bis(arylimino)sulfur compounds in the presence of acid acceptors.[106]

The facile cleavage of the silicon–nitrogen bonds of certain silicon amines has also been used to prepare alkylimino sulfur difluorides.[104] The only known dialkylaminosulfur trifluoride, dimethylaminosulfur

$$CH_3N[Si(CH_3)_3]_2 \xrightarrow{SF_4} CH_3N{=}SF_2 + (CH_3)_3SiF$$

$$C_2H_5NHSi(CH_3)_3 \xrightarrow{SF_4} C_2H_5N{=}SF_2 + (CH_3)_3SiF + C_2H_5NH_2{\cdot}HF$$

trifluoride, is similarly prepared from dimethylaminotrimethylsilane.[104, 107]

$$(CH_3)_2N{-}Si(CH_3)_3 \xrightarrow{SF_4} (CH_3)_2NSF_3$$

This compound could not be isolated from the reaction of dimethylamine with sulfur tetrafluoride.[104]

Diphenylamine combines with sulfur tetrafluoride in the presence of trimethylamine to give, after hydrolysis, N,N′-sulfinylbis(diphenylamine) in small yield. Presumably the corresponding sulfur difluoride is an intermediate.[106]

$$2\,(C_6H_5)_2NH \xrightarrow{SF_4} [(C_6H_5)_2NSF_2N(C_6H_5)_2] \xrightarrow{H_2O} (C_6H_5)_2N\overset{\overset{O}{\|}}{S}N(C_6H_5)_2$$

[105] R. Cramer, *J. Org. Chem.*, **26,** 3476 (1961). See ref. 51.

[106] R. D. Peacock and I. N. Rozhkov, *J. Chem. Soc., A*, **1968,** 107.

[107] G. C. Demitras and A. G. MacDiarmid, *Inorg. Chem.*, **6,** 1903 (1967).

Trifluoromethyliminosulfur difluoride, $CF_3N{=}SF_2$, is frequently encountered as a product of the reaction of sulfur tetrafluoride with amines and other nitrogen-containing substances,[108, 109] especially with compounds containing carbon–nitrogen multiple bonds,[34] in reactions carried out at 150–350° (see pp. 48–49). Trifluoromethyliminosulfur difluoride is thermally stable up to 600°.[101]

Cyanamide reacts at 40° with sulfur tetrafluoride in the presence of sodium fluoride as a hydrogen fluoride acceptor to give N-cyanoiminosulfur difluoride and bis(iminosulfurdifluoride)difluoromethane.[110]

$$H_2NC{\equiv}N \xrightarrow{SF_4} \underset{(65\%)}{F_2S{=}N{-}C{\equiv}N} + \underset{(25\%)}{F_2S{=}N{-}CF_2{-}N{=}SF_2}$$

Compounds with Carbon—Nitrogen Multiple Bonds

Compounds having carbon-nitrogen multiple bonds react with sulfur tetrafluoride at elevated temperatures, often with the formation of iminosulfur difluorides ($RN{=}SF_2$). This class of compounds is also formed in the reaction of primary amines and sulfur tetrafluoride (see p. 46). Most aromatic heterocyclic systems, however, resist cleavage to iminosulfur difluorides unless cesium fluoride is present.

Organic isocyanates react to give iminosulfur difluorides and carbonyl fluoride. Thus, when sulfur tetrafluoride and phenyl isocyanate are allowed to react at 150°, phenyliminosulfur difluoride is formed in 88% yield.[34]

$$C_6H_5NCO + SF_4 \xrightarrow{125^\circ} C_6H_5N{=}SF_2 + COF_2$$

Sulfur tetrafluoride converts cyano compounds to iminosulfur difluorides of the general type $RCF_2{=}SF_2$. Trifluoroacetonitrile at 350° gave pentafluoroethyliminosulfur difluoride in 70% yield. In contrast,

$$CF_3CN \xrightarrow{SF_4,\ 350^\circ} CF_3CF_2N{=}SF_2$$

acetonitrile reacts at 350° with the formation of 1,1,2-trifluoroethyliminosulfur difluoride in very low yield. The fluorination of a carbon-hydrogen bond in this reaction is one of the few examples of a hydrogen replacement occurring with sulfur tetrafluoride.[34]

$$CH_3CN \xrightarrow{SF_4} CH_2FCF_2N{=}SF_2$$

At 180–250° sulfur tetrafluoride reacts with benzonitrile to give α,α-difluorobenzyliminosulfur difluoride in 48% yield. Inorganic thiocyanates

[108] B. Marinier and J. L. Boivin, *Can. J. Chem.*, **42**, 1759 (1964).

[109] J. E. Griffiths and D. F. Sturman, *Spectrochim. Acta*, **25A**, 1355 (1969).

[110] O. Glemser and U. Bierman, *Inorg. Nucl. Chem. Lett.*, **3**, 223 (1967) [*C.A.*, **67**, 99612 (1967)].

are also converted to iminosulfur difluoride derivatives in good yield. For example, sodium thiocyanate yields N-trifluoromethyliminosulfur difluoride as follows.[34]

$$\mathrm{NaSCN} \xrightarrow[200\text{–}350^\circ]{SF_4} \underset{(69\%)}{CF_3N{=}SF_2}$$

Sodium cyanide and cyanogen bromide are similarly transformed to trifluoromethyliminosulfur trifluoride.[34]

Bis(iminosulfur difluoride) derivatives are similarly prepared. For example, treatment of cyanogen with sulfur tetrafluoride at 350° for 8 hours afforded 1,1,2,2-tetrafluoroethane-1,2-bis(iminosulfur difluoride) and cyanodifluoromethyliminosulfur difluoride in 26 and 8% yields, respectively. As mentioned earlier (see p. 48), cyanamide in the presence

$$N{\equiv}C{-}C{\equiv}N \xrightarrow{SF_4} NCCF_2N{=}SF_2 + F_2S{=}NCF_2CF_2N{=}SF_2$$

of sodium fluoride is converted to N-cyanoiminosulfur difluoride, with 1,1-difluoromethane-1,1-*bis*(iminosulfur difluoride) as a by-product.[110] The reaction of cyanuric acid and sulfur tetrafluoride gave trifluoromethyliminosulfur difluoride,[34] whereas cyanuric fluoride, which can be prepared from cyanuric chloride by halogen exchange with sulfur tetrafluoride, was stable.[12] However, in the presence of excess cesium fluoride, cyanuric fluoride was converted to N-trifluoromethyliminosulfur difluoride in 96% yield.[12]

$$\text{(HO)}_3C_3N_3 \xrightarrow{SF_4} 3\ CF_3N{=}SF_2$$

$$F_3C_3N_3 \xrightarrow{SF_4,\ CsF} 3\ CF_3N{=}SF_2$$

Thiocarbonyl Compounds

The thiocarbonyl group is fluorinated by sulfur tetrafluoride in a manner similar to the carbonyl group except that formal oxidation-reduction of the sulfur atom occurs and results in formation of free sulfur and complete utilization of the fluorine atoms in the sulfur tetrafluoride. A mechanism can be written for this reaction that parallels that proposed for carbonyl compounds (see p. 8).[33] Instead of eliminating SOF_2, the thiocarbonyl reaction would, by analogy, eliminate sulfur monofluoride,

S_2F_2, which disproportionates at room temperature to give sulfur tetrafluoride and sulfur.[33]

Ethylene trithiocarbonate reacted smoothly with sulfur tetrafluoride at 110° in the absence of catalyst to give 2,2-difluoro-1,3-dithiolane in 82% yield.[33]

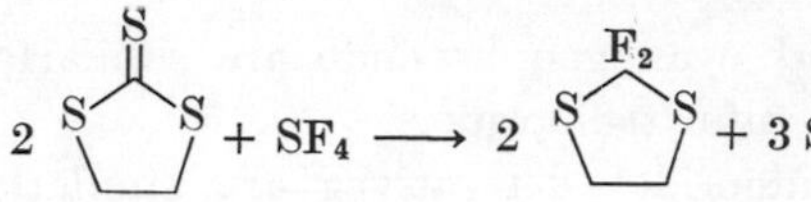

$$2\ \text{(ethylene trithiocarbonate, C=S)} + SF_4 \longrightarrow 2\ \text{(2,2-difluoro-1,3-dithiolane, CF}_2\text{)} + 3\,S$$

Fluorination of thiuram sulfides with sulfur tetrafluoride also proceeds with ease, affording dialkyltrifluoromethylamines.[33] For example, bis(pentamethylene)thiuram disulfide was converted at 100° to the moisture-sensitive 1-trifluoromethylpiperidine in 70% yield.

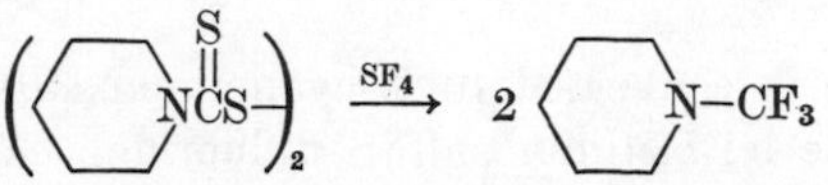

$$\left(\text{C}_5\text{H}_{10}\text{NC(=S)S}-\right)_2 \xrightarrow{SF_4} 2\ \text{C}_5\text{H}_{10}\text{N}-CF_3$$

Reactions of carbon disulfide with sulfur tetrafluoride under a variety of conditions have been studied.[33, 100, 111] At 450°, in the presence of arsenic trifluoride, carbon tetrafluoride is obtained in nearly quantitative yield.[33]

No reactions of sulfur tetrafluoride with thiones have been reported.

Haloalkenes

Although under carefully controlled conditions elemental fluorine can be added to double bonds with some selectivity,[112] it has proved advantageous to moderate the reactivity of fluorine by a heavy metal that acts as an effective carrier; thus metal fluorides in high oxidation states are frequently selective fluorinating agents. Such reagents generated *in situ* or prior to reaction with or without isolation have been used effectively for selective additions of fluorine.[113]

The selective addition of fluorine to the double bond of a halogen-containing olefin[73, 113] employed lead tetrafluoride which was generated *in situ* from lead dioxide and hydrogen fluoride. Other metal fluorides generated *in situ* from oxides and hydrogen fluoride have been used for fluorination of double bonds. Although these procedures sometimes give good yields, they have not found widespread synthetic application, probably because it is necessary to use large excesses of hydrogen fluoride

[111] C. W. Tullock, U.S. Pat. 2,972,637 (1961) [*C.A.*, **55**, 16420 (1961)].

[112] R. F. Merritt and T. E. Stevens, *J. Amer. Chem. Soc.*, **88**, 1822 (1966).

[113] A. L. Henne and T. P. Waalkes, *J. Amer. Chem. Soc.*, **67**, 1639 (1945).

if satisfactory yields are to be obtained. Procedures making use of preformed metal fluorides usually involve elemental fluorine or chlorine trifluoride, both of which require special handling techniques.

Recently some of these methods have been modified and improved so that they are mild enough to give selective fluorine addition to double bonds of complex molecules. Thus pregnenolone acetate in dry methylene chloride with excess lead tetraacetate and anhydrous hydrogen fluoride for 15 minutes at $-75°$ gave $5\alpha,6\alpha$-difluoro-3β-hydroxypregnan-20-one acetate in 27% yield along with recovered starting material in 63% yield.[114]

$COCH_3$ CH_3CO_2 $\xrightarrow{Pb(OAc)_4,\ HF}$ $COCH_3$ CH_3CO_2 F F

The observation that metal fluorides can be prepared by the reaction of the corresponding oxides with sulfur tetrafluoride is the basis for an improved modification which avoids the use of hydrogen fluoride in the *in situ* preparation of lead tetrafluoride.[115] The anticipated advantage depended on the fact that water would not be a reaction product, thus eliminating the need for a large excess of fluorinating agent to drive the formation of the metal fluoride to completion. A mixture of lead dioxide and a fivefold excess of sulfur tetrafluoride with respect to olefin gave yields of fluorinated olefins ranging from a few percent to 95% at 40–100°. The yields obtained with the new procedures were significantly higher and fewer by-products were produced. The advantages are particularly apparent with reactive olefins and those containing iodine. Thus reaction of sulfur tetrafluoride with trifluoroiodoethylene in the presence of lead dioxide gave pentafluoroiodoethane in 65% yield,[115] and 1-chloro-1,2-difluoroethylene afforded the corresponding fluorinated product in 96% yield.[115]

Because *cis*- and *trans*-1,2-dichloroethylene both yielded a *dl*- and *meso*-1,2-dichloro-1,2-difluoroethane mixture, it was concluded that the

$$CF_2{=}CFI \xrightarrow{SF_4,\ PbO_2} CF_3CF_2I$$

$$CHF{=}CClF \xrightarrow{SF_4,\ PbO_2} CHF_2CF_2Cl$$

[114] A. Bowers, P. G. Holton, E. Denot, M. C. Loza, and R. Urquiza, *J. Amer. Chem. Soc.*, **84,** 1050 (1962).

[115] E. R. Bissell and D. B. Fields, *J. Org. Chem.*, **29,** 1591 (1964).

two fluorine atoms cannot add simultaneously and that a cyclic intermediate cannot be involved.[115] The possibility that the reaction proceeds by addition of sulfur tetrafluoride to the double bond was deemed unlikely since the haloalkylsulfur trifluorides that would be formed by such an addition are stable at temperatures considerably higher than those employed in the sulfur tetrafluoride-lead dioxide fluorinations. For example, heptafluoroisopropylsulfur trifluoride and bisperfluoroalkylsulfur difluoride are prepared by the addition of sulfur tetrafluoride to hexafluoropropene in the presence of cesium fluoride at 150°,[116] whereas the reaction of sulfur tetrafluoride and hexafluoropropene in the presence of lead dioxide at 100° gives octafluoropropane in 63% yield.[115] Similarly,

$$3\ CF_3CF{=}CF_2 + 2\ SF_4 \xrightarrow[150^\circ]{CsF} (CF_3)_2CFSF_3 + [(CF_3)_2CF]_2SF_2$$

$$CF_3CF{=}CF_2 \xrightarrow[100^\circ]{SF_4,\ PbO_2} CF_3CF_2CF_3$$

fluorinated vinyl ethers at 100 and 120° gave the corresponding sulfur trifluoride adducts.[117] By using inert solvents and careful temperature

$$n\text{-}C_3H_7OCF{=}CF_2 \xrightarrow{SF_4} \underset{(63\%)}{n\text{-}C_3H_7OCF_2CF_2SF_3}$$

control, it should be possible to employ the sulfur tetrafluoride-lead dioxide method to fluorinate sensitive nonhalogenated olefins and acetylenes.[118] The ever-increasing interest in selective fluorination of molecules of biological importance makes additional studies along these lines imperative. Fluorination of steroid 5-enes with lead tetraacetate and hydrogen fluoride results in *cis* addition, suggesting a cyclic mechanism;[114] however, in similar fluorinations of halogenated olefins the data speak against a cyclic mechanism.[115]

Organo-arsenic, -boron, -phosphorus, and -silicon Compounds

Sulfur tetrafluoride replaces oxygen doubly bonded to either phosphorus or arsenic with fluorine. Thus triphenylphosphine oxide was converted to difluorotriphenylphosphorane in 67% yield.[66] Hydroxyl groups bonded

$$(C_6H_5)_3P{=}O \xrightarrow[14\ hr]{50\text{–}150^\circ} (C_6H_5)_3PF_2 + SOF_2$$

to these metalloids are also replaced by fluorine, and this occurs at a lower temperature than that required for replacement of doubly bonded oxygen. Reaction of phenylarsenic acid with sulfur tetrafluoride at 70° for 10 hours

[116] R. M. Rosenberg and E. L. Muetterties, *Inorg. Chem.*, **1**, 756 (1962).
[117] W. R. Hasek, U.S. Pat. 2,928,970 (1960) [*C.A.*, **54**, 18361 (1960)].
[118] Ref. 16, pp. 108–109.

gave phenylarsenic tetrafluoride in good yield.[66] In addition to its direct fluorinating ability, sulfur tetrafluoride can act as an oxidative fluorinating

$$C_6H_5AsO_3H \xrightarrow{SF_4} C_6H_5AsF_4$$

agent toward triaryl-phosphines and -arsines. Triphenylphosphine was converted to difluorotriphenylphosphorane by this method at 150° with benzene as solvent.[66] A similar reaction occurs with tris(dimethylamino)-

$$(C_6H_5)_3P \xrightarrow{SF_4} (C_6H_5)_3PF_2 + S$$

phosphine with partial cleavage of the phosphorus-nitrogen bond and fluorine transfer.[119] Below 120° sulfur tetrafluoride cleaves the phosphorus-

$$[(CH_3)_2N]_3P \xrightarrow{SF_4} [(CH_3)_2N]_3PF_2 + [(CH_3)_2N]_2PF_3$$

nitrogen bonds of N,N,N′,N′-tetramethylphosphonic diamide to give methylphosphonyl difluoride.[119] This selective formation of a product containing an oxygen doubly bonded to phosphorus is noteworthy because

$$[(CH_3)_2N]_2P(O)CH_3 \xrightarrow[2\text{ hr, }60°]{SF_4} F_2P(O)CH_3 + (CH_3)_2NSF_3$$

previous work showed that under more vigorous conditions such groups are converted to difluorophosphoranes.[66]

Sulfur tetrafluoride converts silicon-chlorine bonds into silicon-fluorine groups, whereas silicon-oxygen bridges are cleaved.[120] Silicon-nitrogen

$$(CH_3)_3SiCl \xrightarrow[20°]{SF_4} \underset{(90\%)}{(CH_3)_3SiF} + SCl_2 + Cl_2$$

$$(CH_3)_3SiOSi(CH_3)_3 \xrightarrow[20°]{SF_4} \underset{(40\%)}{2\ (CH_3)_3SiF} + SOF_2$$

bonds in certain silicon amines can be cleaved at low temperatures by sulfur tetrafluoride to produce the corresponding iminosulfur difluoride in good yield together with a silicon fluoride derivative.[104] When sulfur

$$CH_3N[Si(CH_3)_3]_2 \xrightarrow{SF_4} CH_3N{=}SF_2 + (CH_3)_3SiF$$

tetrafluoride and dimethylaminotrimethylsilane are allowed to react at low temperatures, dimethylaminosulfur trifluoride is formed. Attempts to isolate this compound by the direct action of sulfur tetrafluoride on dimethylamine were unsuccessful.[104]

[119] D. H. Brown, K. D. Crosbie, J. I. Darragh, D. S. Ross, and D. W. A. Sharp, *J. Chem. Soc., A*, **1970**, 914.

[120] R. Mueller and D. Mross, *Z. Anorg. Allg. Chem.*, **324**, 78 (1963).

Sulfur tetrafluoride converts aliphatic boronic acid esters either to the corresponding olefin or to the alkyl fluoride.[121]

$$[n\text{-}C_4H_9CH(C_2H_5)CH_2O]_3B \xrightarrow[-70 \text{ to } 25^\circ]{SF_4} n\text{-}C_4H_9C(C_2H_5){=}CH_2 \quad (85\%)$$

$$(n\text{-}C_8H_{17}O)_3B \xrightarrow[-70 \text{ to } 25^\circ]{SF_4} n\text{-}C_8H_{17}F \quad (44\%)$$

Miscellaneous Reactions

Other functional groups are known which react with sulfur tetrafluoride. These reactions are summarized in Table XV. Sulfur tetrafluoride converts iodoso compounds, by brief contact at -20 to $-10°$, and their bis(trifluoroacetoxy) derivatives, on heating at 70°, to the corresponding difluoroiodoso derivatives in good to excellent yields.[122] The reaction is

$$RI{=}O \xrightarrow{SF_4} RIF_2 + SOF_2$$

$$RI(OCOCF_3)_2 \xrightarrow{SF_4} RIF_2 + 2\,SOF_2 + 2\,CF_3COF$$

general and has been used for the synthesis of aromatic, heterocyclic, and aliphatic difluoroiodides. The use of trichlorotrifluoroethane as a diluent permits the recovery of difluoroiodobenzene as a crystalline solid.

Dimethyl sulfoxide reacts with sulfur tetrafluoride at 20° to give bis(monofluoromethyl) ether,[123] which is also the product of the reaction of sulfur tetrafluoride with *sym*-trioxane.[34] Reaction of sulfur tetra-

$$(CH_3)_2SO \xrightarrow{SF_4} [(CH_3)_2SF_2] \xrightarrow{\text{Glass or } (CH_3)_2SO} CH_2FOCH_2F + CH_2F_2 + SOF_2 + HF$$

fluoride with organomagnesium halides gives dialkyl and diaryl sulfides and alkyl and aryl halides.[124] Thus phenylmagnesium bromide afforded diphenyl sulfide and bromobenzene. Similarly, the reaction with n-butylmagnesium bromide led to di-n-butyl sulfide and n-butyl bromide.

$$C_6H_5MgBr \xrightarrow[-78^\circ]{SF_4} [(C_6H_5)_2SF_2] \xrightarrow{MgBr_2} (C_6H_5)_2S + Br_2$$

[121] A. Dornow and M. Siebrecht, *Chem. Ber.*, **95**, 763 (1962).

[122] V. V. Lyalin, V. V. Orda, L. A. Alekseeva, and L. M. Yagupol'skii, *Zh. Org. Khim.*, **1970**, 329 [*C.A.*, **72**, 110915 (1970)].

[123] J. I. Darragh, A. M. Noble, D. W. A. Sharp, and J. M. Winfield, *J. Inorg. Nucl. Chem.*, **32**, 1745 (1970).

[124] R. A. Bekker, B. L. Dyatkin, and I. L. Knunyantes, *Zh. Vses. Khim. Obshchest.*, **14**, 599 (1969) [*C.A.*, **72**, 55552 (1970)].

Equimolar amounts of sulfur tetrafluoride and boron trifluoride in nitromethane react with *m*-xylene at 25° to give the crystalline triarylsulfonium fluoroborate in 30% yield.[54b]

$$m\text{-}C_6H_4(CH_3)_2 \xrightarrow{SF_4\cdot BF_3} [(CH_3)_2C_6H_3]_3S^+BF_4^-$$

Trifluoromethyl hypofluorite and sulfur tetrafluoride react under the influence of ultraviolet irradiation or heat to give trifluoromethoxysulfur pentafluoride.[125] Similarly, bis(trifluoromethyl) peroxide gave a mixture

$$CF_3OF \xrightarrow[h\nu \text{ or heat}]{SF_4} \underset{(35\%)}{CF_3OSF_5}$$

$$CF_3OOCF_3 \xrightarrow[500 \text{ mm}]{SF_4,\ h\nu} \underset{(10\%)}{CF_3OSF_5} + \underset{(10\%)}{(CF_3O)_2SF_4}$$

of trifluoromethoxysulfur pentafluoride and *cis*-bis(trifluoromethoxy)-tetrafluorosulfur.

Treatment of hydroxyl-containing polymers (*e.g.*, cellophane and cellulose acetate) with sulfur tetrafluoride has improved their water resistance.[126] This property is especially desirable for films intended for wrappings.

Treatment of unsaturated elastomers such as natural rubber, polychloroprene, and butadiene-styrene copolymers with sulfur tetrafluoride has been reported to reduce the coefficient of friction.[127, 128]

PREPARATION AND PROPERTIES OF SULFUR TETRAFLUORIDE

Preparation. *Caution! Sulfur tetrafluoride and sulfur dichloride are toxic chemicals that should be handled only in a well-ventilated fume hood.*

The most convenient laboratory synthesis of sulfur tetrafluoride is that based on the reaction of sulfur dichloride with finely divided sodium

$$3\ SCl_2 + 4\ NaF \xrightarrow{CH_3CN} SF_4 + S_2Cl_2 + 4\ NaCl$$

fluoride at 50–70° in a polar aprotic medium such as acetonitrile; yields range from 60 to 80% based on sulfur dichloride. The reaction may be carried out in well-dried Pyrex glass apparatus. Details of the preparation

[125] L. C. Duncan and G. H. Cady, *Inorg. Chem.*, **3**, 850 (1964).

[126] A. K. Schneider and J. C. Thomas, U.S. Pat. 2,983,626 (1961) [*C.A.*, **55** 17106 (1961)].

[127] G. J. Tennenhouse, Brit. Pat. 1,034,028 (1966) [*C.A.*, **65**, 9148 (1966)].

[128] Fr. Pat. 1,425,629 (1966) [*C.A.*, **65**, 13930 (1966).

have been published in *Inorganic Syntheses*.[129] After distillation at atmospheric pressure through an efficient low-temperature still, the product is a water-white to very pale yellow liquid containing 5–10% thionyl fluoride. Thionyl fluoride does not interfere with most fluorinations and usually is not removed. At least 100 g of sulfur tetrafluoride should be prepared to minimize the degree of hydrolysis to thionyl fluoride by adventitious moisture.

Sulfur tetrafluoride is also produced on heating sulfur dichloride (or its precursors, sulfur and 2 molar equivalents of chlorine) with sodium fluoride at 200–250° in a sealed autoclave. The only other practical laboratory route to sulfur tetrafluoride is the oxidative fluorination of sulfur with iodine pentafluoride at 200–300°.[6]

$$5\ S + 4\ IF_5 \rightarrow 5\ SF_4 + 2\ I_2$$

The preferred preparation is carried out in a 2-liter, round-bottomed, four-necked Pyrex glass flask fitted with a thermometer, a 500-ml dropping funnel, a paddle stirrer, and an efficient water-cooled condenser which exits to a trap cooled with solid carbon dioxide and acetone. The flask is charged with finely divided sodium fluoride (420 g, 10.0 mol) and 1 liter of dry acetonitrile. Freshly distilled sulfur dichloride (515 g, 5.0 mol) is added to the stirred slurry over 30 minutes; during addition the temperature rises from 25 to about 40°. The resulting mixture is stirred at 50° for 1 hour and at 70° for 1.5 hours. The crude product is purified by distillation through an efficient low-temperature still. After very slow removal of the low-boiling forerun (which may be yellow), there is obtained 120–160 g of essentially colorless sulfur tetrafluoride, bp −38 to −35°, which contains 5–10% thionyl fluoride. The product may be stored temporarily in glass vessels at −78° or indefinitely in stainless steel cylinders at room temperature.

The purity of sulfur tetrafluoride may be estimated from infrared absorption data. Characteristic absorption bands are as follows, the *italicized* values being those most useful for this analysis: for sulfur tetrafluoride, a moderately intense sharp band at *1744* cm^{-1}, a moderately intense band at 1281 cm^{-1}, strong triplet bands centered at *889* cm^{-1} and *867* cm^{-1}, and a strong band at 728 cm^{-1}; for thionyl fluoride, strong bands at 1546 cm^{-1} and 1480 cm^{-1}, a strong doublet at *1333* cm^{-1}, and strong bands at *806* cm^{-1} and *748* cm^{-1}. Purity also may be determined by mass spectral analysis.

There are several considerations of major importance in this synthesis.

[129] F. S. Fawcett and C. W. Tullock, *Inorg. Syn.*, 7, 119 (1963).

The sodium fluoride must be finely divided to be reactive; samples with a major portion of the particles below 8 μ in diameter are suitable. A simple means of determining whether or not the sodium fluoride is sufficiently reactive is to determine its bulk density by filling a 100-ml graduated cylinder with the sodium fluoride to a constant level by persistent tapping and weighing the sodium fluoride. Samples with bulk densities ranging from 0.85 to 0.90 g/ml usually are very reactive; those with bulk densities of 1.35 to 1.67 g/ml give significantly poorer results. The larger sodium fluoride particles lower the yields of product and catalyze the decomposition of sulfur dichloride to sulfur monochloride and chlorine; the latter must be removed in a separate step.

Freshly distilled sulfur dichloride should be employed. Technical grade material often contains as much as 25% sulfur monochloride (S_2Cl_2) and chlorine. Sulfur dichloride may be purified by distilling it through a short Vigreux column at 115 mm and collecting the fraction boiling at 20–25°. Distillation of sulfur dichloride at atmospheric pressure results in significant disproportionation to chlorine and sulfur monochloride.

If chlorine is present in the sulfur tetrafluoride, it should be removed to prevent tar formation and chlorinated by-products. Liquid sulfur tetrafluoride which contains chlorine is yellow or greenish yellow. Chlorine also may be detected by the development of an iodine color on passage of a test sample through aqueous potassium iodide solution.

Small amounts of chlorine may be removed by distillation. The chlorine, which codistils with varying amounts of sulfur tetrafluoride and thionyl fluoride, is removed as a yellow forerun, bp −44 to −38°. If this forerun is removed very slowly in an efficient low-temperature still, essentially colorless sulfur tetrafluoride is obtained. A fractionating column with a rectifying section 12 mm in diameter and 280 mm in length packed with 2.4-mm ($\frac{3}{32}$-in) i.d. glass helices and cooled with solid carbon dioxide-acetone refrigerant has been found satisfactory. Pure sulfur tetrafluoride also can be obtained by a slow distillation through a Podbielniak column. Last traces of chlorine may be removed chemically by 24-hour storage of sulfur tetrafluoride at room temperature in a stainless steel cylinder containing finely powdered sulfur (excess sulfur is required to ensure formation of the high-boiling S_2Cl_2 rather than the much more volatile SCl_2) followed by redistillation.

If small amounts of chlorine must be removed, distillation from liquid mercury through a low-temperature still is satisfactory. Large amounts of chlorine are best removed by treatment of the sulfur tetrafluoride with mercury in an autoclave at room temperature prior to distillation.[130]

[130] G. E. Arth and J. Fried, U.S. Pat. 3,046,094 (1962) [*C.A.*, **57**, 9460 (1962)].

Sulfur tetrafluoride may be purchased in the U.S.A. from several companies.*

Properties. Sulfur tetrafluoride is colorless in the gaseous, liquid, and solid states. The Trouton constant of the liquid indicates that it is associated. It is thermally stable up to 500°. The physical properties of sulfur tetrafluoride are summarized in the accompanying tabulation.[4, 8, 131]

PHYSICAL PROPERTIES OF SULFUR TETRAFLUORIDE

Melting point $= -121.0 \pm 0.5°$
Boiling point $= -38°$
Density at 200°K = 1.9191 g/cm³
Density at T°K (170–200°K) $= 2.5471 - 0.00314T$
Critical temperature = 91°
Surface tension at 200°K = 25.70 dynes/cm
Heat of formation $= -171.7 \pm 2.5$ kcal/mole; -176 kcal/mole
Molar heat of vaporization = 6.32 kcal
Trouton's constant = 27.1
Vapor pressure (mm) at T°K (160–224°K), $\log P = 8.8126\text{-}1381/T$
Coefficient of cubical expansion (cm³/cm³/°C) (170–200°K) = 0.00170
Dipole moment $= 0.632 \pm 0.003$ Debye; 1.0 ± 0.1 Debye
S-F bond energy *ca.* 78 kcal/mole

Sulfur tetrafluoride is hydrolyzed rapidly and exothermally by aqueous media at all pH values.

Structure. Sulfur tetrafluoride has the structure of a distorted trigonal bipyramid ($C_{2}v$ symmetry) in which two fluorine atoms and the unshared electron pair are in the equatorial positions and the other fluorine atoms are in the polar positions (see Fig. 1). The equatorial bonds have the length of the normal S-F bond (1.545 Å); the longer polar bonds (1.645 Å) may have a larger contribution from the more diffuse and higher energy *d* orbitals. The equatorial F-S-F angle is only 101°33′ compared to 120° in a true trigonal bipyramid. The structure has been established by a

* Air Products and Chemicals, Inc., P.O. Box 538, Allentown, Pa. 18105; Chemical Procurement Laboratories, Inc., 18-17 130th St., College Point, N.Y. 11356; Columbia Organic Chemicals Company, Inc., 912 Drake St., Columbia, S.C. 29205; K and K Laboratories, Inc., 121 Express St., Plainview, N.Y. 11803; Matheson Gas Products, 932 Paterson Plank Rd., East Rutherford, N.J. 07073; Pfaltz and Bauer, Inc., 126-02 Northern Blvd., Flushing, N.Y. 11368; PRC, Incorporated, P.O. Box 1466, Gainesville, Fl. 32601; Research Organic/Inorganic Chemical Cc., 11686 Sheldon, Sun Valley, Ca. 91352; Scientific Gas Products of New England, 150 Charles St., Malden, Mass. 02148; Scientific Gas Products Inc., 513 Raritan Center, Edison, N.J. 08817.

[131] The Matheson Gas Data Book, *Sulfur Tetrafluoride*, 4th ed., pp. 459–462, Matheson Company, Inc., East Rutherford, N.J., 1966.

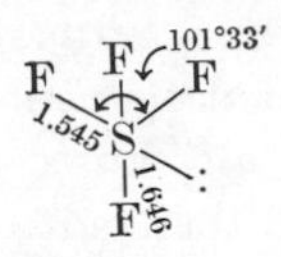

FIGURE 1. Structure of Sulfur Tetrafluoride.

study of the vibrational,[132] NMR,[133, 134] electron diffraction,[135, 136] and microwave spectra.[137]

Inorganic Complexes. The unshared electron pair in the sulfur tetrafluoride molecule correctly suggests that it should have basic character. The following Lewis acid addition compounds, in the approximate order of decreasing stability, have been prepared: $SF_4 \cdot SbF_5$, $SF_4 \cdot AsF_5$, $SF_4 \cdot IrF_5$, $SF_4 \cdot BF_3$, $SF_4 \cdot PF_5$, $(SF_4)_2 \cdot GeF_4$, and $SF_4 \cdot AsF_3$.[138–143] Early investigators suggested that these adducts were simple Lewis acid-Lewis base complexes of the type $SF_4 \rightarrow BF_3$. However, infrared data on the solid sulfur tetrafluoride-boron trifluoride adduct were obtained which suggest that it should be regarded as $SF_3^+BF_4^-$.[144] Nothing is known, however, of the character of these complexes in solution. All adducts prepared thus far are completely dissociated in the gas phase.

Sulfur tetrafluoride also acts as a weak electron-pair acceptor. Vapor pressure measurements indicate that 1:1 adducts of sulfur tetrafluoride with pyridine and with triethylamine are produced.[145, 146] The adduct $(CH_3)_4N^+SF_5^-$ is produced when sulfur tetrafluoride is bubbled into a solution of tetramethyl ammonium fluoride in dimethylformamide at room temperature.[147] The $CsSF_5$ adduct is produced when sulfur tetrafluoride and cesium fluoride are heated at 110° in a closed reactor.[148]

[132] R. E. Dodd, L. A. Woodward, and H. L. Roberts, *Trans. Faraday Soc.*, **52**, 1052 (1956).
[133] F. A. Cotton, J. W. George, and J. S. Waugh, *J. Chem. Phys.*, **28**, 994 (1958).
[134] E. L. Muetterties and W. D. Phillips, *J. Amer. Chem. Soc.*, **81**, 1084 (1959).
[135] K. Kimura and S. H. Bauer, *J. Chem. Phys.*, **39**, 3171 (1963).
[136] V. C. Ewing and L. E. Sutton, *Trans. Faraday Soc.*, **59**, 1241 (1963).
[137] W. M. Tolles and W. D. Gwinn, *J. Chem. Phys.*, **36**, 1119 (1962).
[138] W. C. Smith and E. L. Muetterties, U.S. Pat. 3,000,694 (1961) [*C.A.*, **56**, 3123 (1962)].
[139] N. Bartlett and P. L. Robinson, *J. Chem. Soc.*, **1961**, 3417.
[140] N. Bartlett and P. L. Robinson, *Chem. Ind.* (London), **1956**, 1351; *Proc. Chem. Soc.*, **1957**, 230.
[141] A. L. Oppegard, W. C. Smith, E. L. Muetterties, and V. A. Engelhardt, *J. Amer. Chem. Soc.*, **82**, 3835 (1960).
[142] P. L. Robinson and G. J. Westland, *J. Chem. Soc.*, **1956**, 4481.
[143] F. A. Cotton and J. W. George, *J. Inorg. Nucl. Chem.*, **7**, 397 (1958).
[144] F. Seel and O. Detmer, *Z. Anorg. Allg. Chem.*, **301**, 113 (1959).
[145] E. L. Muetterties, W. D. Phillips, and W. C. Smith, U.S. Pat. 2,897,055 (1959) [*C.A.*, **54**, 1564 (1960)].
[146] E. L. Muetterties, *J. Amer. Chem. Soc.*, **82**, 1082 (1960).
[147] R. Tunder and B. Siegel, *J. Inorg. Nucl. Chem.*, **25**, 1097 (1963).
[148] C. W. Tullock, D. D. Coffman, and E. L. Muetterties, *J. Amer. Chem. Soc.*, **86**, 357 (1964).

Toxicity and Hazards. Sulfur tetrafluoride has the same order of toxicity as phosgene based on inhalation tests with mice.[8, 149] For example, a concentration of 20 ppm in air causes death to mice after a 4-hour exposure; an autopsy indicated that pulmonary edema was the probable cause of death. This toxicity, together with the ease of hydrolysis and resulting formation of hydrogen fluoride, requires that reasonable care be taken when working with sulfur tetrafluoride. Its pungent irritating odor, which resembles that of sulfur dioxide, makes it easy to detect. Small to moderate amounts of sulfur tetrafluoride are easily destroyed by passage through excess aqueous alkali.[131]

Sulfur tetrafluoride may be handled in well-dried Pyrex glass apparatus up to about 30° with only nominal attack. Stainless steel, copper, and nickel are all inert to sulfur tetrafluoride at both ordinary and elevated temperatures. Hence apparatus fabricated from these materials is preferred where entended usage is contemplated. For reactions at elevated temperatures, pressure reactors lined with stainless steel or Hastelloy* have been used satisfactorily. Care must be taken to avoid overloading these reactors with the liquid because sulfur tetrafluoride has a vapor pressure of 10 atm even at 25°. Stainless steel cylinders used for storage of sulfur tetrafluoride should be equipped with a safety rupture disk assembly and should have a capacity of at least 1 cm^3 for each gram of sulfur tetrafluoride.[131, 149]

EXPERIMENTAL CONSIDERATIONS

Caution! Sulfur tetrafluoride and hydrogen fluoride are toxic and corrosive chemicals that should be handled with utmost care in a well-ventilated fume hood. Gloves and safety goggles should be used.

Reactions with sulfur tetrafluoride are generally carried out in stainless steel or Hastelloy-lined shaker tubes of 80–1000-ml capacity. Liquid or solid reactants are placed in the shaker tube under a nitrogen atmosphere, the head is screwed into place, the tube is cooled to solid carbon dioxide temperature, the nitrogen is removed with a vacuum pump, and gaseous reactants (HF, BF_3, SF_4, etc.) are condensed in the shaker tube. After having been heated for the prescribed period, the shaker tube is allowed to cool. If the gaseous products are of interest, they are condensed in an evacuated stainless steel cylinder at liquid nitrogen temperature; otherwise, excess sulfur tetrafluoride and volatile by-products are vented

* Hastelloy is a chemically resistant alloy of nickel, iron, and molybdenum manufactured by the Union Carbide Corporation, 270 Park Avenue, New York, N.Y. 10017.

[149] New Product Information Bulletin 2b, *Sulfur Tetrafluoride Technical*, Organic Chemicals Department, E.I. du Pont de Nemours and Co., Inc., 1961.

from the tube. Liquid or solid products are recovered when the tube is opened, and pure products are obtained by the usual processes of distillation, chromatography,* recrystallization, or sublimation. When necessary, hydrogen fluoride is removed from the crude mixture by either (1) pouring the crude mixture into water and recovering the product by filtration, extraction, or steam distillation, or (2) pouring the crude mixture into a suspension of sodium fluoride ($NaF + HF \rightarrow NaHF_2$) in an inert solvent followed by filtration and fractional distillation. Sometimes removal of hydrogen fluoride from a solid or high-boiling liquid is accomplished by vaporizing the hydrogen fluoride and absorbing it in solid sodium fluoride or sodium hydroxide.

Fluorinations with sulfur tetrafluoride occur under a great variety of conditions. Many proceed satisfactorily simply on heating the reactant with the requisite amount or a moderate excess of sulfur tetrafluoride in a closed metal pressure vessel. Temperatures ranging from 10 to 350° and higher have been employed; however, temperatures in the range 10–200° usually are sufficient. Carbonyl compounds generally require a coreactant such as hydrogen fluoride or boron trifluoride. Hydrogen fluoride is preferred because it gives higher yields under milder conditions and side reactions are minimized. Anhydrous hydrogen fluoride may be added as such or produced *in situ* by the addition of small amounts of water or ethanol. The presence of chlorine or sulfur chlorides in sulfur

$$SF_4 + H_2O \rightarrow 2\,HF + SOF_2$$

tetrafluoride is a common cause of intractable products; these impurities can be removed by distillation or by treatment with sulfur or mercury (p. 57).

Halogenated hydrocarbon solvents such as methylene chloride or chloroform commonly are added to the reaction mixture to improve the solubility of the organic reactant and to moderate the vigor of the fluorination reaction.

Anhydrous hydrogen fluoride may be used as a solvent in special cases. In addition to its catalytic and solvent actions, hydrogen fluoride also protects certain functional groups during fluorinations. For example, primary and secondary amino groups of amino acids, which normally are very reactive toward sulfur tetrafluoride, are protected by protonation by excess hydrogen fluoride, thus permitting conversion of amino acids to trifluoromethyl amines.[83]

* When chromatography is required, as it often is for steroids and other sensitive compounds, alumina or Florisil is usually employed. Florisil is a trademark for a highly selective adsorbent of extremely white, hard granular or powdered magnesia-silica gel (magnesium silicate) manufactured by the Floridin Company, 2 Gateway Center, Pittsburgh, Pa., 15222.

Where circumstances warrant, sodium fluoride may be added to remove hydrogen fluoride as produced. For example, the replacement of hydroxyl groups by fluorine with sulfur tetrafluoride does not require hydrogen fluoride catalyst, although it is produced in the reaction. Replacement of hydroxyl groups by fluorine in compounds containing acid-sensitive ether groups has been successfully carried out simply by adding sodium fluoride to the reactants.[150] Sodium fluoride is also beneficial where the conversion of carboxylic acid to carboxylic acid fluoride is desired; removal of hydrogen fluoride by sodium fluoride as the former is produced effectively prevents subsequent conversion of the acid fluoride to the trifluoromethyl derivative.[150] Treatment of the sodium salt of a carboxylic acid with sulfur tetrafluoride, with concomitant formation of sodium fluoride, accomplishes the same result.[7] Cesium fluoride is required as a catalyst in the addition of sulfur tetrafluoride to perfluoroolefins.[116]

EXPERIMENTAL PROCEDURES

p-Fluoro(heptafluoroisopropyl)benzene.[24] *p*-Fluoro-α,α-bis(trifluoromethyl)benzyl alcohol (15.9 g, 0.061 mol) was placed in a 140-ml Hastelloy-lined pressure vessel. After cooling and evacuating the vessel, 16 g (0.15 mol) of sulfur tetrafluoride was added, and the reactants were heated at 150° for 8 hours. The product was treated with excess sodium fluoride powder or pellets to remove hydrogen fluoride and was suction-filtered, using a small amount of methylene chloride or trichlorofluoromethane as a rinse. Distillation gave 11.2 g (70%) of *p*-fluoro(heptafluoroisopropyl)benzene, bp 126°.

1,4,4,4-Tetrafluoro-3-trifluoromethyl-1,2-butadiene.[14] 1,1,1-Trifluoro-2-trifluoromethyl-3-butyn-2-ol (20 g, 0.104 mol) was placed in a 300-ml stainless steel pressure reactor. The reactor was closed, cooled to −78°, and evacuated. Sulfur tetrafluoride (11.25 g, 0.104 mol) was introduced through a vacuum manifold system. The reactor was then allowed to stand overnight at ambient temperature. At the end of this period (about 16 hours), the excess pressure was released and the liquid contents were slurried with sodium fluoride, filtered, and distilled, giving 9.0 g (45%) of 1,4,4,4-tetrafluoro-3-trifluoromethyl-1,2-butadiene as a water-white liquid, bp 34°.

2,2,2-Trinitrofluoroethane.[28] Sulfur tetrafluoride (103 ml at −78°) was condensed into an evacuated 300-ml stainless steel bomb containing 100 g (0.55 mol) of trinitroethanol. The mixture was allowed to stand at ambient temperature for 4 days. Excess sulfur tetrafluoride

[150] C. W. Tullock, unpublished results from this laboratory.

and volatile by-products were then vented through an aqueous alkali trap. The residue was dissolved in 300 ml of methylene chloride, and the solution was washed six times with 50-ml portions of water. The solvent was removed by distillation. The residue was sublimed onto a −78° cold finger condenser at 0.1 mm to give 64 g (63%) of 2,2,2-trinitrofluoroethane, mp 34–35°.

3,5,7-Tribromo-2-fluorocycloheptatrienone.[7] 3,5,7-Tribromo-2-hydroxycycloheptatrienone (1.18 g, 0.0033 mol) and 5 ml of benzene were charged to a 15-ml platinum tube under nitrogen. The tube was cooled to solid carbon dioxide temperature, the nitrogen was removed with a vacuum pump, and 2.24 g (0.02 mol) of sulfur tetrafluoride was condensed into the tube. The tube was sealed under nitrogen and heated to 60° for 8 hours with shaking. The tube was cooled to room temperature and the excess sulfur tetrafluoride and other volatile by-products were vented. The residue was diluted with 25 ml of benzene and washed with water and saturated salt solution. After drying over anhydrous magnesium sulfate, the benzene solution was evaporated under reduced pressure to give 0.68 g (57%) of 3,5,7-tribromo-2-fluorocycloheptatrienone as a crystalline solid, mp 134–142°.

5α,6,6-Trifluoropregnane-3β,17α-diol-20-one Diacetate[52] (Equation on p. 23). A mixture of 13.7 g (0.0304 mol) of 5α-fluoropregnane-3β, 17α-diol-6,20-dione diacetate, 1.5 ml of water, and 100 ml of methylene chloride was placed in a 400-ml shaker tube. The reactor was sealed, cooled in a bath of solid carbon dioxide and acetone, and evacuated. Sulfur tetrafluoride (160 g, 1.48 mol) was condensed into the reactor, and the reactor was allowed to warm to 20°, at which temperature it was agitated for 10 hours. The volatile products were vented, and the contents were poured with caution into ice water. The phases were separated, and the methylene chloride solution was washed with water, 5% sodium bicarbonate solution, water, and saturated salt solution. The methylene chloride solution was dried over magnesium sulfate and then evaporated under reduced pressure to leave a tan, crystalline residue. Recrystallization from methanol-methylene chloride gave 8.75 g (61%) of product as thick, colorless blades: mp 249–250°; $[\alpha]^{24}_{D}$ −13° (chloroform).

The product from reaction of sulfur tetrafluoride with steroid ketones usually must be purified by column chromatography, *e.g.*, on Florisil or alumina. Less reactive ketones may require relatively more hydrogen fluoride and longer reaction times. With more sensitive ketones, a reaction temperature of about 10° may be preferable.

17,20;20,21-Bis(methylenedioxy)-6α-difluoromethyl-3β-hydroxy-5α-pregnan-11-one Acetate.[55] The following unusually mild conditions were devised for the reaction of a steroid aldehyde containing an acid-sensitive bis(methylenedioxy) protective group. The role of tetrahydrofuran may be that of a base which competes with the bis(methylenedioxy) group for the strong acid.

A mixture of 0.94 g (2.03 mmol) of 17,20;20,21-bis(methylenedioxy)-6α-formyl-3β-hydroxy-5α-pregnan-11-one acetate (**14**, p. 25), 0.05 ml of water, 0.25 ml of tetrahydrofuran, 20 ml of methylene chloride, and 46 g (0.426 mol) of sulfur tetrafluoride was agitated for 16 hours at 15° in a 100-ml stainless steel autoclave. After the autoclave was vented, the contents were diluted with methylene chloride and washed with excess aqueous potassium bicarbonate. The organic fraction was dried (sodium sulfate) and concentrated to dryness. Chromatography on Florisil with increasing percentages of acetone in hexane gave crystalline material from the 2–5% acetone-hexane eluates; recrystallization of the fractions melting between 232 and 238° afforded 0.59 g of product (60%), mp 233–238°. Further recrystallization from acetone and hexane gave an analytical sample: mp 238–242°; $[\alpha]_D$ −40° (chloroform).

1,1-Difluorocyclohexane.[50] A 2-liter Hastelloy-C pressure vessel containing a mixture of 245 g (2.5 mol) of cyclohexanone and 300 ml of methylene chloride was cooled in a bath of solid carbon dioxide and acetone and evacuated. Hydrogen fluoride (60 g, 3.0 mol) followed by sulfur tetrafluoride (216 g, 2.0 mol) was added to the vessel, which was then sealed and rocked at 30° for 48 hours. The vessel was cooled to ambient temperature and vented slowly. The residual liquid was washed successively with water, 10% aqueous sodium bicarbonate, and saturated salt solution, and then dried over sodium sulfate. The organic layer was flash-distilled in a rotary evaporator. The distillate was fractionated in a spinning-band column, giving 210 g (70%) of 1,1-difluorocyclohexane; bp 99–100°; n^{25}_D 1.3904.

1,1,1-Trifluoroheptane.[75] Heptanoic acid (26 g, 0.20 mol) was placed in a 145-ml Hastelloy-C pressure vessel. The air in the vessel was displaced with nitrogen, and the head of the vessel was secured in place. The vessel was cooled in a bath of acetone and solid carbon dioxide, and the vessel was evacuated with a vacuum pump to a pressure of 0.5–1.0 mm. Sulfur tetrafluoride (65 g, 0.57 mol, 95% pure) was transferred to the cold vessel.

The pressure vessel was heated with agitation at 100° for 4 hours and at 130° for 6 hours. The vessel was cooled to room temperature, and volatile by-products were vented. The crude, fuming, liquid product was poured into a stirred suspension of 10 g of finely divided sodium fluoride in 60 ml of pentane and filtered. The filtrate was fractionated through a

15-cm Vigreux column. 1,1,1-Trifluoroheptane (24.6 g, 70%) was collected; bp 100–101°; n^{25}D 1.3449.

24,24,24-Trifluoro-5β-cholan-3α-ol Acetate (Equation on p. 32).[30] A mixture of 1.0 g (2.39 mmol) of lithocholic acid acetate, 20 ml of methylene chloride, and 0.75 ml of water was placed in a 100-ml stainless steel autoclave. The autoclave was sealed and cooled in a bath of solid carbon dioxide and acetone and evacuated. Sulfur tetrafluoride (46 g, 0.426 mol) was added. The autoclave was rocked for 16 hours at 20°, after which the volatiles were vented through caustic solution. Remaining volatile material was removed under vacuum. The residue was taken up in 200 ml of methylene chloride and washed with 10% aqueous potassium bicarbonate. The methylene chloride solution was dried over sodium sulfate and concentrated to dryness under reduced pressure, leaving 1.20 g of brown gum which was chromatographed on 60 g of Florisil. Elution with a petroleum ether (bp 40–60°)-methylene chloride mixture gave 273 mg of crystalline solid. Recrystallization from acetone-water afforded 268 mg (25%) of 24,24,24-trifluoro-5β-cholan-3α-ol acetate as colorless crystals: mp 147–149°; [α]D +44° (chloroform).

3α-Hydroxy-5β-cholanyl Fluoride Acetate.[30] A mixture of 5.00 g (12 mmol) of lithocholic acid acetate, 20 ml of methylene chloride, and 46 g (0.426 mol) of sulfur tetrafluoride was agitated for 16 hours at 25° in a 100-ml stainless steel autoclave. The autoclave was vented and its contents taken up in methylene chloride, washed with 10% aqueous potassium bicarbonate, dried over sodium sulfate, and concentrated to dryness under reduced pressure. This left 5.0 g of crystalline residue. Recrystallization from methylene chloride-petroleum ether afforded 3α-hydroxy-5β-cholanyl fluoride acetate (4.75 g, 95%): mp 157–158°; [α]D +44°.

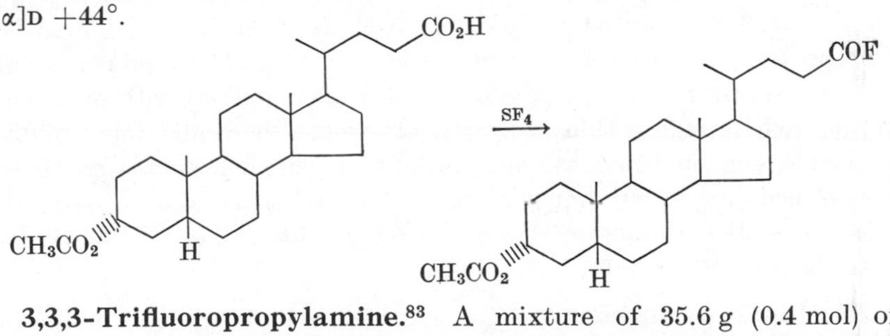

CO_2H $\xrightarrow{SF_4}$ COF

CH_3CO_2 H CH_3CO_2 H

3,3,3-Trifluoropropylamine.[83] A mixture of 35.6 g (0.4 mol) of β-alanine, 50 g (2.5 mol) of hydrogen fluoride, and 100 g (0.93 mol) of sulfur tetrafluoride was heated at 120° for 8 hours with good agitation in a 400-ml shaker tube lined with stainless steel or Hastelloy. The reaction product was poured into a polyethylene dish and warmed on a steam bath to expel hydrogen fluoride. The residue was stirred with water and filtered to separate insoluble materials from the amine hydrofluoride. The filtrate was evaporated to give solid amine hydrofluoride from which tarry contaminants were removed by washing with acetone. The amine was liberated with 40% aqueous potassium hydroxide, separated, and dried over sodium hydroxide. Distillation gave 18.5 g (41%) of product, bp 67.5–68°; n^{25}D 1.3270.

2,4-Bis(hexafluoroisopropylidene)-1,3-dithietane.[86] A 400-ml stainless steel autoclave containing 60 g (0.149 mol) of 2,4-bis(dicarbethoxymethylene)-1,3-dithietane was cooled to −80° and charged with 10 g (0.5 mol) of hydrogen fluoride and 125 g (1.16 mol) of sulfur tetrafluoride. The autoclave was heated with agitation for 2 hours at 125°, for 2 hours at 150°, and for 4 hours at 200°. It was cooled to 25° and volatile products were vented. The reactor was then cooled to −80° and recharged with 35 g (1.75 mol) of hydrogen fluoride and 125 g (1.16 mol) of sulfur tetrafluoride. The recharged tube was heated at 150° for 2 hours and at 200° for 4 hours, cooled to 25°, and vented. The crude product was poured onto ice. The crystalline solid was separated by filtration and washed with water and 10% aqueous sodium carbonate. The product was steam-distilled and filtered from the aqueous distillate. The moist crystals were dissolved in boiling methylene chloride, anhydrous magnesium sulfate was added, and the hot solution was filtered. Cooling the filtrate gave 40 g (69%) of 2,4-bis(hexafluoroisopropylidene)-1,3-dithietane in two crops: mp 84.5–85.5°; bp 173°.

3,4-Dichloro-2,2,5,5-tetrafluoro-2,5-dihydrofuran.[97] A 133-ml stainless steel autoclave, containing 57.4 g (0.33 mol) of dichloromaleic anhydride, was cooled in solid carbon dioxide-acetone and evacuated. Sulfur tetrafluoride (89 g, 0.823 mol) was charged to the autoclave. The sealed reactor was heated at 300° for 10 hours; then it was allowed to cool and the gases were vented into a caustic trap. The liquid residual mixture was poured into a stainless steel beaker, flushed with nitrogen, and poured through a thin layer of glass wool. The liquid was rapidly distilled through a 105-cm column packed with glass helices. The distillate was washed with dilute aqueous potassium carbonate. Careful redistillation gave 31.9 g (46%) of 3,4-dichloro-2,2,5,5-tetrafluoro-2,5-dihydrofuran, bp 74°, n^{25}D 1.3611.

4,6-Difluoropyrimidine.[100] A mixture of 20 g (0.13 mol) of 4,6-dichloropyrimidine (containing a small amount of 2,4,6-trichloropyrimidine) and 60 g (0.55 mol) of sulfur tetrafluoride was heated in a 300-ml stainless steel bomb with rocking for successive 3-hour periods at 50, 100, and 150°. The liquid product in the bomb was dissolved in diethyl ether and fractionated to give 11.2 g of a yellow liquid, bp 35–40° (55 mm). Redistillation gave a main fraction, bp 50° (100 mm). Mass spectrometric analysis showed the product to be 4,6-difluoropyrimidine containing a trace of chlorodifluoropyrimidine. The yield was about 70%.

Cyanuric Fluoride.[100] A mixture of 23.1 g (0.13 mol) of cyanuric chloride and 81 g (0.75 mol) of sulfur tetrafluoride was heated in a 300-ml stainless steel bomb with rocking at 150° for 2 hours, at 200° for 4 hours, and at 250° for 6 hours. A brown liquid product (24.5 g) was obtained. Distillation gave 6.8 g (40%) of cyanuric fluoride, bp 70–73°.

Bis(phenylimino)sulphur.[151] Aniline (56 g, 0.6 mol) in 100 ml of methylene chloride was cooled in solid carbon dioxide/acetone, and 11 g (0.1 mol) of sulfur tetrafluoride was passed in. The mixture was allowed to warm to 24° and the aniline hydrofluoride was removed by filtration. The solvent was boiled off, finally under vacuum on a steam bath. Petroleum ether was added to precipitate tar, and the decanted solution was distilled to give 10.0 g (94%) of bis(phenylimino)sulfur, bp 105° (0.1 mm).

Phenyliminosulfur Difluoride.[34] Phenyl isocyanate (59.5 g, 0.50 mol) and sulfur tetrafluoride (61 g, 0.55 mol; 10% excess) were heated at 100° for 4 hours, at 150° for 6 hours, and then at 100° for 4 hours. The volatile product (41 g) was shown by mass spectrometric analysis to contain 80 mole% carbonyl fluoride (COF_2), 6 mole% thionyl fluoride (SOF_2), 1 mole% carbon dioxide, and unreacted sulfur tetrafluoride. The crude liquid product (77.5 g) was distilled to give 71.2 g (88% yield) of phenyliminosulfur difluoride, bp 36–36.5° (2 mm).

α,α-Difluorobenzyliminosulfur Difluoride.[34] Benzonitrile (20.6 g, 0.20 mol) and sulfur tetrafluoride (44 g, 0.40 mol; 100% excess) were heated at 180° for 2 hours and then at 250° for 16 hours. The bomb was opened and 41 g of a clear, light-green liquid product was recovered. Distillation gave 28.0 g of a pale-yellow product, bp 33° (3 mm). Redistillation gave 20.5 g (48%) of colorless product, bp 55.0° (11 mm).

Trifluoromethyliminosulfur Difluoride.[109] A 6.3-mmol sample of cesium fluoride was transferred to a 10-ml stainless steel cylinder in a nitrogen-filled dry box. The salt was vacuum-dried for 2 hours at 190° before admitting 1.89 mmol of cyanuric fluoride (vp at 0°, 22.0 mm) and 7.59 mmol of purified sulfur tetrafluoride.* Material transfers were made in an all-glass vacuum system fitted with mercury cutoff valves. Contents of the steel reaction cylinder were heated at 155° for 18 hours, after which the volatile materials were separated in a vacuum system. On the basis of molecular weight, vapor pressure, and infrared data, the

* Impurities had previously been removed from the commercial sulfur tetrafluoride by forming the sulfur tetrafluoride-boron trifluoride complex from which sulfur tetrafluoride was liberated by treatment with a deficiency of diethyl ether at 0°.

[151] M. S. Raasch, *J. Org. Chem.*, **37**, 1347 (1972). See ref. 105.

main product was identified as 5.47 mmol of trifluoromethyliminosulfur difluoride accompanied by a small amount (0.0586 g) of a mixture containing cyanuric fluoride, sulfur tetrafluoride-silicon tetrafluoride, and an unidentified fluorocarbon material. Trifluoromethyliminosulfur difluoride was obtained in 96% yield.

Lowering the reaction time to 13 hours reduced the yield to 65–70%.

1-(Trifluoromethyl)piperidine.[33] A 145-ml Hastelloy shaker vessel charged with 47 g (0.15 mol) of bis(pentamethylene)thiuram disulfide and 49 g (0.45 mol) of sulfur tetrafluoride was heated for 6 hours at 100°. A mixture (69 g) of magenta-colored liquid and yellow solid was obtained. Filtration gave 23 g (88%) of sulfur. Distillation of the filtrate through an oven-dried Vigreux column gave 34 g (70%) of 1-(trifluoromethyl)-piperidine, bp 44° (68 mm); some decomposition occurred during distillation. The amine is quite sensitive to traces of moisture and decomposes fairly rapidly in the presence of glass.

1-Chloro-1,1,2,2-tetrafluoroethane.[115] A 25-ml stainless steel pressure vessel was charged with 3.59 g (0.015 mol) of lead dioxide, 1.17 g (0.010 mol) of 1-chloro-1,2-difluoroethylene, and 5.40 g (0.05 mol) of sulfur tetrafluoride. Lead dioxide was weighed directly into the vessel before closing. Sulfur tetrafluoride and the olefin, which were gaseous at room temperature, were measured gasometrically and condensed into the pressure vessel, which had been previously cooled in liquid nitrogen and evacuated. The reaction vessel was allowed to warm to room temperature and was then rocked at 100° for 2 hours. The vessel was cooled to room temperature and the volatile contents were passed through a wash tower containing 300 ml of 30% aqueous sodium hydroxide cooled to 0°. Material passing the sodium hydroxide wash was condensed in a trap cooled with solid carbon dioxide and acetone. The total amount of material thus obtained was measured either gasometrically or by transferring it to a suitable container. Product purity was determined from infrared absorption data, with medium-strong bands at 7.25, 7.30, and 7.40 μ; strong bands at 8.55, 8.75, 8.80, 9.25, and 9.30 μ; and strong triplet bands centered at 7.95, 10.00, 12.02, and 14.75 μ. 1-Chloro-1,1,2,2-tetrafluoroethane was obtained in 96% yield. If desired, an analytically pure sample of product may be recovered by vapor-liquid phase chromatography. The analytical column is a 6-mm o.d. copper or stainless steel tubing packed with silica gel heated to 100°; length, 229 cm.

Phenyltetrafluorophosphorane.[66] A 145-ml Hastelloy shaker vessel was charged with 29.5 g (0.25 mol) of phenylphosphoric acid and 108 g (1.00 mol) of sulfur tetrafluoride. It was heated successively at 100° for 2 hours, at 120° for 4 hours, and at 150° for 10 hours. The volatile reaction products were vented. The residual light-brown, liquid product (41.8 g) was treated with 20 g of sodium fluoride suspended in 100 ml of petroleum ether (bp 30–60°) to remove hydrogen fluoride. Filtration under nitrogen and distillation at atmospheric pressure through a Vigreux column gave 27.0 g (58%) of phenyltetrafluorophosphorane, bp 133–134°.

p-(Difluoroiodo)toluene.[122] A 1.2-g portion of *p*-iodosotoluene was suspended in 20 ml of methylene chloride in a dry quartz reaction vessel. The suspension was cooled to −20°. Sulfur tetrafluoride was bubbled in with stirring by a magnetic agitator. After 10 minutes the reaction was complete, the precipitate having dissolved. Dry nitrogen was bubbled through the reaction mixture and then the methylene chloride was removed by distillation to give a 100% yield of a residue of pure *p*-(difluoroiodo)toluene, mp 108–109°.

Difluoroiodides are extremely sensitive to atmospheric moisture and hydrolyze on storage with formation of iodoso derivatives. They must therefore be handled in a dry box and stored in platinum vessels.

TABULAR SURVEY

The following tables list the reactions of sulfur tetrafluoride reported through 1971. Compounds are divided among the tables according to the functional group undergoing reaction as follows: I, Alcohols; II, Aldehydes; III, Ketones; IV, Carboxylic Acids; V, Acyl Halides; VI, Esters; VII, Amides; VIII, Anhydrides; IX, Organic Halides (Excluding Haloalkenes or Acyl Halides); X, Amines; XI, Compounds with Carbon-Nitrogen Multiple Bonds; XII, Thiocarbonyl Compounds; XIII, Haloalkenes; XIV, Organo-arsenic, -boron, -phosphorus, and -silicon Compounds; XV, Miscellaneous Compounds.

In the several cases in which a compound has two or more different functional groups *which undergo reaction*, it appears in each appropriate table. Thus a hydroxyacid will appear in Table I (Alcohols) if its hydroxyl group is replaced by a fluorine atom, and in Table IV (Carboxylic Acids) if its carboxyl group is converted to CF_3 or COF.

Within each table, compounds are arranged by their empirical formulas according to the method of *Chemical Abstracts*, *i.e.*, in the order of increasing number of carbon atoms and then in the order of increasing number of the other elements in alphabetical order. The first reference listed for an entry gives the best experimental procedure. Other references are given in numerical order. "No reaction" indicates reported unreactivity with recovery of the starting material. When rearrangement, elimination, and substitution products occur, they are included in the tables.

Yields are reported whenever available. If there is more than one reference, the higher yield is reported. In such cases the yield is from the first reference listed.

The only uncommon abbreviation used is *hν* for ultraviolet irradiation.

TABLE I. ALCOHOLS

	Alcohol	Conditions	Product(s) and Yield(s) (%)	Refs.
C_2	$(O_2N)_2CFCH_2OH$	25°, 20 hr	$(O_2N)_2CFCH_2F$ (I, 25), $(O_2N)_2CFCH_2OSF_3$ (II, 28), $[(O_2N)_2CFCH_2O]_2SO$ (III, 15)	12
		100°, 20 hr	I (30), II (9),[a] III (—)	12
	$(O_2N)_3CCH_2OH$	Room temp, 4 d	$(O_2N)_3CCH_2F$ (63)	28
	CH_2OHCO_2H	160°, 5 hr	CH_2FCF_3 (43)	7
C_3	$CHF_2CF_2CH_2OH$	45–80°, 6 hr	$CHF_2CF_2CH_2F$ (65)	21
	$CH_3C(NO_2)_2CH_2OH$	85–90°, 20 hr	$CH_3C(NO_2)_2CH_2F$ (46)	28
	$CH_2OHC(NO_2)_2CH_2OH$	85–90°, 8 hr	$CH_2FC(NO_2)_2CH_2F$ (62)	28
C_4	$(CF_3)_2C(OH)CCl_3$	300°, 15 hr	$(CF_3)_2C(CCl_2F)Cl$ (53), $(CF_3)_2C(CClF_2)Cl$ (10)	25
	$CClF_2C(OH)(CF_3)CH_3$	50°, 16 hr	$CF_3C(CH_3)ClCOF$ (—)[b]	25
		50°, prolonged treatment	$(CF_3)_2C(CH_3)Cl$ (—)	25
	$(CClF_2)_2C(OH)CH_3$	90°, 16 hr	$CClF_2CCl(CF_3)CH_3$ (65)	38, 25
	$CCl_2FC(OH)(CClF_2)CH_3$	90°, 16 hr	$CCl_2FCCl(CF_3)CH_3$ (—)	38
	$(CCl_3)_2C(OH)CH_3$	90°, 16 hr	$CCl_3CCl(CCl_2F)CH_3$ (—)	38
	$(CF_3)_2C(OH)CH_3$	90°, 18 hr	$(CF_3)_2C{=}CH_2$ (72)	25
	$CF_3C(OH)(CHF_2)CH_3$	90–92°, 16 hr	$(CF_3)(CHF_2)C{=}CH_2$ (55)	25
	$(CHF_2)_2C(OH)CH_3$	95°, 16 hr	$(CHF_2)_2C{=}CH_2$ (38)	25
	$(CH_3)_2C(NO_2)CH_2OH$	110°, 8 hr	No reaction	28
C_5	$(CF_3)_2C(OH)C{\equiv}CCl$	Room temp, 16 hr	$(CF_3)_2C{=}C{=}CClF$ (13), $(CF_3)_2CFC{\equiv}CCl$ (30)	14, 152
	$CClF_2C(OH)(CF_3)C{\equiv}CH$	Room temp, 16 hr	$CClF_2C(CF_3)C{=}C{=}CHF$ (30)	14, 152
	$(CClF_2)_2C(OH)C{\equiv}CH$	Room temp, 16 hr	$(CClF_2)_2C{=}C{=}CHF$ (58)	14, 152
	$(CF_3)_2C(OH)C{\equiv}CH$	20–25°, 16–18 hr	$(CF_3)_2C{=}C{=}CHF$ (45)	14, 152
	$(CHF_2)_2C(OH)C{\equiv}CCl$	Room temp, 16 hr	$(CHF_2)_2CFC{\equiv}CCl$ (84)	14, 152

	$CHF_2C(OH)(CClF_2)C{\equiv}CH$	Room temp, 16 hr	$CHF_2C(CClF_2){=}C{=}CHF$ (26), $(CHF_2)(CClF_2)CFC{\equiv}CH$ (26)	14, 152
	$CHF_2C(CF_3)(OH)C{\equiv}CH$	Room temp, 16 hr	$CHF_2C(CF_3){=}C{=}CHF$ (32), $CHF_2CF(CF_3)C{\equiv}CH$ (4)	14, 152
	$CClF_2C(OH)(CF_3)CH{=}CH_2$	Room temp, 17 hr	$CClF_2C(CF_3){=}CHCH_2F$ (—)	27
	$(CHF_2)_2C(OH)C{\equiv}CH$	Room temp, 16 hr	$(CHF_2)_2C{=}C{=}CHF$ (32), $(CHF_2)_2CFC{\equiv}CH$ (26)	14
	$(CF_3)_2C(OH)CH{=}CH_2$	Room temp, 3 d	$(CF_3)_2C{=}CHCH_2F$ (47)	27, 14
		Room temp, 17 hr	" (59)	27, 14
	$(CH_3)_2C(OH)C{\equiv}CH$	Exothermic reaction	$CH_2{=}C(CH_3)C{\equiv}CH$ (—), $(CH_3)_2CFC{\equiv}CH$ (—)	14
C_6	$CCl_2FC(CF_3)(OH)C(CH_3){=}CH_2$	Room temp, 17 hr	$CCl_2FC(CF_3){=}C(CH_3)CH_2F$ (—)	27
C_7	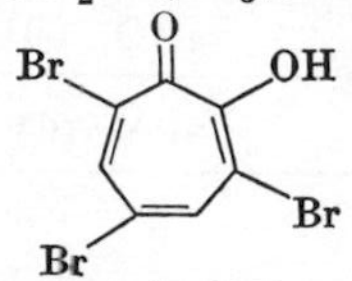 Br, O, OH, Br, Br (tropolone structure)	60°, 8 hr, C_6H_6	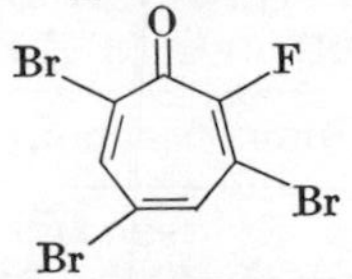 Br, O, F, Br, Br (tropone structure) (57)	7
	$C_6F_5CH_2OH$	85°, 20 hr	$C_6F_5CH_2F$ (80)	37
	$CHF_2(CF_2)_5CH_2OH$	60–80°, 10 hr	$CHF_2(CF_2)_5CH_2F$ (100)	21
	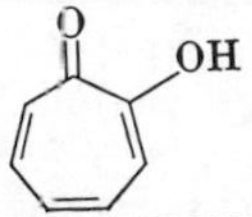 O, OH (tropolone structure)	60°, 10 hr, C_6H_6	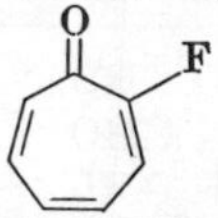 O, F (tropone structure) (28)	7
C_8	$C_6F_5CHOHCClF_2$	50–90°, 18 hr	$C_6F_5CHFCClF_2$ (90)	23, 153
	$C_6F_5CHOHCF_3$	50°, 2 hr	$C_6F_5CHFCF_3$ (12)	23, 153, 22
		50–90°, 12–15 hr	" (90)	23, 22, 153
	$C_6F_5CHOHCO_2H$	180°, 20 hr	$C_6F_5CHFCF_3$ (35)	37

Note: References 152–183 are on p. 124.

[a] The reduced yield was attributed to water-washing during workup.

[b] The product was not positively identified as 1-chloro-1-trifluoromethylpropionyl fluoride.

TABLE I. ALCOHOLS (*Continued*)

	Alcohol	Conditions	Product(s) and Yield(s) (%)	Refs.
C_8	$CF_3(CF_2)_6CH_2OH$	60–80°, 8 hr	$CF_3(CF_2)_6CH_2F$ (54)	21
(*contd.*)	$C_6F_5CHOHCH_3$	100°, 20 hr	*trans*-$C_6F_5CH{=}CHCH(CH_3)C_6F_5$ (80)	37
C_9	p-$FC_6H_4C(CF_3)_2OH$	150°, 8 hr	p-$FC_6H_4C(CF_3)_2F$ (70)	24
	$C_6H_5C(CF_3)_2OH$	150°, 8 hr	$C_6H_5C(CF_3)_2F$ (91)	24, 36
	o-$H_2NC_6H_4C(CF_3)_2OH$	150°, 8 hr	o-$H_2NC_6H_4C(CF_3)_2F$ (8)	24
	p-$H_2NC_6H_4C(CF_3)_2OH$	150°, 8 hr	p-$H_2NC_6H_4C(CF_3)_2F$ (19)	24
C_{10}	p-$CH_3C_6H_4C(CF_3)_2OH$	150°, 8 hr	p-$CH_3C_6H_4C(CF_3)_2F$ (64)	24
C_{11}	p-$(CH_3)_2NC_6H_4C(CF_3)_2OH$	150°, 8 hr	p-$(CH_3)_2NC_6H_4C(CF_3)_2F$ (62)	24
C_{12}	p-$(CH_3)_2CHC_6H_4C(CF_3)_2OH$	150°, 8 hr	p-$(CH_3)_2CHC_6H_4C(CF_3)_2F$ (54)	24

TABLE II. ALDEHYDES

	Aldehyde	Conditions[a]	Product(s) and Yield(s) (%)	Refs.
C_1	$(CH_2O)_n$[b]	150°, 6 hr	CH_2F_2 (49), FCH_2OCH_2F (21)	7
C_2	Cl_3CCHO	150°, 6 hr	$Cl_2CFCHClF$ (62)	67
	CH_3CHO	50°, 14 hr	CH_3CHF_2 (35)	7, 76
C_5	$H(CF_2)_4CHO$	100°, 10 hr	$H(CF_2)_4CHF_2$ (55)	7
	5-nitrofuran-2-CHO (O_2N, O, CHO)	65°, 8 hr	5-nitro-2-(difluoromethyl)furan (O_2N, O, CHF_2) (28)	41
	5-formyluracil (HN, O, CHO, N-H)	50–100°, 25 hr[c]	5-(difluoromethyl)uracil (HN, O, CHF_2, N-H)	42, 154
C_6	C_6H_5CHO	150°, 6 hr[d]	$C_6H_5CHF_2$ (81)	7, 76

C_7	C_6F_5CHO	100°, 24 hr[e]	$C_6F_5CHF_2$ (80)	37
	n-$C_6H_{13}CHO$	60°, 8 hr	n-$C_6H_{13}CHF_2$ (43)	7
C_8	p-$C_6H_4(CHO)_2$	150°, 8 hr	p-$C_6H_4(CHF_2)_2$ (88)	7, 76
		150°, 13 hr, 1300 psi	" (80)	49
	p-$OHCC_6H_4CO_2H$	150°, 13 hr 1100 psi	p-$F_2CHC_6H_4CF_3$ (47) p-$F_2CHC_6H_4COF$ (19)[f]	49
C_{22}	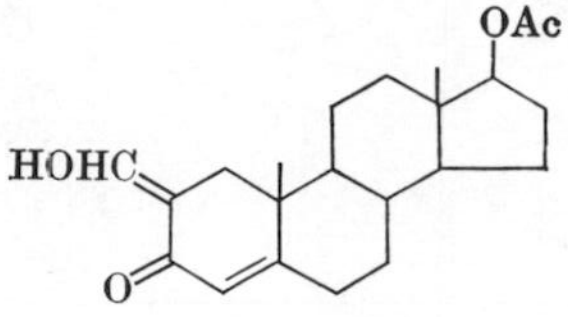 HOHC, O, OAc	20°, 16 hr, CH_2Cl_2[c]	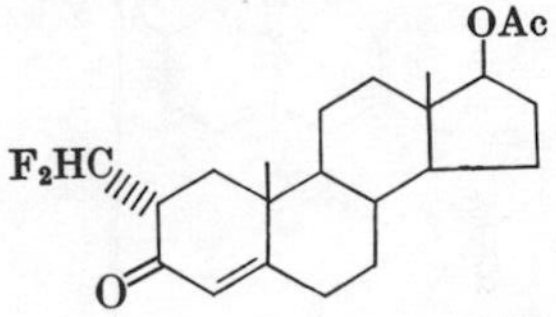 F_2HC, O, OAc (26)[g]	30, 74
	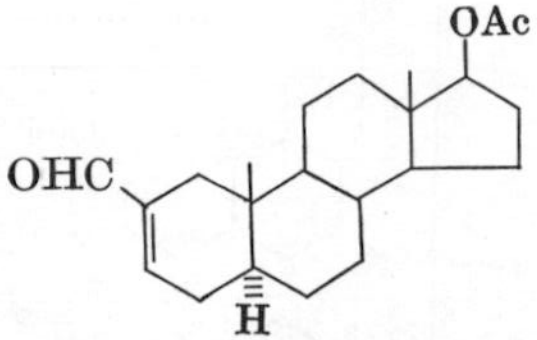 OHC, H, OAc	15°, 16 hr, $CHCl_3$[h]	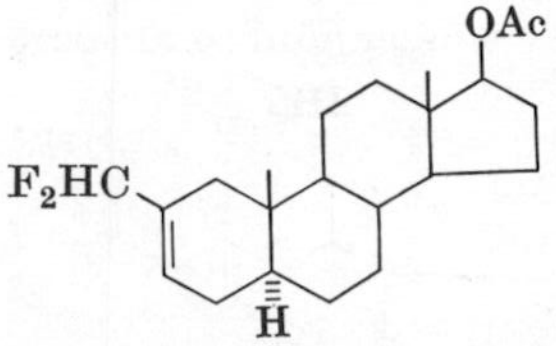 F_2HC, H, OAc (28)	56

Note: References 152–183 are on p. 124.

[a] Unless indicated otherwise, all aldehyde reactions were carried out without added hydrogen fluoride.
[b] α-Polyoxymethylene. See Table XV for reaction with $(CH_2O)_3$.
[c] Water was added to generate hydrogen fluoride.
[d] The molar ratio of aldehyde to sulfur tetrafluoride was 1:2; with a ratio of 1:1 no product was isolated.
[e] Hydrogen fluoride was added.
[f] The acid fluoride was obtained as a higher-boiling fraction in one preparation of p-difluoromethylbenzotrifluoride.
[g] Isolation of the product was complicated by the presence of a chlorine-containing impurity that was not characterized.
[h] Ethanol (2%) was added to generate hydrogen fluoride.

TABLE II. ALDEHYDES (*Continued*)

Aldehyde	Conditions[a]	Product(s) and Yield(s) (%)	Refs.
C_{24} 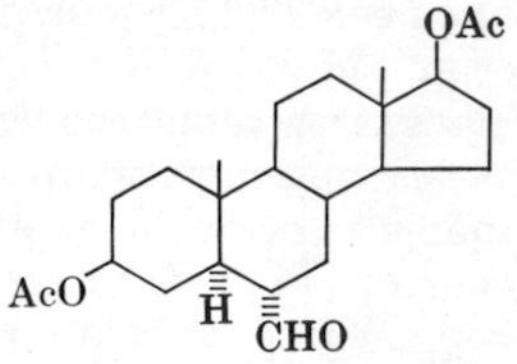 (AcO, OAc, H, CHO)	15°, 16 hr[c]	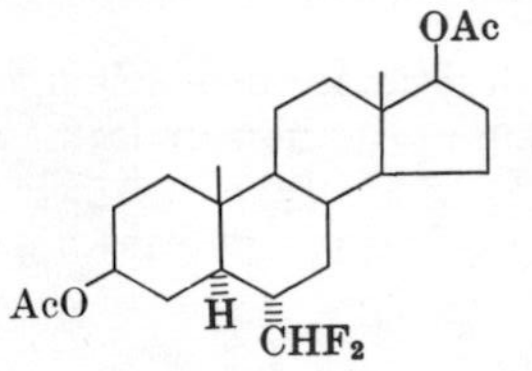 (AcO, OAc, H, CHF_2) (25)	74
C_{26} 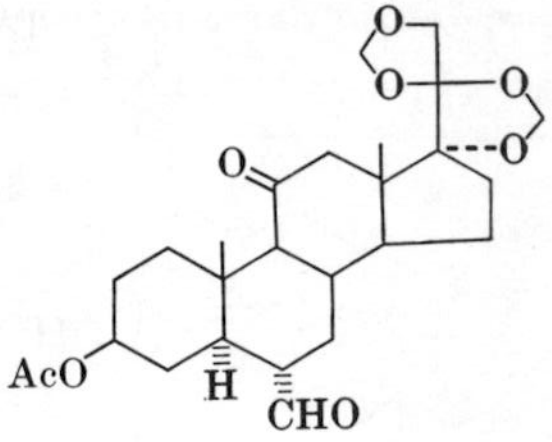 (AcO, H, CHO)	15°, 16 hr, CH_2Cl_2[i]	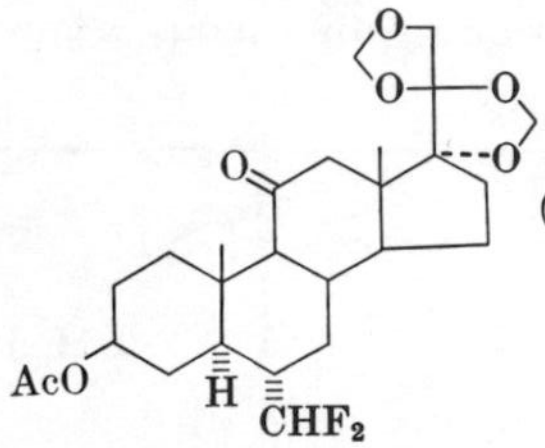 (AcO, H, CHF_2) (60)	55, 74, 155
	15°, overnight, CH_2Cl_2[j]	Mixture of products: 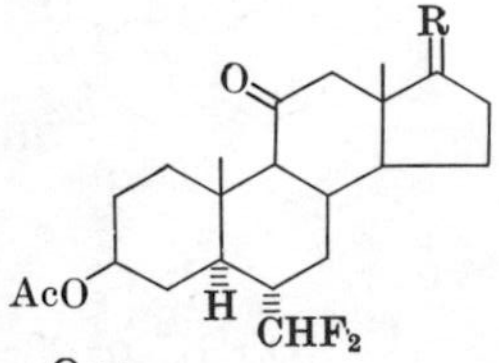 (AcO, H, CHF_2, R) where R is 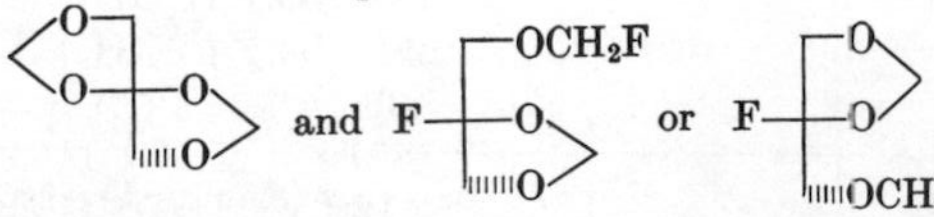 ... and F— ... OCH_2F or F— ... OCH_2F	55

Note: References 152–183 are on p. 124.

[c] Water was added to generate hydrogen fluoride.

[i] Tetrahydrofuran (0.25 ml), water (0.05 ml), and steroid (0.94 g) were used.

TABLE III. Ketones

	Ketone	Conditions[a]	Product(s) and Yield(s) (%)	Refs.
C_3	CD_3COCD_3	110°, 16 hr[b]	$CD_3CF_2CD_3$ (60–70)	156
	CH_3COCH_3	110°, 16 hr[b]	$CH_3CF_2CH_3$ (60)	7, 76, 156
C_4	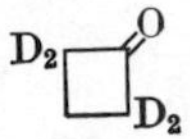 D_2, O, D_2	30°, 20 hr	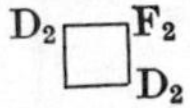 D_2, F_2, D_2 (60)	57
	$(CH_2)_3CO$	30°, 20 hr	$(CH_2)_3CF_2$ (60)	57
C_5	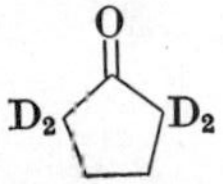 O, D_2, D_2	43°, 22 hr	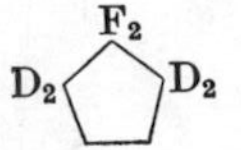 F_2, D_2, D_2 (ca. 50)	58
	$(CH_2)_4CO$	43°, 22 hr	$(CH_2)_4CF_2$ (50)	58
		30°, 120 hr, CH_2Cl_2	$(CH_2)_4CF_2$ (39)	50
C_6	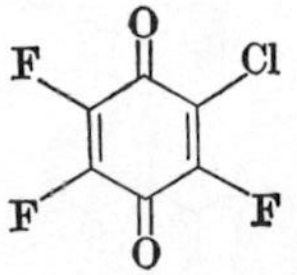 O, O, F, Cl, F, F	125°, 20 hr	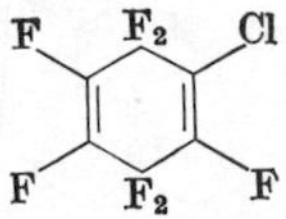 F, F_2, Cl, F, F_2, F (ca. 70)	64
	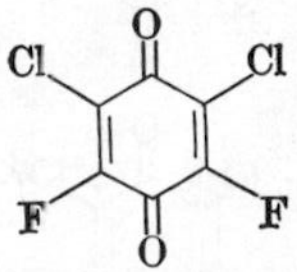 O, O, Cl, Cl, F, F	100°, 17 hr	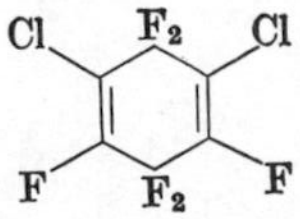 Cl, F_2, Cl, F, F_2, F (ca. 70)	64

Note: References 152–183 are on p. 124.

[a] Unless indicated otherwise, all reactions were run with hydrogen fluoride catalyst added as such or generated *in situ* from water.

[b] No catalyst was added.

TABLE III. KETONES (*Continued*)

	Ketone	Conditions[a]	Product(s) and Yield(s) (%)	Refs.
C_6 (*contd.*) 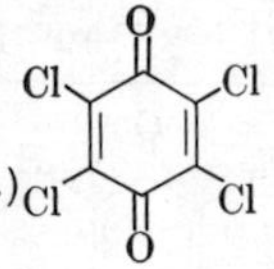	2,3,5,6-Tetrachloro-1,4-benzoquinone	270°, 2.5 hr	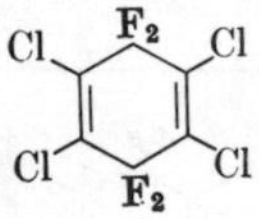 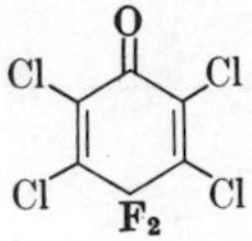 1,2,4,5-Tetrachloro-3,3,6,6-tetrafluorocyclohexa-1,4-diene (75), 2,3,5,6-tetrachloro-4,4-difluorocyclohexa-2,5-dienone (2)	7
	2,3,5,6-Tetrafluoro-1,4-benzoquinone	160°, 19 hr	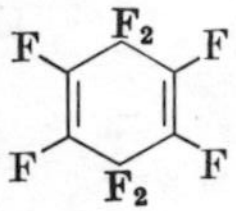 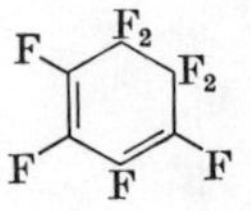 Octafluorocyclohexa-1,4-diene (74), octafluorocyclohexa-1,3-diene (6)	25
	1,4-Benzoquinone	200°, 4 hr[c]	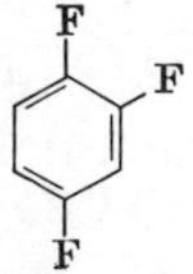 1,2,4-Trifluorobenzene (30)	7
	2,5-Dihydroxy-1,4-benzoquinone	60°, 8 hr	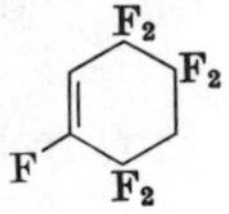 1,3,3,4,4,6,6-Heptafluorocyclohexene (40)	7
	O_2N-furan-$COCH_3$ (5-nitro-2-acetylfuran)	75°, 10 hr[d]	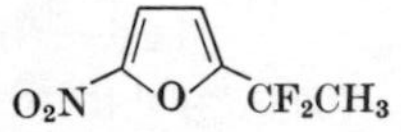 O_2N-furan-CF_2CH_3 (25)	41
		55–60°, 10 hr	" (34)	41
	1,2-Cyclohexanedione	50°, 24 hr	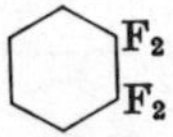 1,1,2,2-Tetrafluorocyclohexane (14)	61

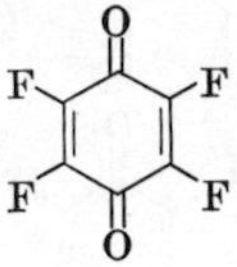

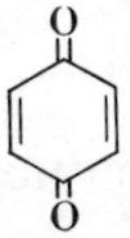

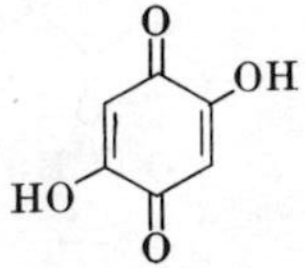

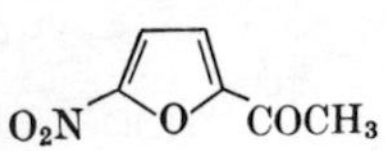

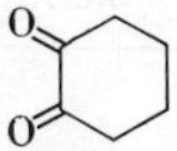

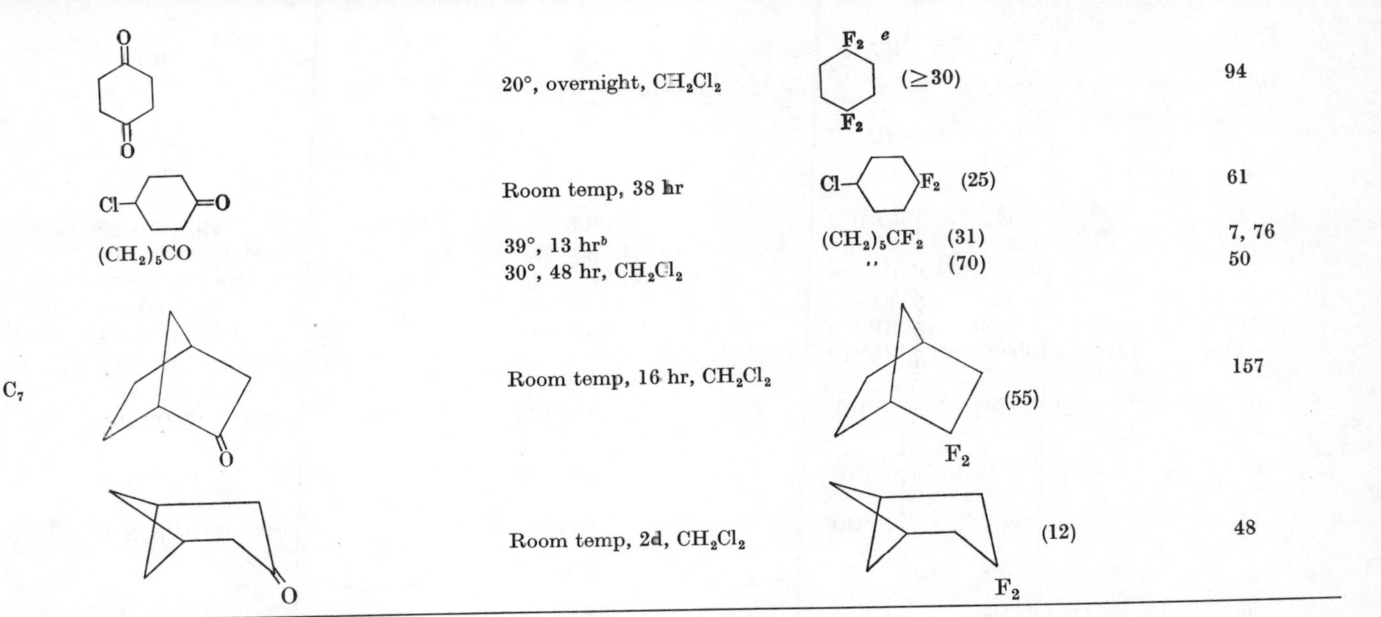

	Reactant	Conditions[a]	Product (Yield %)	Refs.
	1,4-cyclohexanedione	20°, overnight, CH_2Cl_2	F_2 / F_2 [e] ($\geq$30)	94
	4-chlorocyclohexanone	Room temp, 38 hr	Cl– / F_2 (25)	61
	$(CH_2)_5CO$	39°, 13 hr[b]	$(CH_2)_5CF_2$ (31)	7, 76
		30°, 48 hr, CH_2Cl_2	" (70)	50
C_7		Room temp, 16 hr, CH_2Cl_2	F_2 (55)	157
		Room temp, 2d, CH_2Cl_2	F_2 (12)	48

Note: References 152–183 are on p. 124.

[a] Unless indicated otherwise, all reactions were run with hydrogen fluoride catalyst added as such or generated *in situ* from water.

[b] No catalyst was added.

[c] Without catalyst no reaction occurred at 200° and charring occurred at 250°.

[d] At 75° with no catalyst, starting ketone (50%) was recovered; at 40° with catalyst, 70% of the ketone was recovered.

[e] 1,1,4,4-Tetrafluorocyclohexane, a volatile, low-melting (*ca.* 30°) solid, has been obtained in the authors' laboratory, in contrast to a report[61] that reactions of 1,4-cyclohexanedione with sulfur tetrafluoride under "various conditions" were unsuccessful.

[f] Conditions were not given.

[g] The product "seems to be" this.

TABLE III. KETONES (*Continued*)

	Ketone	Conditions[a]	Product(s) and Yield(s) (%)	Refs.
C_7 (*contd.*)	2-Methylcyclohexanone	[f]	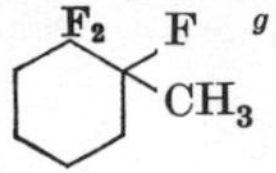 F_2 F CH_3 [g]	61
	3-Methylcyclohexanone	60°, 10 hr	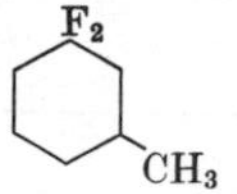 F_2 CH_3 (35)	61
	4-Methylcyclohexanone	70°, 6 hr	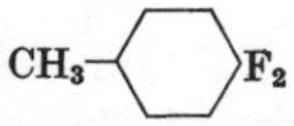 CH_3— F_2 (25)	61
	$(CH_2)_6CO$	30°, 24 hr	$(CH_2)_6CF_2$ (79)	50
	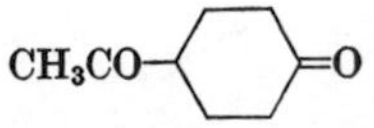 CH_3CO— =O	30°, 18 hr	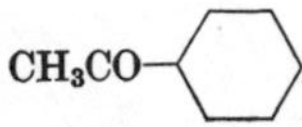 CH_3CO— F_2 (57)	50
	$CH_3COCH_2CH_2CO_2C_2H_5$	95°, 10 hr	$CH_3CF_2CH_2CH_2CO_2C_2H_5$ (16)	7
C_8	$C_6F_5COCF_3$	170°, 20 hr	$C_6F_5CF_2CF_3$ (85)	37
	$C_6F_5COCH_3$	115°, 17 hr[b]	$C_6F_5CF_2CH_3$ (79)	37
	m-$FC_6H_4COCF_3$	150°, 8 hr, $FCCl_3$[b]	*m*-$FC_6H_4CF_2CF_3$ (38)	24
	p-$FC_6H_4COCF_3$	150°, 8 hr, $FCCl_3$[b]	*p*-$FC_6H_4CF_2CF_3$ (77)	24
	$C_6H_5COCF_3$	100°, 8 hr[b]	$C_6H_5CF_2CF_3$ (65)	7
	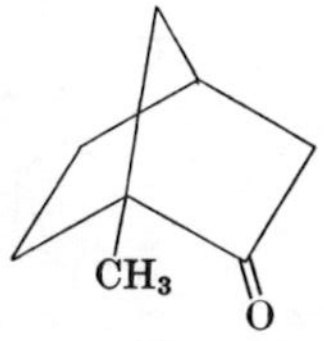 CH_3 O	Room temp, 72 hr, CH_2Cl_2	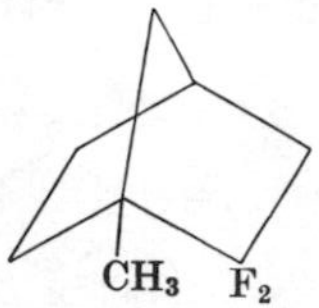 CH_3 F_2 (28)	157

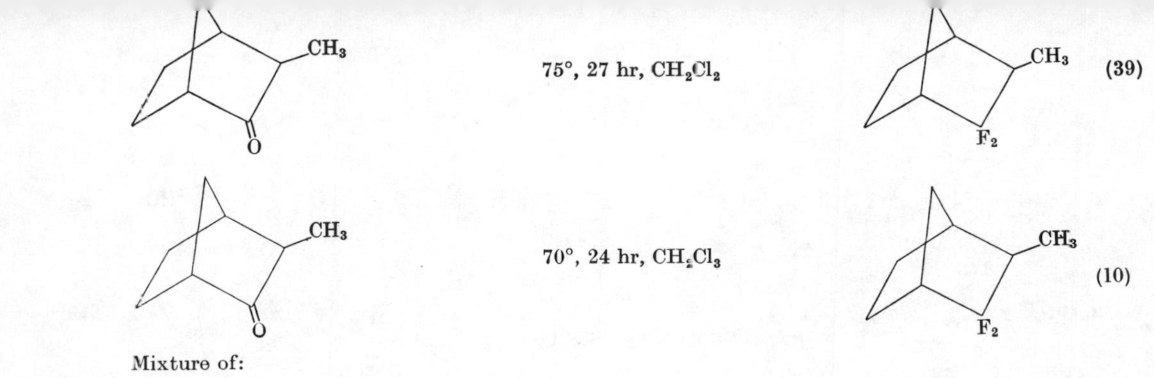

Reactant	Conditions	Product (Yield, %)	Refs.
	75°, 27 hr, CH_2Cl_2	(39)	157
	70°, 24 hr, CH_2Cl_3	(10)	157
Mixture of: ... and ...	60°, 24 hr, CH_2Cl_2	, (30, total)	157

Note: References 152–183 are on p. 124.

[a] Unless indicated otherwise, all reactions were run with hydrogen fluoride catalyst added as such or generated *in situ* from water.
[b] No catalyst was added.
[f] Conditions were not given.
[g] The product "seems to be" this.

TABLE III. KETONES (*Continued*)

Ketone	Conditions[a]	Product(s) and Yield(s) (%)	Refs.
C_8 (*contd.*) 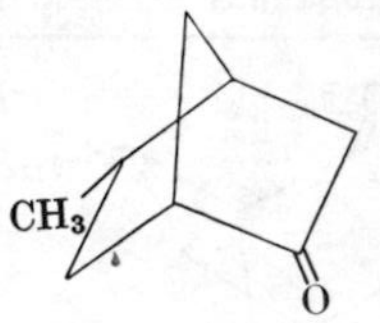 CH_3, O	55°, 20 hr, CH_2Cl_2	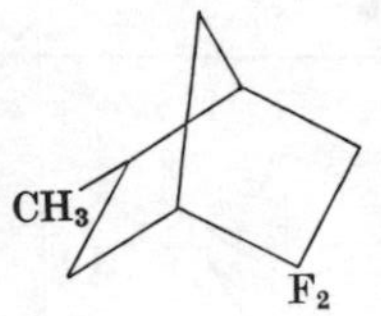 CH_3, F_2 (53)	157
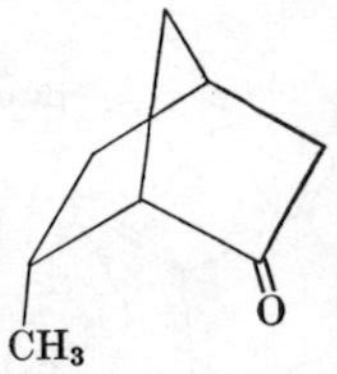 CH_3, O	70°, 19 hr, CH_2Cl_2	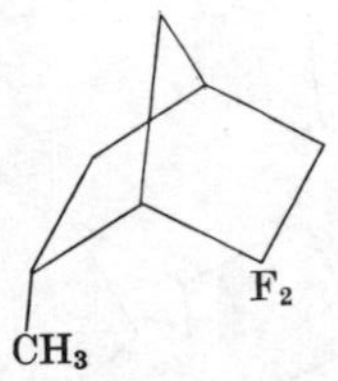 F_2, CH_3 (28)	157
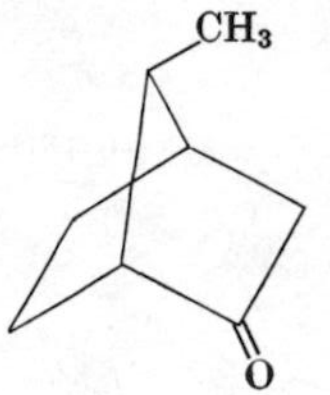 CH_3, O	Room temp, 23 hr, CH_2Cl_2	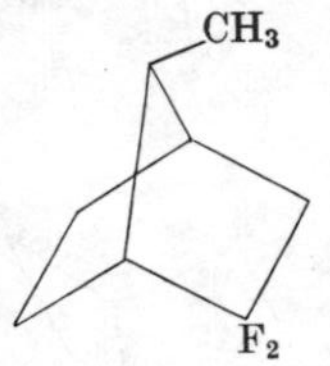 CH_3, F_2 (10)	157
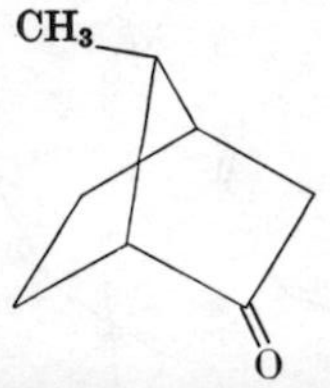 CH_3, O	Room temp, 24 hr, CH_2Cl_2	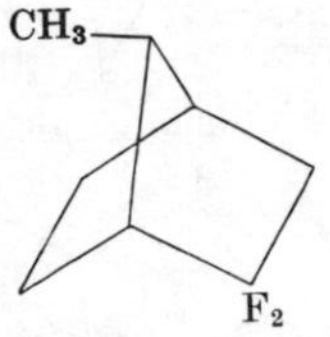 CH_3, F_2 (52)	157

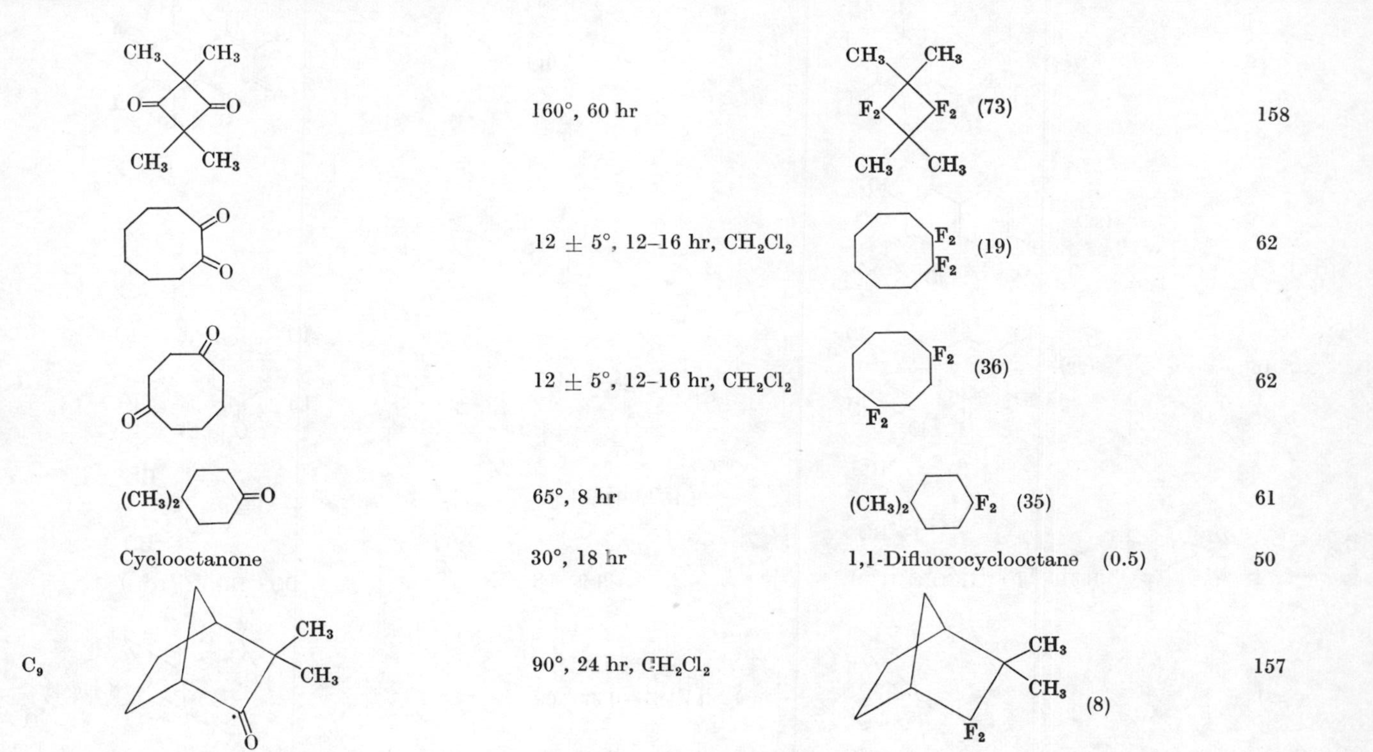

	Reactant	Conditions[a]	Product (Yield, %)	Refs.
		160°, 60 hr	(73)	158
		12 ± 5°, 12–16 hr, CH_2Cl_2	(19)	62
		12 ± 5°, 12–16 hr, CH_2Cl_2	(36)	62
		65°, 8 hr	(35)	61
	Cyclooctanone	30°, 18 hr	1,1-Difluorocyclooctane (0.5)	50
C_9		90°, 24 hr, CH_2Cl_2	(8)	157

Note: References 152–183 are on p. 124.

[a] Unless indicated otherwise, all reactions were run with hydrogen fluoride catalyst added as such or generated *in situ* from water.

TABLE III. KETONES (*Continued*)

Ketone	Conditions[a]	Product(s) and Yield(s) (%)	Refs.
C_9 (*contd.*) 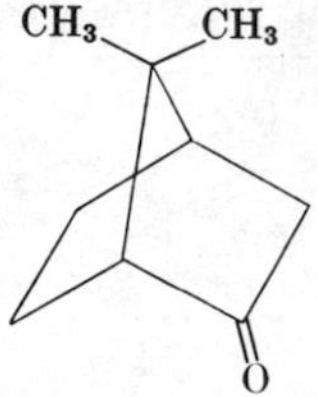	80°, 22 hr, CH_2Cl_2	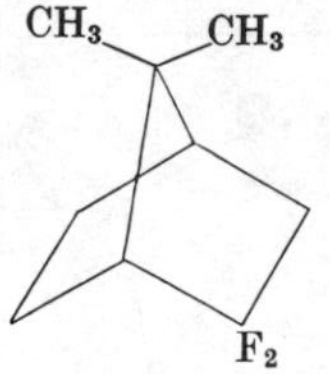 (8)	157
$(C_2H_5O_2CCH_2)_2CO$	80°, 6 hr[b]	$(C_2H_5O_2CCH_2)_2CF_2$ (29)	7
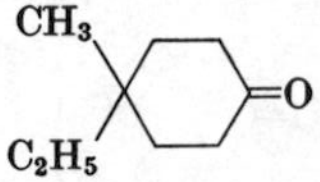	Room temp, 21 hr	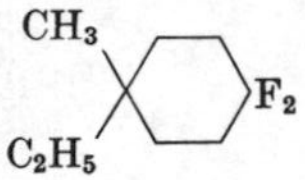 (57)	61
C_{10} 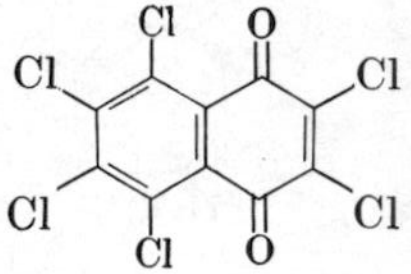	130°, 11 hr	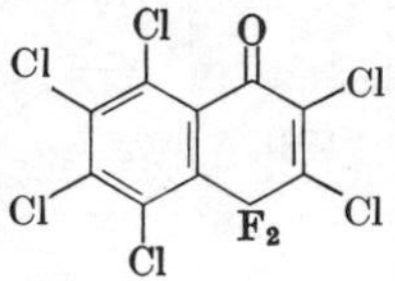 (35), 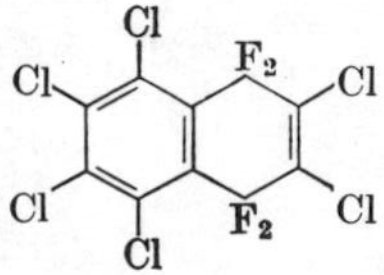 (38)	64
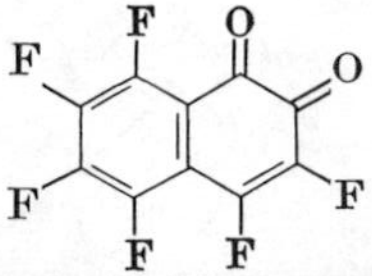	140°, 20 hr	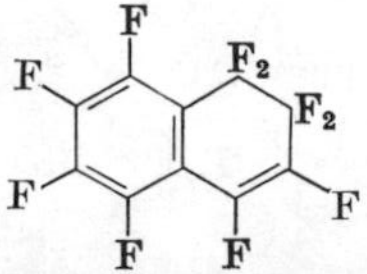 (43),	64

			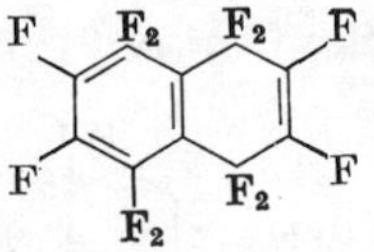 (34),	64
			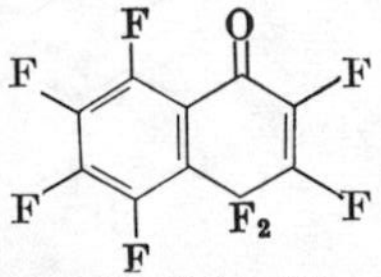 (6)	
C_{10}	m-$FC_6H_4CO(CF_2)_2CF_3$	150°, 8 hr[b]	m-$FC_6H_4(CF_2)_3CF_3$ (78)	24
	p-$FC_6H_4CO(CF_2)_2CF_3$	150°, 8 hr[b]	p-$FC_6H_4(CF_2)_3CF_3$ (63)	24
	$C_6H_5CO(CF_2)_2CF_3$	150°, 8 hr[b]	$C_6H_5(CF_2)_3CF_3$ (84)	24
	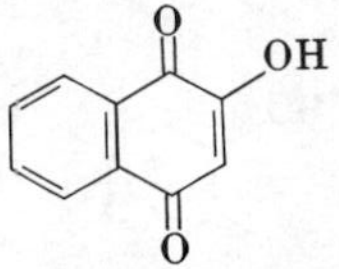	140°, 1.5 hr	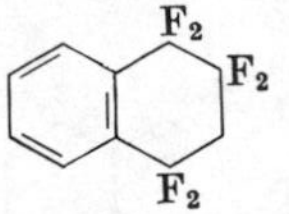 (36)	7
	4-$(CH_3)_2NC_6H_4COCF_3$	150°, 8 hr	4-$(CH_3)_2NC_6H_4CF_2CF_3$	24
	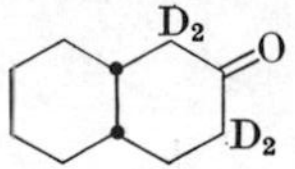 [h]	12 ± 5°, 12–16 hr, CH_2Cl_2	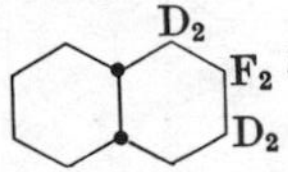 [h] (30)	159
	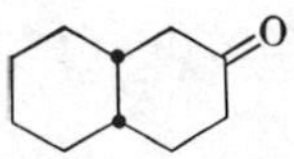 [h]	12 ± 5°, 12–16 hr, CH_2Cl_2	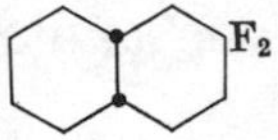 [h] (32)	159

Note: References 152–183 are on p. 124.

[a] Unless indicated otherwise, all reactions were run with hydrogen fluoride catalyst added as such or generated *in situ* from water.

[b] No catalyst was added.

[h] The *trans* isomer was present as an impurity.

TABLE III. KETONES (*Continued*)

	Ketone	Conditions[a]	Product(s) and Yield(s) (%)	Refs.
C_{10} (*contd.*)	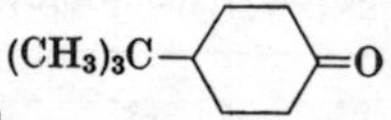 $(CH_3)_3C$-cyclohexanone (=O)	Room temp, 20 hr	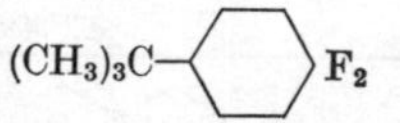 $(CH_3)_3C$-cyclohexane F_2 (19)	61
C_{11}	$C_6H_5COCH{=}CHCO_2CH_3$	160°, 10 hr[b]	$C_6H_5CF_2CH{=}CHCO_2CH_3$ (25)	7
	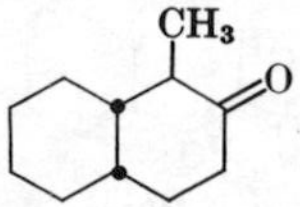 CH_3, O	12 ± 5°, 12–16 hr, CH_2Cl_2	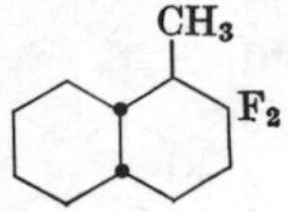 CH_3, F_2 (32)	159
	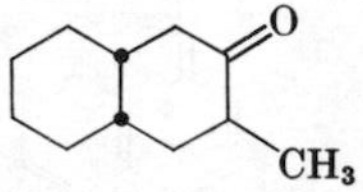 O, CH_3	12 ± 5°, 12–16 hr, CH_2Cl_2	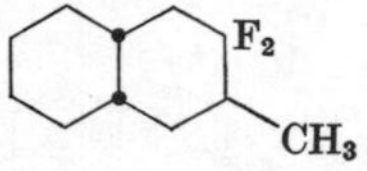 F_2, CH_3 (30)	159
	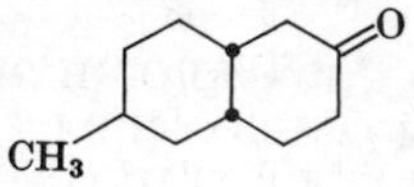 O, CH_3	12 ± 5°, 12–16 hr, CH_2Cl_2	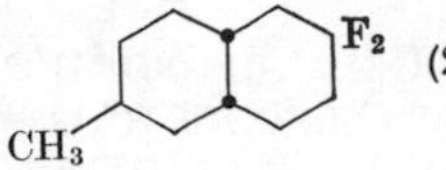 F_2, CH_3 (28)	159
	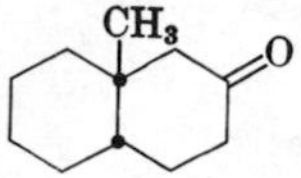 CH_3, O [h]	12 ± 5°, 12–16 hr, CH_2Cl_2	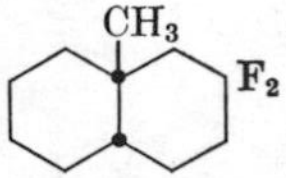 CH_3, F_2 [h] (42)	159
	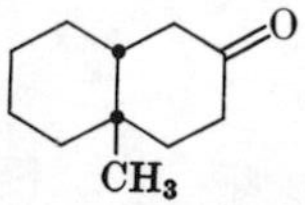 O, CH_3	12 ± 5°, 12–16 hr, CH_2Cl_2	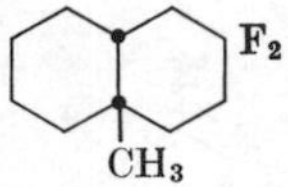 F_2, CH_3 (48)	159
C_{12}	p-$(CH_3)_2NC_6H_4CO(CF_2)_2CF_3$	150°, 8 hr	p-$(CH_3)_2NC_6H_4(CF_2)_3CF_3$ (73)	24

	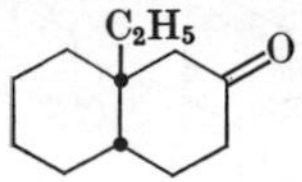 C_2H_5, O	$12 \pm 5°$, 12–16 hr, CH_2Cl_2	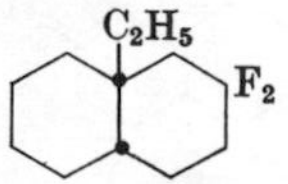 C_2H_5, F_2 (42)	159
	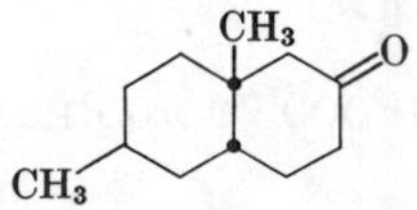 CH_3, O, CH_3	$12 \pm 5°$, 12–16 hr, CH_2Cl_2	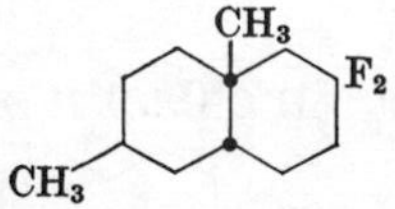 CH_3, F_2, CH_3 (40)	159
	$(CH_2)_{11}CO$	30°, 120 hr	$(CH_2)_{11}CF_2$ (23)	50
C_{13}	S, –CO–$(CF_2)_3$CO–, S	185–250°, 67 hr	S, –$(CF_2)_5$–, S,	160
			F_2, F_2, F_2, S, O, F, F, S [i] (2)	
	$C_6H_5COC_6H_5$	100–180°, 8 hr	$C_6H_5CF_2C_6H_5$ (97)	7, 76
C_{14}	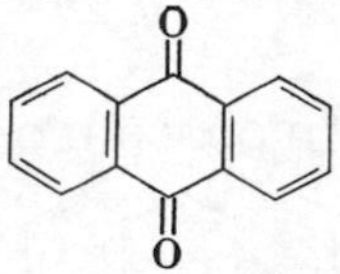 O, O	255°, 8 hr	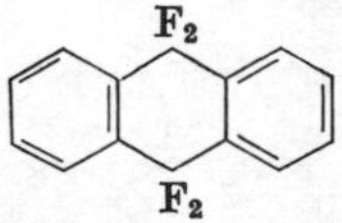 F_2, F_2 (78)	7

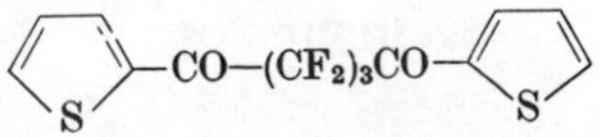

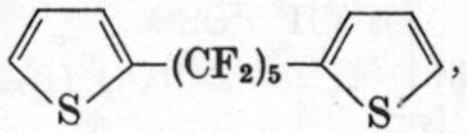

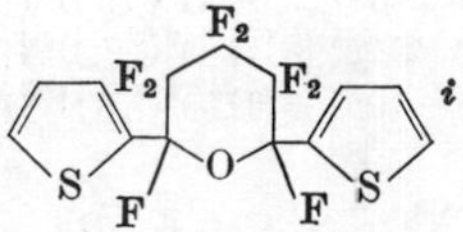

Note: References 152–183 are on p. 124.

[a] Unless indicated otherwise, all reactions were run with hydrogen fluoride catalyst added as such or generated *in situ* from water.

[b] No catalyst was added.

[h] The *trans* isomer was present as an impurity.

[i] This is the "probable" structure.

TABLE III. KETONES (*Continued*)

	Ketone	Conditions[a]	Product(s) and Yield(s) (%)	Refs.
C_{14} (*contd.*)	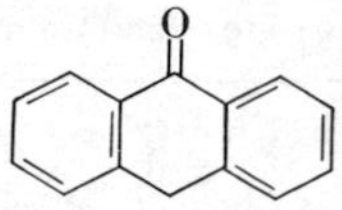	69°, 16 hr, CH_2Cl_2	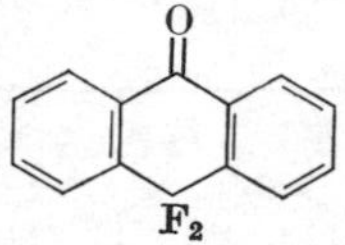 F_2 (85)[j]	65
	$C_6H_5COCOC_6H_5$	180°, 5 hr[b]	$C_6H_5CF_2CF_2C_6H_5$ (34)	7, 76
		175–220°, 45 hr[b]	" (35)	161
	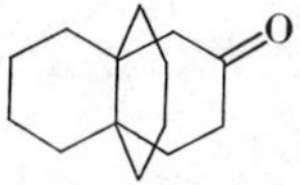	Room temp, 15 hr[b]	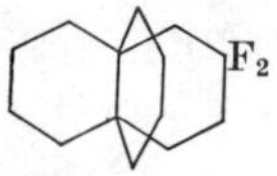 F_2 (44)	162
C_{15}	$C_6H_5COCOCOC_6H_5$	120°, 8 hr	$C_6H_5(CF_2)_3C_6H_5$ (50)	7
C_{16}	$C_6H_5COCF_2CF_2COC_6H_5$	125–220°, 51 hr[b]	$C_6H_5(CF_2)_4C_6H_5$ (77.5)	161
C_{17}	$C_6H_5CO(CF_2)_3COC_6H_5$	125–220°, 51 hr[b]	$C_6H_5(CF_2)_5C_6H_5$ (82.5)	161
		125–220°, 2 hr[b]	" (24), $C_6H_5CO(CF_2)_4C_6H_5$ (39.5)	161
	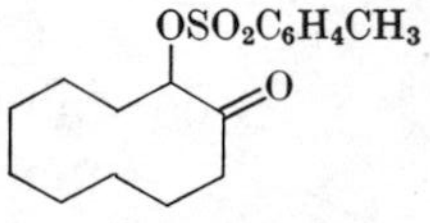 $OSO_2C_6H_4CH_3$	19–23°, 10 hr, CH_2Cl_2	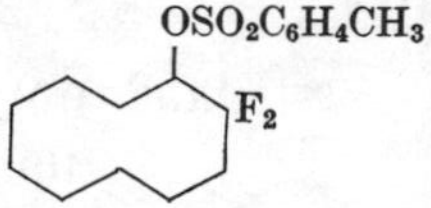 $OSO_2C_6H_4CH_3$ F_2 (14)	59
C_{18}	$C_6H_5CO(CF_2)_4COC_6H_5$	125–220°, 20 hr[b]	$C_6H_5(CF_2)_6C_6H_5$ (78)	161
C_{19}	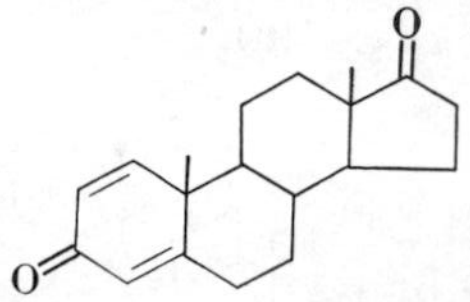	40°, 10 hr, $CHCl_3$[k]	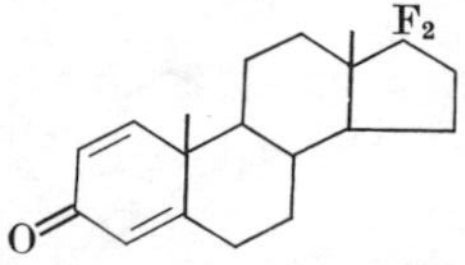 F_2 (3)	43, 163

Reactant	Conditions	Product(s) and Yield(s) (%)	Refs.
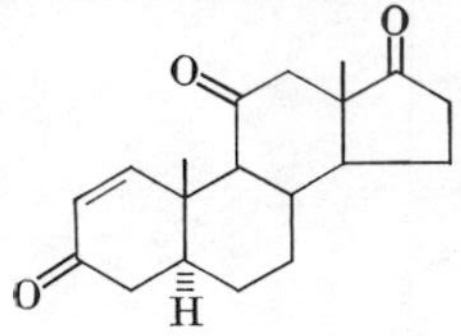 [structure]	20°, 16 hr, CH_2Cl_2	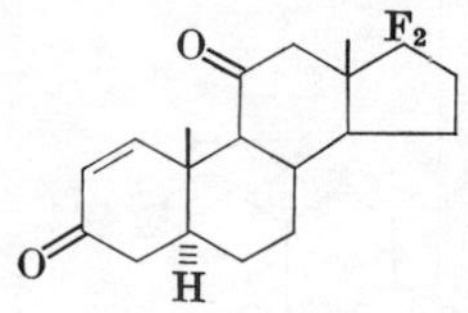 [structure] (66), [structure] (7)	30, 74
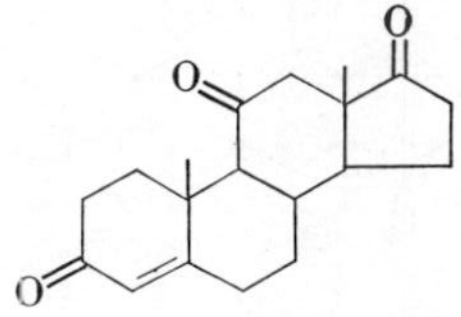 [structure]	40°, 10 hr, $CHCl_3$[k]	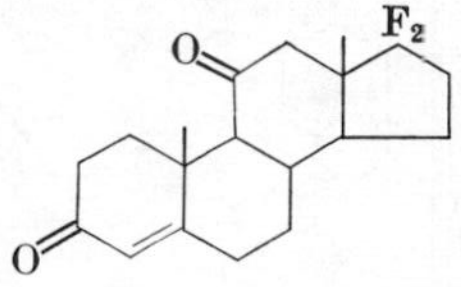 [structure] (—)	164, 163
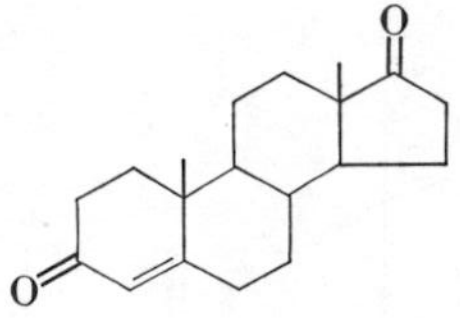 [structure]	20°, 16 hr, CH_2Cl_2	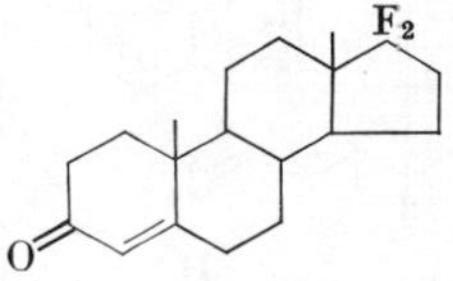 [structure] (63)	30, 74
	40°, 15 hr[m]	" (3)	43
	40°, 10 hr, $CHCl_3$[k]	" (10)	43

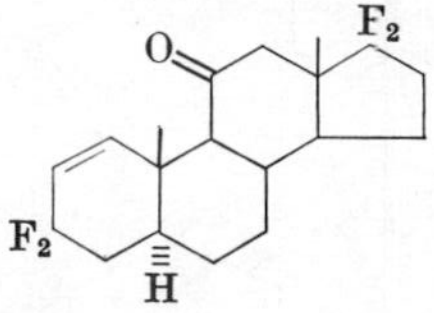

Note: References 152–183 are on p. 124.

[a] Unless indicated otherwise, all reactions were run with hydrogen fluoride catalyst added as such or generated *in situ* from water.

[b] No catalyst was added.

[j] The yield of 10,10-difluoroanthrone was reduced to 29%, and a 47% yield of 10,10′-bianthrone was obtained when the reaction was carried out in the presence of a radical scavenger and an antioxidant.

[k] Boron trifluoride was added to ethanol-free chloroform.

[l] This product was not completely characterized.

[m] Ethanol was added to generate hydrogen fluoride *in situ*.

TABLE III. KETONES (*Continued*)

Ketone	Conditions[a]	Product(s) and Yield(s) (%)	Refs.
C_{19} (*contd.*) 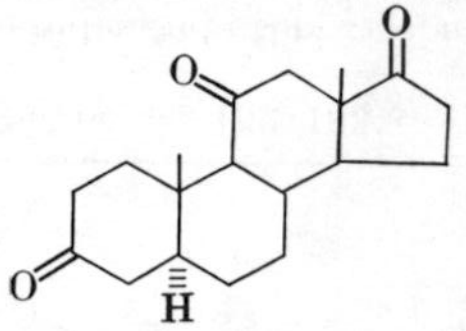	40°, 15 hr, $CHCl_3$[n]	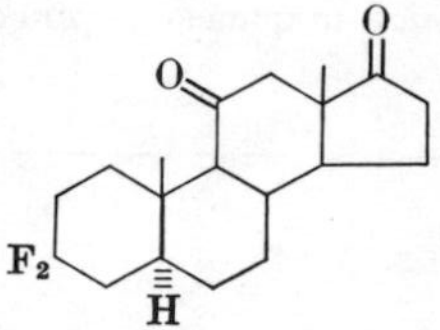 (—)	164, 163
		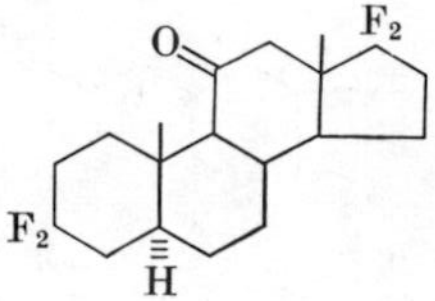 (—)	164, 163
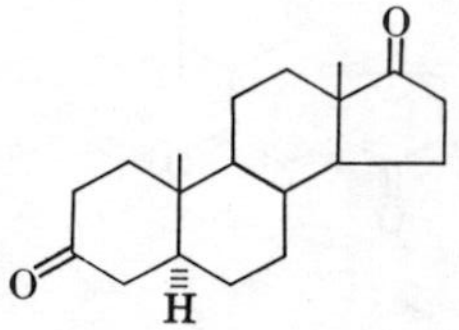	40°, 15 hr, $CHCl_3$[m]	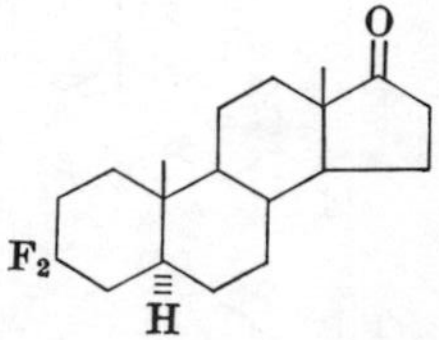 (37),	43, 163
		(11)	

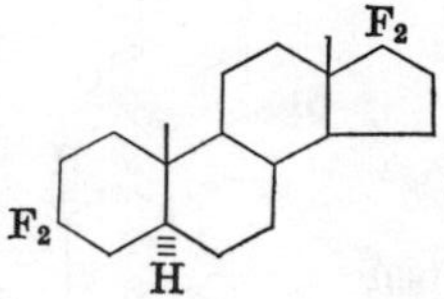

C_{21}	20°, 16 hr, CH_2Cl_2	(37)	30, 74
	20°, 16 hr, CH_2Cl_2	(6), (52)	52, 165
	15°, 16 hr, CH_2Cl_2	(36) [o]	30, 74
	40°, 10 hr, $CHCl_3$ [k]	,, (2)	43, 163

Note: References 152–183 are on p. 124.

[a] Unless indicated otherwise, all reactions were run with hydrogen fluoride catalyst added as such or generated *in situ* from water.

[k] Boron trifluoride was added to ethanol-free chloroform.

[m] Ethanol was added to generate hydrogen fluoride *in situ*.

[n] Hydrogen fluoride (1%) in chloroform was used.

[o] About half of the progesterone was recovered; about 40% of the product was obtained in chromatographic fractions mixed with progesterone.

TABLE III. Ketones (*Continued*)

Ketone	Conditions[a]	Product(s) and Yield(s) (%)	Refs.
C_{21} (*contd.*) [structure: 5α-pregnane-3,11,20-trione]	40°, 15 hr, $CHCl_3$[m]	[structure: 3,3-difluoro-5α-pregnane-11,20-dione] (43),	164, 163
		[structure: 3,3,20,20-tetrafluoro-5α-pregnan-11-one] (9)	
[structure: 5β-pregnane-3,11,20-trione]	40°, 15 hr, $CHCl_3$[m]	[structure: 3,3-difluoro-5β-pregnane-11,20-dione] (18),	164

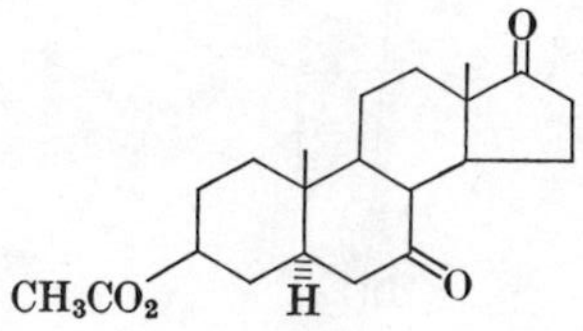

CH_3
CF_2
F_2
H
(6)

O
CH_3CO_2
H
O

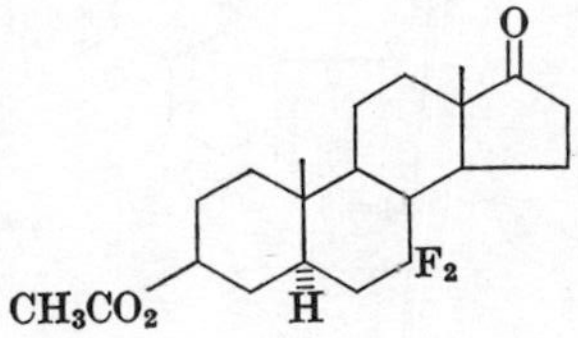

20°, 30 hr, CH_2Cl_2

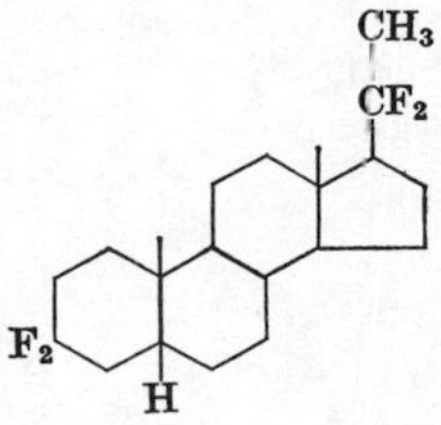

O
CH_3CO_2
H
F_2
(15), 166

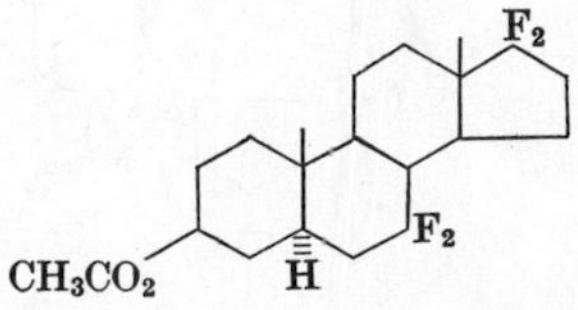

F_2
CH_3CO_2
H
F_2
(41)

Note: References 152–183 are on p. 124.

[a] Unless indicated otherwise, all reactions were run with hydrogen fluoride catalyst added as such or generated *in situ* from water.

[m] Ethanol was added to generate hydrogen fluoride *in situ*.

TABLE III. Ketones (*Continued*)

Ketone	Conditions[a]	Product(s) and Yield(s) (%)	Refs.
C_{21} (*contd.*) 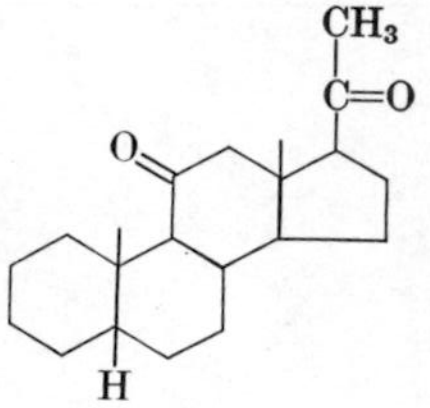 [structure: CH_3, C=O, O, H]	20°, 16 hr, CH_2Cl_2	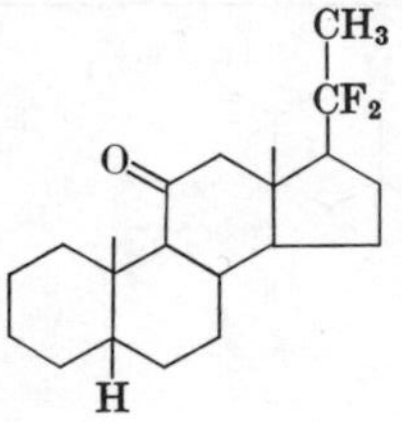 [structure: CH_3, CF_2, O, H] (50)	30, 74
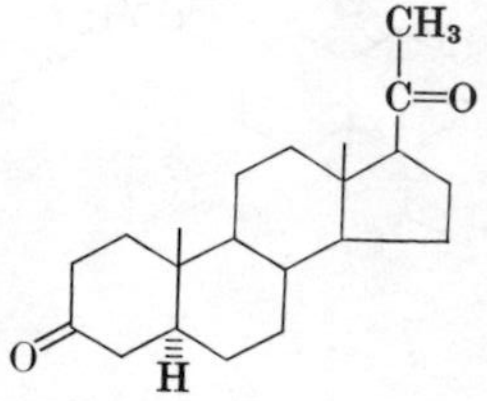 [structure: CH_3, C=O, O, H]	40°, 15 hr, $CHCl_3$[m]	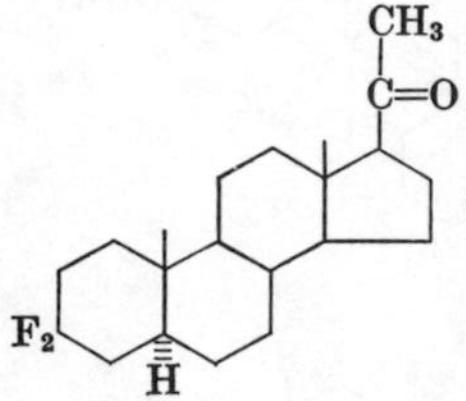 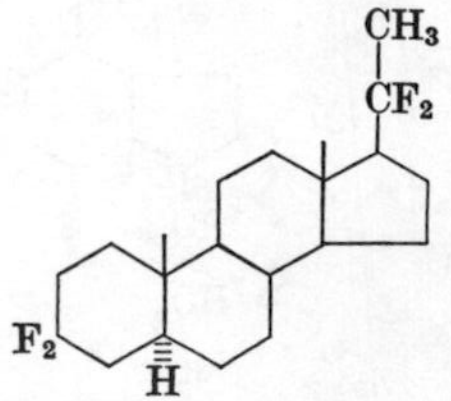 [structure: CH_3, C=O, F_2, H] (33),	43, 163
		[structure: CH_3, CF_2, F_2, H] (5)	

Formula	Reactant	Conditions[a]	Product(s) and Yield(s) (%)	Refs.
		40°, 15 hr, $CHCl_3$[m]	(29), (7)	43, 163
		20°, 16 hr, CH_2Cl_2	(82)	30, 74
C_{22}		20°, 10 hr, CH_2Cl_2	(85)	53, 44

Note: References 152–183 are on p. 124.

[a] Unless indicated otherwise, all reactions were run with hydrogen fluoride catalyst added as such or generated *in situ* from water.

[m] Ethanol was added to generate hydrogen fluoride *in situ*.

TABLE III. KETONES (*Continued*)

	Ketone	Conditions[a]	Product(s) and Yield(s) (%)	Refs.
C_{23} 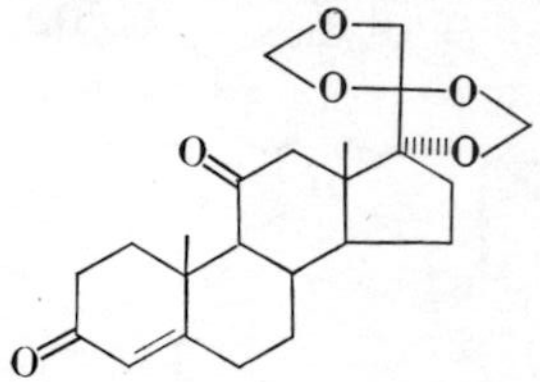	O, O, O, O, O	20°, overnight, CH_2Cl_2, tetrahydrofuran	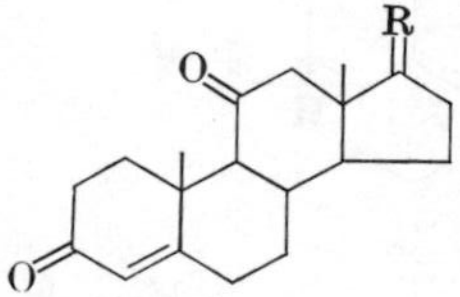 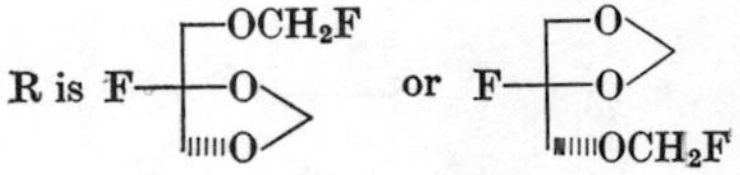 R, O, O (ca. 20); R is F—(—OCH_2F, O, O) or F—(O, O, OCH_2F)	55
	CH_3, C=O, CH_3CO_2, F, O	20°, 16 hr[p]	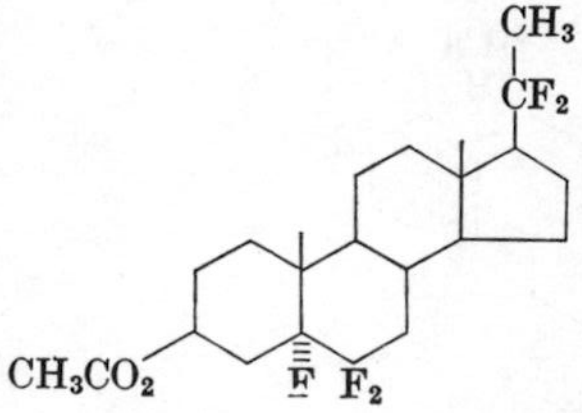 CH_3, CF_2, CH_3CO_2, F, F_2 (29)	52
		20°, 10 hr[q]	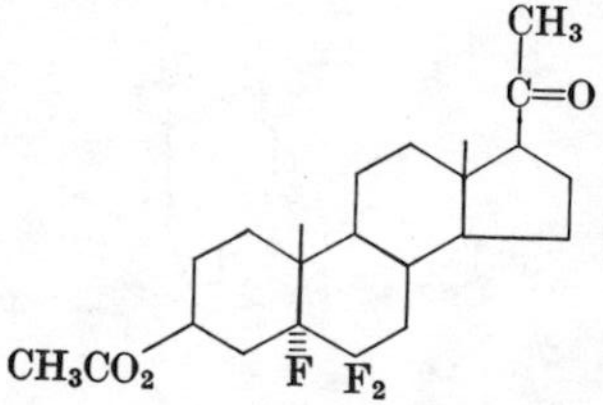 CH_3, C=O, CH_3CO_2, F, F_2 (58)	52

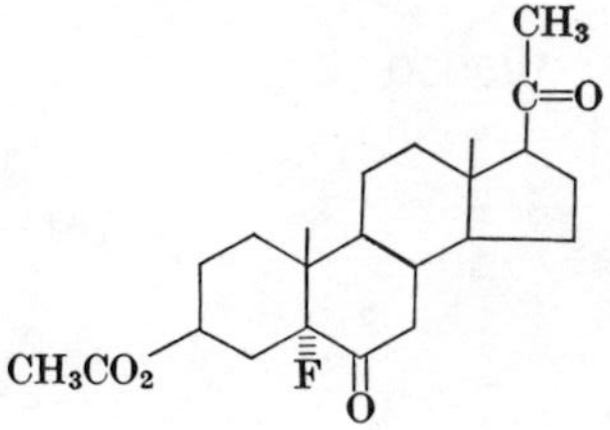

CH_3CO_2, F, O, $OCOCH_3$	20°, 15 hr, CH_2Cl_2[q]	CH_3CO_2, F, F_2, $OCOCH_3$	(82) 52, 165
CH_3CO_2, H, O, CH_3–C=O	20°, 16 hr	CH_3CO_2, H, F_2, CH_3–C=O	(70), 166
		CH_3CO_2, H, F_2, CH_3–CF_2	(26)

Note: References 152–183 are on p. 124.

[a] Unless indicated otherwise, all reactions were run with hydrogen fluoride catalyst added as such or generated *in situ* from water.

[p] Hydrogen fluoride, 24 mole % relative to sulfur tetrafluoride, was used as the catalyst.

[q] Hydrogen fluoride, 11 mole % relative to sulfur tetrafluoride, was used as the catalyst.

TABLE III. KETONES (*Continued*)

	Ketone	Conditions[a]	Product(s) and Yield(s) (%)	Refs.
C_{23} (*contd.*)	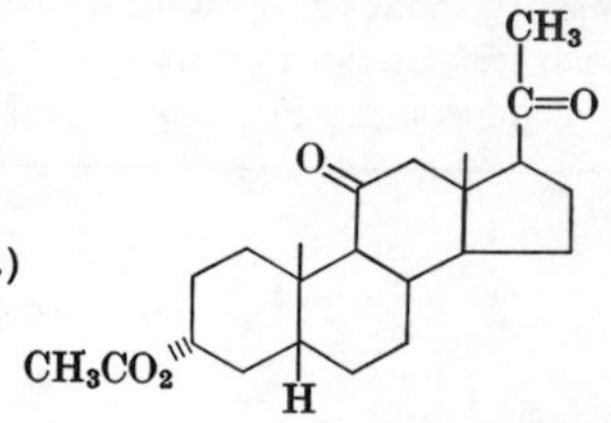 CH_3CO_2, O, CH_3, C=O, H	20°, 16 hr, CH_2Cl_2	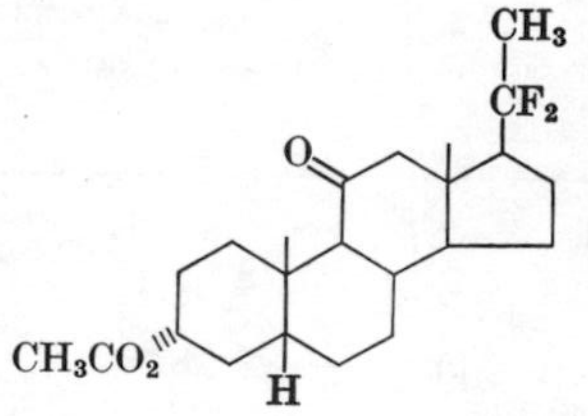 CH_3CO_2, O, CH_3, CF_2, H (58)	30, 74
	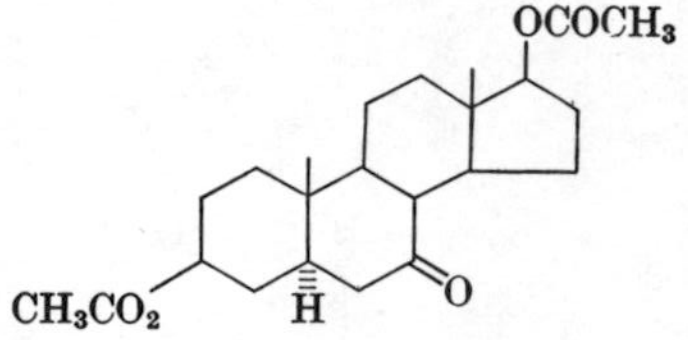 $OCOCH_3$, CH_3CO_2, H, O	20°, 16 hr, CH_2Cl_2 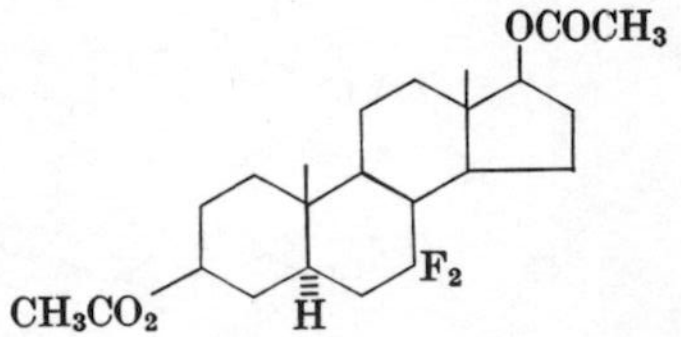	$OCOCH_3$, CH_3CO_2, H, F_2 (87)	166
	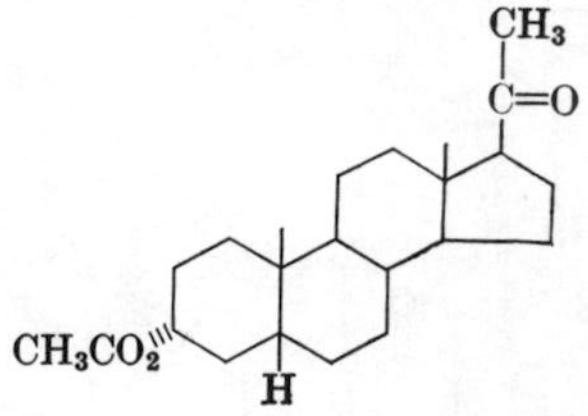 CH_3, C=O, CH_3CO_2, H	20°, 16 hr, CH_2Cl_2	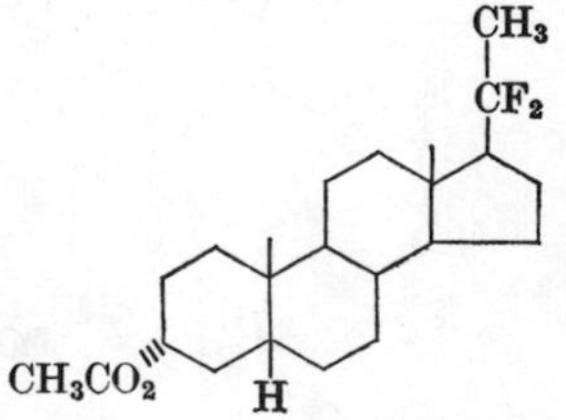 CH_3, CF_2, CH_3CO_2, H (23)	30, 74
C_{24}	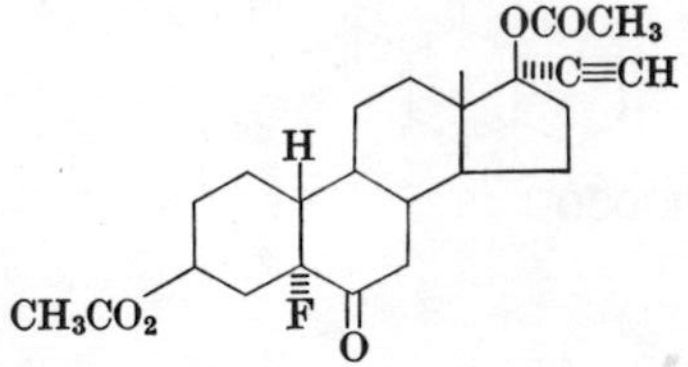 $OCOCH_3$, C≡CH, H, CH_3CO_2, F, O	18–22°, 10 hr, CH_2Cl_2, tetrahydrofuran 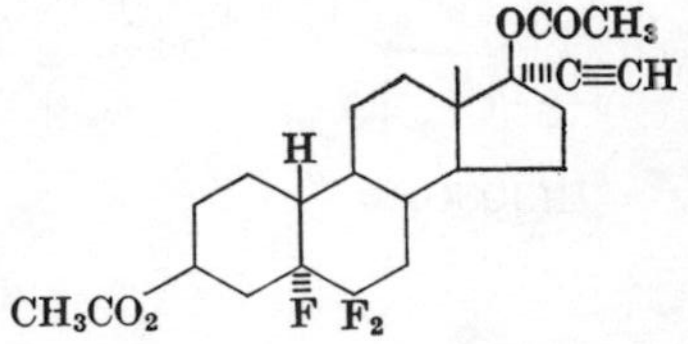	$OCOCH_3$, C≡CH, H, CH_3CO_2, F, F_2 (32)	44

Formula	Reactant	Conditions	Product (Yield, %)	Refs.
C_{25}	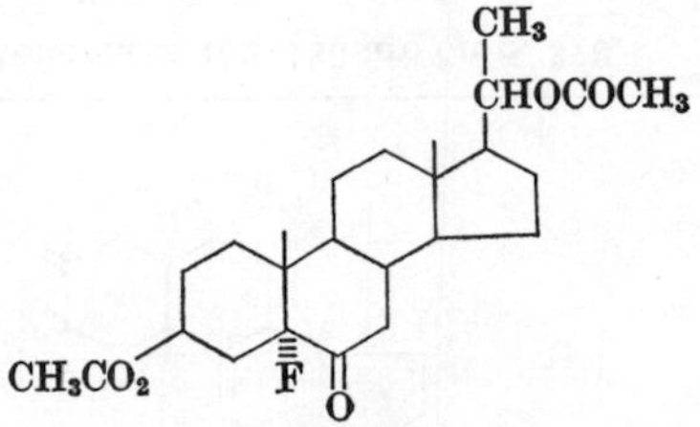 CH_3, $CHOCOCH_3$, CH_3CO_2, F, O	20°, 10 hr, CH_2Cl_2[r]	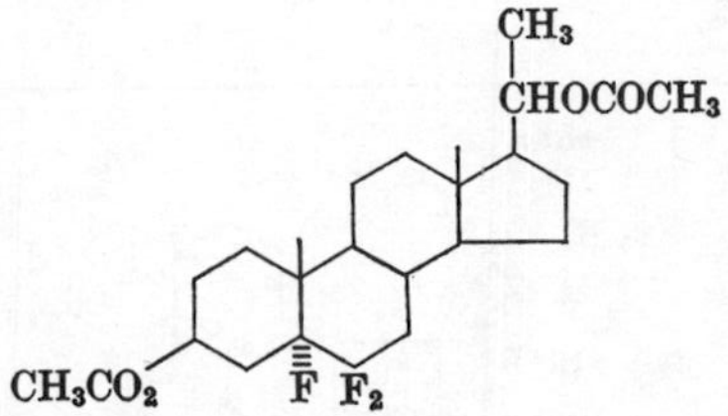 CH_3, $CHOCOCH_3$, CH_3CO_2, F F_2 (68)	52, 165
	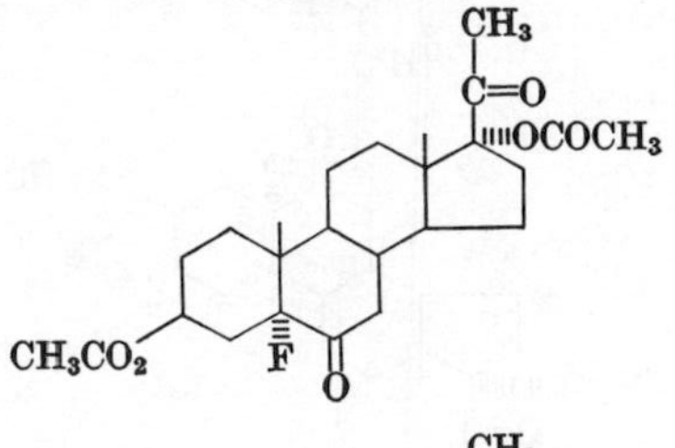 CH_3, C=O, $OCOCH_3$, CH_3CO_2, F, O	20°, 10 hr, CH_2Cl_2[q]	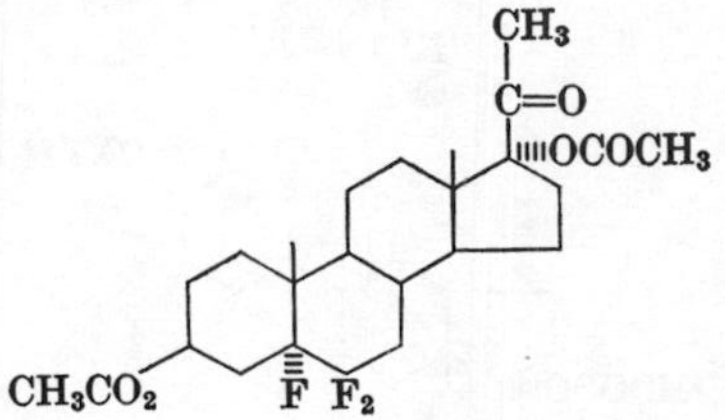 CH_3, C=O, $OCOCH_3$, CH_3CO_2, F F_2 (61)	52, 165
C_{25}	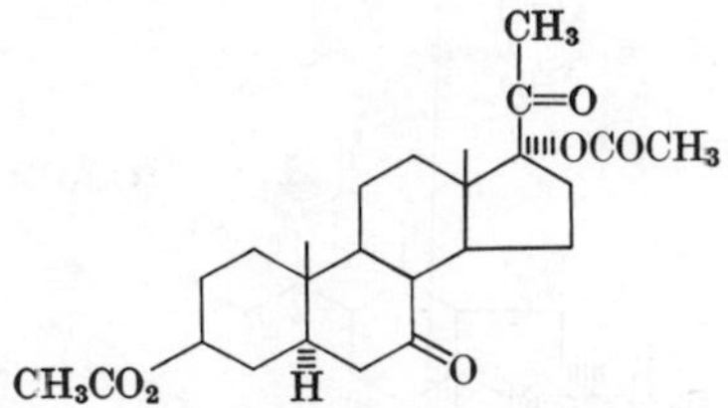 CH_3, C=O, $OCOCH_3$, CH_3CO_2, H, O	20°, 8 hr, CH_2Cl_2	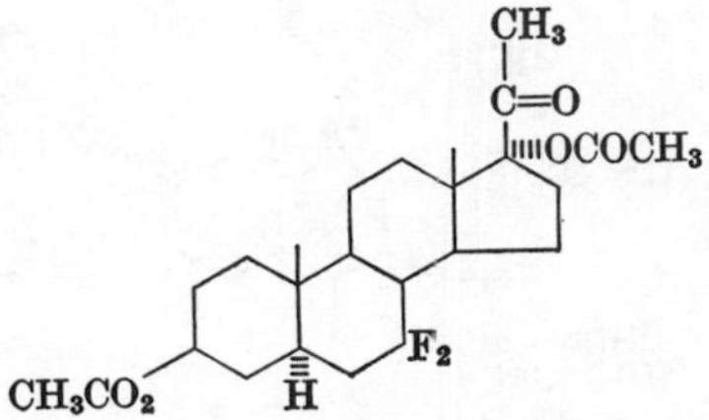 CH_3, C=O, $OCOCH_3$, CH_3CO_2, H, F_2 (53)	166

Note: References 152–183 are on p. 124.

[a] Unless indicated otherwise, all reactions were run with hydrogen fluoride catalyst added as such or generated *in situ* from water.

[q] Hydrogen fluoride, 11 mole % relative to sulfur tetrafluoride, was used as the catalyst.

[r] Hydrogen fluoride, 7 mole % relative to sulfur tetrafluoride, was used as the catalyst.

TABLE III. Ketones (*Continued*)

Ketone	Conditions[a]	Product(s) and Yield(s) (%)	Refs.
C_{26} CH₃CO₂ … F, O; CH₃, C=O, OCOCH₃, CH₃	20°, 10 hr, CH_2Cl_2	CH₃CO₂ … F F₂; CH₃, C=O, OCOCH₃, CH₃ (32–51)	46
C_{26} CH₃CO₂ … Cl, O; CH₃, C=O, OCOCH₃	20°, 18 hr, CH_2Cl_2	CH₃CO₂ … Cl F₂; CH₃, C=O, OCOCH₃ (—)	46
C_{27} O … H; C_8H_{17}	20°, 16 hr, CH_2Cl_2	F₂ … H; C_8H_{17} (78)	30, 74
	40°, 15 hr, $CHCl_3$[m]	" (32)	43

Note: References 152–183 are on p. 376.

[a] Unless indicated otherwise, all reactions were run with hydrogen fluoride catalyst added as such or generated *in situ* from water.

[m] Ethanol was added to generate hydrogen fluoride *in situ*.

TABLE IV. Carboxylic Acids

	Carboxylic Acid	Conditions	Product(s) and Yield(s) (%)	Refs.
C_2	CD_3CO_2D	—	CD_3CF_3 (60–70)	156, 167
	$CHBr(Cl)CO_2H$	100–120°, 10 hr	$CHBr(Cl)CF_3$ (—)	168
	$CH_2(NO_2)CO_2H$	150°, 8 h[a]	$CH_2NO_2CF_3$ (18)	79
	CH_3CO_2H	—	CH_3CF_3 (60–70)	156, 167
	CH_2OHCO_2H	160°, 5 hr	CH_2FCF_3 (48), CH_2FCOF (18)	7
	$CH_2(CO_2H)SO_3H$	180°, 6 hr	$CH_2(CF_3)SO_2F$ (41)	7
	$CH_2(NH_2)CO_2H$	120°, 8 hr	$CH_2(NH_2)CF_3$ (24)	83
C_3	$HC{\equiv}CCO_2H$	30–55°, 3 hr	$HC{\equiv}CCOF$ (28)	7
		120°, 3 hr	$HC{\equiv}CCF_3$ (60)	
	$CHF_2CF_2CO_2H$	250°, 8 hr[b]	$CHF_2CF_2CF_3$ (56)	7
	$CH(CO_2H)_2Br$	100–120°, 10 hr	$CH(CF_3)_2Br$ (—)	169
	$O_2NCH{=}CHCO_2H$	80°, 8 hr[a]	$O_2NCH{=}CHCF_3$ (64)	79
	$CH_2(NO_2)CHClCO_2H$	60°, 8 hr[a]	$CH_2(NO_2)CHClCF_3$ (15)	79
	$CH_2{=}CHCO_2H$	130°, 8 hr	$CH_2{=}CHCF_3$ (45)	7, 76
	$CH_2(CO_2H)_2$	40°, 16 hr	$CH_2(COF)_2$ (70)	7
		150°, 8 hr	$CH_2(CF_3)_2$ (57)	7
	$CH_2(NO_2)CH_2CO_2H$	120°, 8 hr[a,c]	$CH_2(NO_2)CH_2CF_3$ (10)	79
	$CH_3CH_2CO_2H$	150°, 8 hr	$CH_3CH_2CF_3$ (89)	7
	DL—$CH_3CHNH_2CO_2H$	120°, 8 hr	DL—$CH_3CHNH_2CF_3$ (29)	83
	$CH_2(NH_2)CH_2CO_2H$	120°, 8 hr	$CH_2(NH_2)CH_2CF_3$ (41)	83

Note: References 152–183 are on p. 124.

165161

[a] This temperature gave the maximum yield in an 8-hr reaction period.

[b] Boron trifluoride catalyst was used; without catalyst at 180° only HCF_2CF_2COF was obtained.

[c] Yields of 13% were obtained in 12 hours. Longer reaction times gave lower yields.

TABLE IV. Carboxylic Acids (*Continued*)

	Carboxylic Acid	Conditions	Product(s) and Yield(s) (%)	Refs.
C_4	$HO_2CC{\equiv}CCO_2H$	70°, 6 hr[d]	$FCOC{\equiv}CCOF$ (51)	7
		170°, 8 hr[e]	$CF_3C{\equiv}CCF_3$ (80)	7, 170
	(ring structure: O, HN, CO_2H, O, N–N, H)	50°, 24 hr	(ring structure: O, HN, CF_3, O, N–N, H) (53)	154, 82
	trans-$HO_2CCH{=}CHCO_2H$	130°, 9 hr	*trans*-$CF_3CH{=}CHCF_3$ (95)	7
	$CF(NO_2)_2CH_2CH_2CO_2H$	100°, 8 hr[a]	$CF(NO_2)_2CH_2CH_2CF_3$ (87)	79
	$O_2NCH{=}C(CH_3)CO_2H$	60°, 8 hr[a]	$O_2NCH{=}C(CH_3)CF_3$ (37)	79
	$C(NO_2)_3CH_2CH_2CO_2H$	120°, 8 hr[a]	$C(NO_2)_3CH_2CH_2CF_3$ (55)	79
	$CH_2BrCHBrCH_2CO_2H$	140°, 8 hr	$CH_2BrCHBrCH_2CF_3$ (54)	7
	$CH(NO_2)_2CH_2CH_2CO_2H$	80°, 8 hr[a]	$CH(NO_2)_2CH_2CH_2CF_3$ (90)	79
	$CH_2{=}C(CH_3)CO_2H$	130°, 8 hr	$CH_2{=}C(CH_3)CF_3$ (54)	7
	$HO_2CCH_2CH_2CO_2H$	150°, 8 hr	$CF_3CH_2CH_2CF_3$ (41)	7
	$HO_2CCH_2OCH_2CO_2H$	130°, 7 hr	$CF_3CH_2OCH_2CF_3$ (35), (ring structure: CH_2CF_2, O, O, CH_2CF_2) (14)	7
	$O_2N(CH_2)_3CO_2H$	120°, 8 hr[d]	$O_2N(CH_2)_3CF_3$ (20)	79
	$CH_2(NH_2)CH_2CH_2CO_2H$	120°, 8 hr	$CH_2(NH_2)CH_2CH_2CF_3$ (7)	83
	$(CH_3)_2C(NH_2)CO_2H$	120°, 8 hr	$(CH_3)_2C(NH_2)CF_3$ (35)	83
	DL-$CH_3CH_2CH(NH_2)CO_2H$	120°, 8 hr	DL-$CH_3CH_2CH(NH_2)CF_3$ (4)	83
C_5	(ring structure: O_2N–, O, –CO_2H)	120°, 7 hr	(ring structure: O_2N–, O, –CF_3) (37)	47

F_2, F_2 (cyclobutane ring) CO_2H	130°, 16 hr	F_2, F_2 (cyclobutane ring) CF_3 (67)	171
H, N, N (imidazole ring) CO_2H, CO_2H	120°, 8 hr	H, N, N (imidazole ring) CF_3, CF_3 (80)	83
O, HN, O, N, H (pyrimidine ring) CO_2H	100°, 16 hr	O, HN, O, N, H (pyrimidine ring) CF_3 (81, 77)	154, 172
$CF_3CF(NO_2)CH_2CH_2CO_2H$	100°, 8 hr[a]	$CF_3CF(NO_2)CH_2CH_2CF_3$ (73)	79
NH_2, N, O, N, H (pyrimidine ring) CO_2H	100°, 16 hr	NH_2, N, O, N, H (pyrimidine ring) CF_3 (—)	154
$CH_2{=}C(CO_2H)CH_2CO_2H$	160°, 10 hr	$CH_2{=}C(CF_3)CH_2CF_3$ (26), $CH_2{=}C(COF)CH_2CF_3$ (41)	7
$CH_3C(NO_2)_2CH_2CH_2CO_2H$	120°, 8 hr[a]	$CH_3C(NO_2)_2CH_2CH_2CF_3$ (76)	99
$HO_2C(CH_2)_3CO_2H$	120°, 12 hr	$CF_3(CH_2)_3CO_2H$ (27)	173
$CH_2(NO_2)C(CH_3)_2CO_2H$	100°, 8 hr[a]	$CH_2(NO_2)C(CH_3)_2CF_3$ (5)	79
L-$CH_2(CO_2H)CH_2CH(NH_2)CO_2H$	120°, 8 hr	L-$CH_2(CF_3)CH_2CH(NH_2)CF_3$ (12)	83
DL-$(CH_3)_2CHCH(NH_2)CO_2H$	120°, 8 hr	DL-$(CH_3)_2CHCH(NH_2)CF_3$ (4)	83
DL-$CH_3(CH_2)_2CH(NH_2)CO_2H$	120°, 8 hr	DL-$CH_3(CH_2)_2CH(NH_2)CF_3$ (3)	83
DL-$CH_3S(CH_2)_2CH(NH_2)CO_2H$	120°, 8 hr	DL-$CH_3S(CH_2)_2CH(NH_2)CF_3$ (2)	83

Note: References 152–183 are on p. 124.

[a] This temperature gave the maximum yield in an 8-hr reaction period.

[d] Methylcyclohexane was used as a solvent.

[e] Titanium tetrafluoride was used as a catalyst.

TABLE IV. Carboxylic Acids (*Continued*)

	Carboxylic Acid	Conditions	Product(s) and Yield(s) (%)	Refs.
C_6	2,2,4,4-F_2,F_2-3-F_2-cyclobutyl–CH_2CO_2H	160°, 16 hr	2,2,4,4-F_2,F_2-3-F_2-cyclobutyl–CH_2CF_3 (51)	7
	2-pyridyl–CO_2H	120°, 8 hr	2-pyridyl–CF_3 (53)	83
	3-pyridyl–CO_2H	100–120°, 10 hr	3-pyridyl–CF_3 (25)	83
	4-pyridyl–CO_2H	120°, 8 hr	4-pyridyl–CF_3 (57)	83
	$CH_3(CF_3)C(NO_2)CH_2CH_2CO_2H$	100°, 8 hr	$CH_3(CF_3)C(NO_2)CH_2CH_2CF_3$ (78)	79
	$HO_2C(CH_2)_2$–hydantoin (5-substituted; HN, NH, O, O)	100°, 3 hr	$CF_3(CH_2)_2$–hydantoin (5-substituted; HN, NH, O, O) (89)	84
	$HO_2CCH(CH_3)$–hydantoin (5-substituted; HN, NH, O, O)	100°, 3 hr	$CF_3CH(CH_3)$–hydantoin (5-substituted; HN, NH, O, O) (32)	84
	$CH_2(CO_2H)CH{=}CHCH_2CO_2H$	130°, 10 hr	$CH_2(CF_3)CH{=}CHCH_2CF_3$ (58)	7
	cyclobutane–$(CO_2H)_2$	150°, 6 hr	cyclobutane–$(CF_3)_2$ (43)	7

$CH_2(CO_2H)CH(CO_2H)CH_2CO_2H$	130°, 10 hr	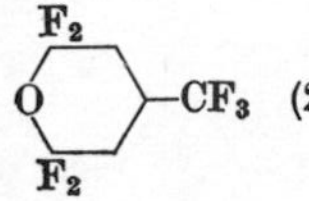 (F_2, O, F_2 ring)–CF_3 (20)	7
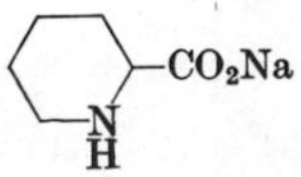 (N–H ring)–CO_2Na	120°, 8 hr	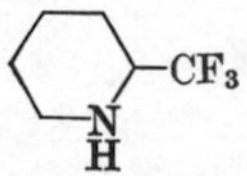 (N–H ring)–CF_3 (9.6)	83
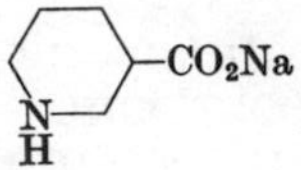 (N–H ring)–CO_2Na	120°, 8 hr	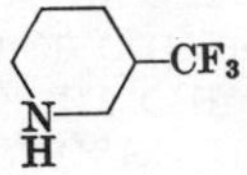 (N–H ring)–CF_3 (40)	83
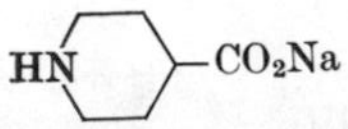 (HN ring)–CO_2Na	120°, 8 hr	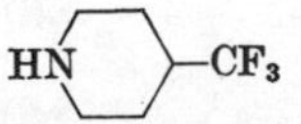 (HN ring)–CF_3 (34)	83
$CH_2(CO_2H)(CH_2)_3CO_2H$	150°, 12 hr	$CH_2(CF_3)(CH_2)_3CO_2H$ (11)	173
	130°, 7 hr	$CH_2(CF_3)(CH_2)_3CF_3$ (19), $CH_2(CF_3)(CH_2)_3CO_2H$ (39)	7
L-$CH(CH_3)_2CH_2CH(NH_2)CO_2H$	120°, 8 hr	L-$CH(CH_3)_2CH_2CH(NH_2)CF_3$ (22)	83
DL-$CH_3(CH_2)_3CH(NH_2)CO_2H$	120°, 8 hr	DL-$CH_3(CH_2)_3CH(NH_2)CF_3$ (14)	83
3-NO_2-2-$BrC_6H_3CO_2H$	140–145°, 7 hr	3-NO_2-2-$BrC_6H_3CF_3$ (82)	80
$C_6H_5CO_2H$	120°, 6 hr	$C_6H_5CF_3$ (22), C_6H_5COF (41)	7
p-$NH_2C_6H_4CO_2H$	125°, 10 hr	p-$NH_2C_6H_4CF_3$ (37)	83
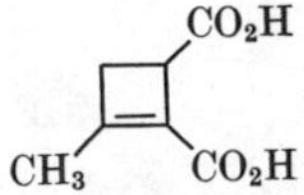 (cyclobutene: CO_2H, CH_3, CO_2H)	120°, 4 hr	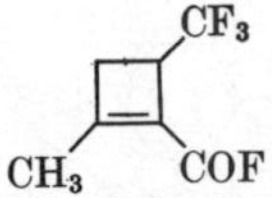 (cyclobutene: CF_3, CH_3, COF) (30) 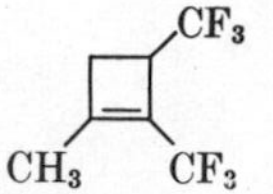 + (cyclobutene: CF_3, CH_3, CF_3) (31)	7
$(O_2N)_2C(CH_2CH_2CO_2H)_2$	100°, 8 hr	$(O_2N)_2C(CH_2CH_2CF_3)_2$ (62)	79
n-$C_6H_{13}CO_2H$	130°, 6 hr	n-$C_6H_{13}CF_3$ (80)	7, 75, 76

Note: References 152–183 are on p. 124.

TABLE IV. CARBOXYLIC ACIDS (*Continued*)

	Carboxylic Acid	Conditions	Product(s) and Yield(s) (%)	Refs.
C_8	p-$HO_2CC_6F_4CO_2H$ (2,3,5,6-tetrafluoro)	200°, 24 hr	p-$F_3CC_6F_4CF_3$ (54)	174
	o-$C_6F_4(CO_2H)_2$	200°, 20 hr	o-$C_6F_4(CF_3)_2$ (65)	37
	$C_6F_5CH_2CO_2H$	180°, 20 hr	$C_6F_5CH_2CF_3$ (88)	37
	$C_6F_5CH(OH)CO_2H$	180°, 20 hr	$C_6F_5CHFCF_3$ (35)	37
	Pyrazinetetracarboxylic acid (HO_2C, CO_2H, HO_2C, CO_2H)	150°, 6 hr	Tetrakis(trifluoromethyl)pyrazine (CF_3, CF_3, CF_3, CF_3) (20)	7
	4-Cl-C_6H_3-1,3-$(CO_2H)_2$	150°, 8 hr	4-Cl-C_6H_3-1,3-$(CF_3)_2$ (62)	7
	$C_6H_5COCO_2H$	100°, 6 hr	$C_6H_5CF_3$ (13), C_6H_5COF (59)	7
	p-$OHCC_6H_4CO_2H$	150°, 13 hr	p-$CHF_2C_6H_4CF_3$ (47)	49
	o-$HO_2CC_6H_4CO_2H$	120°, 6 hr	o-$CF_3C_6H_4CF_3$ (43), o-$CF_3C_6H_4COF$ (23)	7, 76
	p-$HO_2CC_6H_4CO_2H$	120°, 6 hr	p-$CF_3C_6H_4CF_3$ (76), p-$CF_3C_6H_4COF$ (3)	7, 76
	4-CH_3-C_6H_{10}-CO_2H	70°, 10 days	4-CH_3-C_6H_{10}-CF_3	77

	$CH_2(CO_2C_2H_5)(CH_2)_3CO_2H$	130°, 7 hr	$CH_2(CO_2C_2H_5)(CH_2)_3CF_3$ (14), $CH_2(CO_2H)(CH_2)_3CF_3$ (13)	7
	DL-$CH_3(CH_2)_5CH(NH_2)CO_2H$	120°, 8 hr	DL-$CH_3(CH_2)_5CH(NH_2)CF_3$ (39)	83
C_9	$C_6H_5C{\equiv}CCO_2Na$	45°, 6 hr	$C_6H_5C{\equiv}CCOF$ (71)	7
	CO_2H, CO_2H, CO_2H	100–150°, 11 hr	CF_3, COF, COF (69)	85
	p-$HO_2CC_6H_4CO_2CH_3$	130°, 7 hr	p-$CF_3C_6H_4CO_2CH_3$ (63)	7
	DL-$C_6H_5CH_2CH(NH_2)CO_2H$	120°, 8 hr	DL-$C_6H_5CH_2CH(NH_2)CF_3$ (4)	83
	$CH_2(CO_2H)(CH_2)_6CO_2H$	150°, 12 hr	$CH_2(CF_3)(CH_2)_6CO_2H$ (10)	173
	$CH_2(CO_2H)(CH_2)_5CH(CO_2H)SO_3H$	150°, 8 hr	$CH_2(CF_3)(CH_2)_5CH(CF_3)SO_2F$ (33)	7
	$(CH_3)_3CCH_2CH(CH_3)CH_2CO_2H$	120°, 6 hr	$(CH_3)_3CCH_2C(CH_3)CH_2CF_3$ (64)	7
C_{10}	HO_2C, CO_2H, HO_2C, CO_2H	150°, 6 hr	CF_3, CF_3, CF_3, CF_3 (77)	7
	CO_2H, CO_2H, CO_2H, CO_2H	100–160°, 8 hr	CF_3, COF, COF, CF_3 (76)	85
	N, CO_2H	120°, 8 hr	N, CF_3 (72)	83

Note: References 152–183 are on p. 124.

TABLE IV. CARBOXYLIC ACIDS (*Continued*)

	Carboxylic Acid	Conditions	Product(s) and Yield(s) (%)	Refs.
C_{10} (*contd.*)	(−)-$C_6H_5CD(C_2H_5)CO_2H$	35–40°, 9 days	(−)-$C_6H_5CD(C_2H_5)CF_3$ (57)	78
	$C_6H_5CH_2CH(CH_3)CO_2H$	30°, 4.5 days	$C_6H_5CH_2CH(CH_3)CF_3$ (62)	78
	(+)-$C_6H_5CH(C_2H_5)CO_2H$	35–40°, 9 days	(+)-$C_6H_5CH(C_2H_5)CF_3$ (54)	78
	$CH_2(CO_2H)(CH_2)_7CO_2H$	120°, 6 hr	$CH_2(CF_3)(CH_2)_7CF_3$ (27), $CH_2(CF_3)(CH_2)_7COF$ (45), $CH_2(COF)(CH_2)_7COF$ (21)	7
	$C_6H_{11}(CH_2)_3CO_2H$	120°, 10 hr	$C_6H_{11}(CH_2)_3CF_3$ (80)	7
	1-$C_{10}H_7CO_2H$	180°, 24 hr[f]	1-$C_{10}H_7CF_3$ (—)	175
	2-$C_{10}H_7CO_2H$	180°, 24 hr[f]	2-$C_{10}H_7CF_3$ (—)	175
	$(CH_3)_3C$–C_6H_{10}–CO_2H (cyclohexane ring, hashed bond to CO_2H)	70°, 10 days	$(CH_3)_3C$–C_6H_{10}–CF_3 (cyclohexane ring, hashed bond to CF_3) (76)	77
	$(CH_3)_3C$–C_6H_{10}–CO_2H (cyclohexane ring)	70°, 10 d	$(CH_3)_3C$–C_6H_{10}–CF_3 (cyclohexane ring) (57), $(CH_3)_3C$–C_6H_{10}–COF (cyclohexane ring) (10)	77
	$CH_2(CO_2H)(CH_2)_9SO_3H$	130°, 8 hr	$CH_2(CF_3)(CH_2)_9SO_2F$ (42)	7
C_{12}	n-$C_{11}H_{23}CO_2H$	130°, 6 hr	n-$C_{11}H_{23}CF_3$ (88)	7
C_{15}	$CH_3CH(CO_2H)NHC_{12}H_{25}$-$n$	120°, 8 hr	$CH_3CH(CF_3)NHC_{12}H_{25}$-$n$ (61)	83
C_{18}	n-$C_{17}H_{35}CO_2H$	130°, 6 hr	n-$C_{17}H_{35}CF_3$ (93)	7

	Compound	Conditions	Product(s) and Yield(s) (%)	Refs.
C_{26}	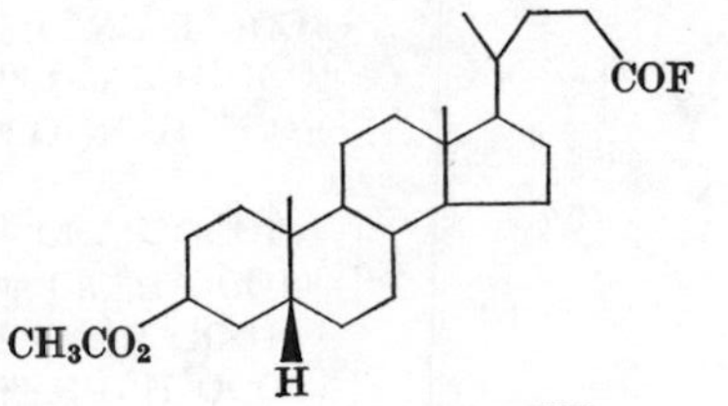 CH_3CO_2 ... COF (steroid structure)	20°, 16 hr	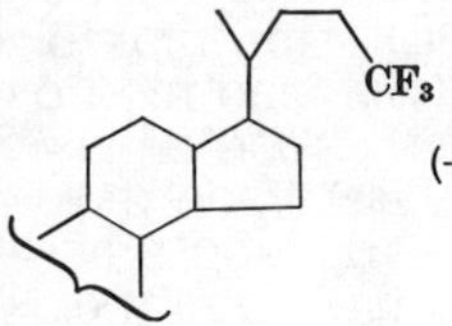 ... CF_3 (—)	74
	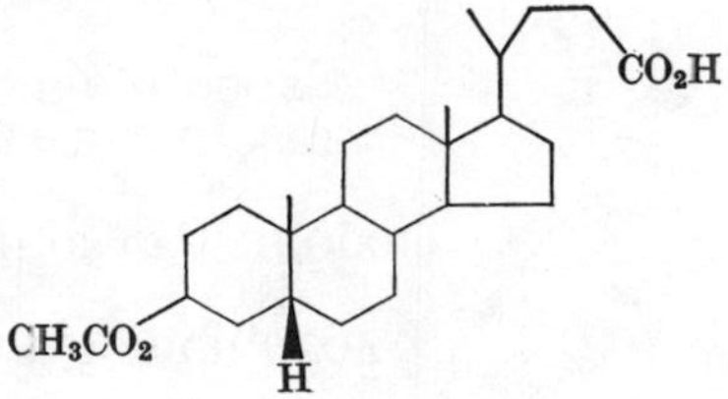 CH_3CO_2 ... CO_2H (steroid structure)	20°, 16 hr, CH_2Cl_2	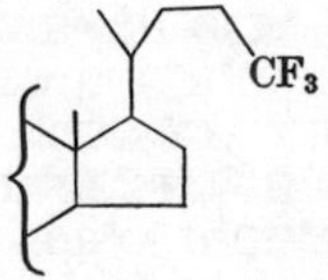 ... CF_3 (42)	74, 30
		20°, 16 hr, tetramethylene sulfone, H_2O	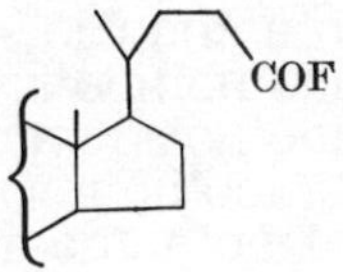 ... COF (82)	74, 30
		25°, 16 hr	" (95)	30

Note: References 152–183 are on p. 124.

[f] The sulfur tetrafluoride pressure was adjusted to 900 psi.

TABLE V. ACID HALIDES

	Acid Halide	Conditions	Product(s) and Yield(s) (%)	Refs.
C_1	$COCl_2$	250°, 4 hr[a]	CF_4 (90), COF_2 (9)	7
		250°	No reaction	7
C_2	CH_3OCOF[b]	100–175°, 6 hr	CH_3OCF_3 (29)	90
C_3	$ClCF_2CH_2OCOF$[b]	100–175°, 6 hr	$ClCF_2CH_2OCF_3$ (38)	90
	CCl_3CH_2OCOF[b]	100–175°, 6 hr	$Cl_3CCH_2OCF_3$ (72)	90
	$BrCH_2CH_2OCOF$[b]	100–175°, 6 hr	$BrCH_2CH_2OCF_3$ (15)	90
	$ClCH_2CH_2OCOF$[b]	100–175°, 6 hr	$ClCH_2CH_2OCF_3$ (24–42)	90
C_4	$CHF_2CF_2CH_2OCOF$[b]	100–175°, 6 hr	$CHF_2CF_2CH_2OCF_3$ (64)	90
	$FOCOCH_2CH_2OCOF$	100–175°, 6 hr	$CF_3OCH_2CH_2OCF_3$ (54), $CF_3OCH_2CH_2OCOF$ (3)	90, 176
	$CF_3OCH_2CH_2OCOF$	100–175°, 6 hr	$CF_3OCH_2CH_2OCF_3$ (53), $CF_3OCH_2CH_2F$ (18)	90
	$H(CF_2CF_2)_2CH_2OCOF$[b]	100–175°, 6 hr	$H(CF_2CF_2)_2CH_2OCF_3$ (35), $H(CF_2CF_2)_2CH_2OCOF$ (29)	90
C_7	2,4-$Br_2C_6H_3OCOF$	100–175°, 6 hr	2,4-$Br_2C_6H_3OCF_3$ (32)	89
	p-BrC_6H_4OCOF[b]	100–175°, 6 hr	*p*-$BrC_6H_4OCF_3$ (27), *p*-BrC_6H_4OCOF (18)	89
	m-BrC_6H_4OCOF[b]	100–175°, 6 hr	*m*-$BrC_6H_4OCF_3$ (18)	89
	o-ClC_6H_4OCOF[b]	100–175°, 6 hr	*o*-$ClC_6H_4OCF_3$ (17)	89
	m-ClC_6H_4OCOF[b]	100–175°, 6 hr	*m*-$ClC_6H_4OCF_3$ (46)	89
	p-ClC_6H_4OCOF[b]	100–175°, 6 hr	*p*-$ClC_6H_4OCF_3$ (58), *p*-ClC_6H_4OCOF (11)	89
	o-$O_2NC_6H_4OCOF$	100–175°, 6 hr	*o*-$O_2NC_6H_4OCF_3$ (69)	89
	m-$O_2NC_6H_4OCOF$[b]	100–175°, 6 hr	*m*-$O_2NC_6H_4OCF_3$ (76)	89
	p-$O_2NC_6H_4OCOF$[b]	100–175°, 6 hr	*p*-$O_2NC_6H_4OCF_3$ (81)	89
	m-FC_6H_4OCOF[b]	100–175°, 6 hr	*m*-$FC_6H_4OCF_3$ (32)	89
	p-FC_6H_4OCOF[b]	100–175°, 6 hr	*p*-$FC_6H_4OCF_3$ (42)	89

	C_6F_5COCl	85–95°, 20 hr	$C_6F_5CF_3$ (95)	37
		120–150°, 22 hr	C_6H_5COF (51), m-$ClC_6H_4CF_3$ (—), charring	7
	C_6H_5COF	120°, 6 hr	$C_6H_5CF_3$ (41)	7
	C_6H_5OCOF	100–175°, 6 hr	$C_6H_5OCF_3$ (67)	89, 91
		,, [b]	,, (46–62)	89
C_8	$H(CF_2CF_2)_3CH_2OCOF$[b]	100–175°, 6 hr	$H(CF_2CF_2)_3CH_2OCF_3$ (51)	90
	m-$FOCOC_6H_4OCOF$[b]	100–175°, 6 hr	m-$CF_3OC_6H_4OCF_3$ (17)	89
	p-$FOCOC_6H_4OCOF$[b]	100–175°, 6 hr	p-$CF_3OC_6H_4OCF_3$ (56)	89
	m-$O_2NC_6H_4N(COF)_2$	75–140°, 8 hr	m-$O_2NC_6H_4N(CF_3)_2$ (5)	63
	p-$O_2NC_6H_4N(COF)CF_3$	75–140°, 8 hr	p-$O_2NC_6H_4N(CF_3)_2$ (5)	63
	m-$O_2NC_6H_4N(COF)CF_3$	75–140°, 8 hr	m-$O_2NC_6H_4N(CF_3)_2$ (39)	63
	m-$FC_6H_4N(COF)CF_3$	75–140°, 8 hr	m-$FC_6H_4N(CF_3)_2$ (12)	63
	p-$FC_6H_4N(COF)CF_3$	75–140°, 8 hr	p-$FC_6H_4N(CF_3)_2$ (12)	63
	$C_6H_5N(COF)_2$	75–140°, 8 hr	$C_6H_5N(CF_3)_2$ (2)	63
	$C_6H_5N(COF)CF_3$	75–140°, 8 hr	$C_6H_5N(CF_3)_2$ (5)	63
	m-$CH_3C_6H_4OCOF$[b]	100°, 2 hr	m-$CH_3C_6H_4OCF_3$ (9)	89
	p-$CH_3C_6H_4OCOF$[b]	100–175°, 6 hr	p-$CH_3C_6H_4OCF_3$ (27)	89
C_9	m-$CH_3C_6H_4N(COF)_2$	75–140°, 8 hr	m-$CH_3C_6H_4N(CF_3)_2$ (4)	63
	p-$CH_3C_6H_4N(COF)_2$	75–140°, 8 hr	p-$CH_3C_6H_4N(CF_3)_2$ (2)	63
C_{12}	$(p$-$FOCOC_6H_4)_2O$[b]	75–140°, 8 hr	$(p$-$CF_3OC_6H_4)_2O$ (5)	89
	$(p$-$FOCOC_6H_4)_2SO_2$[b]	75–140°, 8 hr	$(p$-$CF_3OC_6H_4)_2SO_2$ (56)	89
C_{22}	[steroid structure: O=, OCOF]	20°, 65 hr	[steroid structure: O=, OCF$_3$, F$_2$] (50)	93

Note: References 152–183 are on p. 124.

[a] The experiment was carried out in the presence of methanol and titanium tetrafluoride.

[b] The acyl fluoride was prepared *in situ* from the corresponding alcohol and carbonyl fluoride by heating for 2–4 hours at 100–140°.

TABLE V. ACID HALIDES (*Continued*)

	Acid Halide	Conditions	Product(s) and Yield(s) (%)	Refs.
C_{23}	$OCOCH_3$, H, CH_3CO_2, F, OCOF	20°, 65 hr	H, CH_3CO_2, F, OCF_3 (17)	93
C_{24}	$OCOCH_3$, CH_3CO_2, F, OCOF	20°, 65 hr	CH_3CO_2, F, OCF_3 (19)	93
C_{27}	$OCOCH_3$, CH_3, CH_3CO_2, F, OCOF	20°, 65 hr	CH_3CO_2, F, OCF_3 (9)	93

Note: References 152–183 are on p. 124.

TABLE VI. Esters

	Ester	Conditions	Product(s) and Yield(s) (%)	Refs.
C_2	HCO_2CH_3	200°, 6 hr	CH_3COF (—), CH_3F (high), CHF_3 (high), HCF_2OCH_3 (low)	7
C_5	$CF_3CO_2CH_2OCOCH_3$	100–175°, 6 hr	$CF_3CF_2OCH_2F$ (60)	90
C_6	F_2 F_2 CO_2CH_3	140°, 16 hr	F_2 F_2 CF_3 (10)	7
C_7	$CF_3CO_2CH(CH_3)CH_2OCOCF_3$	—	$CF_3CO_2CH(CH_3)CH_2OC_2F_5$ (—), $C_2F_5OCH(CH_3)CH_2OC_2F_5$ (—)	176
	$(HCF_2CF_2CH_2O)_2CO$	150–250°, 10 hr	$(HCF_2CF_2CH_2O)_2CF_2$ (23)	90
C_8	m-$O_2NC_6H_4OCOCF_3$	100–175°, 6 hr	m-$O_2NC_6H_4OCF_2CF_3$ (83)	89
	p-$O_2NC_6H_4OCOCF_3$	100–175°, 6 hr	p-$O_2NC_6H_4OCF_2CF_3$ (69)	89
	$C_6H_5OCOCF_3$	100–175°, 6 hr	$C_6H_5OCF_2CF_3$ (61)	89
	$C_6H_5CO_2CH_3$	360°, 6 hr	$C_6H_5CF_3$ (55), C_6H_5COF (trace)	7
		250°, 6 hr	No reaction	7
C_9	$CF_3CO_2CH_2CH_2CH(OCOCF_3)C_2H_5$	—	$CF_3CF_2OCH_2CH_2CH(O_2CCF_3)C_2H_5$ (—), $CF_3CO_2CH_2CH_2CH(OC_2F_5)C_2H_5$ (—)	176
C_{10}	p-$O_2NC_6H_4OCO(CF_2)_2CF_3$	100–175°, 6 hr	p-$NO_2C_6H_4O(CF_2)_3CF_3$ (30)	89
	$[H(CF_2CF_2)_2CH_2O]_2CO$	150–250°, 10 hr	$[H(CF_2CF_2)_2CH_2O]_2CF_2$ (34)	90
	$C_6H_5OCO(CF_2)_2CF_3$	100–175°, 6 hr	$C_6H_5O(CF_2)_3CF_3$ (—)	89
	p-$C_6H_4(CO_2CH_3)_2$	130°, 8 hr	p-$CF_3C_6H_4CF_3$ (16), p-$CF_3C_6H_4COF$ (26), p-$C_6H_4(COF)_2$ (4), CH_3F (high)	7
C_{12}	$CF_3CO_2CH(C_6H_5)CH_2OCOCF_3$	—	$C_2F_5OCH(C_6H_5)CH_2OCOCF_3$ (—), $CF_3CO_2CH(C_6H_5)CH_2OC_2F_5$ (—)	176

Note: References 152–183 are on p. 124.

TABLE VI. ESTERS (*Continued*)

	Ester	Conditions	Product(s) and Yield(s) (%)	Refs.
C_{14}	$C_2F_5CF(CF_3)CO_2CH_2CH(CH_2OCH_3)$-$OCOCF(CF_3)C_2F_5$	—	$C_2F_5CF(CF_3)CO_2CH_2$-$CH(CH_2OCH_3)OCF_2CF(CF_3)C_2F_5$ (—)	176
	$(CF_3)_2CFCO_2$–(cyclohexane ring)–$OCOCCF(CF_3)_2$ (—)		$(CF_3)_2CFCO_2$–(cyclohexane ring)–$OCF_2CF(CF_3)_2$	176
	$C_6H_5OCOCO_2C_6H_5$	25°, 65 hr	Polymer (—)	87
		100°, 6 hr	Polymer (—)	87
C_{16}	$(C_2H_5O_2C)_2C{=}C\langle S_2 \rangle C{=}C(CO_2C_2H_5)_2$ (1,3-dithietane ring)	125–200°, 8 hr	$(CF_3)_2C{=}C\langle S_2 \rangle C{=}C(CF_3)_2$ (1,3-dithietane ring) (69)	86
C_{19}	$C_6H_5OCH_2CH(OCOR_f)CH_2OCOR_f$ (—) R_f = $-CF(CF_3)CF_2CF_3$		$C_6H_5OCH_2CH(OCOR_f)CH_2OCF_2R_f$ (—), $C_6H_5OCH_2CH(OCF_2R_f)CH_2OCOR_f$ (—)	176

Note: References 152–183 are on p. 124.

TABLE VII. AMIDES

	Amide	Conditions	Product(s) and Yield(s) (%)	Refs.
C_1	NH_2CONH_2	150–250°, 14 hr	$CF_3N{=}SF_2$ (6)	108
C_2	CH_3CONH_2	20°, 3 d	$(CH_3CONH)_2S$ (—)	106
C_7	$C_6H_5CONH_2$	150°, 8 hr	$C_6H_5CF_3$ (13)	7
C_8	Phthalimide	100°, 10 hr	o-$CF_3C_6H_4COF$ (58)	7
	$C_6H_5CONHCH_3$	60°, 4 hr	C_6H_5COF (48)	7
C_9	$C_6H_5CON(CH_3)_2$	130°, 6 hr	$C_6H_5CF_2N(CH_3)_2$ (17), C_6H_5COF (1)	7

Note: References 152–183 are on p. 124.

TABLE VIII. ANHYDRIDES

	Anhydride	Conditions	Product(s) and Yield(s) (%)	Refs.
C_1	CO	250°, 7 hr	CF_4 (95)	7
	CO_2	500°, 2 hr	CF_4 (80), COF_2 (10)	7
C_4	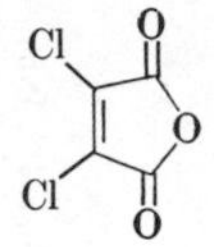 [dichloromaleic anhydride: Cl, Cl, O, O, O ring]	300°, 10 hr	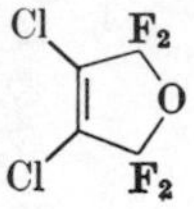 [Cl, Cl, F_2, F_2, O ring] (I, 46)	7
		290°, 13 hr[a]	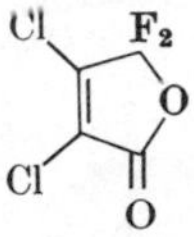 [Cl, Cl, F_2, O ring, C=O] (II, 51)	98
		300°, 51 hr[b]	I (39), II (25), $CF_3ClC{=}CCF_3Cl$ (III, 11) *cis* and *trans*	98
		300°, 63 hr[b]	I (37), III (38)	98
	Maleic anhydride	150°, 13 hr	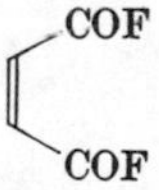 [cis-CH=CH with COF, COF] (71)	7
	$(CH_3CO)_2O$	300°, 10 hr	CH_3CF_3 (50)	7
C_8	Phthalic anhydride	180°, 18 hr	o-$C_6H_4(COF)_2$ (93)	7
		350°, 11 hr	o-$C_6H_4(CF_3)_2$ (45)	7

Note: References 152–183 are on p. 124.

[a] The mole ratio of sulfur tetrafluoride to anhydride was 2.4:1.
[b] The mole ratio of sulfur tetrafluoride to anhydride was 6.3:1.

TABLE IX. Organic Halides (Excluding Haloalkenes and Acyl Halides)[a,b]

	Halide	Conditions	Product(s) and Yield(s) (%)
C_1	CBr_4	150–225°, 3 hr	CBr_2F_2 (27), CBr_3F (16)
		150–325°, 3 hr	$CBrF_3$ (77), CF_4 (5)
	CCl_4	100–225°, 8 hr	CCl_2F_2 (17), CCl_3F (66)
		100–325°, 4 hr	CCl_2F_2 (10), $CClF_3$ (40)
C_2	$CCl_2{=}CCl_2$	200–400°, 10 hr	$CClF_2CClF_2$ (35), CF_3CClF_2 (35)
	CCl_3CCl_3	150–350°, 3 hr	CCl_2FCCl_2F (—)
	$CCl_2{=}CHCl$	200–400°[c], 4 hr	$CClF_2CClF_2$ (25), CF_3CClF_2 (25)
	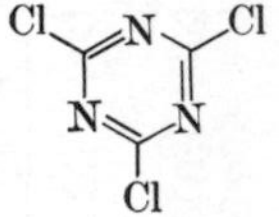 [structure]	150–250°, 12 hr	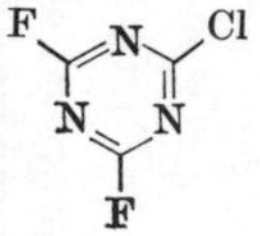 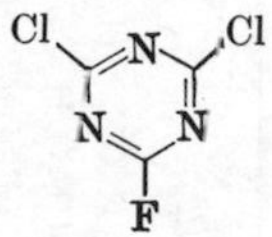 [structure] (39), [structure] (29)
		150–250°, 12 hr	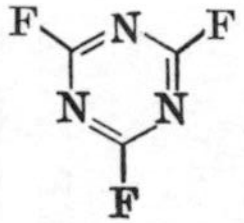 [structure] (40)
C_4	$CF_3CCl_2CCl_2CF_3$	150–325°, 3 hr	$CF_3CClFCCl_2CF_3$ (61)
	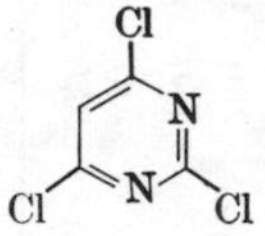 [structure]	175–225°, 7 hr	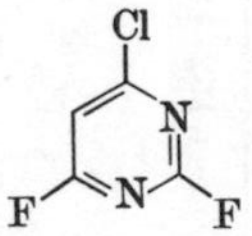 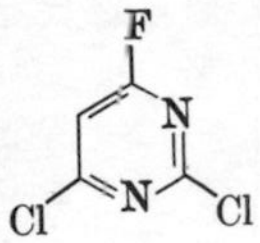 [structure] (—), [structure] (—)
	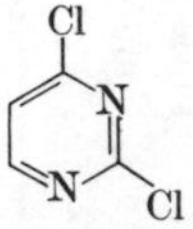 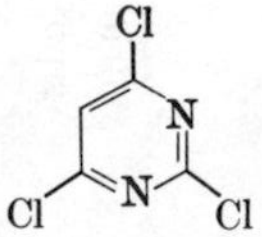 [structure] + [structure] (trace)	50–150°, 9 hr	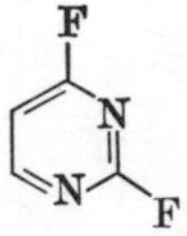 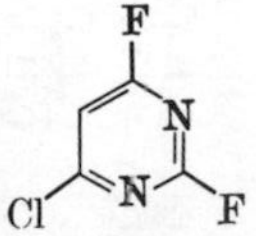 [structure] (30), [structure] (trace)

	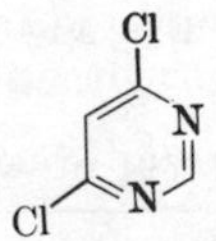 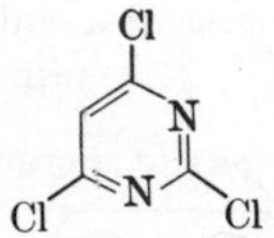 Cl, N, N, Cl + Cl, N, N, Cl, Cl (trace)	50–150°, 9 hr	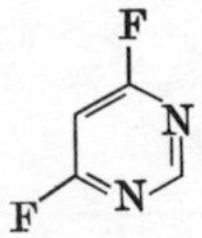 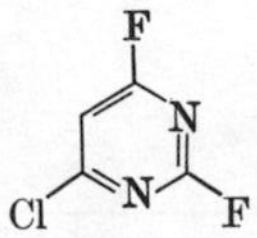 F, N, N, F (30), F, N, N, Cl, F (trace)
C_5	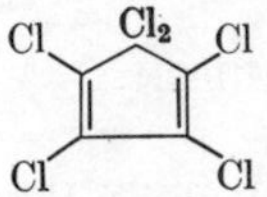 Cl, Cl_2, Cl, Cl, Cl	180–330°, 10 hr	Pentachlorotrifluorocyclopentenes (38)
C_6	Hexachlorobenzene	200–400°, 8 hr	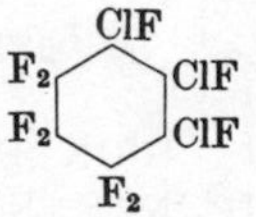 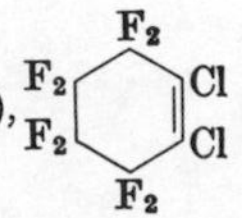 ClF, F_2, ClF, F_2, ClF, F_2 (−), F_2, F_2, Cl, F_2, Cl, F_2 (20)

Note: References 152–183 are on p. 124.

[a] This table covers halogen exchange by sulfur tetrafluoride with or without the addition of fluorine to double bonds.

[b] All the examples in this table were reported by Tullock and co-workers.[100]

TABLE X. AMINES

	Amine	Conditions	Product(s) and Yield(s) (%)	Refs.
C_1	$NCNH_2$	25–40°, 84 hr	$NCN{=}SF_2$ (65), $F_2S{=}NCF_2N{=}SF_2$ (25)	110
	CH_3NH_2	−78°, 0.5 hr[a],	$CH_3N{=}S{=}NCH_3$ (72)	102
		−78°, 2 hr[b]	$CH_3N{=}SF_2$ (56)	103
C_2	$C_2H_5NH_2$	"Low temp"[c]	$C_2H_5N{=}SF_2$ (—)	104
C_5	$(CH_3)_2NSi(CH_3)_3$	"Low temp"[c]	$(CH_3)_2NSF_3$ (75)	107
	$C_2H_5NHSi(CH_3)_3$	"Low temp"[c]	$C_2H_5N{=}SF_2$ (—)	104
C_6	$C_6F_5NH_2$	20°, 1 hr[d]	$C_6F_5N{=}S{=}NC_6F_5$ (90)	106
	p-$ClC_6H_4NH_2$	−10° [d,e]	p-$ClC_6H_4N{=}S{=}NC_6H_4Cl$-$p$ (40)	
	$C_6H_5NH_2$	−78°, CH_2Cl_2[f]	$C_6H_5N{=}S{=}NC_6H_5$ (94)	151
		Reflux, 1 hr	" (37)	105
C_7	$(C_2H_5)_2NSi(CH_3)_3$	"Low temp"[e]	$(C_2H_5)_2NSF_3$	177
	$CH_3N[Si(CH_3)_3]_2$	"Low temp"[c]	$CH_3N{=}SF_2$ (—)	104
C_8	(cyclohexane ring)$NSi(CH_3)_3$	"Low temp"[c]	(cyclohexane ring)NSF_3 (60)	178
C_{12}	$C_6H_5NHC_6H_5$	(1) 25°, 3 d, (2) hydrolysis	$[(C_6H_5)_2N]_2SO$ (36)	106

Note: References 152–183 are on p. 124.

[a] Excess methylamine was used.
[b] A deficiency of methylamine was used.
[c] Reactants were mixed in the cold and allowed to warm to room temperature.
[d] Sodium fluoride was used as a hydrogen fluoride acceptor.
[e] Petroleum ether was the solvent.
[f] Excess aniline was used as the hydrogen fluoride acceptor.
[g] Triethylamine was used as the solvent and hydrogen fluoride acceptor.

TABLE XI. COMPOUNDS HAVING CARBON-NITROGEN MULTIPLE BONDS[a]

	Reactant	Conditions	Product(s) and Yield(s) (%)
C_1	BrCN	150–200°, 16 hr	$CF_3N{=}SF_2$ (37)
	$F_2S{=}NCN$	100–160°, 14 hr	$F_2S{=}NCF_2N{=}SF_2$ (50)
	H_2NCN[b]	Room temp −40°, 84 hr[b]	$F_2S{=}NCN$ (65), $F_2S{=}NCF_2N{=}SF_2$ (25)
	KCNO	350°	$CF_3N{=}SF_2$ (trace)
	NaCN	200–300°, 12 hr	$CF_3N{=}SF_2$ (29)
	NaSCN	200–350°, 12 hr	$CF_3N{=}SF_2$ (69)
C_2	CF_3CN	350°, 14 hr	$CF_3CF_2N{=}SF_2$ (70)
	CH_3CN[c]	160–260°, 13.5 hr	$CH_2FCF_2N{=}SF_2$ (7)
	$(CN)_2$	350°, 8 hr	$NCCF_2N{=}SF_2$ (8), $F_2S{=}N(CF_2)_2N{=}SF_2$ (26)
C_3	OH N N HO N OH	100–300°, 14 hr	$CF_3N{=}SF_2$ (9)
C_5	$(CH_3)_3CCN$	280°	$(CH_3)_3CCF_2N{=}SF_2$ (—)[c]
C_7	C_6H_5CN[d]	180–250°, 18 hr	$C_6H_5CF_2N{=}SF_2$ (48)
	C_6H_5NCO	100–150°, 14 hr	$C_6H_5N{=}SF_2$ (88)

Note: References 152–183 are on p. 124.

[a] With the exceptions of $F_2S{=}NCN$ and H_2NCN, which were reported by Glemser and Bierman,[110] all the examples in this table were reported by Smith and co-workers.[34]

[b] Anhydrous sodium fluoride was included in the reaction to bind hydrogen fluoride.

[c] No reaction occurred at 230°.

[d] The reaction is incomplete at 220°; decomposition takes place at 280°.

TABLE XII. THIOCARBONYL COMPOUNDS

	Thiocarbonyl Compound	Conditions	Product(s) and Yield(s) (%)	Refs.
C_1	$Cl_2C{=}S$	150°, 4 hr[a]	CCl_2F_2 (—), $CClF_3$ (—)	179
	CS_2	150–180°, 10 hr[b]	$CF_3S_3CF_3$ (28), $CF_3S_2CF_3$ (—)	33
		200–475°, 13 hr[c]	CF_4 (100)	33
	$CS_2 + Br_2$	150–225°, 6 hr[c]	$CBrF_3$ (38), CF_4 (8)	110, 111
		200–325°, 3 hr	$CBrF_3$ (87), CF_4 (12)	100, 111
	$CS_2 + Cl_2$	75–225°, 4 hr	CCl_2F_2 (14), CCl_3F (54)	100
		75–325°, 3 hr	CCl_2F_2 (57), $CClF_3$ (19)	100
C_3	$(CF_3S)_2CS$	130–200°, 11 hr	$CF_3S_{1-4}CF_3$ (—), $CF_3SCF_2SCF_3$ (trace)	179
	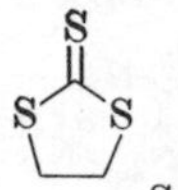 1,3-dithiolane-2-thione (ring S–C(=S)–S)	110°, 8 hr	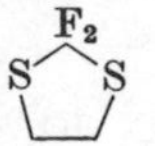 2,2-difluoro-1,3-dithiolane (ring S–CF_2–S) (82)	33
C_5	$C_2H_5OC(=S)SC_2H_5$	50–120°, 8 hr	$C_2H_5OCF_3$ (—), CH_3CH_2F (—)	179
C_6	$[(CH_3)_2NC(=S)S]_2$	120–150°, 8 hr[c]	$(CH_3)_2NCF_3$ (—)[d]	33
C_{10}	$[(C_2H_5)_2NC(=S)S]_2$	120°, 8 hr	$(C_2H_5)_2NCF_3$ (58)	33
C_{12}	 $(C_5H_{10}NC(=S)S)_2$ (piperidine ring)	100°, 6 hr	 $C_5H_{10}NCF_3$ (piperidine ring) (70)	33

Note: References 152–183 are on p. 124.

[a] Antimony trichloride–antimony pentachloride (1:1) was used as catalyst.
[b] Boron trifluoride was the catalyst.
[c] Arsenic trifluoride was used as catalyst.
[d] According to mass spectrometric analysis of the volatile product, this is the sole organic product.

TABLE XIII. Haloalkenes[a,b]

	Haloalkene	Conditions[c]	Product(s) and Yield(s) (%)
C_2	$CF_2{=}CBrF$		CF_3CF_2Br (57)
	$CF_2{=}CClF$		CF_3CF_2Cl (60)
	$CF_2{=}CCl_2$		CF_3CCl_2F (88)
	$CCl_2{=}CClF$		CCl_2FCClF_2 (77)
	$CCl_2{=}CCl_2$		CCl_2FCCl_2F (77)
	$CF_2{=}CFI$		CF_3CF_2I (65)
	$CF_2{=}CF_2$		CF_3CF_3 (2), $CF_3CF_2CF_3$ (6), $CF_3CF_2CF_2CF_3$ (7)
	$CHF{=}CClF$		CHF_2CClF_2 (96)
	$CCl_2{=}CHCl$		$CCl_2FCHClF$ (85)
	$CF_2{=}CHI$		CF_3CHFI (54)
	CF_2CHF		CF_3CHF_2 (41)
	trans-$CHCl{=}CHCl$		$CHClFCHClF$-*dl* (54), -*meso* (26)
	cis-$CHCl{=}CHCl$		$CHClFCHClF$-*dl* (69), -*meso* (21)
	$CCl_2{=}CH_2$		CCl_2FCH_2F (59)
	$CF_2{=}CH_2$		CF_3CH_2F (28)
	$CHBr{=}CH_2$		$CHBrFCH_2F$ (20)
C_3	$CF_3CF{=}CF_2$		$CF_3CF_2CF_3$ (63)
	$CHF_2CBr{=}CHF$	100°, 5 hr[d]	$CHF_2CBrFCHF_2$ (86)
	$CF_2{=}CFOCH_3$	100–120°, 8 hr[e]	$F_3SCF_2CF_2OCH_3$ (41)
C_4	(cyclobutene ring: F_2, F_2, F, F)		(cyclobutane ring: F_2, F_2, F_2, F_2) (23)
	$CF_3CF{=}CFCF_3$		$CF_3(CF_2)_2CF_3$ (—)
C_5	(cyclopentene ring: Cl, Cl, F_2, F_2, F_2)		(cyclopentane ring: ClF, ClF, F_2, F_2, F_2) (47)[f]
	n-$C_3H_7OCF{=}CF_2$	100–120°[e]	n-$C_3H_7OCF_2CF_2SF_3$ (63)

Note: References 152–183 are on p. 124.

[a] This table covers the addition of fluorine or the elements of sulfur tetrafluoride to double bonds.

[b] With the exceptions of $CHF_2CBr{=}CHF$ reported by Regan,[180] and $CF_2{=}CFOCH_3$ and n-$C_3H_7OCF{=}CF_2$ reported by Hasek,[117] all the examples in this table were reported by Bissell and Fields.[115]

[c] Unless otherwise denoted, reactions were run for 2 hr at 100° with a 3:1 mixture of sulfur tetrafluoride and lead dioxide as the fluorinating agent.

[d] A 1.5:1 mixture of sulfur tetrafluoride and lead dioxide was used as the fluorinating agent.

[e] This reaction did not include lead dioxide.

[f] A mixture of *cis* and *trans* isomers was obtained.

TABLE XIV. Organo–Arsenic, –Boron, –Phosphorus, and –Silicon Compounds

	Reactant	Conditions	Product(s) and Yield(s) (%)	Refs.
		Organoarsenic Compounds		
C_6	$C_6H_5AsO(OH)_2$	70°, 10 hr	$C_6H_5AsF_4$ (45)	66
C_{18}	$(C_6H_5)_3As$	50–130°, 14 hr	$(C_6H_5)_3AsF_2$ (46)	66
		Organoboron Compounds		
C_{15}	$[CH_3CH(CO_2C_2H_5)O]_3B$	—	$CH_3CHFCO_2C_2H_5$ (16)	121
C_{18}	Tris(cyclohexyl) borate	1. −70° to rt, 12 hr 2. Reflux, 1 hr	$C_6H_{11}F$ (59)	121
	$(n\text{-}C_6H_{13}O)_3B$	1. −70° to rt, 12 hr 2. Reflux, 1 hr	$n\text{-}C_6H_{13}F$ (31)	121
C_{24}	$[n\text{-}C_4H_9CH(C_2H_5)CH_2O]_3B$	−70° to rt, 2 d	$n\text{-}C_4H_9C(C_2H_5){=}CH_2$ (85)	121
	$(n\text{-}C_8H_{17}O)_3B$	−70° to rt, 24 hr	$n\text{-}C_8H_{17}F$ (44)	121
C_{36}	$(n\text{-}C_{12}H_{25}O)_3B$	−70° to rt, 24 hr	$n\text{-}C_{12}H_{25}F$ (52)	121
	$[i\text{-}C_4H_9CH(CH_3)(CH_2)_3CH(CH_3)CH_2O]_3B$	−70° to rt, 24 hr	$i\text{-}C_4H_9CH(CH_3)(CH_2)_3C(CH_3){=}CH_2$ (82)	121
		Organophosphorus Compounds		
C_3	$(CH_3S)_3P$	Exothermic at 25°	CH_3SPOF_2 (trace), $(CH_3S)_2PSF$ (trace), SPF_3 (—), $(CH_3)_2S_X$ (77)	119
C_5	$CH_3PO[N(CH_3)_2]_2$	60°, 2 hr	CH_3POF_2 (45), $(CH_3)_2NSF_3$ (trace)	119
C_6	$C_6H_5POF_2$	100–150°, 12 hr	$C_6H_5PF_4$ (62)	66
	$C_6H_5PO(OH)_2$	100–150°, 16 hr	$C_6H_5PF_4$ (58)	66
	$[(CH_3)_2N]_3P$	—	$[(CH_3)_2N]_3PF_2$ (51), $[(CH_3)_2N]_2PF_3$ (trace)	119
C_{12}	$(C_6H_5)_2PO(OH)_2$	50–150°, 14 hr	$(C_6H_5)_2PF_3$ (42)	66
C_{18}	$(C_6H_5)_3P$	50–150°, 14 hr	$(C_6H_5)_3PF_2$ (69)	66
	$(C_6H_5)_3PO$	50–150°, 14 hr	$(C_6H_5)_3PF_2$ (67)	66
		Organosilicon Compounds		
C_3	$(CH_3)_3SiCl$	−40° warmed to 20°[a]	$(CH_3)_3SiF$ (90)	120

	$(CH_3)_3SiCl$	20°, 16 hr	$(CH_3)_3SiF$ (90)	120
	$(CH_3)_3SiOH$	20°, 22 hr	$(CH_3)_3SiF$ (70)	120
C_5	$CH_3CH_2NHSi(CH_3)_3$	"Low temp"	$CH_3CH_2N{=}SF_2$ (—), $(CH_3)_3SiF$ (—), $CH_3CH_2NH_2 \cdot HF$ (—)	104
			$(CH_3)_2NSF_3$ (—)	
	$(CH_3)_2NSi(CH_3)_3$	"Low temp"	$(CH_3)_2NSF_3$ (—)	104
	$(CH_3)_3SiOSi(CH_3)_3$	−50° to +10°, 15 hr	$(CH_3)_2SiF_2$ (79)	120
C_6	$(CH_3)_2Si(OC_2H_5)_2$	20°, 18 hr	$(CH_3)_2SiF_2$ (77)	120
C_7	$CH_3N[Si(CH_3)_3]_2$	"Low temp"	$CH_3N{=}SF_2$ (—), $(CH_3)_3SiF$ (—)	104
C_9	$(CH_3)_3SiOC_6H_5$	[b]	$(C_6H_5O)_4S$ (—)	181
		[c]	$(C_6H_5O)SF_3$ (—), $(C_6H_5O)_2SF_2$ (—), $(C_6H_5O)_3SF$ (trace)	181
C_{12}	$(C_6H_5)_2Si(OH)_2$	25–28°, 24 hr	$(C_6H_5)_2SiF_2$ (43)	120
		−50° to +10°, 15 hr	" (79)	120

Note: References 152–183 are on p. 124.

[a] The sulfur tetrafluoride/boron trifluoride complex in excess was used as the fluorinating agent.

[b] The mole ratio of sulfur tetrafluoride to silane was 1:3.

[c] The mole ratio of sulfur tetrafluoride to silane was 1:<3.

TABLE XV. MISCELLANEOUS COMPOUNDS

	Reactant	Conditions	Product(s) and Yield(s) (%)	Refs.
C_1	CF_3OF	Room temp, $h\nu$, gas phase, 3 d	CF_3OSF_5 (I, —), $(CF_3O)_2SF_4$ (II, 8)	125
		150°	I (—), II (—)	
C_2	CF_3OOCF_3	100°, 1 wk	I (35)	
		Room temp, $h\nu$, gas phase	CF_3OSF_5 (~10), $(CF_3O)_2SF_4$ (~10)	182
	$CH_2(CO_2H)SO_3H$	180°, 6 hr	$CH_2(CF_3)SO_2F$ (41)	7
	CH_3SOCH_3	−196 to 20°[a]	CH_2FOCH_2F (—)	123
	$(CH_3)_2SO(NH)$	0°, 12 hr	$[(CH_3)_2SON]_3S^+H_2F_3^-$ (34)	182
C_3	C_3F_7IO	−10°[b]	$C_3F_9IF_2$ (75)	122
	$(CH_2O)_3$[c]	140°, 12 hr	CH_2F_2 (—), CH_2FOCH_2F (—)	123
C_4	n-C_4H_9MgBr	−78°	$(n\text{-}C_4H_9)_2S$ (82)	124
C_5	(2-pyridyl)IO	−10°[b]	(2-pyridyl)IF_2 (100)	122
C_6	C_6F_5IO	−10°[b]	$C_6H_5IF_2$ (66)	122
	m-FC_6H_4IO	−10°[b]	m-$FC_6H_4IF_2$ (100)	122
	p-FC_6H_4IO	−10°	p-$FC_6H_4IF_2$ (100)	122
	o-$O_2NC_6H_4IO$	−10°[b]	o-$O_2NC_6H_4IF_2$ (100)	122
	C_6H_5MgBr	−78°	$C_6H_5SC_6H_5$ (58)	124
	C_6H_5IO	−20°, 10 min[b]	$C_6H_5IF_2$ (82)	122
	$[(CH_3)_3SiN]SF_2$	Room temp, 24 hr	$N{\equiv}SF_2N{=}SF_2$ (56)	183

C_7	n-C_3F_7I	70°, 2 hr	n-$C_3F_7IF_2$ (—)	122
	p-$CH_3C_6H_4IO$	−20°, 10 min[b]	p-$CH_3C_6H_4IF_2$ (100)	122
	m-Xylene	25°, 1 hr[d]	$[2,4\text{-}(CH_3)_2C_6H_3]_3S^+BF_4^-$ (30)	54b
C_9	p-$CH_3C_6H_4SO_2N{=}S(CH_3)_2(NH)$	−10 to −5°	$[p\text{-}CH_3C_6H_4SO_2N{=}S(CH_3)_2N]_3S^+H_2F_3^-$ (—)	182
	$CH_2(CO_2H)(CH_2)_5CH(CO_2H)SO_3H$	150°, 8 hr	$CH_2(CF_3)(CH_2)_5CH(CF_3)SO_2F$ (33)	7
C_{11}	$CH_2(CO_2H)(CH_2)_9SO_3H$	130°, 8 hr	$CH_2(CF_3)(CH_2)_9SO_2F$ (42)	7

Note: References 152–183 are on p. 124.

[a] A vigorous exothermic reaction occurred when a mixture of dimethyl sulfoxide and sulfur tetrafluoride in a 2:1 mole ratio was allowed to warm slowly from −196 to 20°.

[b] The iodoso compound was suspended in methylene chloride in a glass flask and the suspension was cooled by liquid nitrogen. The flask was evacuated and sulfur tetrafluoride was allowed to condense in the flask. The temperature was gradually increased to −10° at which temperature the precipitate dissolved. The reaction was then worked up.

[c] See Table II, reaction with $(CH_2O)_n$.

[d] m-Xylene was added to a solution of the 1:1 adduct of sulfur tetrafluoride and boron trifluoride in nitromethane.

REFERENCES TO TABLES

[152] R. E. A. Dear and E. E. Gilbert, U.S. Pat. 3,372,205 (1968) [*C.A.*, **69**, 26748 (1968)].

[153] L. A. Wall and J. P. Antonucci, U.S. Pat. 3,513,206 (1970) [*C.A.*, **73**, 26056 (1970)].

[154] M. P. Mertes and S. E. Saheb, U.S. Pat. 3,324,126 (1967) [*C.A.*, **68**, 39646 (1968)].

[155] D. G. Martin and J. E. Pike, U.S. Pat. 3,251,834 (1966) [*C.A.*, **63**, 13369 (1965)].

[156] H. J. Ache, D. R. Christman, and A. P. Wolf, *Radiochim. Acta*, **12**, 121 (1969).

[157] J. D. Roberts, J. B. Grutzner, M. Jautelat, J. B. Dence, and R. A. Smith, *J. Amer. Chem. Soc.*, **92**, 7107 (1970).

[158] H. G. Gilch, *J. Org. Chem.*, **30**, 4392 (1965).

[159] J. T. Gerig and J. D. Roberts, *J. Amer. Chem. Soc.*, **88**, 2791 (1966).

[160] E. Jones and I. M. Moodie, *J. Chem. Soc., C*, **1969**, 2051.

[161] E. L. Zaitseva and A. Ya. Yakubovich, *Zh. Obshch. Khim.*, **36**, 359 (1966) [*C.A.*, **64**, 15774 (1966)].

[162] H. Gilboa, J. Altman, and A. Loewenstein, *J. Amer. Chem. Soc.*, **91**, 6062 (1969).

[163] J. S. Tadanier and J. W. Cole, U.S. Pat. 3,281,436 (1966) [*C.A.*, **66**, 55657 (1967)].

[164] J. S. Tadanier and J. W. Cole, U.S. Pat. 3,257,424 (1966) [*C.A.*, **65**, 12267 (1966)].

[165] G. A. Boswell, Jr., U.S. Pat. 3,219,673 (1965) [*C.A.*, **64**, 5183 (1966)].

[166] G. A. Boswell, Jr., U.S. Pat. 3,282,969 (1966) [*C.A.*, **66**, 11129 (1967)].

[167] H. J. Ache, D. R. Christman, and A. P. Wolf, *Radiochim. Acta*, **10**, 41 (1968).

[168] A. F. Crowther and H. L. Roberts, Brit. Pat. 908,290 (1962) [*C.A.*, **58**, 5513 (1963)].

[169] A. F. Crowther and H. L. Roberts, Brit. Pat. 955,478 (1964) [*C.A.*, **60**, 15730 (1964)].

[170] R. E. Putnam, R. J. Harder, and J. E. Castle, *J. Amer. Chem. Soc*, **83**, 391 (1961).

[171] R. E. Putnam and J. E. Castle, *J. Amer. Chem. Soc.*, **83**, 389 (1961).

[172] M. P. Mertes and S. E. Saheb, *J. Pharm. Sci.*, **52**, 508 (1963).

[173] P. M. Enriquez and J. T. Gerig, *Biochemistry*, **8**, 3156 (1969).

[174] G. Camaggi and F. Gozzo, *J. Chem. Soc., C*, **1969**, 489.

[175] T. E. Bull and J. Jonas, *J. Chem. Phys.*, **52**, 1978 (1970).

[176] W. A. Sheppard, U.S. Pat. 3,099,685 (1963) [*C.A.*, **60**, 411h (1964)].

[177] S. P. von Halasz and O. Glemser, *Chem. Ber.*, **103**, 594 (1970).

[178] S. P. von Halasz and O. Glemser, *Chem. Ber.*, **104**, 1247 (1971).

[179] W. C. Smith, U.S. Pat. 2,957,001 (1960) [*C.A.*, **55**, 22086 (1961).

[180] B. M. Regan, U.S. Pat. 3,480,683 (1969) [*C.A.*, **72**, 42729 (1970).

[181] J. I. Darragh and W. A. Sharp, *Angew. Chem.*, **82**, 45 (1970).

[182] R. Appel and E. Lassmann, *Chem. Ber.*, **103**, 2548 (1970).

[183] O. Glemser and R. Hoefer, *Angew. Chem.*, **82**, 324 (1970).

CHAPTER 2

MODERN METHODS TO PREPARE MONOFLUOROALIPHATIC COMPOUNDS

Clay M. Sharts
San Diego State University
San Diego, California

and

William A. Sheppard
Central Research Department, E. I. du Pont de Nemours and Co., Inc.
Wilmington, Delaware

CONTENTS

ACKNOWLEDGMENT

In the fall of 1970 one of us (CMS) taught a graduate course in organic fluorine chemistry and an undergraduate course in chemical literature. Much of the recent literature cited in this chapter was first located by students in these two courses.

The text is derived in part from papers written by these people. We acknowledge the papers written by Charles A. Jacks and James A. Carey, Jr. (Fluorination by Alkali Metal Fluorides), Barnard R. McEntyre (Hydrogen Fluoride Addition to Oxiranes), James D. Mavis (Hydrogen Fluoride Addition to Alkenes), Richard Sherman and Michael Wagner (Fluoramine Reagent), Alan Carder (Perchloryl Fluoride), and Mrs. Krishnamurthy (Halogen Fluoride Addition) and express our sincere thanks to these individuals.

We acknowledge literature searches carried out by Joe Barone, Stephen D. Drew, L. E. Elliot, (Miss) Wendy Gniffke, Tony Host, Robert Lowry, Barry Masters, and Fred Roland, and thank these students for their diligence.

Also we thank Miss Kathleen L. Green and Miss Susan A. Vladuchick of Central Research Department, E. I. du Pont de Nemours and Co., for a comprehensive literature search for the sections on halogen fluoride addition and the fluoralkylamine reagent, and for proofreading and final revisions, respectively.

INTRODUCTION

Organic fluorine chemistry has been important since the 1930s. The discovery of Freons® and their use as refrigerants provided the economic incentive for industrial research and development. The development of fluorine chemistry during World War II is well documented.[1–3] The post-World War II emergence of Teflon® as an unusually stable lubricating polymer continued the earlier economic incentive for industrial support of fluorine research. Nevertheless, in the early 1950s fluorine chemistry, both organic and inorganic, was little recognized by the general profession, and few university professors had active research programs involving fluorine chemistry.

Fluorine chemistry emerged in the 1950s because of three major events.

1. The Advanced Research Projects Agency, Defense Department, U.S. Government, financed large-scale research designed to synthesize fluorine-containing rocket propellants. Of enormous significance were research funds provided for synthesis of N-F and O-F compounds.

2. A fluorine atom at a specific site in a steroid molecule was found to enhance greatly beneficial pharmacological effects.[4]

3. New analytical techniques, such as gas chromatography and fluorine nmr spectroscopy, made possible the separation and identification of fluorinated compounds.

[1] *Ind. Eng. Chem.*, **39**, 236 (1947). This issue contains 53 papers reporting fluorine chemistry related to the Manhattan Project.

[2] C. Slesser and S. R. Schram, Eds., *Preparations, Properties and Technology of Fluorine and Organic Fluoro Compounds*, McGraw-Hill, New York, 1951.

[3] J. H. Simmons, *Fluorine Chemistry*, Vols. I and II, Academic Press, New York, 1930.

[4] J. Fried and N. A. Abraham, "Introduction of Fluorine into the Steroid System," in *Organic Reactions in Steroid Chemistry*, J. Fried and J. A. Edwards, Eds., Vol. 1, Chapter 8, Reinhold, New York, 1972.

One fluorine atom introduced into an organic molecule is potentially valuable as a probe (nmr) for biological, mechanistic, and structural studies or to enhance considerably biological activity or chemical reactivity.

In 1944 the first review on the preparation of aliphatic fluorine compounds was published in *Organic Reactions*.[5] Because of the extensive development of the field, the present review must be limited to preparation of monofluoroaliphatic compounds, and even further restricted to in-depth coverage of modern reagents for introducing a single fluorine atom. Use of sulfur tetrafluoride in the preparation of fluoroaliphatic compounds, particularly *gem*-difluoro and trifluoromethyl compounds, is covered in the companion chapter.[6] Reviews that are useful for preparation of monofluoroaliphatic compounds are summarized below[7–15] and where appropriate in the text for each reaction type.

Eight reagents and/or reactions are important for preparation of monofluoroaliphatic compounds.

1. Halogen fluoride addition to alkenes and alkynes.

$$\overset{\delta+}{X}-\overset{\delta-}{F} + \gt C{=}C\lt \longrightarrow -\overset{X}{\underset{|}{\overset{|}{C}}}-\overset{F}{\underset{|}{\overset{|}{C}}}-$$

X = Cl, Br, I

2. Fluoroalkylamine reagent (**FAR**) for replacement of a hydroxyl group.

$$ROH + \underset{\textbf{FAR}}{(C_2H_5)_2NCF_2CHClF} \rightarrow RF + (C_2H_5)_2NCOCHClF + HF$$

[5] A. L. Henne, *Org. Reactions*, **2**, 49 (1944).

[6] G. A. Boswell, W. C. Ripka, R. M. Scribner, and C. W. Tullock, *Org. Reactions*, **21**, 1 (1974).

[7] F. L. M. Pattison, *Toxic Aliphatic Fluorine Compounds*, Elsevier, New York, 1959.

[8] Houben-Weyl, Ed., *Methoden der Organischen Chemie*, Vol. 5, Part 3, Georg Thieme Verlag, Stuttgart, 1962.

[9] J. W. Chamberlain, "Introduction of Fluorine into the Steroid System" in *Steroid Reactions*, C. Djerassi, Ed., Holden Day, Inc., San Francisco, Calif., 1963, p. 155.

[10] N. F. Taylor and P. W. Kent, *Advan. Fluorine Chem.*, **4**, 113 (1965).

[11] F. L. M. Pattison, R. L. Buchanan, and F. H. Dean, *Can. J. Chem.*, **43**, 1700 (1965).

[12] F. L. M. Pattison and R. A. Peters, "Monofluoro Aliphatic Compounds, "in *Handbook of Experimental Pharmacology*, Vol. 20, Part 1, Chapter 8, Springer, New York, 1966.

[13] J. E. G. Barnett, *Advan. Carbohyd. Chem.*, **22**, 177 (1967).

[14] (a) P. W. Kent, *Chem. Ind.* (London), **1969**, 1128; (b) *Carbon-Fluorine Compounds*, A Ciba Foundation Symposium, Elsevier, Excerpta Medica, North-Holland, 1972.

[15] W. A. Sheppard and C. M. Sharts, *Organic Fluorine Chemistry*, W. A. Benjamin, New York, 1969.

3. Nitrosyl fluoride addition to an activated double bond.

$$NOF + \gt C{=}C\lt_{R} \longrightarrow -\underset{F}{\overset{|}{C}}-\underset{NO}{\overset{|}{C}}-R \xrightarrow[\text{(for R = H)}]{+NOF,\ -HF} -\underset{F}{\overset{|}{C}}-\overset{|}{C}{=}NNO_2$$

$$-\underset{F}{\overset{|}{C}}-\underset{NO}{\overset{|}{C}}-R \longrightarrow \text{Dimer}$$

$$-\underset{F}{\overset{|}{C}}-\overset{|}{C}{=}NNO_2 \xrightarrow{\text{Hydrolysis}} -\underset{F}{\overset{|}{C}}-C{=}O$$

4. Hypofluorites for electrophilic fluorination, by either substitution or addition, of activated olefins.

$$RH + FOCF_3 \xrightarrow{h\nu} RF + HF + COF_2$$

$$\gt C{=}C\lt + FOCF_3 \longrightarrow -\underset{F}{\overset{|}{C}}-\underset{OCF_3}{\overset{|}{C}}- + -\underset{F}{\overset{|}{C}}-\underset{F}{\overset{|}{C}}-$$

5. Inorganic and heteroatom fluorides for replacement of halogen, hydroxyl, or ester groups.

$$\begin{matrix}RX\\ \text{or}\\ ROSO_2Y\end{matrix} + MF \rightarrow RF + \begin{matrix}MX\\ \text{or}\\ MOSO_2Y\end{matrix}$$

6. Hydrogen fluoride for addition to alkenes and oxiranes and replacement of sulfonate or carboxylate ester groups.

$$HF + \gt C{=}C\lt \longrightarrow -\underset{|}{\overset{H}{C}}-\underset{|}{\overset{F}{C}}-$$

$$HF + \gt C\overset{O}{\frown}C\lt \longrightarrow -\underset{|}{\overset{OH}{C}}-\underset{|}{\overset{F}{C}}-$$

$$HF + ROSO_2R' \longrightarrow RF + R'SO_3H$$

$$HF + ROCOR' \longrightarrow RF + R'CO_2H$$

7. Perchloryl fluoride for replacement of hydrogen.

$$RH \xrightarrow{\text{Base}} R^-M^+ \xrightarrow{FClO_3} RF + MClO_3$$

8. Condensation of a monofluorinated organic compound (one or two carbon units) with a nonfluorinated compound, usually an active methylene condensation.

$$\begin{array}{l}CO_2C_2H_5\\ |\\ CO_2C_2H_5\end{array} + CH_2FCO_2C_2H_5 \xrightarrow[(C_2H_5)_2O]{C_2H_5ONa} \begin{array}{l}COCHFCO_2C_2H_5\\ |\\ CO_2C_2H_5\end{array}$$

Only the halogen fluoride additions, fluoroalkylamine hydroxyl replacement, nitrosyl fluoride additions, and hypofluorite electrophilic fluorination (methods 1–4) are done in depth; all published examples are given in the tables. The remaining four methods are reviewed critically but not completely; however, references that have appeared since earlier reviews are included.

The fact that methods 5–8 are not fully reviewed should not detract from the importance of these techniques. For example, the use of alkali metal fluorides to substitute a fluorine atom for a halogen atom or a sulfonate ester group is probably the most important fluorination method, but because of excellent reviews the discussion is limited to recent work.[4, 5, 7–15] Recent advances in certain methods of fluorination have made older techniques obsolete. For example, the addition of hydrogen fluoride to alkenes formerly was one of the most important methods to prepare monofluoroaliphatic compounds.

$$>C{=}C< + HF \longrightarrow -\overset{H}{\underset{|}{\overset{|}{C}}}-\overset{F}{\underset{|}{\overset{|}{C}}}-$$

Although potentially important, particularly for synthesis of tertiary fluorides, the reaction has found little synthetic use in recent years and consequently is discussed briefly. This reaction has been adequately reviewed.[16, 17]

Certain methods are not discussed because they are of limited value for preparation of monofluoroaliphatic compounds. For example, aqueous fluorination has been used extensively to form N-F compounds,[18, 19] but formation of a single C—F bond has been limited to α-fluorination of nitro derivatives.[20, 21] The addition of a molecule of fluorine to carbon-carbon double bonds may be accomplished by molecular fluorine, lead tetrafluoride, and phenyliododifluoride.[22] In these additions vicinal difluorides are formed, but monofluorinated products are also formed.

[16] Ref. 15, Sheppard and Sharts, p. 59.

[17] Ref. 8, Houben-Weyl, p. 95–113.

[18] V. Grakauskas and K. Baum, *J. Org. Chem.*, **34,** 2840 (1969).

[19] V. Grakauskas and K. Baum, *J. Org. Chem.*, **35,** 1545 (1970).

[20] H. G. Adolph, R. E. Oesterling, and M. E. Sitzmann, *J. Org. Chem.*, **33,** 4296 (1968).

[21] K. Baum, *J. Org. Chem.*, **35,** 846 (1970).

[22] Ref. 15, Sheppard and Sharts, p. 129.

Some of the monofluoro compounds thus obtained are inaccessible by other methods. For example, the reaction of lead tetraacetate/hydrogen fluoride (*in situ* preparation of a lead(IV) fluorinating agent) with norbornene gave twelve products of which four were monofluorinated and

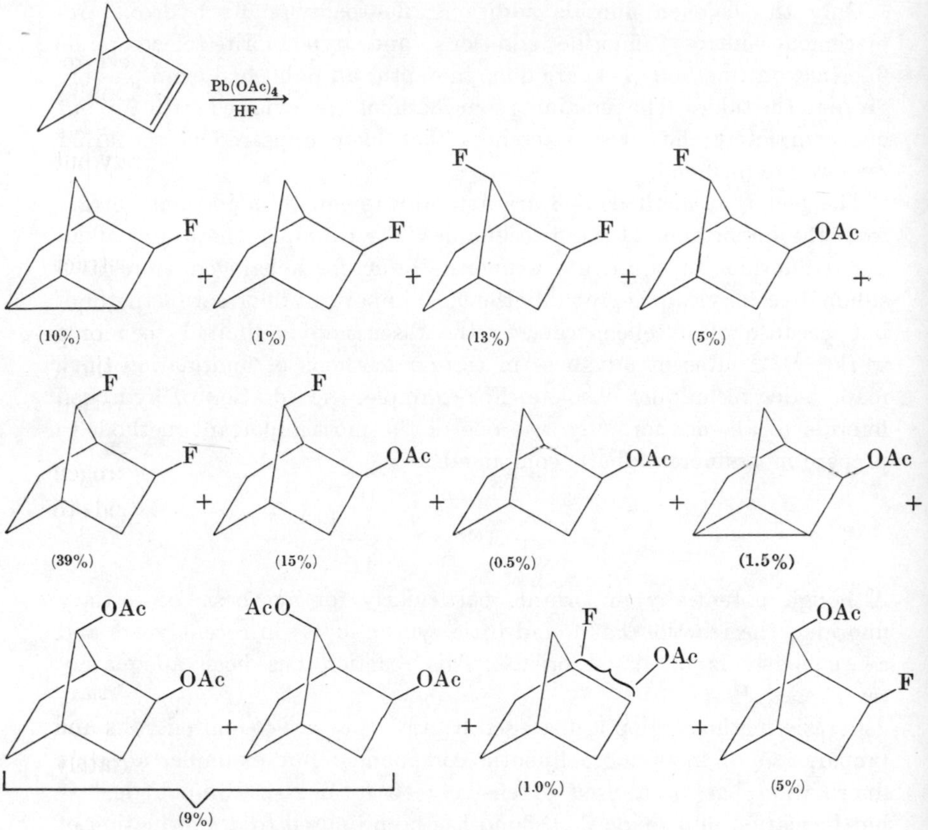

$Pb(OAc)_4$ / HF

(10%) (1%) (13%) (5%)

(39%) (15%) (0.5%) (1.5%)

(9%) (1.0%) (5%)

new.[23] Also the difluoro compounds can give selective elimination of hydrogen fluoride to monofluoro products.

Monofluoro by-products have also been found in fluorination of cholesterol with lead tetraacetate and hydrogen fluoride.[24] Yields of by-products varied with conditions including amount of hydrogen fluoride, solvent, and temperature.

[23] D. D. Tanner and P. Van Bostelen, *J. Am. Chem. Soc.*, **94**, 3187 (1972).

[24] J. Levisalles and J. Molimard, *Bull. Soc. Chim. Fr.*, **1971**, 2037.

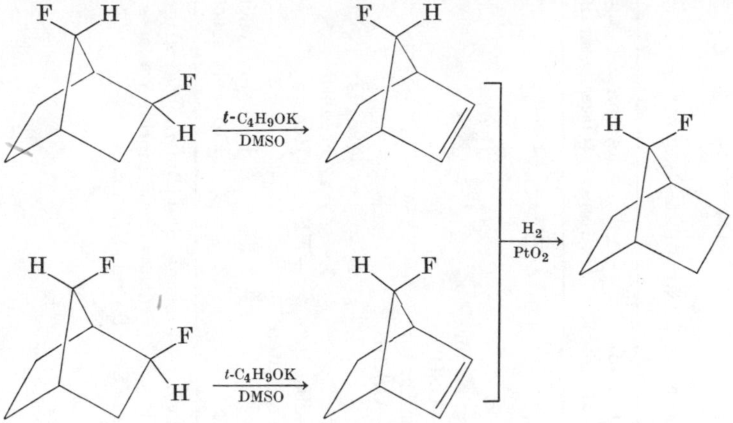

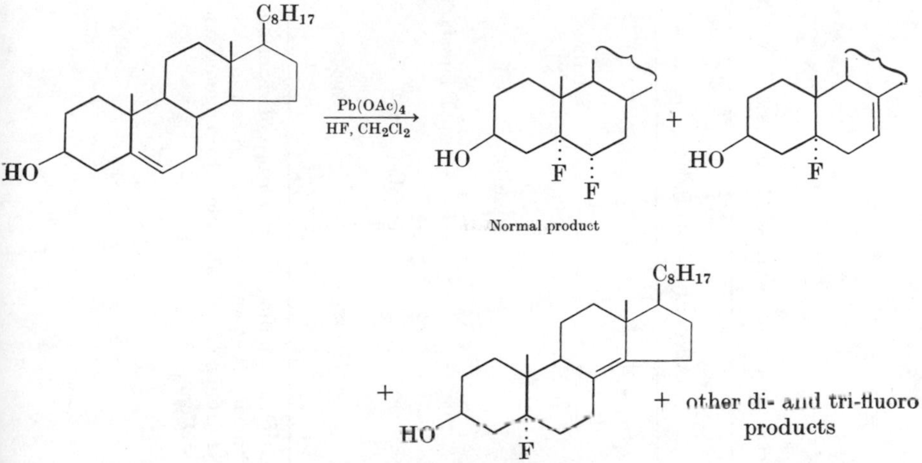

F H F H

F

H $\xrightarrow[\text{DMSO}]{t\text{-}C_4H_9OK}$ H F

$\xrightarrow[PtO_2]{H_2}$

H F H F

F

H $\xrightarrow[\text{DMSO}]{t\text{-}C_4H_9OK}$

C_8H_{17}

HO $\xrightarrow[HF,\ CH_2Cl_2]{Pb(OAc)_4}$ HO F F + HO F

Normal product

C_8H_{17}

\+ HO F + other di- and tri-fluoro products

gem-Difluoro compounds, produced from sulfur tetrafluoride and cycloketones,[6] are a source of 1-fluorocycloalkenes by hydrogen fluoride elimination.[25] This reaction has also been applied to steroids.

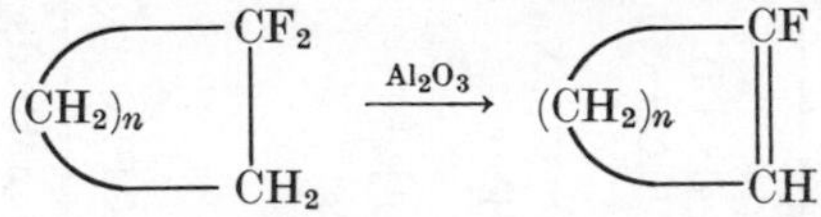

$$(CH_2)_n\,\langle CF_2,\ CH_2\rangle \xrightarrow{Al_2O_3} (CH_2)_n\,\langle CF{=}CH\rangle$$

Table I summarizes the reactivity of functional groups to the various fluorinating agents and indicates utility of reagents in fluorination as well

[25] D. R. Strobach and G. A. Boswell, Jr., *J. Org. Chem.*, **36**, 818 (1971).

TABLE I. REACTIVITY OF FLUORINATING REAGENTS TOWARD VARIOUS FUNCTIONAL GROUPS[a]

Reagent/Functionality	Olefin	Alkynes	Halides	Alcohol	Esters of Alcohols	Oxirane	Anion (Organometallic)
X—F (X = I, Br, Cl)	+XF	+XF	—	—	—	Reacts	Reacts
$(C_2H_5)_2NCF_2CHFCl$	—	—	—	F subst	—	Reacts	Reacts
MF M = Cs^+, K^+, Na^+	—	—	F subst	—	F subst for —OSO_2R	—	—
$(C_4H_9)_4N^+F^-$ [b]	—	—	F subst	—	F subst for —OSO_2R	—	—
R_3PF_2, R_2PF_3 [b]	—	—	—	F subst	—	?	Reacts
$FClO_3$	—	—	—	—	—	?	F subst
HF	+HF	+HF	F subst	—	F subst if activated	+HF	Reacts
FNO	+F—NO	Reacts	—	Reacts	—	?	Probably reacts
$CF_3OF(R_fOF)$	+F—OCF_3	?	—	?	—	?	Probably reacts

[a] Useful methods of monofluorination are indicated by nature of reaction. A dash (—) means that the functional group is usually inert to reagent; "reacts" means that an undesirable reaction occurs. "F subst" means that fluoride replaces the functional group indicated. Carbonyl groups are usually inert to reagents which give fluoride substitution.

[b] Reagent listed in text under MF section.

as the limitations caused by reaction with other functional groups. More extensive evaluations of the methods of fluorination and of the best reagent to achieve a specific objective are given on pp. 260–264.

HALOGEN FLUORIDE ADDITION

Net addition of ClF, BrF, or IF (XF) to a double bond provides a mild method of introducing a single fluorine into organic compounds. The α-halogen substituent is a useful functionality for substitution or elimination.

$$X^+ + R_2C{=}CR_2 \longrightarrow \left[R_2C \overset{\overset{+}{X}}{\cdots} CR_2 \overset{\text{or}}{\longleftrightarrow} R_2C(X){-}\overset{+}{C}R_2 \right] \xrightarrow{F^-} R_2C(X){-}CR_2(F)$$

X = Cl, Br, or i

The Reagents

Iodine fluoride (IF) and bromine fluoride (BrF) are not sufficiently stable for isolation[26] but are effectively prepared *in situ*. Chlorine fluoride (ClF), available commercially, is almost as reactive as elemental fluorine; hence it has been used directly only with perhalogenated or very simple molecules. However, chlorine fluoride is also available *in situ* from a positive chlorine source in the presence of fluoride ion or hydrogen fluoride. The synthesis *in situ* does not mean that the XF reagents are prepared in solution but only that the net result of the reaction is addition of X—F to an unsaturated center. Reagents such as bromine trifluoride and iodine pentafluoride have been used to add XF but only to polyfluoroolefins.

Early attempts to add halogen fluorides to olefins led only to X_2 (or X–OH) addition or allylic substitution.[27,28] However, proper choice of solvent and control of temperature allow the reaction sequence of addition of positive halogen to give a halonium ion followed by fluoride ion attack so that overall XF addition is accomplished in high yield. This technique was primarily developed as a method of preparing fluoro steroids,[29,30] but it has been applied to a wide variety of unsaturated compounds.

The most common positive halogen sources are N-halosuccinimides and -acetamides but several other sources, including the free halogens, have been used. The reagents for XF addition to unsaturated centers for which

[26] H. Schmidt and H. Meinert, *Angew. Chem.*, **72**, 109 (1960).

[27] A. I. Titov and F. L. Maklyaev, *J. Gen. Chem. (USSR), Engl. Transl.*, **24**, 1613 (1954).

[28] E. D. Bergmann and I. Shahak, *J. Chem. Soc.*, **1959**, 1418.

[29] A. Bowers, *J. Amer. Chem. Soc.*, **81**, 4107 (1959).

[30] C. H. Robinson, L. Finckenor, E. P. Oliveto, and D. Gould, *J. Amer. Chem. Soc.*, **81**, 2191 (1959).

experimental details have been given are:

X is Cl: chlorine, hexachloromelamine (HCM), N-chlorosuccinimide (NCS), *t*-butyl hypochlorite;

X is Br: bromine, N-bromoacetamide (NBA), N-bromosuccinimide (NBS), 1-bromo-3,5,5-trimethylhydantoin, and 1,3-dibromo-5,5-dimethylhydantoin (DBH has the advantage of high solubility in a reaction mixture of ether and hydrogen fluoride);

X is I: iodine and N-iodosuccinimide (NIS).

The fluoride ion source is usually hydrogen fluoride, but recently silver(I) fluoride has been claimed to have advantages. Other fluoride sources such as mercury(II) fluoride (HgF_2), zinc fluoride (ZnF_2) or lead(II) fluoride (PbF_2) have not been used successfully. Silver(I) fluoride as a source of fluoride is limited to use with halogens; conditions are critical, including rapid stirring of finely powdered reagent.[31]

The solvents are usually of high ionizing power. Good proton-acceptor solvents, such as tetrahydrofuran or 5% ethanol in chloroform, which assist in ionizing hydrogen fluoride are preferred. A common technique is to dissolve the positive halogen reagent or olefin in methylene chloride, chloroform, or carbon tetrachloride and the hydrogen fluoride in diethyl ether or tetrahydrofuran; two other commonly used solvents are diethylacetic acid and pyridine. Acetonitrile or benzene is used with silver(I) fluoride.

Normally, reactions are run at −80° to prevent side reactions such as hydrogen fluoride addition, carbonium ion rearrangements, or polymerization. The silver(I) fluoride-halogen reagent is normally employed at room temperature.

The preparation of the halogen fluoride reagent is not described here, because the reagent is prepared *in situ*. The experimental details are incorporated in typical experimental procedures beginning on p. 155.

Scope, Limitations, and Side Reactions*

By employing the *in situ* preparation, halogen fluorides have been added to a large number of olefins and to some acetylenes. The olefinic carbons can be substituted with a variety of substituents ranging from strong electron-withdrawing nitrile or ester groups to four alkyl groups or several halogens. Other functionality in the molecule appears limited only by the requirement of low reactivity with positive halogen reagents or hydrogen

[31] L. D. Hall, D. L. Jones, and J. F. Manville, *Chem. Ind.* (London), **1967,** 1787.

* Note added in proof: G. A. Olah, M. Nojima and I. Kerekes, *Synthesis*, **1973,** 780, recently report that halogen fluoride addition to olefins and acetylenes can be accomplished using 70% hydrogen fluoride/pyridine complex with an appropriate positive halogen source in tetramethylene sulfone at room temperature. The preferred halogen sources are iodine, bromine/silver nitrate, N-chlorosuccinimide.

fluoride; hence functions such as carbonyl, hydroxyl, and ether are often present in a range of compounds that include sugars and steroids.

Unsymmetrical olefins might be expected to give a mixture of isomers. Instead, the addition usually occurs with high selectivity to give the Markownikoff product with stereospecific *trans* orientation. The selectivity and stereospecificity are expected from the electrophilic ionic course of the reaction (see p. 146). When the double-bond substituents have equivalent electronegativity, both orientations are found in approximately equal amounts. Further treatment of selectivity and stereospecificity is included in the following discussion of the scope of halogen fluoride addition.

Aliphatic and Alicyclic Olefins

In the first development of the reaction, bromine- and iodine-fluoride addition to cyclohexene was demonstrated to be highly stereospecific to give *trans* adducts.[32]

$$\text{cyclohexene} \underset{Zn\ +\ C_2H_5CO_2H}{\overset{\text{Halogenated amide + HF}}{\rightleftharpoons}} \text{trans-1-X-2-fluorocyclohexane}$$

X = Br or I

In the broadest extension of this reaction to aliphatic olefins,[33–35] Markownikoff orientation was clearly shown. 1-Alkenes of varying chain length with N-bromoacetamide and hydrogen fluoride in ether give the expected 1-bromo-2-fluoroalkanes in 70–80% yield with minor amounts of 2-bromo-1-fluoride. Gaseous alkenes give good yields. Olefins of more than ten carbon atoms give lower yields, presumably because of low solubility in the reaction medium. Branched alkenes give lower yields; for example, 2-ethyl-1-butene is converted to 1-bromo-2-fluoro-2-ethylbutane in only 50% yield. Conjugated olefins such as styrenes give large amounts of tarry by-products unless pyridine is added (it increases the concentration of fluoride ion or decreases proton concentration); 1-bromo-2-fluoro-2-phenylethane is the expected product for an ionic electrophilic addition.

Allyl bromide gives nearly equal amounts of both the Markownikoff and the anti-Markownikoff adducts.

$$CH_2{=}CHCH_2Br \rightarrow \underset{(22\%)}{BrCH_2CHFCH_2Br} + \underset{(32\%)}{FCH_2CHBrCH_2Br}$$

[32] A. Bowers, L. C. Ibáñez, D. Denot, and R. Becerra, *J. Amer. Chem. Soc.*, **82**, 4001 (1960).
[33] F. H. Dean and F. L. M. Pattison, *Can. J. Chem.*, **43**, 2415 (1965).
[34] F. L. M. Pattison, R. L. Buchanan, and F. H. Dean, *Can. J. Chem.*, **43**, 1700 (1965).
[35] F. L. M. Pattison, D. A. V. Peters, and F. H. Dean, *Can. J. Chem.*, **43**, 1689 (1965).

Such a result has precedent from other electrophilic additions and was suggested to be the result of either the steric effect or the statistical advantage of the more exposed primary carbon atoms in a symmetrical dibromoallylcarbonium ion.

$$\left[CH_2 \overset{Br}{\cdots} CH \underset{Br}{\cdots} CH_2 \right]^+$$

The esters of α,β-unsaturated acids give chiefly α-bromo-β-fluorocarboxylic acid esters.[33, 36] Methyl acrylate gives a low yield of the expected methyl 2-bromo-3-fluoropropionate and traces of the unexpected methyl 3-bromo-2-fluoropropionate and methyl 3-bromo-2-ethoxypropionate. However, methyl methacrylate gives only methyl 3-bromo-2-fluoro-2-methylpropionate.

$$CH_2{=}CHCO_2CH_3 \xrightarrow{\text{"BrF"*}} \underset{\text{(Low)}}{FCH_2CHBrCO_2CH_3} + \underset{\text{(Trace)}}{BrCH_2CHFCO_2CH_3} + \underset{\text{(Trace)}}{BrCH_2CH(OC_2H_5)CO_2CH_3}$$

In order to obtain methyl 3-bromo-2-fluoropropionate in quantity, bromine fluoride was added to allyl trichloroacetate.[33] The adduct, obtained in high yield, was readily converted to the propionate derivative that was detected in the methyl acrylate reaction.

$$CH_2{=}CHCH_2OCOCCl_3 \xrightarrow{\text{"BrF"}} BrCH_2CHFCH_2OCOCCl_3 \xrightarrow[\text{2. } CH_3OH,\ (CH_3)_2C(OCH_3)_2]{\text{1. } HNO_3,\ HOAc} BrCH_2CHFCO_2CH_3$$

Steroids

Addition of halogen fluoride to steroid double bonds was independently reported in 1959 by groups headed by Bowers[29] and by Robinson.[30] Robinson's group initially added chlorine and bromine chloride to the 9,11 double bond of corticosteroids and extended the reaction to bromine fluoride addition. Bowers and his associates added bromine fluoride and iodine fluoride to prepare C_6, C_{11}, and C_{16} fluorinated steroids.[32] Subsequent papers and patents describe many examples of halogen fluoride (primarily bromine fluoride) addition to double bonds in steroids. A

[36] (a) A. K. Bose, K. G. Das, and T. M. Jacob, *Chem. Ind.* (London), **1963,** 452. (b) A. K. Bose, K. G. Das, and P. T. Funke, *J. Org. Chem.*, **29,** 1202 (1964).

* "BrF" (or "XF") represents bromine fluoride (halogen fluoride) generated *in situ*.

summary of the parent steroid systems and the mode of addition of halogen fluoride is given in Table II. Other functional groups that may be present in the steroid without adversely affecting addition of halogen fluoride are acetylenic, carbalkoxy, carbonyl, hydroxyl, and halogen groups.

TABLE II. SUMMARY OF HALOGEN FLUORIDE ADDITIONS TO UNSATURATED STEROIDS

Parent Steroid	Double-Bond Position	Halogen Fluoride[a] ClF	BrF	IF
Androstane	5,6		6	
	9,11		11	
Cholestane	8,9		9	
	8,14		8	
Estrane	5,10		5	5
Pregnane	2,3		2	
	4,5		4	
	5,6		6 (or 5)	5
	6,7		6	
	9,11	11	11	11
	13,17		17	
	16,17		16	
Testosterone	6,7		6	

[a] The position of the fluorine atom is indicated by the number.

Because addition to steroids appears to be influenced more by steric than by electronic factors, orientation is more frequently anti-Markownikoff than Markownikoff. The orientation can usually be explained by a careful examination of the steroid three-dimensional structures relative to the required geometry of formation and reaction of the halonium ion intermediate. Some unsaturated centers do not add halogen fluorides; occasionally bromine fluoride does add but iodine fluoride does not. A particularly interesting observation is that bromine fluoride adds normally to a steroidal 5,6 double bond with anti-Markownikoff[32] orientation (5 Br,6 F) while iodine fluoride adds "abnormally" with Markownikoff orientation (5 F,6 I).[37]

Recently the orientation of bromine fluoride addition was shown to depend on reaction conditions; hence either 5α-fluoro-6β-bromo- or 6β-fluoro-5α-bromo- steroids could be formed.[38]

[37] A. Bowers, E. Denot, and R. Becerra, *J. Amer. Chem. Soc.*, **82**, 4007 (1960).

[38] Von U. Kerb and R. Wiechert, *Ann. Chem.*, **752**, 78 (1971).

Sugars

The addition of halogen fluorides to unsaturated carbohydrates by adaptation of the Bowers method was reported in 1963.[39] A *cis* addition, in contrast to the normal *trans* addition for aliphatic olefins, was proposed but the original structural assignment has been corrected by detailed spectral studies,[40] including an X-ray structure proof.[41] Actually the halogen fluoride addition gives a mixture of *cis* and *trans* adducts similar to other electrophilic additions to unsaturated sugars.[42, 43] The most extensive study demonstrated that chlorine, bromine, and iodine monofluorides add to unsaturated carbohydrates including glucals, galactals, arabinals, and xylals.[40, 44a] A mild method using silver(I) fluoride and halogen at room temperature in benzene or acetonitrile was developed and compared to the older halosuccinimide-hydrogen fluoride method and to other electrophilic additions to unsaturated sugars. The silver(I) fluoride-chlorine reaction is sufficiently mild to permit isolation of the chlorine fluoride adduct in good yield, whereas N-chlorosuccinimide-hydrogen fluoride gives no fluorinated product.

The adducts from addition of bromine fluoride to D-glucal triacetate were the same when either silver fluoride-bromine at room temperature or N-bromosuccinimide-hydrogen fluoride at −78° was used as reagent, except that in the reaction at −78° a minor amount of hydrogen fluoride-addition product was obtained. The relative yields of *trans*- and *cis*-bromofluoro adducts were similar to the *trans* and *cis* adducts obtained from electrophilic addition of benzoyl hypobromite.[40] However, the *trans*

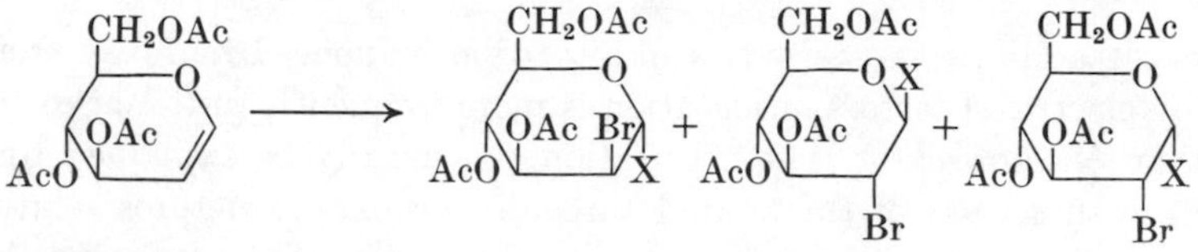

Method	X		Yield, %	
AgF, Br_2, CH_3CN, room t	F	70	21	9
AgF, Br_2, C_6H_6, room t	F	42	42	16
NBS, HF, $(C_2H_5)_2O$, −78°	F	55	30	9
Br_2, AgOBz, C_6H_6, room t	OBz	31	42	26

[39] (a) P. W. Kent, F. O. Robson, and V. A. Welch, *Proc. Chem. Soc.*, **1963**, 24; (b) *J. Chem. Soc.*, **1963**, 3273.

[40] L. D. Hall and J. F. Manville, *Can. J. Chem.*, **47**, 361 (1969).

[41] (a) J. C. Campbell, R. A. Dwek, P. W. Kent, and C. K. Prout, *Chem. Commun.*, **1968**, 34; (b) *Carbohyd. Res.*, **10**, 71 (1969).

[42] (a) R. U. Lemieux and B. Fraser-Reid, *Can. J. Chem.*, **42**, 532 (1964); (b) **43**, 1460 (1965).

[43] R. J. Ferrier, *Advan. Carbohyd. Chem.*, **20**, 77 (1965).

[44] (a) L. D. Hall and J. F. Manville, *Can. J. Chem.*, **47**, 379 (1969).

adduct forms to a greater extent for bromine fluoride addition than for other electrophilic additions. Relative yields of geometric isomers depend on the solvent.

The greater amount of *cis* isomer found for the hydrogen fluoride method has been attributed to isomerization of the *trans diequatorial* (β-gluco) to the more thermodynamically stable *cis* isomer (α-gluco).[40] However, solvent and temperature differences affect the ionic intermediates and change the relative ratios (see p. 147). The chlorine fluoride addition is unusual in that it gives all four possible stereoisomers, both *trans* and both *cis* forms. This is the first example of formation of a β-D-manno isomer by an electrophilic addition,[44a] but it has also been reported in chlorination using chlorine and hydrogen chloride.[44b] In all reactions, including the model tetrahydropyran system, the fluorine appears alpha to the ether linkage because the positive charge in the intermediate halonium ion is better stabilized in the alpha position by delocalization to oxygen[45] (see p. 151).

Other Unsaturated Centers

Acetylenes add bromine fluoride using hydrogen fluoride in ether at −80° to give the bromofluoroalkenes.[46] Presumably addition of a second bromine-fluoride "molecule" is prevented by the electronic and steric effects of the initially added bromine fluoride. The addition is primarily *trans* with fluoride ion entering from the less hindered side. For terminal acetylenes the orientation follows the Markownikoff rule. 1-Hexyne gives a 48% yield of 95% *trans*- and 5% *cis*-1-bromo-2-fluoro-1-hexene. Acetylenes with electronegative substituents gave no adducts (only halogenation in some cases). The silver(I) fluoride-bromine method did not work.[47]

A novel geminal addition of bromine fluoride has been accomplished using ethyl diazoacetate, hydrogen fluoride, and N-bromosuccinimide or bromine.[48]

$$N_2CHCO_2C_2H_5 \xrightarrow[\text{HF}]{\text{NBS or Br}_2} \underset{(59\%)}{BrFCHCO_2C_2H_5} + N_2$$

Perfluoroolefins also add halogen fluorides. For example, iodine fluoride addition occurs from potassium fluoride and iodine.[49]

$$R_fCF{=}CF_2 + KF \xrightarrow[100°]{CH_3CN} R_f\bar{C}FCF_3 \xrightarrow{I_2} R_fCFICF_3$$

$$R_f = \text{fluoroalkyl}$$

[44] (b) K. Igarashi, T. Honma, and T. Imagawa, *J. Org. Chem.*, **35,** 610 (1970).

[45] I. Jenkins, J. G. Moffat, and J. P. Verheyden, Ger. Pat. 2,228,750 (1973). [*C.A.*, **78,** 111702k (1973)].

[46] R. E. A. Dear, *J. Org. Chem.*, **35,** 1703 (1970).

[47] L. D. Hall and J. F. Manville, *Chem. Commun.*, **1968,** 35.

[48] H. M. Machleidt, R. Wessendorf, and M. Klockow, *Ann. Chem.*, **667,** 47 (1963).

[49] C. G. Krespan, *J. Org. Chem.*, **27,** 1813 (1962).

Usually perfluoroolefins require more strenuous conditions such as higher temperature and preformed halogen fluorides (ClF, BrF_3, or IF_5). A nucleophilic addition has been suggested in which fluoride can add to form an intermediate carbanion stabilized by the electronegative halogen substituent. However, the direct addition of iodine fluoride (from IF_5 + $2I_2$) to highly fluorinated olefins, including tetrafluoroethylene and hexafluoropropylene, has recently been suggested to be an electrophilic addition.[50] Since highly fluorinated compounds are beyond the scope of this chapter, the interested reader is referred to a review[51] and to recent references.[50, 52]

Halogen fluoride addition to unsaturated systems containing a heteroatom (*e.g.*, C=O or C=N) is limited to highly fluorinated molecules such as perfluoroketones and nitriles or inorganic compounds such as sulfur dioxide. The addition of chlorine fluoride to cyanogen chloride to give N,N-dichloro(chlorodifluoromethyl)amine[53] is included in Table X as an example of chlorine fluoride addition to a C=N function. Again for additional information the reader is referred to the review[51] and recent references such as chlorine fluoride addition to perfluoroketones,[54–56] to sulfur dioxide,[57] and to other sulfur-oxygen compounds.[58, 59]

Side Reactions

Side reactions are generally minor or nonexistent and are usually encountered only under abnormal reaction conditions. The main side reactions are substitution of hydrogen by halogen or of halogen by fluorine [particularly using silver(I) fluoride], hydrogen fluoride addition, oxidation, haloamination, haloalkylation, resinification, and rearrangement. Most side reactions occur at higher than normal temperatures or with the silver(I) fluoride-halogen reagent.

[50] P. Sartori and A. J. Lehnen, *Chem. Ber.*, **104**, 2813 (1971).

[51] Ref. 15, Sheppard and Sharts, p. 127.

[52] E. S. Lo, J. D. Readio, and H. Iserson, *J. Org. Chem.*, **35**, 2051 (1970).

[53] (a) D. E. Young, L. R. Anderson, and W. B. Fox, *Chem. Commun.*, **1970**, 395; (b) U.S. Pat. 3,689,563 (1972). [*C.A.*, **78**, 29213t (1973)].

[54] C. J. Schack and W. Maya, *J. Amer. Chem. Soc.*, **91**, 2902 (1969).

[55] D. E. Gould, L. R. Anderson, D. E. Young, and W. B. Fox, *J. Amer. Chem. Soc.*, **91**, 1310 (1969).

[56] D. E. Young, L. R. Anderson, and W. B. Fox, *Inorg. Chem.*, **9**, 2602 (1970).

[57] C. V. Hardin, C. T. Ratcliffe, L. R. Anderson, and W. B. Fox, *Inorg. Chem.*, **9**, 1938 (1970).

[58] C. J. Schack, R. D. Wilson, J. S. Muirhead, and S. N. Cohz, *J. Amer. Chem. Soc.*, **91**, 2907 (1969).

[59] C. J. Schack and R. D. Wilson, *Inorg. Chem.*, **9**, 311 (1970).

Allylic Halogenation.[28]

$$C_6H_5CH_2CH_2CH{=}CH_2 \xrightarrow[\text{(No solvent)}]{\text{Room t, AgF + }I_2} C_6H_5CH_2CHICH{=}CH_2$$

Replacement of Halogen by Fluorine.[31] The difluoroacenaphthylene arises from replacement of iodine by fluorine (from silver fluoride) in the initial adduct. Difluoroiodoacenaphthylene must result from dehydrohalogenation of the initial iodine fluoride adduct followed by addition of iodine fluoride.

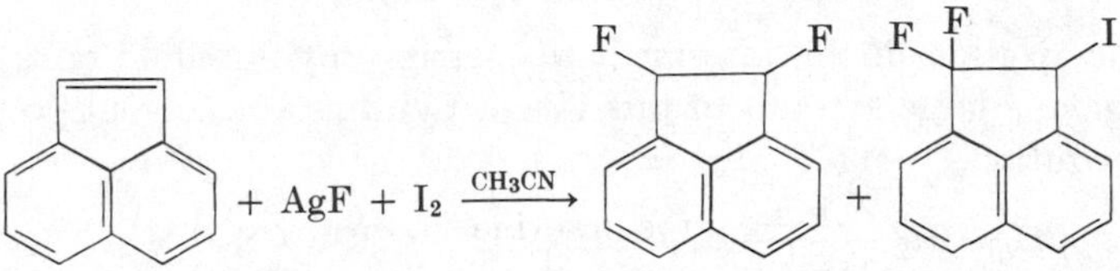

$$+ \text{AgF} + I_2 \xrightarrow{CH_3CN} \quad + $$

Hydrogen Fluoride Addition. Surprisingly the only definite report of this reaction is in the bromine fluoride addition to D-glucal triacetate using N-bromosuccinimide and hydrogen fluoride in an ether at −80°.[29,40]

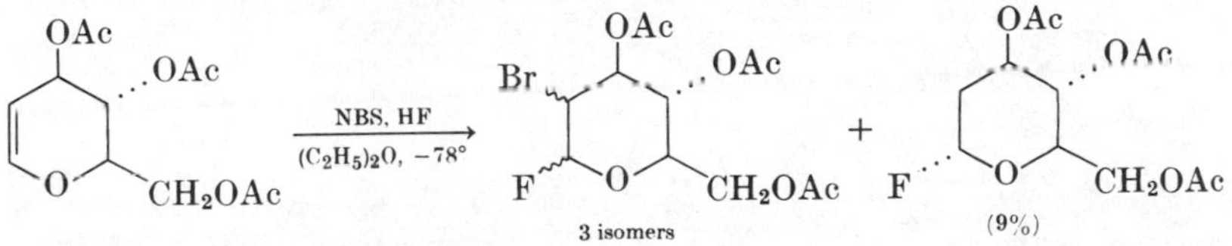

$$\xrightarrow[(C_2H_5)_2O,\ -78°]{\text{NBS, HF}} \quad \text{(3 isomers)} + \quad (9\%)$$

A minor by-product from the reaction of N-bromosuccinimide and hydrogen fluoride with a pregnene must also come from hydrogen fluoride addition.[38] In our opinion, addition of hydrogen fluoride is frequently a side reaction. However, the adducts have not been detected because they are formed in small amounts at −80° and are probably removed at an early step in the workup procedure.

Oxidation.[28]

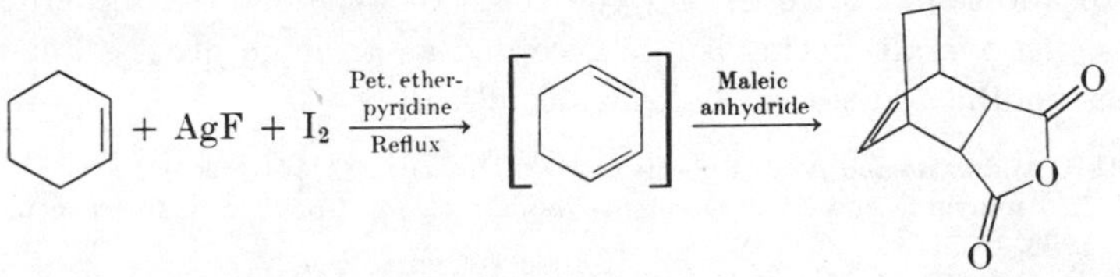

$$+ \text{AgF} + I_2 \xrightarrow[\text{Reflux}]{\text{Pet. ether-pyridine}} [\ ] \xrightarrow{\text{Maleic anhydride}}$$

This side reaction could be the result of an allylic halogenation followed by dehydrohalogenation by pyridine.

Haloamination.[60] (See also example under rearrangements below.)

$$\text{cyclohexene} + AgF + I_2 \xrightarrow[\text{-10°, aq. workup}]{CH_3CN} \text{2-iodo-1-}(NHCOCH_3)\text{cyclohexane} \; (17\%)$$

Polymerization[35] **and Haloalkylation.**[61]

$$C_6H_5CH{=}CH_2 + NBA + HF \rightarrow C_6H_5CHFCH_2Br + tar \quad \text{(Ref. 35)}$$

(a) in diethyl ether, 25%
(b) in diethyl ether-pyridine, 68%

Polymerization of the intermediate, possibly by Friedel-Crafts alkylation, gave a large amount of tar. Use of pyridine as a solvent prevented polymerization.

$$CH_2{=}CH_2 \xrightarrow[C_6H_6-HF]{HCM} ClCH_2CH_2F + \underset{(8\%)}{ClCH_2CH_2C_6H_5} \quad \text{(Ref. 61)}$$

Rearrangements.[62] (Also see p. 149.)

$$\text{norbornene} \xrightarrow[\text{Diethyl ether}]{NBA—HF} \text{bromonortricyclene} \; (33\%) + \text{Br, F-norbornane} \; (20\%) + \text{Br, F-norbornane} \; (34\%)$$

Mechanism

No studies have been directed primarily to the mechanism of halogen fluoride addition. However, on the basis of accumulated experimental results, the overall mechanism is accepted as an ionic process analogous to electrophilic addition of halogens to olefins.[63]

[60] R. H. Andreatta and A. V. Robertson, *Aust. J. Chem.*, **19**, 161 (1966).

[61] I. L. Knunyants and L. S. German, *Bull. Acad. Sci. USSR, Div. Chem. Sci.*, (*Engl. Transl.*, **1966**, 1016.

[62] F. H. Dean, D. R. Marshall, E. W. Warnhoff, and F. L. M. Pattison, *Can. J. Chem.*, **45**, 2279 (1967).

[63] P. B. D. de la Mare and R. Bolton, *Electrophilic Additions to Unsaturated Systems* Elsevier, Co., New York, 1966.

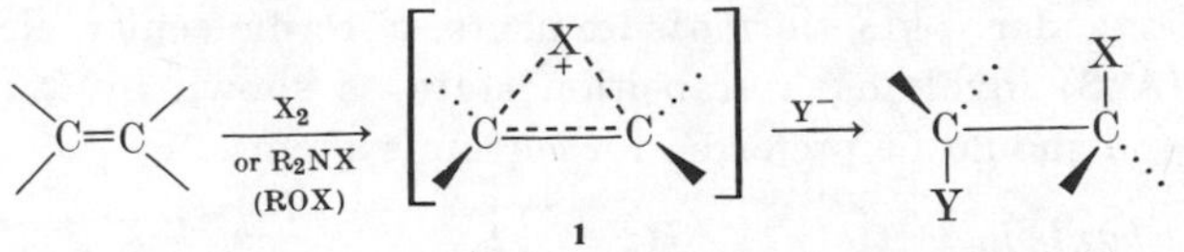

$$>C=C< \xrightarrow[\text{(ROX)}]{X_2 \text{ or } R_2NX} \left[\text{C}\overset{X^+}{\text{------}}\text{C} \right] \xrightarrow{Y^-} \text{X-C-C-Y}$$

1

X = Cl, Br, I

$Y^- = F^-, Cl^-, Br^-, I^-$

Experimental results that provide the basis for the ionic mechanism are summarized by the five following points.

1. Except for steroids, orientation in addition to unsymmetrical olefins is usually Markownikoff, corresponding to the polarity of the double bond. The electronegative fluorine always bonds to the carbon with the highest electron density.

2. Addition is stereospecifically *trans*, and exceptions are readily explained by steric or conjugative interactions.

3. The best yields are obtained in solvents with relatively high ionizing power and with reagents that are good sources of X^+.

4. Intermediate carbonium ion species can be trapped by electrophilic attack on other reagents, such as an aromatic nucleus.[61]

5. If the olefin forms a carbonium ion species that is prone to rearrange, typical Wagner-Meerwein rearrangement products are found.

More extensive mechanistic studies are required to gain a more detailed picture of the reaction. Actually variations in reagents and conditions could produce considerable changes in mechanism. The intermediate **1** could vary from a molecular adduct **1a** through a π complex **1b**. A classical carbonium ion **1c** is generally excluded on the basis of stereospecific *trans* addition, but the halocarbonium ion could closely approach **1c** through a form such as **1d**, which in certain cases appears classical because

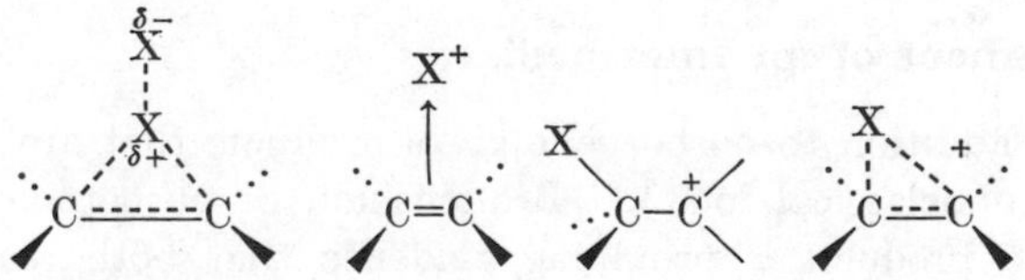

1a 1b 1c 1d

of conjugation with substituents on carbon. Form **1d** is preferred because of the predominance of Markownikoff-oriented products, since a highly symmetrical halonium ion intermediate should favor non-Markownikoff orientation.[63]

Overall, the "halogen fluoride" probably adds by an electrophilic bimolecular mechanism (designated A_E2) as defined originally for electrophilic addition of halogens to olefins.[63] However, without more extensive studies, particularly kinetic measurements, a termolecular electrophilic addition (A_E3) involving a transition state as shown by **2** cannot be excluded and should be preferred for certain systems.

A_E2 Mechanism.

$$\text{Olefin} + X^{+} \text{ (or XB)} \xrightarrow{\text{Slow}} \begin{array}{c}(\text{olefin})^{+}\\ |\\ X\end{array} \xrightarrow[\text{Fast}]{+Y^{-}} \begin{array}{c}\text{olefin}\\ |\quad|\\ X\quad Y\end{array}$$

A_E3 Mechanism.

$$\text{Olefin} + XB + AY \xrightarrow{\text{Slow}} \begin{array}{c}\text{Olefin}\\ |\quad|\\ X\quad Y\end{array} + B^{-} + A^{+}$$

```
              δ+      δ-
              X------B
              :
        δ+    :
        C------Cδ-
        :
        :
  A-----Y
  δ+    δ-
```

2

The molecular addition of halogen fluoride formed in a fast reaction is excluded by the instability of the halogen fluoride reagents and no indication of *cis* addition such as noted for some additions of fluorine to olefins.[63]

In terms of the preceding mechanism discussion, some specific examples of halogen fluoride addition are now analyzed, particularly with regard to the origin of certain by-products.

Rearrangement of an Intermediate

The BrF addition to norbornene gives products that are proposed to arise from nonclassical ion **3**.[62] Bromination of norbornene also gives similar type products[64a] providing evidence that both reactions have similar rate controlling steps to form the common nonclassical carbonium ion intermediate **3** (X = Br) rather than bromonium ion **4** proposed earlier.[64b] The chlorination data are not as definitive since a radical reaction is competitive and complicates the picture.[65] (See Table III; also Table X p. 274; also compare with lead(IV) fluoride fluorination on p. 134).

[64] (a) D. R. Marshall, P. Reynolds-Warnhoff, E. W. Warnhoff, and J. R. Robinson, *Can. J. Chem.*, **49**, 885 (1971); (b) H. Kwart and L. Kaplan, *J. Amer. Chem. Soc.*, **76**, 4072 (1954).

[65] M. L. Poutsma, *J. Amer. Chem. Soc.*, **87**, 4293 (1965); *Science*, **157**, 997 (1967).

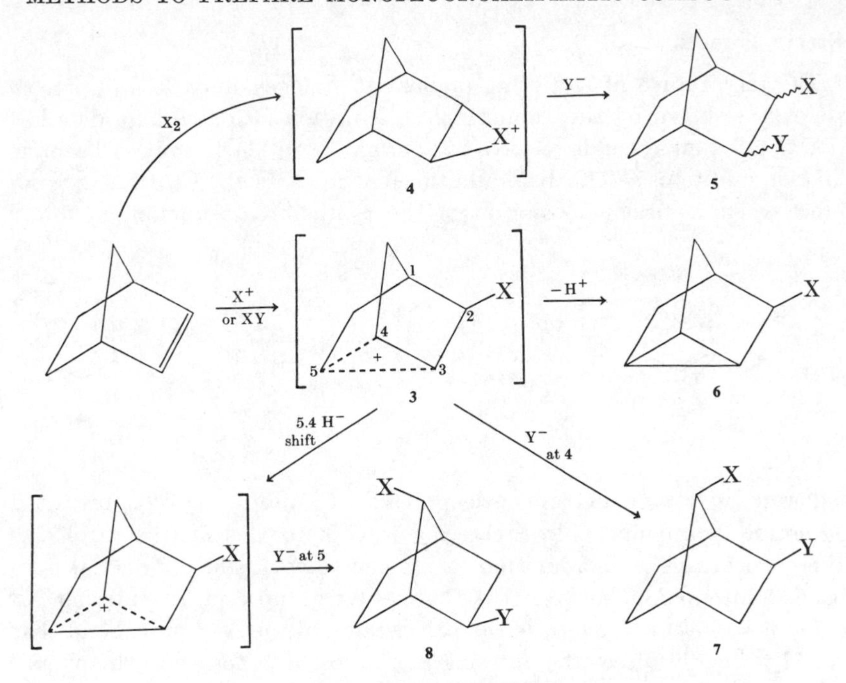

TABLE III. HALOGEN FLUORIDE AND HALOGEN ADDITIONS TO NORBORNENE

Compound No.	Yield, % X = Y = Cl[a]	X = Y = Br[b]	X = Br, Y = F[c]
5	6 *trans*	3.6 *trans*	0
	3.5 *cis*		
6	65	42	33
7	26	33	20
8	0	15	34

[a] Cl_2 in CCl_4 at 0°.[65]
[b] Br_2 in CH_2Cl_2 at 0°.[64a]
[c] NBA in HF, ether at −78°.[62]

Although product **8** could arise from 5,4-hydride shift as shown, the reaction with excess hydrogen fluoride at low temperature could be concerted by the A_E3 type of mechanism.

H^+F^- $^+Br\bar{N}R_2$

Steric Effects

The importance of steric interactions in halogen fluoride addition to steroids is shown by the "abnormal" orientation in iodine fluoride addition to 5,6 double bonds relative to "normal" anti-Markownikoff bromine fluoride additions.[37] The BrX additions where X is Br, Cl, OH, or F all proceed by transient formation of the α-orientated bromonium ion **9**

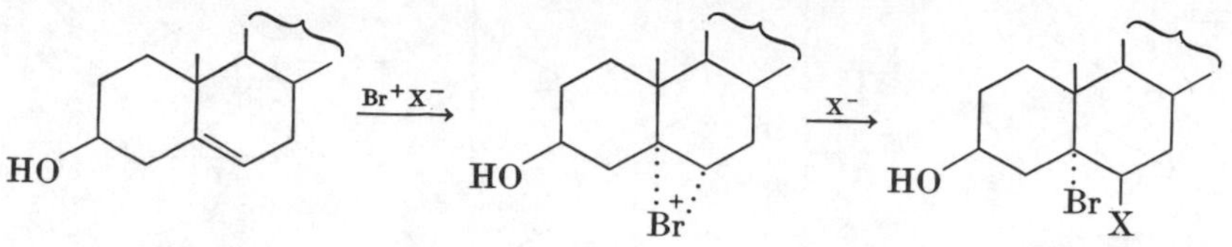

followed by attack of the nucleophile X^- to afford the 6β-substituted 5α-bromo compound **10**. Stereochemical factors clearly control the approach of the bromonium ion (or its equivalent). The 1,3-diaxial nonbonded interaction between the C_{19} methyl group and an entering X^- group is sufficient to override the polarization of the Δ^5 double bond that would normally direct the entering cation to the 5 position. The overall result, therefore, is 5,6 anti-Markownikoff addition of BrX to the double bond.

The 5 F,6 I orientation is unexpected for the bulky iodine atom on the basis of the analysis above; however, a rational explanation has been offered in terms of thermodynamic control.[37]

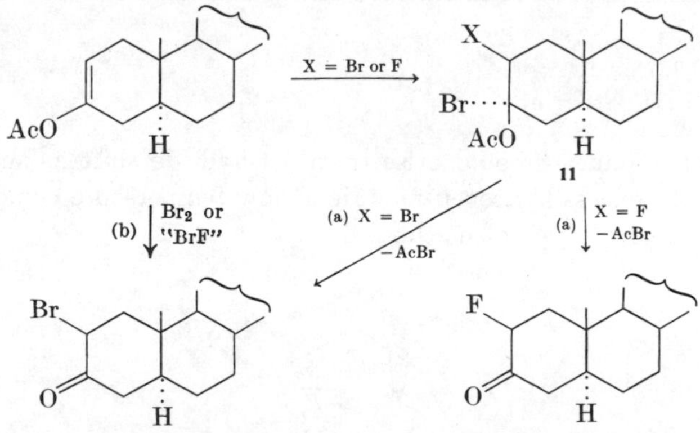

The bromine fluoride addition reaction was also used as a probe to the mechanism of addition of halogens to enol acetates.[32] The bromoketone **13** could form via the adduct **11**, reaction path (a), or directly, reaction path (b). If path (a) were operative, bromine fluoride addition to the enol acetals of a C-3 ketone (5α-allo series) would give 2β-fluoro-3-ketone (**12**) via the adduct **11** (X = F). However, only the 2β-bromoketone (**13**) was formed with either bromine or bromine fluoride, a result that strongly supports path (b) where acetyl bromide is eliminated directly from the bromonium intermediate.

Resonance Effects

Electrophilic additions of bromine and halomethoxyl to unsaturated sugars have been studied in detail.[42] The overall similarity of stereochemistry of the products from halogen fluoride addition relative to other electrophilic additions suggests similar mechanisms. The products are readily explained by significant contribution by form **14b** to the halonium ion intermediate **14a** because of resonance interactions with oxygen (form **14c**). Obviously form **14b** can lead to both *cis* and *trans* adducts. The halogen can greatly influence the relative contribution of each and, as expected, iodine greatly favors form **14a** and gives the

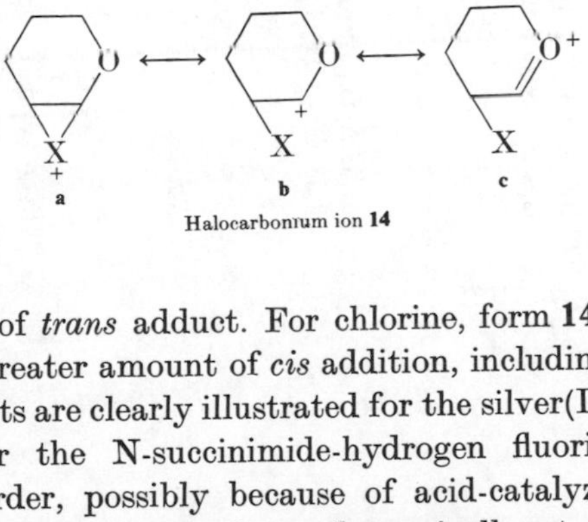

a b c

Halocarbonium ion **14**

greatest amount of *trans* adduct. For chlorine, form **14a** is least favored and results in a greater amount of *cis* addition, including both *cis* stereoisomers. The results are clearly illustrated for the silver(I) fluoride-halogen system. However the N-succinimide-hydrogen fluoride system does not show this order, possibly because of acid-catalyzed isomerization of *trans* adduct to the more thermodynamically stable *cis* adduct.[40] Steric interference from hydrogen bonding of hydrogen fluoride to the ether oxygen may also affect the stereochemistry of the product.[51]

Utility of Products

The halogen fluoride adducts serve as intermediates for the preparation of a variety of monofluoro derivatives. Some examples were selected for Table IV (pp. 162–164) to show the type of conversions and uses that have been made of these derivatives.

TABLE IV. Examples of Utilization of Halogen Fluoride Adducts

Adduct	Chemical Conversion or Utility	Refs.
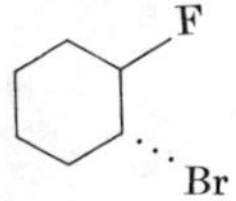 F, Br (cyclohexane)	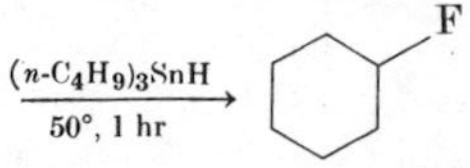 $\xrightarrow[50°,\ 1\ hr]{(n\text{-}C_4H_9)_3SnH}$ F (cyclohexane) (73%) (This method is proposed as a mild two-step method for effecting the indirect addition of HF to olefins.)	66
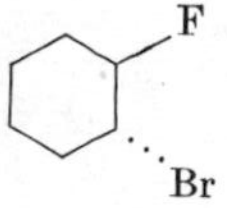 F, Br (cyclohexane)	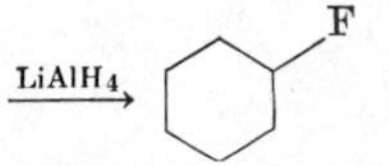 $\xrightarrow{LiAlH_4}$ F (cyclohexane) (45%) [But $LiAlH_4$ reduction of $C_6H_5CHFCHBrC_6H_5$ gives $C_6H_5CH_2CH_2C_6H_5$ (86%)]	67
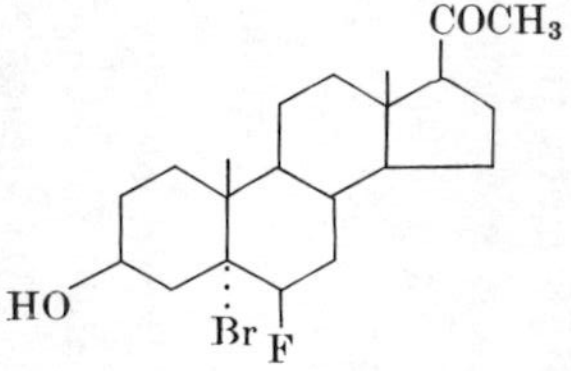 HO, Br, F, $COCH_3$ (steroid)	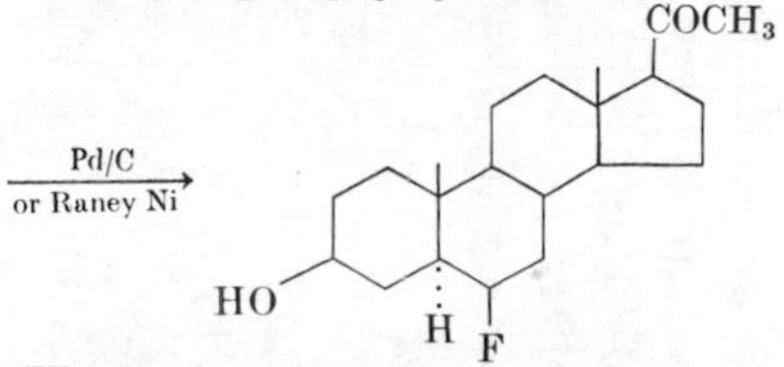 $\xrightarrow[\text{or Raney Ni}]{Pd/C}$ HO, H, F, $COCH_3$ (steroid) (–) (These types of steroids are medicinally useful.)	68
$BrCH_2CHFCH_2OCOCCl_3$	$\xrightarrow[2)\ CH_3OH,\ (CH_3)_2C(OCH_3)_2]{1)\ HNO_3,\ HOAc}$ $BrCH_2CHFCO_2CH_3$ $\xrightarrow{NaOCH_3}$ $CH_2{=}CFCO_2CH_3$ Note that HF is eliminated from certain adducts such as the normal adduct from methyl acrylate. $FCH_2CHBrCO_2CH_3 \xrightarrow{NaOCH_3} CH_2{=}CBrCO_2CH_3$	33, 36*b*

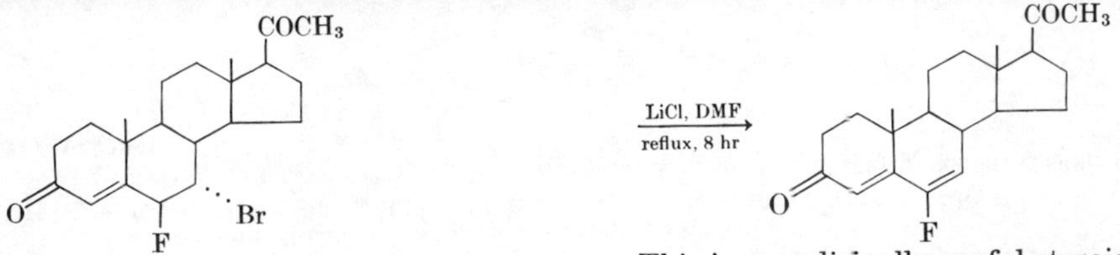

Reactant	Reaction and Product	Ref.
Steroid with $COCH_3$, O, F, Br	$\xrightarrow{\text{LiCl, DMF, reflux, 8 hr}}$ Steroid with $COCH_3$, O, F. This is a medicinally useful steroid.	68
Steroid with COC_2H_5, OH, HO, Br, F	$\xrightarrow{Br_2}$ $COCHBrCH_3$, OH $\xrightarrow{NaOAc}$ Steroid with $COCH(CH_3)OAc$, OH, O, F. This is an adrenocortically active steroid.	69
$RCHFCH_2Br$	$\longrightarrow RCHFCH_2X$	35
	Conditions — **where X is**	
	NaOAc, NaI, $(CH_3)_2NCHO$ — $-OAc$	
	HCO_2Na, NaI, $(CH_3)_2NCHO$ — $-OCHO$	
	$AgNO_3$, CH_3CN — $-ONO_2$	
	KSCN, C_2H_5OH — $-SCN$	
	$LiOCOC_6H_4NO_2$-*p*, $(CH_3)_2NCHO$ — $-OCOC_6H_4NO_2$-*p*	
	$AgOSO_2C_6H_4CH_3$-*p*, CH_3CN — $-OSO_2C_6H_4CH_3$-*p*	
	(In all cases substitution requires more vigorous conditions because of deactivating effect of β-F. Sodium cyanide gave disubstitution.)	
Cyclohexane with Br, F	$\xrightarrow[\text{25°, } CH_2Cl_2]{Cl(CO)Ir[P(CH_3)_3]_2}$ Cyclohexane with F and $Ir[P(CH_3)_3]_2(CO)ClBr$	70
	(Oxidative addition of alkyl halide to a d^8 square planar iridium(I) complex was proposed to proceed with inversion of configuration at carbon. This reaction could not be repeated.[71])	

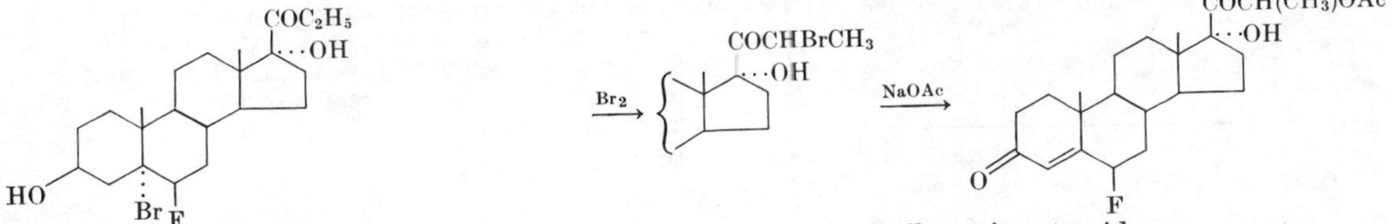

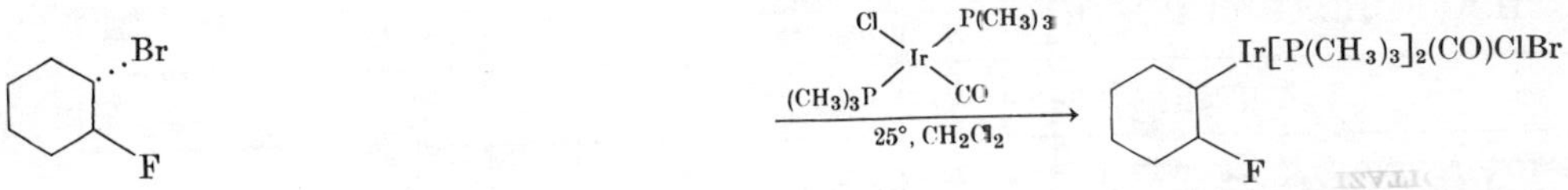

Note: References 66–70 are on p. 164.

TABLE IV. EXAMPLES OF UTILIZATION OF HALOGEN FLUORIDE ADDUCTS (*Continued*)

Adduct	Chemical Conversion or Utility	Refs.
$CH_3(CH_2)_4CHFCH_2Br$	$\xrightarrow[HCON(CH_3)_2]{NaOCOCH_3,\ NaI} CH_3(CH_2)_4CHFCH_2OCOCH_3 \xrightarrow[CH_3CO_2H]{HNO_3} CH_3(CH_2)_4CHFCO_2H$ (The series of reactions was used to prepare a series of monofluoroacids for toxicity studies. Other methods of conversion to the acids were also reported.)	34, 72
$BrCH_2CH_2F$	$\xrightarrow[-60°]{SbF_3,\ SO_2}$ CH_2—CH_2 bridged by $\overset{+}{Br}$	73
ICH_2CH_2F	$\xrightarrow[-60°]{SbF_3,\ SO_2}$ CH_2—CH_2 bridged by $\overset{+}{I}$ (The bridged halonium ions were observed by nmr. Many additional examples were studied.)	

[66] G. L. Grady, *Synthesis*, **1971**, 255.

[67] J. F. King and R. G. Pews, *Can. J. Chem.*, **42**, 1294 (1964).

[68] A. Bowers, U.S. Pat. 3,173,914 (1965) [C.A., **63**, 4372h (1965)].

[69] H.-J. E. Hess, S. K. Figdor, G. M. K. Hughes, and W. T. Moreland, U.S. Pat. 3,042,691 (1962) [C.A., **57**, 15205c (1962)].

[70] J. A. Labinger, R. J. Braus, D. Dolphin, and J. A. Osborn, *Chem. Commun.*, **1970**, 612.

[71] F. R. Jensen and B. Knickel, *J. Amer. Chem. Soc.*, **93**, 6339 (1971).

[72] F. H. Dean, J. H. Amin, and F. L. M. Pattison, *Org. Syn.*, **46**, 37 (1966).

[73] (a) G. A. Olah and J. M. Bollinger, *J. Amer. Chem. Soc.*, **89**, 4744 (1967)(b) **90**, 947 (1968); (c) G. A. Olah, J. M. Bollinger, and J. Brinich, *ibid.*, **90**, 2587 (1968).

Experimental Considerations and Procedures

Many combinations of positive halogen reagent and solvent are possible for halogen fluoride additions. The major division is between use of hydrogen fluoride at −80° and silver(I) fluoride at room temperature. In the standard method using hydrogen fluoride in an ionizing solvent at −80°, the reaction is simply run in a polyethylene bottle equipped with a magnetic stirrer and inlet tube.[29] In general, yields are high for a wide range of olefins. However, handling anhydrous hydrogen fluoride can be hazardous for the inexperienced chemist, and *great care must be taken to avoid contact of hydrogen fluoride with the skin because of serious burns, and to avoid breathing the vapors.* The safety precautions recommended in the section on hydrogen fluoride addition (pp. 220–223) should be followed.

The silver(I) fluoride method appears to be less hazardous, but yields are poor unless the silver(I) fluoride is finely powdered and the reaction mixture vigorously stirred. Since this method was reported only recently, it has not been applied to a range of compounds. Another advantage of the hydrogen fluoride method is that N-halosuccinimides and other N-halo or O-halo sources of positive halogen are generally easier to handle and control than the free halogen required for the silver(I) fluoride procedure. The silver(I) fluoride method should be milder and less likely to cause side reactions promoted by strongly acidic hydrogen fluoride. However, silver(I) fluoride is an active fluorinating agent and can replace other halogens by fluorine.

For chemists with adequate laboratory facilities and experience in the use of hydrogen fluoride, N-haloamides in hydrogen fluoride at −78° are recommended as most likely to give the best results. Where provisions for safe handling of hydrogen fluoride are inadequate, silver(I) fluoride and a halogen at room temperature should be used.

***trans*-1-Fluoro-2-bromocyclohexane.**[32] Dry diethyl ether (98 g) was added to anhydrous hydrogen fluoride (54 g, 2.84 mol) at −80° (acetone-dry ice) in a polyethylene bottle. N-Bromoacetamide (30 g, 0.17 mol) and cyclohexene (16 g, 0.24 mol) were added portionwise and simultaneously over 10 minutes. After an additional 2 hours at −80° with stirring (polyethylene-covered magnet), the solution was kept at 0° for 2 hours. The reaction mixture was added cautiously to excess ice-cold aqueous sodium bicarbonate. Ether extraction and distillation afforded *trans*-1-fluoro-2-bromocyclohexane (15.0 g, 48%), bp 30° (13 mm), n^{20}_{D} 1.4830.

***trans*-1-Fluoro-2-iodocyclohexane.**[32] Dry diethyl ether (270 g) was added to anhydrous hydrogen fluoride (131 g, 6.9 mol) at −80°.

N-Iodosuccinimide (150 g, 0.65 mol) and cyclohexene (45 g, 0.67 mol) were added portionwise and simultaneously during 10 minutes with stirring. After a further 2 hours at $-80°$ and 1 hour at 0° the reaction mixture was added cautiously to excess ice-cold aqueous sodium bicarbonate. The product was extracted from the alkaline solution with ether. The iodine-colored ether solution was washed successively with water, aqueous sodium thiosulfate, and water, and dried over anhydrous sodium sulfate. The ether was removed under reduced pressure. The product was distilled from finely powdered sodium carbonate (2–3 g) or calcium carbonate (in the absence of carbonate the distillation was accompanied by decomposition) to give *trans*-1-fluoro-2-iodocyclohexane (90.4 g, 73%), bp 64° (9 mm), $n^{20}{}_{D}$ 1.5314.

This product still contained a trace of free iodine. It was washed with 10 ml of 2% aqueous sodium thiosulfate, dried, and redistilled over sodium carbonate to afford 76 g of an almost colorless product, $n^{20}{}_{D}$ 1.5318, unchanged after a further fractional distillation. The product was stored at 0–5° in the dark. A sample exposed to sunlight liberated iodine.

1-Bromo-2-fluorohexane.[35] The preparation of 1-bromo-2-fluorohexane in 60–77% yield from 1-hexane, N-bromoacetamide, and hydrogen fluoride in diethyl ether at $-80°$ is described in *Organic Syntheses*.[74] The product contains 5–10% of impurities, mostly 2-bromo-1-fluorohexane. This procedure is generally useful for "bromine fluoride" addition to olefins.

9α-Bromo-11β-fluoro-1,4-pregnadiene-17α,21-diol-3,20-dione 21-Acetate.[30] A solution of 1,4,9(11)-pregnatriene-17α,21-diol-3,20-dione 21-acetate (1.0 g, 2.6 mmol) in 50 ml of diethylacetic acid was stirred in a polyethylene bottle. A solution of hydrogen fluoride in chloroform-tetrahydrofuran (5 ml, 270 mg of hydrogen fluoride per ml of solution) was added, followed by N-bromoacetamide (395 mg, 2.85 mmol). Stirring was continued for 17 hours, and the solution was then poured into 500 ml of 10% aqueous sodium carbonate. The mixture was extracted with methylene chloride. The extracts were washed with water, dried ($MgSO_4$), and the solvents removed by evaporation under reduced pressure to give 1.15 g (91%) of crude product. Crystallization from acetone-hexane yielded pure 9α-bromo-11β-fluoro-1,4-pregnadiene-17α,21-diol-3,20-dione 21-acetate (660 mg, 51%), mp 225–228° dec, $[\alpha]_D$ +123°.

2-Chloro-3-fluoro-2,3-dimethylbutane.[73a] To a solution of anhydrous hydrogen fluoride (25 g, 1.3 mol) and dry ether (50 ml) were alternately added, at $-78°$, tetramethylethylene (4.2 g, 0.05 mol, Aldrich) and *t*-butyl hypochlorite (5.6 g, 0.051 mol, Frinton Laboratories) during

[74] F. H. Dean, J. H. Amin, and F. L. M. Pattison, *Org. Syn.*, **46**, 10 (1966).

5 minutes. The mixture was stirred for 1 hour at −78°, and for an additional hour at 0°, and then poured into ice water containing excess potassium carbonate. An additional 60 ml of ether was added, and the layers were separated. Removal of solvent left an oil which, after flash distillation at 1 mm, partially crystallized. The yield was 3.6 g (46%) of material containing small amounts of an unidentified contaminant. The analytical sample was purified by preparative gas chromatography on an Apiezon L on Chromosorb W column, mp 63–64°.

2-Bromo-2-deoxymannopyranosyl Fluoride Triacetates.[40] (See equation on p. 142). D-Glucal triacetate (1.36 g, 5.0 mmol) in dry acetonitrile (*ca.* 25 ml) was stirred vigorously with powdered silver(I) fluoride (4.0 g, 31 mmol). A solution of bromine (0.85 g, 5.3 mmol, 10% w/v in benzene) was then added dropwise. After completion of the bromine addition (*ca.* 10 minutes), the solution was stirred for a further 20 minutes and filtered from the copious precipitate of silver halide. To this solution was added 5 ml of saturated aqueous sodium chloride, the precipitated silver chloride was removed, the filtrate was concentrated to about 10 ml, and 30 ml of chloroform was added. The chloroform solution was extracted successively with aqueous sodium thiosulfate, aqueous sodium bicarbonate, and water, and dried over sodium sulfate. Evaporation produced a clear syrup (1.8 g, 97%). Inspection of the crude reaction product by proton magnetic resonance showed that no starting material remained. ^{19}F nmr showed the presence of three glycopyranosyl fluorides that were subsequently identified as: 2-bromo-2-deoxy-α-D-mannopyranosyl fluoride triacetate (**A**), 70%; ϕ_c +123.2; $J_{1,F}$ 50.2 Hz; $J_{2,F}$ 2.85 Hz; 2-bromo-2-deoxy-β-D-glycopyranosyl fluoride triacetate (**B**), 21%; ϕ_c +136.0; $J_{1,F}$ 50.3 Hz; $J_{2,F}$ 10.0 Hz; and 2-bromo-2-deoxy-α-D-glycopyranosyl fluoride triacetate (**C**), 9%; ϕ_c +144.9, $J_{1,F}$ 51.5 Hz; $J_{2,F}$ 25.2 Hz.

After workup the reaction mixture was dissolved in boiling diethyl ether. On cooling, the solution deposited crystals (0.98 g), mp 137–142°; addition of light petroleum afforded a second crop of the same material (0.13 g), total yield 1.11 g, 56%. Recrystallization from ethanol afforded pure **A** as prismatic crystals; mp 138–140; $[\alpha]_D^{22}$ −32.0° (c 2.06).

The mother liquors remaining after isolation of **A** were evaporated, and the resulting syrup was dissolved in ethanol and cooled to −5°; after several weeks an impure sample of **C** was obtained (0.12 g, 6%). Two recrystallizations from aqueous ethanol afforded pure **C** as fine needles; mp 120–121°, $[\alpha]_D^{22}$ +134.5° (c 1.80).

The residual syrup from above was shown by nmr to contain mainly the β-D-gluco isomer (**B**) together with smaller quantities of the other isomers (**A**) and (**C**). Although not crystalline, this syrup was sufficiently pure for detailed nmr analysis.

THE FLUOROALKYLAMINE REAGENT

The one-step conversion of a hydroxy derivative to the corresponding fluoro compound by α,α-difluorosubstituted amines (usually 2-chloro-1,1,2-trifluorotriethylamine, known as the Yarovenko reagent or the fluoroamine reagent), is a valuable synthetic method. Recent reviews on

$$ROH + (C_2H_5)_2NCF_2CHFCl \rightarrow RF + (C_2H_5)_2NCOCHFCl + HF$$

this subject are of limited value because one is in Czechoslovakian[75] and the other deals only with steroids.[4] α-Fluoroalkylamines have also been used for difluorination of the carbonyl group,[75b] but only monofluorination is reviewed in this chapter.

Preparation of the Reagent

The fluoroalkylamine reagent, abbreviated in this chapter as **FAR**,* is prepared by addition of diethylamine to the double bond of chlorotrifluoroethylene.[76–79] Many compounds similar to **FAR** have been pre-

$$(C_2H_5)_2NH + CF_2{=}CFCl \rightarrow \underset{\textbf{FAR}}{(C_2H_5)_2NCF_2CHClF}$$

pared from chlorotrifluoroethylene and various amines such as dimethylamine, diisopropylamine, dibutylamine, pyrrolidine, 2-methylpyrrolidine, 2,2-dimethylpyrrolidine, 4-methylpiperazine, morpholine, piperidine, and 2-methylpiperidine.[79]

$$R_1R_2NH + CF_2{=}CFCl \rightarrow R_1R_2NCF_2CHClF$$

Similar reagents have been prepared from tetrafluoroethylene.[77, 80]

* From FluoroalkylAmine Reagent.

[75] (a) F. Liska. *Chem. Listy*, **66**, 189 (1972) [*C.A.*, **76**, 99936p (1972)]; (b) A. V. Fokin, V. I. Zimin, Yu. N. Studnev, and D. A. Sultanbekov, *J. Gen. Chem.* (*USSR*), *Engl. Transl.* **38**, 1459 (1968).

[76] N. N. Yarovenko and M. A. Raksha, *J. Gen. Chem.* (*USSR*), *Engl. Transl.*, **29**, 2125 (1959).

[77] D. C. England, L. R. Melby, M. A. Dietrich, and R. V. Lindsey, Jr., *J. Amer. Chem. Soc.*, **82**, 5116 (1960).

[78] R. L. Pruett, J. T. Barr, K. E. Rapp, C. T. Bahner, J. D. Gibson, and R. H. Lafferty, Jr., *J. Amer. Chem. Soc.*, **72**, 3646 (1950).

[79] D. E. Ayer, U.S. Pat. 3,153,644 (1964) [*C.A.*, **62**, 619e (1965)].

[80] N. N. Yarovenko, M. A. Raksha, V. N. Shemanina, and A. S. Vasilyeva, *J. Gen. Chem.* (*USSR*), *Engl. Transl.*, **27**, 2305 (1957).

Although most of these reagents will replace the hydroxyl group of alcohols by fluorine, none approaches **FAR** in convenience, frequency of use, or general utility.

Three procedures are given in the following numbered paragraphs, after which the advantages and disadvantages of the procedures are described and some pertinent properties of the reagent listed.

1.[79] Chlorotrifluoroethylene (15 ml) was condensed into a pressure tube cooled in a bath of acetone and dry ice. To the pressure tube was added 10.3 ml of diethylamine previously cooled to −40°. The tube was then sealed, placed in an ice bath, and allowed to warm slowly to room temperature. The tube and contents were then kept at room temperature for about 48 hours, cooled, opened, and the contents distilled under reduced pressure with minimum exposure to atmospheric moisture. 2-Chloro-1,1,2-trifluorotriethylamine (**FAR**, 15.7 g, 85%) was formed as a liquid, bp 33–34° (6 mm).

2.[81] At −70°, chlorotrifluoroethylene (128.1 g, 1.40 mol) was liquefied (in an inert atmosphere), and dry diethylamine (72 g, 1.0 mol) and dry methylene chloride (60 ml) were added successively. The mixture was kept in an autoclave at room temperature for 48 hours and fractionated, the product being collected at 45–52° (25 mm) in a receiver cooled with acetone-dry ice mixture. The yield was 80–90%. **FAR** is kept in a closed bottle in a desiccator and without exposure to light. It is stable for several days under these conditions.

3.[76] In a bottle cooled to −5 to −10° was placed 40.0 g (0.56 mol) of diethylamine. Over a period of 10 hours, well-dried chlorotrifluoroethylene (70.0 g, 0.06 mol) was bubbled through the solution. The reaction mixture was then distilled under reduced pressure and the product (70.0 g, 61%) was collected at 33° (6 mm).

Preparations 1 and 2 give higher yields but have the disadvantage of pressurization, including the hazard of sealed glass tubes. Preparation 3 is more convenient, since a gas is simply bubbled through a liquid. Successful preparation of **FAR** depends on the rigorous exclusion of moisture. For fluorination reactions it is best to use freshly prepared **FAR**, although it can be stored for several days without significant loss of fluorinating power. Because of instability in storage it is not available commercially.

The nmr spectrum of **FAR** is useful for characterization and analysis for purity, and as a probe to follow the reaction (see p. 172).[82, 83] The data reported were obtained in carbon tetrachloride solution using as

[81] E. D. Bergmann and A. M. Cohen, *Isr. J. Chem.*, **8**, 925 (1970).

[82] F. J. Weigert, Central Research Department, E. I. du Pont de Nemours and Co., personal communication.

[83] K. Schaumburg, *J. Magn. Resonance*, **7**, 177 (1972).

internal references trichlorofluoromethane for ^{19}F and tetramethylsilane for ^{1}H.[82]

^{19}F —C$\underline{F}$HCl δ_F −150 ppm, J_{HF} 49 Hz (d), J_{FF} 15 (t), 15 (t)
—C$\underline{F}_2$— δ_F −90.1 and −86.8 ppm, J_{FF} 206 Hz (weak, strong, strong, weak), J_{FF} 15 Hz (d), J_{HF} not resolved
^{1}H —$C_2\underline{H}_5$ −1.19 ppm, J 7 Hz (t), −3.19 ppm, J 7 Hz (q), J 0.2 Hz (t)
—C$\underline{H}$FCl −6.50 ppm, J_{HF} 49 Hz (d), J_{HF} 5.5 Hz (t), J_{HF} 4 Hz (t)

Scope, Typical Reactions and Limitations

FAR, 2-chloro-1,1,2-trifluorotriethylamine, is a mild fluorinating reagent which replaces hydroxyl groups by fluorine.

$$ROH + \underset{\textbf{FAR}}{(C_2H_5)_2NCF_2CHClF} \rightarrow RF + (C_2H_5)_2NCOCHClF + HF$$

The reaction with primary alcohols usually proceeds in high yield with limited side reactions. Tertiary alcohols rarely give tertiary fluorides; in most cases, side reactions predominate.

Three commonly encountered side reactions of **FAR** with alcohols are elimination of water, esterification of the hydroxyl group, and carbonium ion rearrangement of the carbon skeleton of the alcohol. The structure of the alcohol determines which side reactions predominate. This will become clear in the discussion of steroidal alcohols.

$$>CH{-}\overset{|}{\underset{|}{C}}{-}OH + (C_2H_5)_2NCF_2CHClF \longrightarrow >C{=}C< + (C_2H_5)_2NCOCHClF + 2\,HF$$

$$ROH + HOH^* + (C_2H_5)_2NCF_2CHClF \longrightarrow ROCOCHClF + (C_2H_5)_2NH{\cdot}HF + HF$$

$$2\,R_2CH{-}\overset{|}{\underset{|}{C}}{-}OH + (C_2H_5)_2NCF_2CHClF \longrightarrow R_2C{=}C< + R_2CH\overset{|}{\underset{|}{C}}COCHClF + (C_2H_5)_2NH{\cdot}HF + HF$$

A major advantage of **FAR** for fluorination of alcohols is that the reagent is unreactive to most functional groups such as carbon-carbon double and triple bonds, ketones, aldehydes, esters, amides, and nitriles. It does react, however, with primary and secondary amines. Numerous efforts to obtain fluorinated amines and amino acids from hydroxy amines

* The water is probably the result of the reaction of hydrogen fluoride with the glass reaction vessel.

and hydroxy amino acids have failed; however, when the N-function was protected, fluorination succeeded.[81, 84]

Aliphatic and Simple Alicyclic Alcohols

The high yields of fluorides from simple primary aliphatic alcohols are illustrated by the reaction of **FAR** with 1-butanol and 3-bromo-1-butanol.

$$CH_3(CH_2)_2CH_2OH \xrightarrow{\text{FAR}} \underset{(67\%)}{CH_3(CH_2)_2CH_2F} \qquad \text{(Ref. 76)}$$

$$CH_3CHBrCH_2CH_2OH \xrightarrow[\text{Room t, 15 hr}]{\text{FAR}/(C_2H_5)_2O} \underset{(73\%)}{CH_3CHBrCH_2CH_2F} \qquad \text{(Ref. 85)}$$

Secondary alcohols usually react cleanly with **FAR** to give the corresponding fluorides. With cyclic secondary alcohols the elimination side reaction often predominates.

$$BrCH_2CH_2CHOHCH_3 \xrightarrow[\text{Room t, 15 hr}]{\text{FAR}/(C_2H_5)_2O} \underset{(68\%)}{BrCH_2CH_2CHFCH_3} \qquad \text{(Ref. 85)}$$

$$\underset{(70\%\ \textit{trans},\ 30\%\ \textit{cis})}{(CH_3)_3C\text{-}C_6H_{10}\text{-}OH} \xrightarrow[0°,\ 14\ \text{hr}]{\text{FAR}/(C_2H_5)_2O} \underset{(9\%;\ 90\%\ \textit{cis},\ 10\%\ \textit{trans})}{(CH_3)_3C\text{-}C_6H_{10}\text{-}F}$$

$$+ \underset{(68\%)}{(CH_3)_3C\text{-}C_6H_9} \qquad \text{(Ref. 86)}$$

Both primary and secondary alcohols of higher molecular weight give good yields of the corresponding fluorides. As illustrated with methyl ricinoleate, **FAR** does not react with carbon-carbon double bonds.[87]

$$CH_3(CH_2)_5CHOHCH_2CH{=}CH(CH_2)_7CO_2CH_3 \xrightarrow[0°,\ 12\ \text{hr}]{\text{FAR}/CH_2Cl_2}$$

$$\underset{(90\%)}{CH_3(CH_2)_5CHFCH_2CH{=}CH(CH_2)_7CO_2CH_3}$$

[84] M. Hudlicky and B. Kakac, *Collect. Czech. Chem. Commun.*, **31,** 1101 (1966).
[85] M. Hudlicky and I. Lejhancova, *Collect. Czech. Chem. Commun.*, **31,** 1416 (1966).
[86] E. L. Eliel and R. J. L. Martin, *J. Amer. Chem. Soc.*, **90,** 682 (1968).
[87] E. D. Bergmann and A. Cohen, *Isr. J. Chem.*, **3,** 71 (1965).

Although a tertiary alcohol usually cannot be converted to the corresponding unrearranged tertiary fluoride, an exception occurs in the bicyclooctane series.[88]

OH → FAR → F (64%)

Steroids

The fluoroalkylamine reagent has been used more extensively for the fluorination of steroids than for any other class of compounds. The yields vary from very high to very low depending on the type of hydroxyl group, the stereochemistry of the hydroxyl group and neighboring carbon atoms, and the presence of allylic or homoallylic double bonds. As might be expected, hydroxymethyl groups on the steroid skeleton can usually be substituted by fluorine in good yield with negligible side reactions.[89] 19-Hydroxy-

$COCH_3$, CH_2OH → FAR, CH_2Cl_2, 40° → $COCH_3$, CH_2F (46%)

steroids are exceptions; as might be expected of these highly substituted neopentyl alcohols, rearrangements are often observed (see p. 167).

A few steroids with a hydroxyl group attached directly to a steroid ring can be converted by **FAR** to the corresponding fluoride in good yield without significant side reactions.[89–91] As illustrated in this example the

HO → **FAR**/THF, room t, 20 min; Or **FAR**/CH_2Cl_2, 0°, 3 hr → F (90–96%)

[88] J. Kopecky, J. Smejkal, and M. Hudlicky, *Chem. Ind.* (London), **1969,** 271.

[89] D. E. Ayer, *Tetrahedron Lett.*, **1962,** 1065.

[90] (a) L. H. Knox, E. Velarde, S. Berger, D. Cuadriello, and A. D. Cross, *Tetrahedron Lett.*, **26,** 1249 (1962); (b) *J. Org. Chem.*, **29,** 2187 (1964).

[91] J.-C. Brial and M. Mousseron-Canet, *Bull. Soc. Chim. Fr.*, **1968,** 3321.

configuration is retained at C_3. The β-hydroxyl group is converted to a β-fluoro group. However, inversion of configuration is usual for fluoroalkylamine substitution of hydroxyl by fluoride. Configuration may be retained when attack from the backside of carbon is sterically hindered, either by the conformation or by a bulky substituent or when a stabilized intermediate carbonium ion is formed.

1-Hydroxysteroids. Few 1-fluorosteroids have been prepared. Reaction of **FAR** with 17β-acetoxy-5α-androst-2-en-1α-ol gave 1-fluoro-5α-androst-2-en-17β-ol acetate of uncertain configuration at C-1.[92a]

OAc OH H $\xrightarrow[\text{Room t, 30 min}]{CH_2Cl_2}$ F H (53%)

1α-Hydroxy-5α-androstane-3,17-dione with **FAR** gave no fluoro product but only retropinacol rearrangement of the C-19 methyl group.[92b]

3-Hydroxysteroids. The reaction of **FAR** with many 3-hydroxysteroids has been studied. The products formed depend on several factors including solvent, time of reaction, and temperature. For example, 3β-hydroxyandrost-5-en-17-one is converted at 0–5° in dichloromethane or at room temperature in tetrahydrofuran to the corresponding 3β-fluoro derivative (no inversion!) in over 90% yield (see example on p. 162).

In contrast, 3β-hydroxy-5α-androstan-17-one (**15**) in tetrahydrofuran gives only an elimination product **16**, but in methylene chloride at 0°, 42% of the 3α-fluoro product is formed in addition to 34% of the elimination product **16**.[90b] Also, ester and amide products can be formed in significant amounts under other conditions.

Experimental evidence indicates that side reactions are more likely to occur with the 3α-hydroxy- than with the 3β-hydroxy-steroids. In the reaction of **FAR** with 3α-hydroxy-17β-acetoxy-5(10)-estrene in methylene chloride under mild conditions (0° for 20 minutes), seven products were isolated.[93] The products were typical of those expected if a carbonium ion intermediate was involved (see discussion of mechanism on pp. 168–171 and Table IX). In general, rearrangements and side reactions occur more

[92] (a) T. J. Foell and L. L. Smith, *J. Med. Chem.*, **9**, 953 (1966); (b) L. L. Smith, T. J. Foell, and D. M. Teller, *J. Org. Chem.*, **30**, 3781 (1965).

[93] M. Mousseron-Canet and J.-L. Borgna, *Bull. Soc. Chim. Fr.*, **1969**, 613.

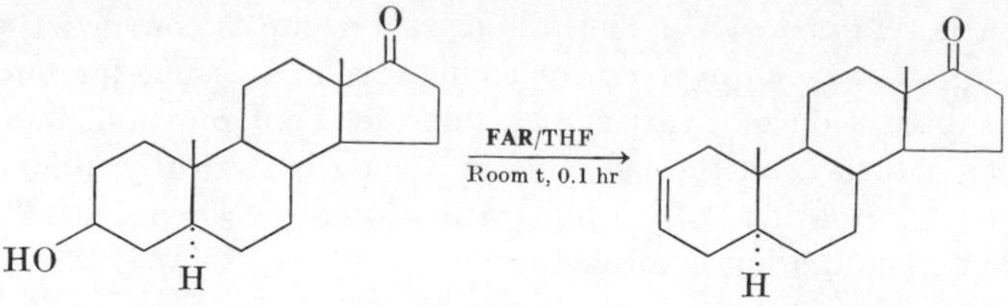

FAR/THF, Room t, 0.1 hr

15 → **16** (100%)

15 → **FAR**/CH_2Cl_2, 0°, 16 hr → (42%) + **16** (34%)

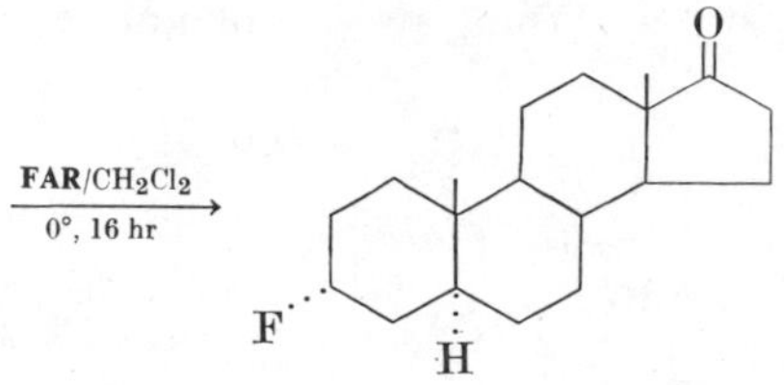

frequently if the A and B rings are fully saturated. Unsaturation at the Δ^5 position reduces side reactions.

6-Hydroxysteroids. The fluorination of 6-hydroxysteroids gives variable results. Usually a fair yield of the desired fluoro compound is obtained along with the usual side-reaction products. A very interesting example is the selective reaction of **FAR** with the 6β-hydroxyl group of 21-acetoxy-16α-chloro-5α,6β,11β,17α-tetrahydroxypregnane-3,20-dione (no inversion).[94]

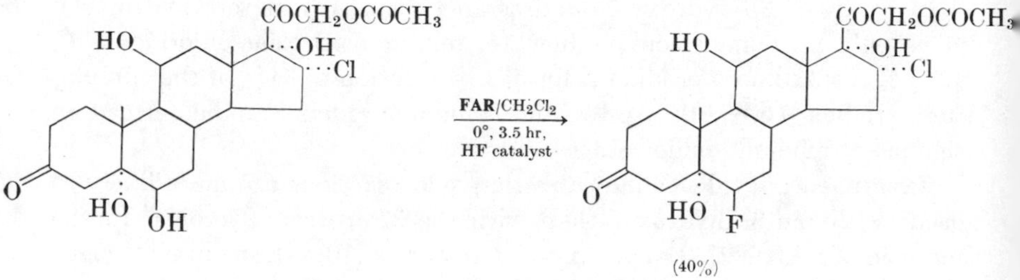

$COCH_2OCOCH_3$ — **FAR**/CH_2Cl_2, 0°, 3.5 hr, HF catalyst → (40%)

This example illustrates the stability of tertiary hydroxyl groups (C_5 and C_{17}) toward **FAR**. The secondary hydroxyl at C_{11} reacts sluggishly. In general, 6-fluorosteroids are more easily prepared by addition of bromine fluoride to Δ^5 (or Δ^6) unsaturation followed by subsequent removal of bromine (Table IV, pp. 152–153).

[94] G. B. Spero, J. E. Pike, F. H. Lincoln, and J. L. Thompson, *Steroids*, **11**, 769 (1968).

11-Hydroxysteroids. With 11-hydroxysteroids, elimination products predominate. For example, reaction of 11α-hydroxypregn-4-ene-3,20-dione with **FAR** in tetrahydrofuran gave only the elimination product.[90a] The 19-methyl group in the 11-hydroxysteroids is sterically im-

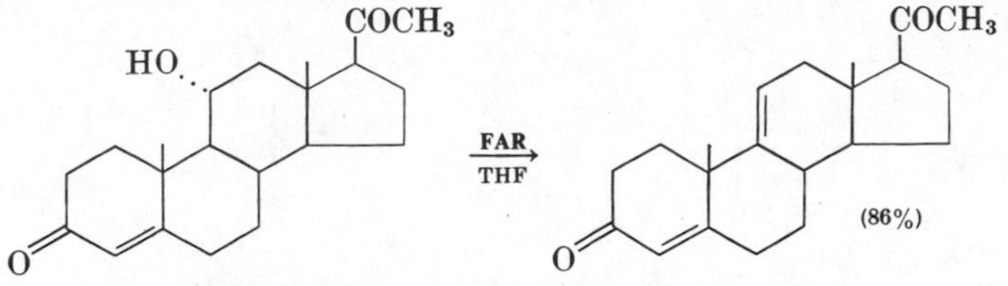

COCH3 HO O FAR THF COCH3 O (86%)

portant for, in contrast with the above, 11α-hydroxy-19-norsteroids give the 11β-fluoro-19-norsteroids in yields up to 45%.[95]

An 11β-hydroxyl group reacts less readily than a 6β-hydroxyl group (see the example under 6-fluoro).[96] When conditions are made sufficiently vigorous, both groups undergo reaction. The 6β-hydroxyl gives the fluoride and the 11β-hydroxyl eliminates; other side reactions also occur.[96]

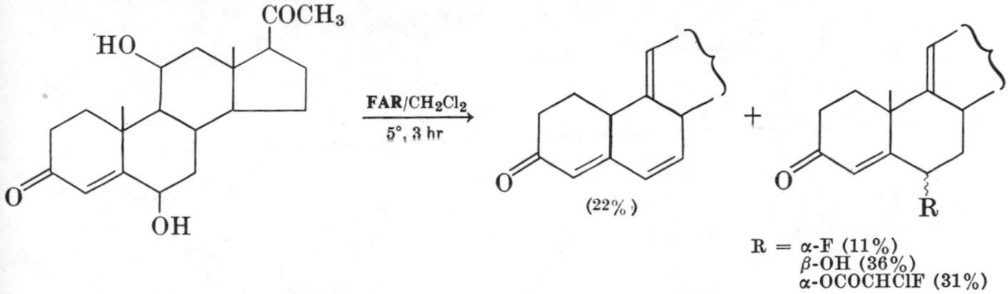

COCH3 HO O OH FAR/CH2Cl2 5°, 3 hr O (22%) + O R

R = α-F (11%)
β-OH (36%)
α-OCOCHClF (31%)

As a general statement, **FAR** is not satisfactory to convert 11-hydroxysteroids to fluorides. The sterically hindered 11-hydroxy group is too unreactive. When reaction occurs, elimination predominates.

15-Hydroxysteroids. The 15α-hydroxysteroids are converted by **FAR** into the corresponding 15β-fluorosteroids in reasonable yield, for in this position no conjugating double bonds or hindering methyl groups can interfere with fluorination. However, the usual elimination reaction products are also formed.[96]

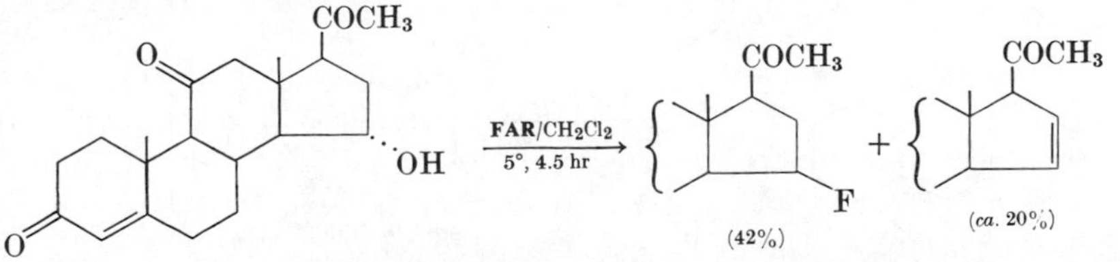

COCH3 O OH O FAR/CH2Cl2 5°, 4.5 hr COCH3 F (42%) + COCH3 (*ca.* 20%)

[95] E. J. Bailey, H. Fazakerley, M. E. Hill, C. E. Newall, G. H. Phillipps, L. Stephenson, and A. Tulley, *Chem. Commun.*, **1970**, 106.

[96] D. E. Ayer, U.S. Pat. 3,056,807 (1962) [*C.A.*, **58**, 9197g (1963)].

16-Hydroxysteroids. By the action of **FAR,** both 16α- and 16β-hydroxysteroids are converted with inversion to the 16β- and 16α-fluorosteroids, respectively. Yields are high.[94]

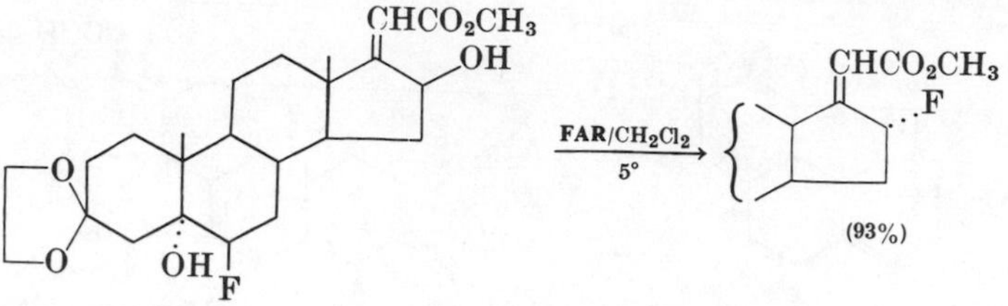

CHCO$_2$CH$_3$ OH → FAR/CH$_2$Cl$_2$, 5° → CHCO$_2$CH$_3$ F (93%)

In competition, the 16α-hydroxyl group is more reactive than the 11β-hydroxyl group.[96] The limited experimental data suggest that **FAR** is the preferred reagent for introducing a 16-fluoro group.

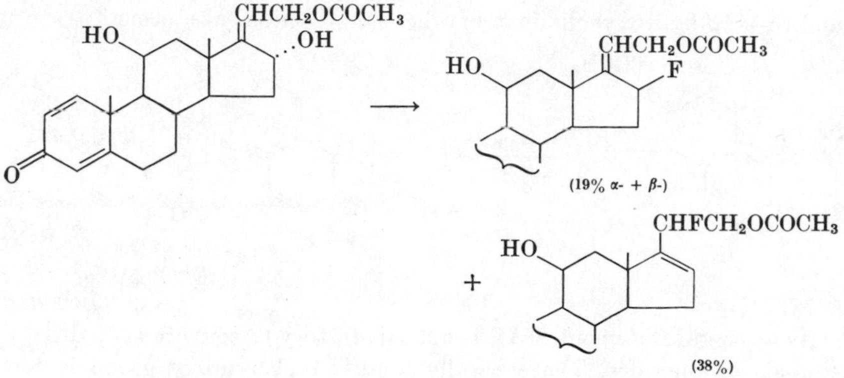

CHCH$_2$OCOCH$_3$ HO OH → HO CHCH$_2$OCOCH$_3$ F (19% α- + β-) + HO CHFCH$_2$OCOCH$_3$ (38%)

17-Hydroxysteroids. Only rarely can a 17-hydroxysteroid be converted into the corresponding 17-fluorosteroid. With testosterone a low yield of fluoride is obtained along with elimination and rearrangement products.[90b]

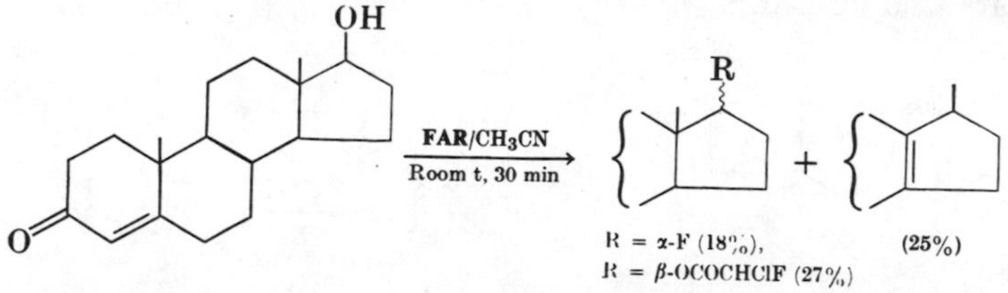

OH → FAR/CH$_3$CN, Room t, 30 min → R + (25%)

R = α-F (18%),
R = β-OCOCHClF (27%)

If the hydroxyl group at C_{17} is tertiary, it is relatively inert to reaction with **FAR**. When the temperature is raised, only rearrangement occurs.[90b]

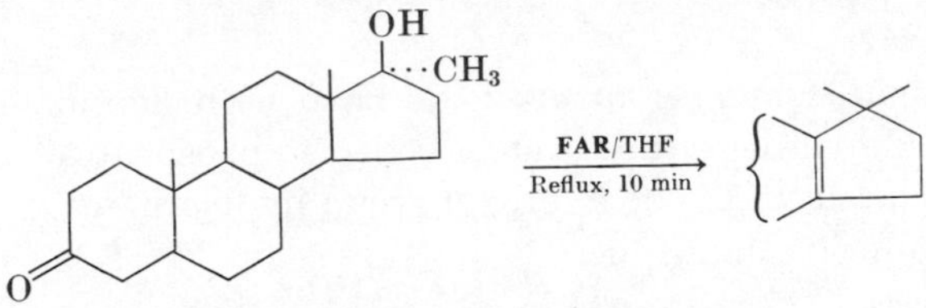
OH
CH3
O
FAR/THF
Reflux, 10 min

19-Hydroxysteroids. The usual reaction of these steroids is rearrangement of the carbon skeleton, as might be expected of the highly substituted neopentyl system. There are a few exceptions, one of which is illustrated in the accompanying equation.[97]

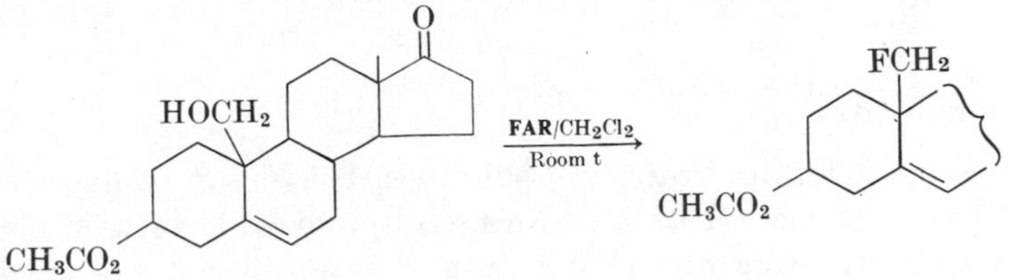
O
HOCH2
CH3CO2
FAR/CH2Cl2
Room t
FCH2
CH3CO2

An example of the type of skeletal rearrangement to be expected is illustrated by the following equation.[98]

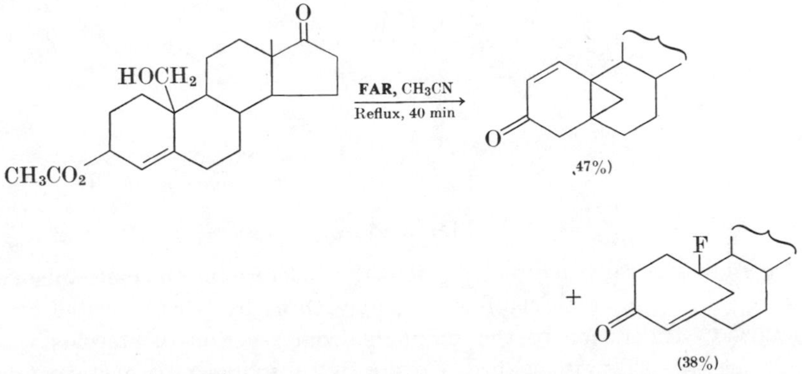
O
HOCH2
CH3CO2
FAR, CH3CN
Reflux, 40 min
O
(47%)
+
F
O
(38%)

In general, 19-hydroxysteroids should not be expected to give 19-fluorosteroids, although rearranged products containing fluorine are sometimes formed.[99]

[97] A. Bowers, U.S. Pat. 3,101,357 (1963) [*C.A.*, **60**, 3048 (1964)].

[98] L. H. Knox, E. Velarde, and A. D. Cross, *J. Amer. Chem. Soc.*, **85**, 2533 (1963); **87**, 3727 (1965).

[99] P. H. Bentley, M. Todd, W. McCrae, M. L. Maddox, and J. A. Edwards, *Tetrahedron*, **28**, 1411 (1972).

Alkaloids

Only a limited number of alkaloids have been fluorinated by **FAR**. Morphine, dihydromorphine, and codeine phosphate sesquihydrate undergo replacement of one hydroxyl group by fluorine.[100]

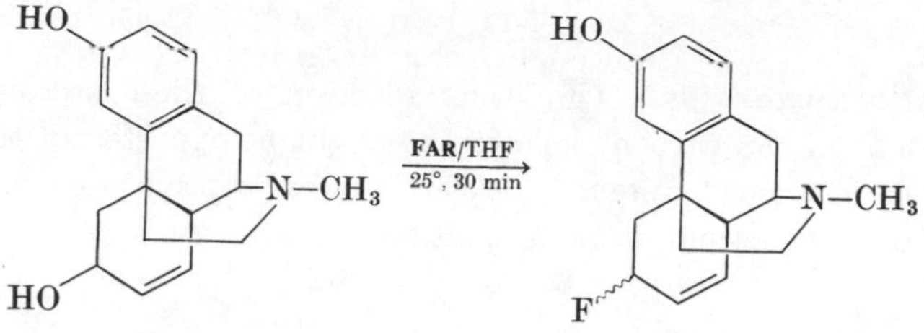

HO ... N—CH_3 $\xrightarrow[25°,\ 30\ min]{FAR/THF}$ HO ... N—CH_3

HO ... F

Morphine

Carbohydrates

Attempts to replace hydroxyl groups in carbohydrates by fluorine using **FAR** have not succeeded.[101–103] No useful products have been obtained when reactions were carried out with ring-substituted carbohydrates. With hydroxymethyl compounds the fluorochloroacetate derivatives were obtained. In some cases a chloro compound was isolated.[101]

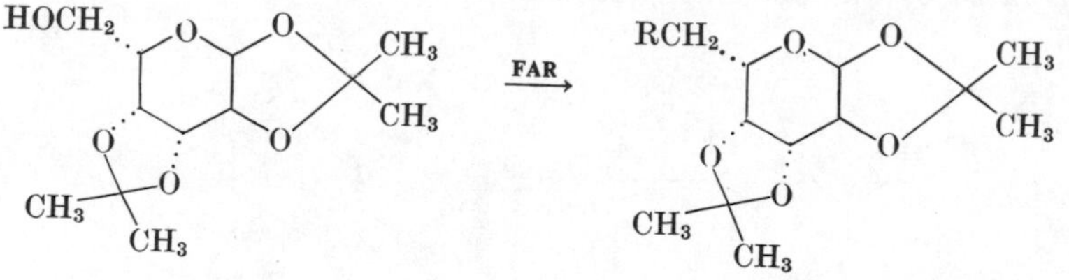

$HOCH_2$ O O CH_3 CH_3 O O O CH_3 CH_3 $\xrightarrow{FAR}$ RCH_2 O O CH_3 CH_3 O O O CH_3 CH_3

R = $CHClFCO_2$ + R = Cl

Mechanism

No studies have been directed primarily to determine the mechanism of reaction of the fluoroalkylamine reagent with hydroxyl-substituted compounds. Consideration of the products from reaction of steroids with **FAR** suggests that either an S_N1 or an S_N2 mechanism is operative depending on steroid structure and reaction conditions. For most steroids, inversion of configuration occurs at the hydroxyl-substituted carbon atom. For these steroids the S_N2 mechanism is favored.

The reaction of octyl (+)-phenylhydroxyacetate with **FAR** gave octyl (−)-phenylfluoroacetate; an S_N2 mechanism is highly probable.[81]

[100] D. E. Ayer, U.S. Pat. 3,137,701 (1964) [*C.A.*, **61**, 4409h (1964)].
[101] P. W. Kent, *Chem. Ind.* (London), **1969**, 1128.
[102] K. R. Wood, D. Fisher, and P. W. Kent, *J. Chem. Soc., C*, **1966**, 1994.
[103] L. D. Hall and L. Evelyn, *Chem. Ind.* (London), **1968**, 183.

On the basis of the stereochemistry of reactions of **FAR** with secondary alcohols, the S_N2 mechanism is also reasonable for reaction of **FAR** with primary alcohols.

The S_N2 mechanism, as suggested in the literature, is illustrated for 1-butanol.[89, 90b]

$$(C_2H_5)_2\ddot{N}-CF_2-CHClF \rightleftharpoons (C_2H_5)_2\overset{+}{N}=CF-CHClF \; F^- \xrightarrow{H\ddot{O}-CH_2C_3H_7\text{-}n}$$

$$(C_2H_5)_2\ddot{N}-CF(\overset{+}{O}H-CH_2C_3H_7\text{-}n)-CHClF \; F^- \longrightarrow (C_2H_5)_2\ddot{N}-CF(OCH_2C_3H_7\text{-}n)-CHClF + HF \longrightarrow$$

$$(C_2H_5)_2\overset{+}{N}=C(\ddot{O}CH_2C_3H_7\text{-}n)-CHClF \; F^- \; (\text{or } N:) \longrightarrow (C_2H_5)_2\ddot{N}-C(=O)CHClF + FCH_2C_3H_7\text{-}n$$

Note that this is a postulated mechanism and the existence of any of the separate steps can be questioned. A concerted mechanism is also reason-

$$(C_2H_5)_2\ddot{N}-CF(\ddot{O}H-CH_2C_3H_7\text{-}n)-CHClF \longrightarrow \left[(C_2H_5)_2N-C(CHClF)(:\ddot{F}:)-\ddot{O}-CH_2C_3H_7\text{-}n\right] \longrightarrow$$

$$(C_2H_7)_2N-C(CHClF)=O + CH_2C_3H_7\text{-}n \; (F)$$

able, although an intermediate must be formed when other nucleophiles (:Nuc⁻) intercept the carbon skeleton of the alcohol.

$$\left[(C_2H_5)_2N-C(FCHCl)(F)-\ddot{O}-CH_2C_3H_7 \; :Nuc^-\right] \longrightarrow (C_2H_5)_2N-C(CHClF)=O \; F^- + NucCH_2C_3H_7\text{-}n$$

where :Nuc⁻ is Cl^-, Br^-, or I^- or some neutral species (*e.g.*, ROH). The best mechanism may be the preceding where :Nuc⁻ is F^-.

If the reaction of an alcohol and **FAR** is carried out in the presence of chloride ion, a chloro compound may be formed. Here :Nuc$^-$ is Cl$^-$ and

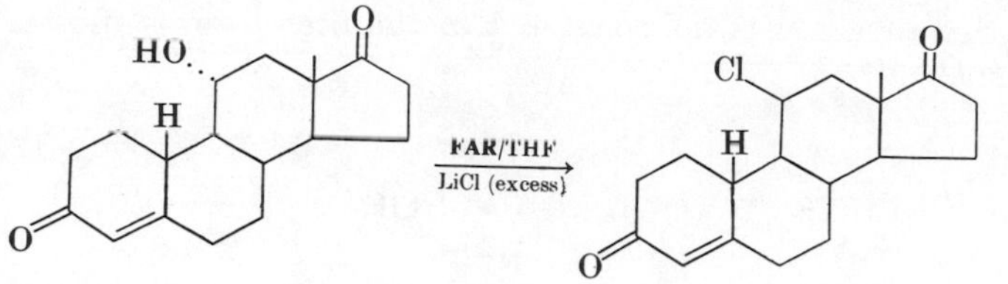

$$\xrightarrow[\text{LiCl (excess)}]{\text{FAR/THF}}$$

inversion occurs at carbon atom 11, the point of substitution.[95]

For many compounds an S_N1 mechanism with a carbonium ion intermediate (or incipient carbonium ion intermediate) is the only way to account for the product distribution. In the accompanying formulation

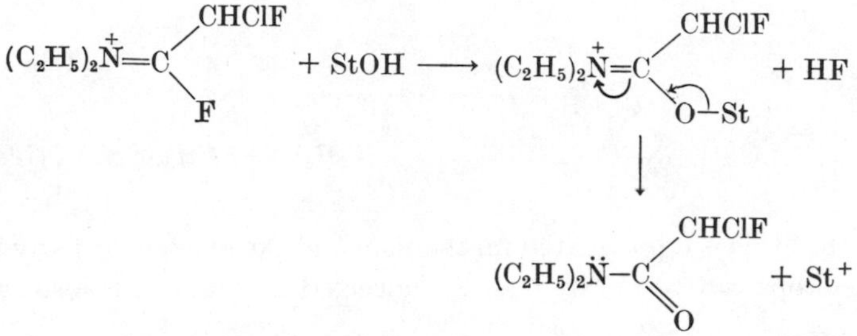

$$(C_2H_5)_2\overset{+}{N}{=}C(CHClF)F + StOH \longrightarrow (C_2H_5)_2\overset{+}{N}{=}C(CHClF){-}O{-}St + HF$$

$$\downarrow$$

$$(C_2H_5)_2\ddot{N}{-}C(CHClF){=}O + St^+$$

the symbol StOH means a hydroxyl-substituted steroid. A free carbonium ion is written, although a concerted reaction may occur with an incipient carbonium ion. An example is the reaction of 17α-acetoxy-3α-hydroxy-5(10)-estrene with **FAR**.[93] The product mixture obtained (Table XI) suggests that at least two carbonium ions are intermediates:

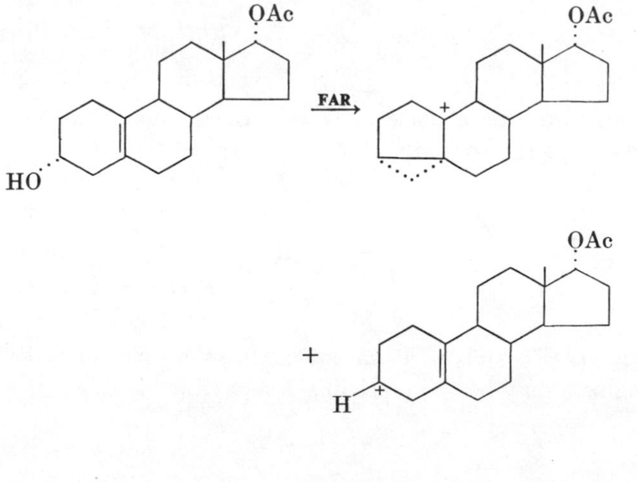

$$\xrightarrow{\text{FAR}}$$

+

The reaction of 3β,19-dihydroxyandrost-5-en-17-one 3-acetate with **FAR** under different conditions and with different workup procedures gave several products four of which were identified.[104] Each of the

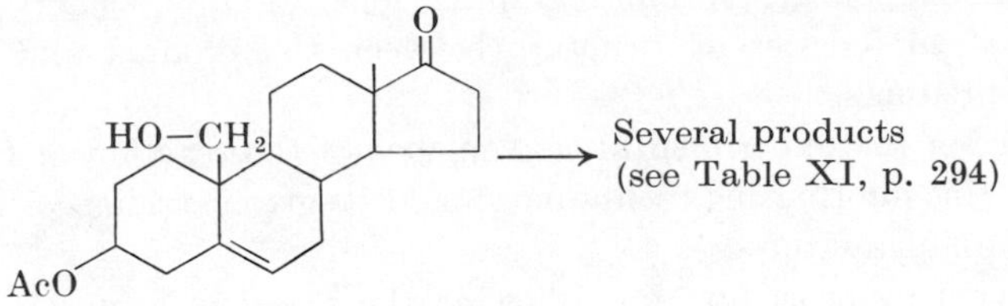

HO—CH_2 … AcO → Several products (see Table XI, p. 294)

identified products can be considered to be derived from one of the appropriate carbonium ion (or incipient carbonium ion) intermediates, **17a–e**.

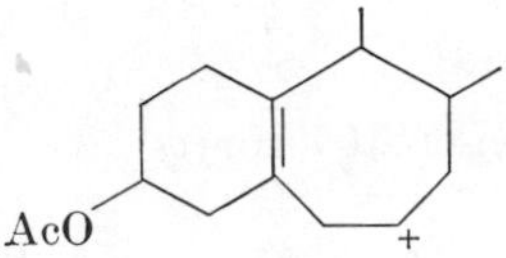

AcO **17a** AcO **17b** AcO **17c**

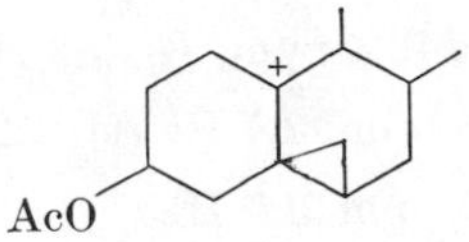

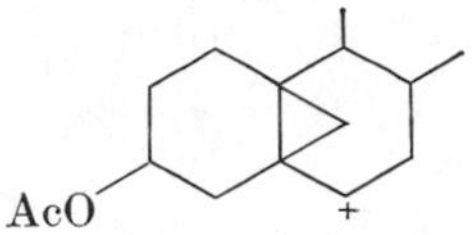

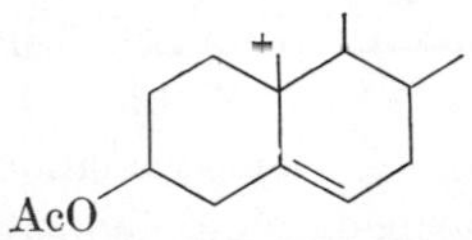

AcO **17d** AcO **17e**

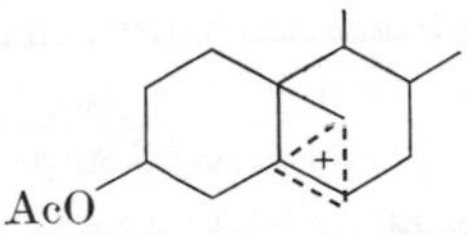

In reactions of **FAR** with steroidal alcohols, stereochemistry frequently determines the composition of products. Detailed analyses of such requirements in steroids have been published.[91, 105]

Experimental Considerations and Procedures

The fluorination of alcohols by **FAR** is one of the simplest, most convenient, and safest fluorination techniques. The alcohol is dissolved in a nonreacting solvent, typically diethyl ether, dichloromethane, or acetonitrile. At 0–25°, a 50–100% excess of **FAR** is added slowly enough to control a possible exothermic reaction. The reaction mixture is allowed to stand at 0–25° for 3–24 hours and worked up.

Over 90% of the fluorinations with **FAR** have been carried out as described above. Occasionally temperatures as low as −20° or as high as the boiling point of the reaction mixture have been used.

[104] L. H. Knox, E. Velarde, S. Berger, I. Delfin, R. Grezemkovsky, and A. D. Cross, *J. Org. Chem.*, **30**, 4160 (1965).

[105] M. Mousseron-Canet and J.-C. Lanet, *Bull. Soc. Chim. Fr.*, **1969**, 1745.

The major difficulty is usually encountered in the workup. When side reactions do not interfere, excess reagent is easily removed by distillation and the residual product purified by distillation or crystallization. Where side reactions occur, chromatographic separation is usually required because of similarities in structure between the product and the side-reaction products.

Because of the experimental convenience of the reaction of **FAR** with alcohols, this fluorinating technique should be given serious consideration when planning syntheses.

A convenient procedure for following the reaction is by ^{19}F nmr.[82, 83] The CF_2 resonance for **FAR** (see p. 160) disappears and the $-CHClF$ changes to that of the amide product. The nmr of $(C_2H_5)_2NCOCHClF$ was obtained in carbon tetrachloride solution using as internal references trichlorofluoromethane for ^{19}F and tetramethylsilane for ^1H.[82]

^{19}F CHClF δ_F − 142.4 ppm, J_{HF} 53 Hz (d)

^1H C_2H_5 −1.13 ppm, J 7 Hz (t); 1.21 ppm J 7 Hz (t)

−3.37 ppm, J 7 Hz (q)

CHClF −6.47 ppm, J_{HF} 50 Hz

3-Bromo-1-fluorobutane.[85] To a solution of 3-bromo-1-butanol (95 g, 0.62 mol) in 250 ml of absolute ether, 2-chloro-1,1,2-trifluorotriethylamine (120 g, 0.63 mol) in 25 ml of ether was added with stirring during 1 hour. The mixture, which warmed spontaneously and separated into two layers, was allowed to stand at room temperature for 15 hours, after which it became homogeneous. The solution was washed with 50 ml of water, with 20% aqueous potassium hydroxide until neutral, with water, dried (anhydrous $MgSO_4$), and the solvent removed by distillation through a 20-cm column filled with Raschig rings. The residue composed of 3-bromo-1-fluorobutane and the diethylamide of chlorofluoroacetic acid was distilled at 10 mm into a receiver cooled to −75°. There was obtained a liquid (85 g) boiling over the range 25–87°, and diethylamide of chlorofluoroacetic acid (95.5 g, 90%) distilling at 87–95° (10 mm), n^{20}D 1.4560; reported bp 105° (27 mm), n^{20}D 1.4490.

Distillation of the lower-boiling fraction through a 20-cm column gave 3-bromo-1-fluorobutane (70.4 g, 73.2%) distilling over the range 113–121° (742 mm). The pure compound distilled at 116.5° (742 mm), n^{20}D 1.4328. Only a single component was detected by gas-liquid chromatography.

Hexadecyl Fluoride (Cetyl Fluoride).[81] **FAR** (14.3 g, 0.075 mol) was added to cetyl alcohol (12.1 g, 0.050 mol) dissolved in 20 ml of dry methylene chloride at 0°. The reaction flask was stoppered and kept at

0° for 24 hours. The mixture was washed successively with cold 10% aqueous potassium carbonate and water, dried ($MgSO_4$), and distilled under reduced pressure. The first fraction was N,N-diethylchlorofluoroacetamide, bp 65–67° (0.5 mm). Cetyl fluoride distilled at 130° (0.1 mm); the yield was 8.5 g (70%).

Methyl 3-Ethylenedioxy-5α-hydroxy-6β,16α-difluoro-11-keto-*trans*-17(20)-pregnen-21-oate.[94] To a solution of methyl 3-ethylenedioxy-5α,16β-dihydroxy-6β-fluoro-11-keto-*trans*-17(20)-pregnen-21-oate (13.6 g, 30.1 mmol) in 550 ml of methylene chloride at ice-bath temperature was added 13.6 ml of **FAR**. After standing at 5° for 3.5 hours, the solution was washed with saturated sodium bicarbonate and water, dried over anhydrous sodium sulfate, and evaporated to dryness. The residual oil was dissolved in a minimum amount of ethylene dichloride and allowed to evaporate to dryness on the steam bath in a stream of air. The evaporation was repeated several times until crystals formed. (In some instances where crystallization failed at this stage, chromatography over Florisil gave crystalline product.) The crystalline residue was recrystallized from methanol to give two crops of product (12.8 g, 94%), of mp 204–216°. The analytical sample melted at 215–217°; $[\alpha]_D$ −24° (acetone); λ_{max}^{alc} 213 nm, (ε 13,750).

3β-Fluoroandrost-5-en-17-one.[90b] A mixture of 3β-hydroxyandrost-5-en-17-one (2.9 g, 10 mmol), **FAR** (2.85 g, 15 mmol), and dry methylene chloride (25 ml) was heated under reflux for 15 minutes and then evaporated under reduced pressure on the steam bath. The residue was absorbed from hexane containing a little benzene onto Florisil (90 g). The crystalline fractions eluted with hexane and hexane-ether (4:1) were combined (2.43 g) and recrystallized from hexane yielding 3β-fluoroandrost-5-en-17-one (1.67 g, 57.5%); mp 155–157°, raised to 157–158° by recrystallization from hexane; $[\alpha]_D$ −13°, γ_{max} 1743 cm^{-1} (17-ketone).

Further elution with diethyl ether afforded the ether (ROR) from the starting alcohol (0.27 g, 4.6%), mp 276–278°, raised to 278–280° by recrystallization from acetone; $[\alpha]_D$ −5°; γ_{max} 1745 and 1100 cm^{-1}.

MONOFLUOROSUBSTITUTION OF ORGANIC HALIDES, ESTERS, OR ALCOHOLS BY FLUORIDES

The fluorination of organic halides and similar derivatives was first reviewed in *Organic Reactions* by Henne[5] at a time when the majority of fluorinations were multisubstitutions of chloride by fluoride. Initially most monofluorosubstitutions consisted of halide replacement using silver(I) or mercury(II) fluorides.[106] Recent advances have been made in monofluorination, particularly in the development of new reagents for selective substitution of alcohol functions and in the use of aprotic solvent systems. Reviews on the preparation of monofluoroaliphatic derivatives have outlined these developments in comprehensive and critical fashion.[8,15] Specific surveys are available on the preparation of fluorosteroids,[4,9,10] fluoro sugars,[13,101] and monofluoroalkanoic acids.[7,11,12] Although this review emphasises recent work, the scope and details of the reaction will be covered thoroughly, Work published since 1967 is emphasized in the tabular entries, but some earlier references are included to provide a complete overview.

Monofluorosubstitution is fundamentally a simple reaction. A fluoride (metallic or nonmetallic) reacts with an organic halide, ester (usually sulfonate), or alcohol to substitute fluorine for the original group.

Halide: $RX + MF \rightarrow RF + MX$
(R = 1°, 2° organic groups;
X = Cl, Br, I;
M = alkali ions or R_4N^+)

Sulfonic ester: $ROSO_2R' + MF \rightarrow RF + MOSO_2R'$
(R = 1°, 2° organic groups;
$R' = p\text{-}CH_3C_6H_4, CH_3$;
M = alkali metal or R_4N^+)

Alcohol: $ROH + (C_6H_5)_xPF_{5-x} \rightarrow RF + (C_6H_5)_xPOF_{3-x} + HF$
(R = 1°, 2° organic groups; $x = 1\text{–}3$)

Availability and Preparation of Reagents

Listed below are the various fluorides that have been successfully used for monosubstitution on at least a few organic halides, sulfonate esters, or alcohols.

Alkali metals:	LiF, NaF, <u>KF</u>, <u>CsF</u>
Transition metals:	<u>AgF</u>, HgF_2, $AgBF_4$
Ammonium compounds:	NH_4F, $(CH_3)_4NF$, $(C_2H_5)_4NF$, $(n\text{-}C_4H_9)_4NF$
Hydrogen fluoride:	HF, NaF·HF, KF·HF
Groups IV, V, and VI:	SiF_4, Na_2SiF_6, AsF_3, $(C_6H_5)_3PF_2$, $(C_6H_5)_2PF_3$, $C_6H_5PF_4$, KSO_2F, SF_4

Two reagents not included in this review are sulfur tetrafluoride and sodium hexafluoroaluminate. Sulfur tetrafluoride monofluorinates certain

[106] A. M. Lovelace, D. A. Rausch, and W. Postelnek, *Aliphatic Fluorine Compounds*, ACS Monograph No. 138, Reinhold, New York, 1958.

halides and alcohols, as reviewed in the companion chapter.[6] Another extensive review includes substitution of chlorine by sodium hexafluoroaluminate.[107] This reagent (Na_3AlF_6) may have significant importance for industrial fluorinations, but it does not appear to be useful in the laboratory. Alkali metal-hydrogen fluoride complexes, such as KHF_2, are included in the following section on hydrogen fluoride.

The underlined compounds in the preceding list are the preferred fluorinating agents that have been most widely used. The one exception is tetra-*n*-butylammonium fluoride which has only recently been employed but appears to have excellent potential. The majority of the reagents can be obtained from commercial suppliers.[108] Thorough drying and pulverizing of commercial alkali fluorides is the primary prerequisite for a successful fluorination. For example, finely divided potassium fluoride is usually dried for 24 hours at 140° under reduced pressure (<1 mm).

Procedures for preparing most of the fluorinating reagents have been compiled.[109] Experimental details for preparation of tetra-*n*-butylammonium fluoride, triphenyldifluorophosphorane, diphenyltrifluorophosphorane, and phenyltetrafluorophosphorane follow.

Tetra-*n*-butylammonium Fluoride.[110] Aqueous hydrofluoric acid (10%) was added slowly to aqueous 0.15 *M* tetra-*n*-butylammonium hydroxide (100 ml) until the pH value began to fall. The reaction was completed by titration with 3% hydrofluoric acid to pH 7. The resulting light-brown solution was diluted with water to a final volume of 300 ml (*ca.* 0.5 M). When cooled in ice, fine crystals of clathrate $(n\text{-}C_4H_9)_4NF \cdot 32.8\ H_2O$ were deposited. The product was filtered, washed with cold water, and dried in air (yield 72 g; mp 23–25°). A second crop of clathrate (32 g) was obtained by concentration of mother liquors, dilution to about 0.5 *M*, and cooling to 0°. The clathrate could be stored in a polyethylene container at 0° for several weeks.

For exchange reactions, the calculated quantity of clathrate was heated to 50° at 15 mm in a round-bottomed flask attached to a rotary evaporator. The melted solid gave a viscous syrup as water was lost. Finally a colorless glass was dislodged from the walls of the flask with a glass rod to give soft white granules of a gel-like material. This material was stored over diphosphorus pentoxide under reduced pressure for 24 hours in a desiccator within a dry box. The resulting granules were

[107] B. Cornils, *Chem. Ztg. Chem. App.*, **91,** 629 (1967).

[108] M. Hudlicky, *Organic Fluorine Chemistry*, Plenum Press, New York, 1971, pp. 9–17.

[109] M. Hudlicky, *Chemistry of Organic Fluorine Compounds*, MacMillan, New York, 1962, pp. 49–63.

[110] P. W. Kent and R. C. Young, *Tetrahedron*, **27,** 4057 (1971).

pulverized with a glass rod and the product further desiccated over diphosphorus pentoxide to constant weight (4 days). The final product was extremely hygroscopic and was always manipulated in a dry box.

Difluorotriphenylphosphorane.[111,112] Distilled triphenylphosphine (60.0 g, 0.17 mol), sulfur tetrafluoride (53.8 g, 0.50 mol), and dry benzene (90 ml) were heated in a 100-ml autoclave at 150° for 10 hours with stirring. After the conclusion of reaction, the autoclave was cooled to room temperature and the valve was opened to discharge the unreacted sulfur tetrafluoride. The autoclave was again heated to 40–50° under reduced pressure (20–30 mm) to remove benzene completely. Pale-yellow crystals were removed from the autoclave under a dry nitrogen atmosphere.

The crystalline substance was washed in anhydrous diethyl ether, sublimed at 160–170° (10^{-4} mm), and recrystallized from dimethylformamide. Colorless prisms (63.4 g), mp 134–142°, were obtained. The yield based on triphenylphosphine was 92.5%. Although the substance was not completely pure, its purity was sufficient for fluorinating alcohols.

Trifluorodiphenylphosphorane.[113] (A) Diphenylphosphinic acid (80.0 g, 0.37 mol) and sulfur tetrafluoride (99.6 g, 0.82 mol) were heated at 150° for 12 hours in a 100-ml autoclave lined with Hastelloy C. After the reaction was complete, unchanged sulfur tetrafluoride was vented safely. Any dissolved gas was completely eliminated under reduced pressure, and the remaining dark-brown liquid was distilled to furnish a yellow liquid (81.8 g, 93.8%), bp 116–117° (3 mm).

(B) To chlorodiphenylphosphine (49.0 g, 0.22 mol), antimony(III) fluoride (56.0 g, 0.31 mol) was added in five portions with vigorous stirring. A slightly exothermic reaction occurred, but cooling was not necessary. After warming the reaction mixture at 60° for 5 hours, it was fractionated under reduced pressure in an atmosphere of nitrogen. Antimony(III) chloride distilled first at 120° (18 mm); then 22.8 g (43.4%) of diphenyltrifluorophosphorane distilled at 135° (4 mm).

Phenyltetrafluorophosphorane. (A)[114] A 1-l four-necked flask was equipped with a reflux condenser with a drying tube, a thermometer, a mechanical stirrer, and a hose-connected solid addition funnel. The reaction system was flushed with dry nitrogen. The funnel was charged

[111] W. C. Smith, *J. Amer. Chem. Soc.*, **82**, 6176 (1960).

[112] Y. Kobayashi and C. Akashi, *Chem. Pharm. Bull.* (Tokyo), **16**, 1009 (1968).

[113] (a) Y. Kobayashi, C. Akashi, and K. Morinaga, *Chem. Pharm. Bull.* (Tokyo), **16**, 1784 (1968); (b) Y. Kobayashi, I. Kumadaki, A. Ohsawa, and M. Honda, *Chem. Pharm. Bull.*, (Toyko), **21**, 867 (1973).

[114] R. Schmutzler, *Inorg. Chem.*, **3**, 410 (1964).

with antimony(III) fluoride (787 g, 4.4 mol), and phenyldichlorophosphine (537 g, 3.0 mol) was added to the flask. Antimony(III) fluoride was added in small portions with stirring at a rate sufficient to maintain an internal reactant temperature of 40–50° for the slightly exothermic reaction. Addition required 2 hours, after which time the black mixture was heated and stirred at 60° for 1 hour. Phenyltetrafluorophosphorane (520 g, 0.80 mol, 94%), was separated by distillation at 60–80° (60 mm); most of the product distilled at 65° (60 mm). Redistillation of the product through a 25-cm glass-helix packed column gave a colorless liquid, bp 134.5–136°, which fumed when exposed to the atmosphere.

(B)[111] A shaker vessel charged with phenylphosphonic acid (38.5 g, 0.205 mol) and sulfur tetrafluoride (108 g, 1.00 mol) was heated at 100° for 2 hours, at 120° for 4 hours, and finally at 150° for 10 hours. The volatile reaction products (*caution: thionyl fluoride, hydrogen fluoride, and unreacted sulfur tetrafluoride are highly toxic gases*) were vented safely from the reactor to an isolated area. The light-brown liquid product (41.8 g) was treated with 20 g of anhydrous granular sodium fluoride suspended in 100 ml of petroleum ether (bp 30–60°) to remove hydrogen fluoride. The insoluble solids were separated by filtration under a nitrogen atmosphere. The filtrate was distilled at atmospheric pressure through a Vigreux column to give phenyltetrafluorophosphorane (27.0 g, 0.145 mol, 58%), bp 133–134° (760 mm).

The substitution reactions of the listed fluorides are usually carried out in an aprotic solvent, although liquid organic compounds frequently serve as their own solvents. Most solvents are commercially available, but require *thorough removal of water*. Distillation of the solvent from a solid drying agent is the most commonly used procedure. The solvents most commonly used as media for fluorinations are acetonitrile, diethyl ether (ether), diethylene glycol (diglycol), 2,2′-dimethoxydiethyl ether (diglyme), dimethoxyethane (glyme), N,N-dimethylformamide (DMF), dimethyl sulfoxide (DMSO), dimethyl sulfone (DMS), ethylene glycol (glycol), formamide, hexamethylphosphortriamide (HMPT), N-methylpyrrolidone (NMP), N-methylacetamide (NMA) and tetramethylene sulfone (TMS).

Studies have shown that hexamethylphosphortriamide often leads to a high yield of side-reaction elimination products rather than monofluorination.[115]

Scope and Limitations

The usual classes of organic compounds that are fluorinated by the fluorides are summarized in Table V. The prominence of potassium fluoride

[115] J. F. Normant and H. Deshayes, *Bull. Soc. Chim. Fr.*, **1967**, 2455.

TABLE V. Common Classes of Compounds Undergoing Monofluorosubstitution at Carbon with a Metallic Fluoride or Related Compound

Class of Compound	Typical Fluorinating Reagent	Typical Solvent	Conditions and Other Comments
Primary aliphatic halides	KF	Diglycol, glycol, DMSO, DMF; often no solvent is used	100–200° as required
Secondary aliphatic halides	KF	Diglycol, DMSO, DMF, glycol	Poor yields common
	AgF	CH_3CN	30–60°, fair to good yields
	$AgBF_4$	Ether, toluene	0–36°, yields can be high
Sulfonic acid esters	KF	Diglycol, DMSO, glycol, no solvent	Low yields are common
	AgF	CH_3CN	30–60°
	$(n\text{-}C_4H_9)_4NF$	CH_3CN	30–60°, good yields
Alcohols	$(C_6H_5)_2PF_3$, $(C_6H_5)_3PF_2$	CH_3CN, CCl_4	150°, autoclave
Trimethylsilyl ethers	$C_6H_5PF_4$	No solvent	0° to room t, low to high yields

in Table V reflects the fact that it often represents the best compromise between cost and effectiveness among the alkali metal fluorides. For many substitutions, cesium fluoride gives superior yields, but the much higher cost precludes its use except when absolutely required. A cheaper alternative to potassium fluoride is sodium fluoride, but the lower cost is far outweighed by the much lower yields common with sodium fluoride.

The main side reaction is attack of a second mole of potassium fluoride on the fluoroalkane to eliminate hydrogen fluoride and form the corresponding alkene. This elimination is possible because alkali metal fluorides are powerful bases toward hydrogen fluoride.

$$KF + HF = K^+HF_2^-$$

$$\text{H}\!-\!\overset{|}{\text{C}}\!-\!\overset{|}{\text{C}}\!-\!\text{Cl} + 2\,KF \longrightarrow \text{C}\!=\!\text{C} + KHF_2 + KCl$$

Silver(I) fluoride is not as strong a base as potassium fluoride. Consequently silver(I) fluoride frequently succeeds as a fluorinating agent whereas potassium fluoride gives elimination. Additionally, silver(I) fluoride may react by a different mechanism (see p. 188) because of the covalent bond in silver chloride, silver bromide, and silver iodide. (The solubility of silver(I) fluoride in water implies the presence of an ionic bond, Ag^+F^-.)

Substitution by Fluoride

Primary Halides. The usual reagent for the substitution is potassium fluoride in a solvent such as diglycol, dimethyl sulfoxide, or glycol, although cesium fluoride usually gives better yields.*

Elimination of hydrogen halide is the predominant side reaction because fluoride ion is a strong base in aprotic systems.[116, 117] If elimination by potassium fluoride is a problem in a fluorination, then silver(I) fluoride or silver tetrafluoroborate could be substituted. The cost of silver fluoride is high, but more discouraging is the fact that two moles of reagent are required for each halogen substituted. The reason for this is the stability of mixed halogen silver salts.

$$RX + 2\,AgF \rightarrow RF + AgF{\cdot}AgX$$

The yields in the following examples of halide substitution are not as high

* Note added in proof: Potassium fluoride complexed with 1,4,7,10,13,16-hexaoxacyclooctadecane (18-Crown-6 ether) in acetonitrile or benzene, has recently been described as a source of "naked" fluoride which is an effective nucleophile for replacement of aliphatic halides in synthesis of aliphatic fluorides. It is also a potent base for promoting elimination to olefins. C. L. Liotta and H. P. Harris, *J. Amer. Chem. Soc.*, **96**, 2250 (1974).

[116] J. Hayami, N. Ono, and A. Kaji, *Tetrahedron Lett.*, **1968**, 1385.

[117] J. Hayami, N. Ono, and A. Kaji, *Nippon Kagaku Zasshi*, **92**, 87 (1971) [*C.A.*, **76**, 24604v (1972)].

as are sometimes found (see Table XII) but are representative.

$$CH_2{=}CH(CH_2)_8CH_2Br \xrightarrow[180°]{KF,\ Diglycol} CH_2{=}CH(CH_2)_8CH_2F \quad (46\%) \qquad \text{(Ref. 118)}$$

$$ClCH_2(CH_2)_{16}CH_2Cl \xrightarrow[175°]{KF,\ Diglycol} FCH_2(CH_2)_{16}CH_2F \quad (40\%) \qquad \text{(Ref. 118)}$$

CH_3O, CH_3, O, O, N, C_2H_5, $CH_2OCONHCH_3$, CH_2Br $\xrightarrow[2\ hr,\ reflux]{AgF,\ CH_3CN}$ $CH_2OCONHCH_3$, CH_2F (30%) (Ref. 119)

The reaction of 4-chlorobutyryl chloride with potassium fluoride is an unusual elimination that provides a convenient entry into the cyclopropane system.[120] Unfortunately this ring-closure reaction is not general.

$$ClCH_2CH_2CH_2COCl \xrightarrow[reflux]{KF,TMS} \triangleright\text{—}COF \quad (70\%)$$

No products were identified from reaction of 5-chloropentanoyl chloride with potassium fluoride. With 6-bromohexanoyl chloride the 6-fluorohexanoyl fluoride and 5-hexenoyl fluoride were formed; however, a small yield of cyclopentanecarbonyl fluoride was obtained.[120]

Secondary Halides. Attempts to substitute fluoride for a secondary halide are frequently unsuccessful and the elimination side reaction often dominates. For the special case of α-halo-esters, -amides, -nitriles, and similar α-halo-substituted compounds, substitution usually occurs in good yield with either potassium fluoride or silver(I) fluoride in a suitable solvent.[121]

$$CH_3(CH_2)_4CHBrCONHC_6H_5 \xrightarrow[180°,\ 2\ hr]{KF,\ Diglycol} CH_3(CH_2)_4CHFCONHC_6H_5 \quad (61\%)$$

Substitutions into secondary ring positions of steroids have been reported with silver(I) fluoride. Retention of configuration has been reported for some substitutions, probably because of steric control.[122]

[118] F. L. M. Pattison and J. J. Norman, *J. Amer. Chem. Soc.*, **78,** 2311 (1957).

[119] G. R. Allen, Jr., J. F. Poletto, and M. J. Weiss, *J. Med. Chem.*, **10,** 14 (1967).

[120] R. E. A. Dear and E. E. Gilbert, *J. Org. Chem.*, **33,** 1690 (1968).

[121] I. Shahak and E. D. Bergmann, *J. Chem. Soc., C*, **1967,** 319.

[122] W. T. Moreland, D. P. Cameron, R. G. Berg, and C. E. Maxwell, *J. Amer. Chem. Soc.*, **84,** 2966 (1962).

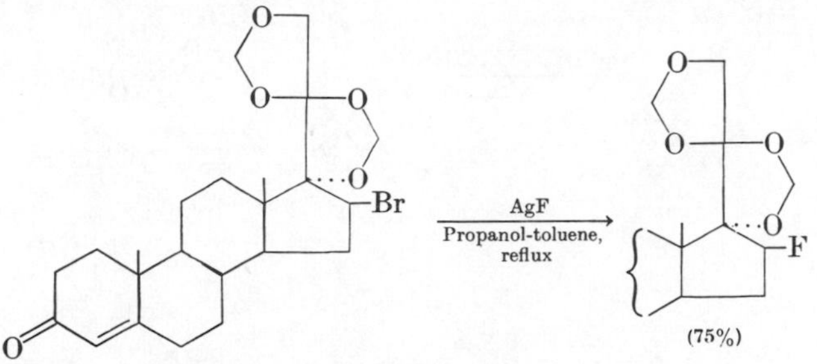

O
O O
...O
Br
AgF
Propanol-toluene,
reflux
O
O O
...O
F
O
(75%)

Substitution of secondary halogen has been accomplished more frequently in carbohydrates. The fluorinating agents of choice have been silver(I) fluoride and silver tetrafluoroborate.[123]

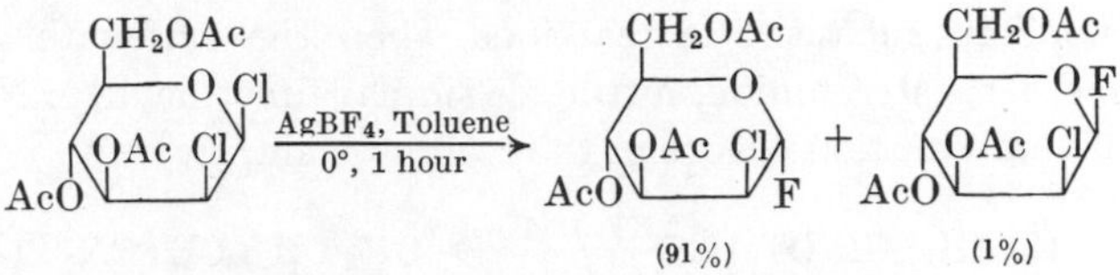

CH_2OAc O Cl OAc Cl AcO $\xrightarrow[\text{0°, 1 hour}]{AgBF_4\text{, Toluene}}$ CH_2OAc O OAc Cl AcO F (91%) + CH_2OAc O F OAc Cl AcO (1%)

Interesting data on relative rates of substitution of chlorine by silver(I) fluoride have been observed for some carbohydrate systems.[124] Again note that elimination reactions are expected to accompany any attempted fluorinations of secondary halogens in organic compounds. This is emphasized by recent studies of elimination reactions with tetra-*n*-butylammonium fluoride.[125] In N,N-dimethylformamide, tetra-*n*-butylammonium fluoride gave butenes from 2-iodobutane in 77% yield and from 2-bromobutane and 2-chlorobutane in 37% and 2% yields, respectively. This elimination reaction can be contrasted with the successful fluorinations of carbohydrates by tetra-*n*-butylammonium fluoride in acetonitrile (see Table XII). Clearly, the nature of tetra-*n*-butylammonium fluoride as a fluorination or elimination reagent is not understood. Because the reagent is very hygroscopic,[110] we suggest that the variability of the results is a function of the amount of water absorbed by the reagent.

Sulfonic Acid Esters. Primary sulfonic acid esters, usually as the *p*-toluenesulfonate of an alcohol, readily undergo substitution by inorganic fluorides. Again, potassium fluoride and silver(I) fluoride are the most frequently used reagents; however, recent work has employed cesium fluoride[126] and tetra-*n*-butylammonium fluoride.[127]

[123] K. Igarashi, T. Honma, and J. Irisawa, *Carbohyd. Res.*, **11**, 577 (1969); **13**, 49 (1970).
[124] L. D. Hall and J. F. Manville, *Can. J. Chem.*, **47**, 379 (1969).
[125] R. A. Bartsch, *J. Org. Chem.*, **35**, 1023 (1970).
[126] J. H. Bateson and B. E. Cross, *Chem. Commun.*, **1972**, 649.
[127] A. B. Foster, R. Hems, and J. M. Webber, *Carbohyd. Res.*, **5**, 292 (1967).

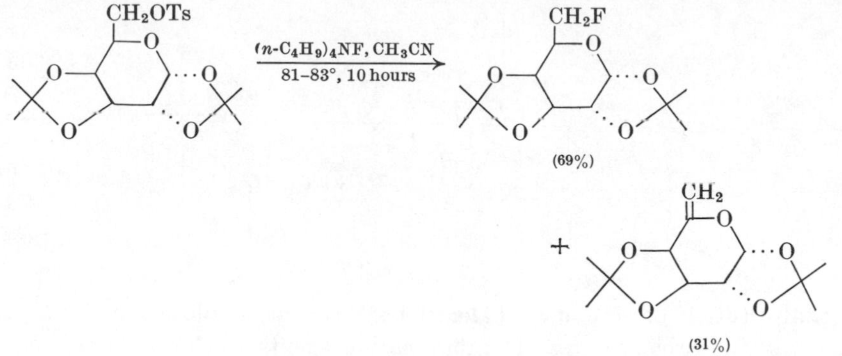

CH_2OTs $\xrightarrow[81-83°,\ 10\ hours]{(n\text{-}C_4H_9)_4NF,\ CH_3CN}$ CH_2F (69%) + CH_2 (31%)

Sulfonate esters of secondary alcohols, including sulfonates of carbohydrates, undergo substitution reactions when the sulfonate group is alpha to an ether, ester, amide, nitrile, or similar functionality. Note that inversion of configuration occurs in the second example.

$$CH_3(CH_2)_2CH(OTs)CN \xrightarrow[Distil]{KF,\ Diglycol} CH_3(CH_2)_2CHFCN \quad \text{(Ref. 128)}$$

(49%)

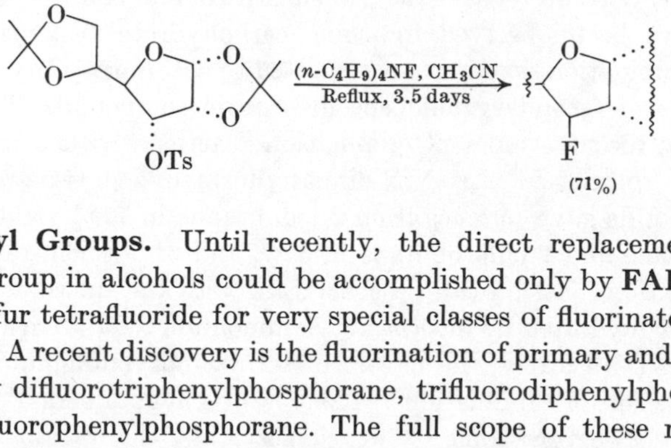

OTs $\xrightarrow[Reflux,\ 3.5\ days]{(n\text{-}C_4H_9)_4NF,\ CH_3CN}$ F (71%) (Ref. 127)

Hydroxyl Groups. Until recently, the direct replacement of the hydroxyl group in alcohols could be accomplished only by **FAR** (p. 158), and by sulfur tetrafluoride for very special classes of fluorinated tertiary alcohols.*[,6] A recent discovery is the fluorination of primary and secondary alcohols by difluorotriphenylphosphorane, trifluorodiphenylphosphorane, and tetrafluorophenylphosphorane. The full scope of these reagents is currently under evaluation. For example, direct reaction of phenyltetrafluorophosphorane with alcohols led to extensive elimination and formation of isomeric fluorides in contrast to diphenyltrifluorophosphorane.[113b]

[128] E. D. Bergmann and I. Shahak, *Bull. Res. Counc. Isr.*, **10A**, 91 (1961).

* Note added in proof: G. A. Olah, M. Nojima and I. Kerekes, *J. Amer. Chem. Soc.*, **96**, 925 (1974) use selenium tetrafluoride complexed with pyridine in 1,1,2-trichlorotrifluoroethane solvent at 0° to replace hydroxyl by fluoride in aliphatic alcohols. W. J. Middleton reported at 2nd Winter Fluorine Conference, February, 1974, that diethylaminosulfur trifluoride (prepared from sulfur tetrafluoride and diethylaminotrimethylsilane) in trichlorofluoromethane at −78° is also a general reagent for high yield conversion of alcohols to alkyl fluorides (see also L. N. Markovskij, V. E. Pashinnik and A. V. Kirsanov, *Synthesis*, **1973**, 787).

$$n\text{-}C_8H_{17}OH \xrightarrow[\text{Autoclave, 150°}]{(C_6H_5)_2PF_3,\ CH_3CN} n\text{-}C_8H_{17}F \quad \text{(Ref. 113a)}$$

(76%)
[78% with $(C_6H_5)_3PF_2$]

$$ClCH_2CH_2CH_2OH \xrightarrow[\text{Autoclave, 150°}]{(C_6H_5)_3PF_2,\ CH_3CN} ClCH_2CH_2CH_2F \quad \text{(Ref. 113a)}$$

(64%)
[64% with $(C_6H_5)_2PF_3$]

In very recent work, phenyltetrafluorophosphorane was used in a stepwise process to fluorinate alcohols in higher yields. First the alcohol is converted by trimethylchlorosilane to a silyl ether and the ether is then fluorinated.[129, 130] Elimination is a common side reaction; for example, a

$$n\text{-}C_5H_{11}OH + (CH_3)_3SiCl \rightarrow n\text{-}C_5H_{11}OSi(CH_3)_3 + HCl$$

$$n\text{-}C_5H_{11}OSi(CH_3)_3 + C_6H_5PF_4 \rightarrow [n\text{-}C_5H_{11}OPF_3C_6H_5] + (CH_3)_3SiF$$

$$[n\text{-}C_5H_{11}OPF_3C_6H_5] \rightarrow \underset{(50\%)}{n\text{-}C_5H_{11}F} + C_6H_5P(O)F_2 + \underset{(50\%)}{n\text{-}C_3H_7CH{=}CH_2} \quad \text{(Ref. 129)}$$

50% yield of 1-pentene is formed along with the 50% yield of fluoropentane;[130] 2- and 3-pentanol both gave 60% fluorination and 40% elimination.

This new procedure for fluorinating alcohols has been reported by two different groups who disagree on the amount of by-products formed.[129, 130] Recently, the reaction was evaluated much more extensively.[129b,c] Yields of fluoroalkyl products ranged from 15 to 95% for primary and secondary alcohols, and were usually higher, sometimes nearly quantitative, for tertiary alcohols and primary or secondary alcohols β-substituted with electron-attracting groups. However, alcohols substituted with very strong electronegative groups gave only a stable alkoxyfluorophosphorane. Rearranged fluorides are found if the carbonium ion produced from the alcohol is prone to rearrange. Also, alkenes, ethers, and phenylfluorophosphonates are common side products.

Examples using the arylfluorophosphoranes are given in Tables XIII and XIV, but some more extensive recent work[113b, 129b,c] was received too late for inclusion in the Tables.

Aromatic Halides. Aprotic solvents at 200–500° have been used extensively for alkali metal fluoride replacements with activated aromatic halides[131] and polychlorinated aliphatic compounds.[132] Two examples are cited to illustrate these replacements but, since they have not been applied to the synthesis of monofluoroaliphatic compounds, they are not discussed further.

[129] (a) D. U. Robert and J. G. Riess, *Tetrahedron Lett.*, **1972,** 847; (b) D. U. Robert, G. N. Flatau, A. Cambon, and J. G. Riess, *Tetrahedron*, **29,** 1877 (1973); (c) R. Guedj, R. Nabet and T. Wade, *Tetrahedron Lett.*, 907 (1973).

[130] H. Koop and R. Schmutzler, *J. Fluorine Chem.*, **1,** 252, (1972).

[131] Ref. 15, Sheppard and Sharts, p. 90.

[132] J. T. Maynard, *J. Org. Chem.*, **28,** 112 (1963).

2-Chloropyridine $\xrightarrow[200\text{–}210°, 21 \text{ days}]{KF, DMSO}$ 2-Fluoropyridine (50%) (Ref. 133)

Pentachloropyridine $\xrightarrow[400\text{–}500°]{KF}$ Pentafluoropyridine (Ref. 134)

Replacement in aromatic halides has its greatest value for preparing polyfluoroaromatics. For introducing one fluorine atom on an aromatic ring, the Balz-Schiemann[8, 135, 136] reaction is usually the method of choice, particularly the use of the hexafluorophosphate salt.[137, 138]

$R-C_6H_4-NH_2$ $\xrightarrow[2.\ NaBF_4]{1.\ HONO}$ $R-C_6H_4-N_2^+BF_4^-$ $\xrightarrow{\text{Heat in sand}}$ $R-C_6H_4-F$

Miscellaneous Reagents. Arsenic trifluoride, silicon tetrafluoride, and sodium hexafluorosilicate do not have any significant use as laboratory reagents for monofluorinations.[139] Arsenic trifluoride is an effective fluorinating reagent only on perhalo compounds. The silicon fluorides have limited potential for commercial use at high temperatures.

The conversion of 1-methoxy- and 1-methoxy-4-methylbicyclo[2.2.2]-octane to 1-fluoro- and 1-fluoro-4-methylbicyclo[2.2.2]octane by acetyl fluoride is an interesting substitution but does not appear to be generally useful.[140]

1-OCH_3-4-R-bicyclo[2.2.2]octane $\xrightarrow[BF_3 \cdot H_3PO_4]{CH_3COF}$ 1-F-4-R-bicyclo[2.2.2]octane

R = H, CH_3 R = H, 70%; CH_3, 74%

[133] G. C. Finger, L. D. Starr, D. R. Dickerson, H. S. Gutowsky, and J. Hamer, *J. Org. Chem.*, **28,** 1666 (1963).

[134] R. D. Chambers, J. Hutchinson, and W. K. R. Musgrave, *Proc. Chem. Soc.*, **1964,** 83.

[135] (a) A. Roe, *Org. Reactions*, **5,** 193 (1949); (b) H. Suschetzky, *Advan. Fluorine Chem.*, **4,** 1 (196).

[136] A. E. Pavlath and A. J. Leffler, *Aromatic Fluorine Compounds*, ACS Monograph No. 155, Reinhold, New York, 1962, pp. 12–16, 42–45.

[137] Ref. 15, Sheppard and Sharts, p. 168.

[138] K. G. Rutherford and W. Redmond, *Org. Syn.*, **43,** 12 (1960).

[139] Ref. 15, Sheppard and Sharts, p. 163.

[140] Z. Suzuki and K. Morita, *Bull. Chem. Soc. Jap.*, **41,** 1724 (1968).

A nitro group in polynitromethanes is displaced by fluoride ion.[141]

$$C(NO_2)_4 \rightarrow FC(NO_2)_3$$

KF/DMF 30°, 75 min, 57%
RbF/DMF 30°, 60 min, 59%
CsF/DMF 30°, 60 min, 39%
CsF/acetone 50°, 2.3 hr, 50%
CsF/acetone 50°, 8 hr, 49%

$$ClC(NO_2)_3 \xrightarrow[30°,\ 8\ hr]{CsF,\ DMF} FClC(NO_2)_2 \ (41\%)$$

$$FC(NO_2)_3 \xrightarrow[30°,\ 1.75\ hr]{CsF,\ DMF} F_2C(NO_2)_2 \ (47\%)$$

Unfortunately this interesting monofluorosubstitution reaction has limited utility in synthetic schemes.

The Arbusov reaction has been used for preparation of halogenated carbohydrates[13] and has been extended to prepare a fluorosugar.[142]

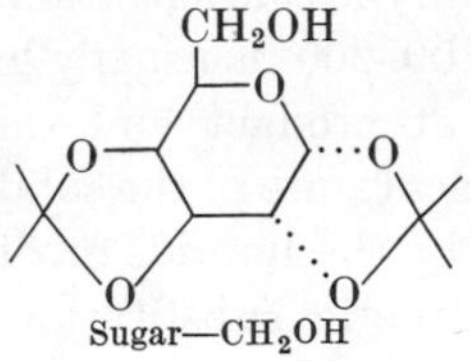

CH_2OH

Sugar—CH_2OH $\xrightarrow{(n\text{-}C_3H_7)_2PCl}$ sugar-$CH_2OP(C_3H_7\text{-}n)_2$

$$\xrightarrow[140°,\ 7\ hr]{CF_2{=}CFCF_3} \text{sugar-}CH_2F + (n\text{-}C_3H_7)_2P(O)CF{=}CFCF_3$$

(60%)

6-Deoxy-6-fluoro-1,2:3,4-di-O-isopropylidenegalactose was also prepared in 19% yield from reaction of 1,2:3,4-di-O-isopropylidenegalactose-6-(N,N-diethyl-P-methyl-phosphonamidate) and ethyl fluoroacetate.[143]

Indirect replacement of hydroxyl groups is accomplished by decomposition of fluoroformates and fluorosulfinates.[144] Fluoroformates can be prepared directly from carbonyl fluoride,[145] although preparation from chloroformates by halogen exchange was used in the original fluoroformate decomposition studies.[146] Decomposition of fluorosulfites was recently developed for synthesis of a series of alkyl fluorides in yields of 40–75% (R = CH_3, C_2H_5, *n*- and *iso*-C_3H_7, *sec*- and *iso*-C_4H_9).[147]

141 G. Kh. Khisamutdinov, V. I. Slovetskii, M. Sh. L'vova, O. G. Usyshkin, M. A. Besprozvannyi, and A. A. Fainzil'berg, *Bull. Acad. Sci. USSR, Div. Chem. Sci., Engl. Transl.*, **1970**, 2397.

142 K. A. Petrov, E. E. Nifant'ev, A. A. Shchegolev, and V. G. Terekhov, *J. Gen. Chem. (USSR), Engl. Transl.*, **34**, 1463 (1964).

143 E. E. Nifant'ev, I. N. Sorochkin, and A. P. Tuseev, *J. Gen. Chem. (USSR), Engl. Transl.*, **35**, 2248 (1965).

144 Ref. 15, Sheppard and Sharts, p. 165.

145 P. E. Aldrich and W. A. Sheppard, *J. Org. Chem.*, **29**, 11 (1964).

146 S. Nakanishi, T. C. Myers, and E. V. Jensen, *J. Amer. Chem. Soc.*, **77**, 3099, 5033 (1955).

147 F. Seel, J. Boudier, and W. Gombler, *Chem. Ber.*, **102**, 443 (1969).

$$\mathrm{ROCOCl} \xrightarrow{\mathrm{MF}} \mathrm{ROCOF} \xrightarrow[\text{Pyridine or } BF_3]{\text{Heat}} \mathrm{RF} + CO_2$$

$$\mathrm{ROH} \xrightarrow{COF_2} \mathrm{ROCOF}$$

$$\mathrm{ROH} \xrightarrow{SOF_2} \mathrm{ROSOF}$$

$$\mathrm{ROSOCl} \xrightarrow{KF(SO_2)} \mathrm{ROSOF} \xrightarrow{80^\circ} \mathrm{RF} + SO_2$$

Mechanism

No definitive mechanistic studies have been reported for the synthetic substitution reactions presented in this section. A major difficulty is that the reactions are often heterogeneous and the concentration of fluoride ion in solution at reaction temperatures of 100–200° is usually low and not readily determined. Furthermore, yields of product and side-reaction products are usually dependent on the physical state of the solid fluorinating agent and on traces of water absorbed by the fluoride or dissolved in the solvent. Reaction conditions for synthetic substitution reactions under discussion here are simply not defined sufficiently for mechanism studies.

The lack of kinetic data on the forward reaction of fluorination does not mean that the forward fluorination reaction is not understood. The reverse reaction, namely, the displacement of fluorine by halogen in organic compounds, has been intensively studied. A thorough review is available.[148] As might be expected, an S_N2 reaction occurs for most primary and secondary fluorides. Where conditions are adjusted to favor carbonium ions, the S_N1 mechanism prevails.

From the accumulated data on reaction conditions, most substitution fluorinations by potassium fluoride, cesium fluoride, and tetra-*n*-butylammonium fluoride must occur by an S_N2 mechanism. Where stereochemistry has been observed, inversion of configuration is usually found.

Tetra-O-acetyl glycopyranosyl bromide (CH_2OAc, OAc, AcO, Br, OAc) $\xrightarrow[\text{Room t, overnight}]{AgF,\ CH_3CN}$ tetra-O-acetyl glycopyranosyl fluoride (CH_2OAc, OAc, AcO, F, OAc) (80%) (Ref. 149)

[148] R. E. Parker, *Advan. Fluorine Chem.*, **3**, 63 (1963).

[149] L. D. Hall, J. F. Manville, and N. S. Bhacca, *Can. J. Chem.*, **47**, 1 (1969).

Retention of configuration has been observed for substitutions in steroids. For this case an S_N2 mechanism is not consistent with the stereochemistry (see example on p. 181), but steric factors are often controlling in steroid reactions and lead to retention of configuration. Furthermore, the initially formed S_N2 product frequently epimerizes to the more stable stereochemical isomer; this corresponds to a product of frontside substitution.

The greater fluorinating power of cesium fluoride and potassium fluoride is due to the large ion size, which increases cation solubility. The effectiveness of $(n\text{-}C_4H_9)_4NF$ is due to both the large ion size and the organic groups. The large ions bring unsolvated fluoride ion into aprotic solvents through ion-pair formation. Unsolvated fluoride ion can then attack carbon without the additional energy required to shed a solvent sheath.

$$CsF + x(\text{solvent}) \rightarrow [Cs(\text{solvent})_x]^+F^-$$

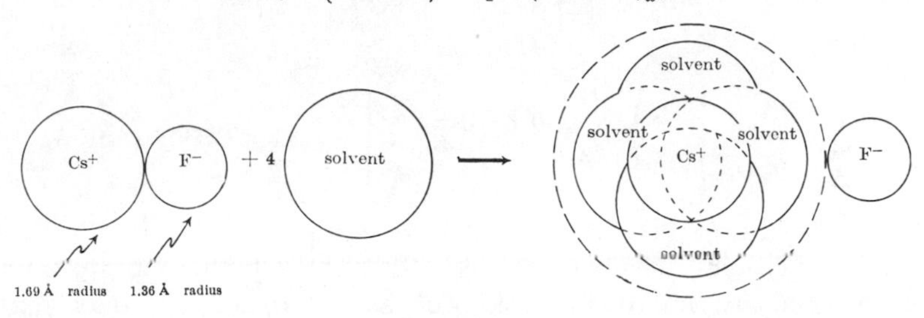

The concept that "naked" fluoride ion is responsible for fluorination is substantiated by the effect of water on the fluorinating ability of alkali metallic fluorides. When fluoride ion is hydrated, the energy is −123 kcal/mol. For chloride ion the value is only −89 kcal/mol. Hence for substitution of chloride by fluoride the heat of hydration of the ions opposes the reaction by 34 kcal and entropy is unfavorable by −15.5 gibb/mole.

$$RCl + [F(H_2O)_{x+y}]^- \rightarrow RF + [Cl(H_2O)_x]^- + y\,H_2O$$

To explain retention of configuration in substitution of tosylate by fluoride in a sugar, the accompanying mechanism has been proposed.[150] The expected inversion is observed in other cases.[151]

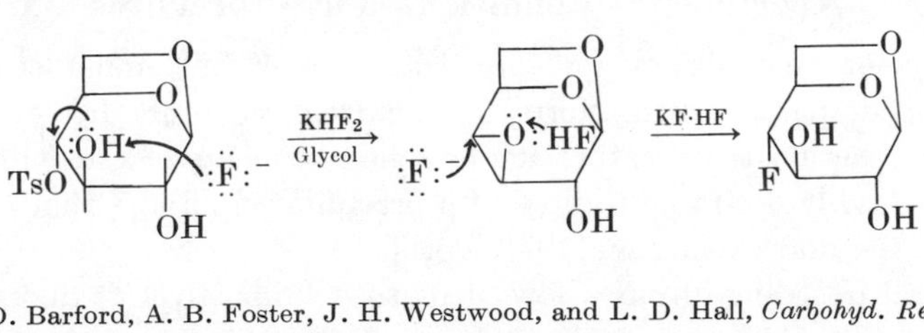

$\xrightarrow[\text{Glycol}]{KHF_2}$ $\xrightarrow{KF\cdot HF}$

150 A. D. Barford, A. B. Foster, J. H. Westwood, and L. D. Hall, *Carbohyd. Res.*, **11**, 287 (1969).
151 L. D. Hall and P. R. Steiner, *Can. J. Chem.*, **48**, 451 (1970).

The effectiveness of silver fluoride toward secondary halides suggests that something more than a simple S_N2 mechanism is involved. Two facts are pertinent: Ag^+ forms complex ions such as $AgCl_2^-$; silver(I) fluoride forms 1:1 complexes with silver chloride, bromide, and iodide. A reasonable postulate is that silver(I) fluoride operates by a push-pull mechanism either in solution or at the solid interface. However, these remain suppositions until more experimental evidence is provided.

R(H)(R')C–Cl - - → Ag^+F^- (attacked by Ag^+F^-) ⟶ [R(H)(R')C···$Cl^{\delta-}$ - - - Ag^+F^- ; $F^{\delta-}$ ··· Ag^+] ⟶

[R(H)(R')C–F Ag^+ $(Cl–Ag–F)^-$] ⟶ RR′CHF + (AgF·AgCl)

The fluorination of alcohols (or silyl ethers) by arylfluorophosphoranes must proceed by initial formation of an intermediate monoalkoxyfluorophosphorane.[113b, 129b] This alkoxyphosphorane can decompose to alkyl fluoride by an intra-molecular fluoride transfer from the phosphorus or may dissociate to a carbonium ion intermediate. A carbonium ion can be attacked by fluoride or some other nucleophile (ether by-product), rearrange, or eliminate a proton (olefin by-product). The course of the reaction is dependent both on the stereochemistry of the intermediate alkoxyphosphorane and the nature of the alcohol. The stereochemistry of the intermediate phosphorane, which is related to the electronegativity of the substituents, must influence the fluorination reaction.[113b]

Experimental Conditions and Procedures

Experimental conditions for the class of reactions under study are among the simplest known. Normally the *dried* reactants are mixed and heated to reaction temperature under reflux or heated in an autoclave. Stirring is highly desirable to break up precipitated halides that coat the surface of the fluorinating metallic fluoride.

Temperatures range from a low of 0° to 60° for silver(I) fluoride and silver tetrafluoroborate reactions, up to 100–200° for most potassium fluoride reactions. The primary experimental precautions are that the fluorinating reagent must be pulverized finely and in most cases all water must be eliminated.

The major hazard is, of course, leakage from a pressurized autoclave upon rupture. Because all inorganic fluorides are toxic, precautions against ingestion or absorption are required. Monofluorinated organic compounds may be toxic. *All products should be handled as if they are toxic until proved otherwise.*

Ethyl 7-Fluoroheptanoate.[152] (A) A mixture of ethyl 7-bromoheptanoate (35 g, 0.15 mol), ethylene glycol (110 g), and anhydrous potassium fluoride (13.55 g, 0.23 mol) was placed in a 250-ml round-bottomed flask equipped with a mercury-sealed stirrer and reflux condenser. The mixture was heated at 130° for 8 hours with vigorous stirring. The reaction mixture was cooled, diluted with 3 volumes of water, and thoroughly extracted with ether. The extract was dried over magnesium sulfate and ether was removed. Fractionation yielded ethyl 7-fluoroheptanoate (10.5 g, 40.3%), bp 39–40° (0.1 mm), n^{25}_{D} 1.4111.

(B) To ethyl 7-bromoheptanoate (11.85 g, 0.05 mol), contained in a 50-ml round-bottomed flask equipped with a reflux condenser and calcium chloride tube, was added pure, dry silver(I) fluoride (13.5 g, 0.11 mol). The mixture was shaken at room temperature for a few minutes and then heated in a water bath at 60° with constant shaking. A yellow precipitate of silver bromide rapidly formed. The temperature was raised slowly to 80° to complete the reaction. The mixture was cooled in an ice bath, diluted with 20 ml of ether, and filtered. The silver residue was washed thoroughly with three additional portions of ether. The combined extracts were washed with water and dried over sodium carbonate. After removal of the ether, fractionation yielded ethyl 7-fluoroheptanoate (3.0 g, 35%).

α-Fluoropentanoyl-N,N-diethylamide.[121] A stirred mixture of α-bromovaleroyl-N,N-diethylamide (22 g, 0.1 mol), dry potassium fluoride (15 g, 0.25 mol), and bis-(2-hydroxyethyl) ether (100 ml) was heated rapidly to 125–130° and kept at this temperature for 2 hours. The product was cooled, diluted with water (200 ml), and extracted with three 150-ml portions of benzene. The benzene was washed with 5% aqueous sodium hydrogen carbonate (100 ml), dried, and concentrated. The α-fluoropentanoyl-N,N-diethylamide which distilled at 132–134° (24 mm) amounted to 11 g (69%).

[152] F. L. M. Pattison, S. B. D. Hunt, and J. B. Stothers, *J. Org. Chem.*, **21,** 883 (1956).

3,4,6-Tri-O-acetyl-2-chloro-2-deoxy-α-D-glucopyranosyl Fluoride (A) and 3,4,6-Tri-O-acetyl-2-chloro-2-deoxy-β-D-glucopyranosyl Fluoride (B).[123] To a solution of silver tetrafluoroborate (136 mg, 0.700 mmol) in anhydrous ether (1.6 ml) cooled to 0° was added at once 3,4,6-tri-O-acetyl-2-chloro-2-deoxy-α-D-glucopyranosyl chloride (200 mg, 0.575 mmol) with stirring. The mixture was stirred for 15 minutes and poured into ice-cold, 10% aqueous sodium bicarbonate. The organic layer was separated and the water layer extracted with dichloromethane. The combined ether and dichloromethane extracts were washed with water, dried with sodium sulfate, and evaporated. The product was fractionated by preparative thin-layer chromatography. From the faster-moving zone (R_f 0.68), 115 mg (0.348 mmol, 60.5%) of **A**, mp 98.5–99.5°, $[\alpha]_D^{21}$ $+152.3 \pm 1.8$ (c 1.064 chloroform) [lit. mp 95–97°, $[\alpha]_D^{25}$ +203° (c 1.13, chloroform)],[44a] was obtained as colorless leaflets after recrystallization from ether and petroleum ether (bp 30–50°). From the slower-moving zone (R_f 0.63), 33.7 mg (0.102 mmol, 17.7%) of **B**, mp 79.5–80.5° $[\alpha]_D^{21}$ $+92.3 \pm 1.3°$ (c 0.997, chloroform [lit. mp 77–78°, $[\alpha]_D^{25}$ +124° (c 1.13, chloroform)],[44] was similarly obtained as colorless leaflets. Both compounds showed only one peak by gas-liquid chromatography and were identical with authentic samples When 200 mg of 3,4,6-tri-O-acetyl-2-chloro-2-deoxy-β-D-glucopyranosyl chloride (200 mg, 0.575 mmol) was treated as above, **A** and **B** were obtained in 62 and 15% yields, respectively.

Different results were obtained in toluene. To a solution of silver tetrafluoroborate (136 mg, 0.700 mmol) and anhydrous toluene (7.4 ml) was added 3,4,6-tri-O-acetyl-2-chloro-2-deoxy-α-D-glucopyranosyl chloride (200 mg, 0.575 mmol) at 0° with stirring. After 1 hour the mixture was treated as above. Compounds **A** and **B** were obtained in 76 and 1% yields, respectively. In a similar manner, **A** and **B** were obtained in 77 and 1% yields from 3,4,6-tri-O-acetyl-2-chloro-2-deoxy-β-D-glucopyranosyl chloride. Quantitative gas-liquid chromatographic analysis showed for the reactions in toluene that **A** and **B** were obtained in the same ratio of 95.5:4.5 from either initial starting material isomer.

1-Ethyl-2-fluoromethyl-3-hydroxymethyl-5-methoxy-6-methylindole-4,7-dione Methylcarbamate.[119] A Soxhlet extractor was charged with 1.3 g (10 mmol) of silver(I) fluoride and 100 ml of acetonitrile. The silver(I) fluoride was extracted for 1 hour. Then 1-ethyl-3-hydroxymethyl-5-methoxy-2-methoxymethyl-6-methylindole-4,7-dione methylcarbamate (245 mg, 0.69 mmol) was added to the reaction flask. The reaction was maintained at reflux for 2 hours, filtered, and the filtrate was evaporated. The residue was dissolved in methylene chloride, washed successively with water and saline, and evaporated. This residue was chromatographed on silica gel, the product being eluted with ether. The first ether fraction (50 ml) was removed by evaporation. Recrystallization of the residue from ether-petroleum ether gave 70 mg (0.21 mmol, 30%) of the 1-ethyl-2-fluoromethyl-3-hydroxymethyl-5-methoxy-6-methylindole-4,7-dione methylcarbamate as yellow crystals, mp 137–138°.

19-Fluoro-Δ^4-10α-androsten-17β-ol-3-one 17-Acetate.[153a] A suspension of lithium fluoride (10 g, 0.38 mmol) in 50 ml of dimethylformamide was heated to boiling. A solution of 17β-19-dihydroxy-Δ^4-10α-androsten-3-one 17-acetate-19-*p*-toluenesulfonate (2 g, 4.0 mmol) in 10 ml of dimethylformamide was added. The mixture was refluxed for 5 hours, cooled, and poured into water. The precipitate that formed was filtered and crystallized from methanol-acetone to give the acetate of 19-fluoro-Δ^4-10α-androsten-17β-ol-3-one (no yield or mp stated).

3,6-Anhydro-5-deoxy-5-fluoro-1,2-O-isopropylidene-α-L-iodofuranose.[151] A mixture of dried 3-O-acetyl-1,2-O-isopropylidene-5,6-di-O-mesyl-α-D-glucofuranose (5 g, 12.5 mmol, mp 135°), anhydrous potassium fluoride (5 g, 84 mmol), and anhydrous ethylene glycol (50 ml) was heated under reflux for 2.5 hours. During this time, white crystals sublimed onto the condenser walls. The discolored solution was poured into ice water and the whole was extracted with chloroform. The combined chloroform extracts were dried over sodium sulfate and concentrated to give crude 3,6-anhydro-5-deoxy-5-fluoro-1,2-O-isopropylidene-α-L-iodofuranose. This product, together with the sublimed crystals from the condenser, was purified by sublimation to give fine needle-like crystals (1.21 g, 6.0 mmol, 48%), mp 96.5–97.5°, $[\alpha]_D^{22}$ +15.4° (c 1.97).

3-O-Benzyl-5-deoxy-5-fluoro-1,2-O-isopropylidene-α-D-xylofuranose.[110] A mixture of 3-O-benzyl-5-deoxy-5-tosyl-1,2-O-isopropylidene-α-D-xylofuranose (4.9 g, 20 mmol), anhydrous tetra-*n*-butylammonium fluoride [desiccated from 60 g (64 mmol) of clathrate, $(n\text{-}C_4H_9)_4NF\cdot 32.8\ H_2O$], and anhydrous acetonitrile was heated under reflux for 6 hours. The resulting brown solution was cooled and poured into 100 ml of diethyl ether. The ethereal solution was washed with three 25-ml portions of water and dried over magnesium sulfate. The material was separated by thin-layer chromatography (developed twice in chloroform-petroleum ether (1:2)). The first product (R_f 0.75) was obtained as a syrup (1.53 g, 9.6 mmol, 48%), n^{19}D 1.4969, $[\alpha]_D^{22}$ −40.5° (c 2.5, ethanol) and was identified as the desired compound.

[153] (a) A. Bowers, U.S. Pat. 3,210,389 (1965) [*C.A.*, **64**, 2143 (1966)]; (b) N. J. M. Birdsall, *Tetrahedron Lett.*, **1971**, 2675.

A second component (R_f 0.75, 0.18 g, 1.2 mmol, 6%) was a syrup, $n^{19}D$ 1.5115, $[\alpha]_D^{22}$ −29.1° (c 1.43, ethanol), identified as 3-O-benzyl-5-deoxy-1,2-O-isopropylidene-β-L-threo-pent-4-enefuranose.

Methyl 12-Fluorostearate.[153b] Methyl 12-tosyloxystearate was obtained in good yield by the tosylation of methyl 12-hydroxystearate. Displacement of the tosylate function by tetra-*n*-butylammonium fluoride in acetonitrile proceeded readily at room temperature to give methyl 12-fluorostearate. It was isolated in 40% yield after several recrystallizations from methanol. No additional experimental details were provided.

***n*-Octyl Fluoride.**[112] A mixture of difluorotriphenylphosphorane (12.63 g, 0.042 mol), *n*-octyl alcohol (2.84 g, 0.022 mol), and acetonitrile (25 ml) was heated at 170° with stirring for 8 hours in an autoclave lined with Hastelloy C. After it had been cooled to room temperature, the autoclave was opened and crystalline phosphorus compounds were removed by filtration. The filtrate was carefully distilled in the presence of sodium fluoride (3.0 g). The main fraction, amounting to approximately 2 g (70%) of *n*-octyl fluoride, was purified using preparative gas chromatography; bp 145–146°, $n^{24.5}D$ 1.3939.

***n*-Pentyl Fluoride.**[113a] In a stainless steel reaction tube, freshly distilled diphenyltrifluorophosphorane (10.2 g, 0.042 mol), *n*-pentyl alcohol (1.67 g, 0.019 mol), and acetonitrile (30 ml) were heated at 150° for 10 hours with stirring. The reaction mixture was cooled to room temperature, transferred to a separatory funnel, and shaken with 30 ml of benzene. The benzene layer was washed with 30% aqueous sodium hydroxide and water, dried over sodium sulfate, and distilled. The product, *n*-pentyl fluoride, bp 62–63°, weighed 1.06 g (0.012 mol, 64%).

HYDROGEN FLUORIDE AS A MONOFLUORINATING REAGENT

Caution! Fluorinations with hydrogen fluoride are potentially hazardous because of the high toxicity of hydrogen fluoride and the grave damage that it causes to human tissues. Safety precautions are mandatory when working with this reagent. (See pp. 220–223.)

Hydrogen fluoride is the most important industrial fluorinating agent.[154–158] It is also a valuable laboratory fluorinating reagent, solvent,

[154] J. M. Hamilton, Jr., *Advan. Fluorine Chem.*, **3**, 117 (1963).

[155] Ref. 8, Houben-Weyl, pp. 119–196.

[156] A. K. Barbour, L. J. Belf, and M. W. Buxton, *Advan. Fluorine Chem.*, **3**, 181 (1963).

[157] Ref. 15, Sheppard and Sharts, pp. 59–71.

[158] M. Hudlicky. *Chemistry of Organic Fluorine Compounds*, Macmillan, New York, 1962, pp. 88–92.

and catalyst,[159–161] but its highly injurious effects on human tissues and its high reactivity with glass have limited its use for laboratory syntheses.

In the 1930s and 1940s, before the development of safer and more convenient reagents, hydrogen fluoride was frequently added to carbon-carbon double and triple bonds.[5,16] These additions follow Markownikoff's

$$\rangle C{=}C\langle \; + \; HF \longrightarrow -\overset{H}{\underset{|}{\overset{|}{C}}}-\overset{F}{\underset{|}{\overset{|}{C}}}-$$

$$-C{\equiv}C- \; + \; 2\,HF \longrightarrow -\overset{H}{\underset{H}{\overset{|}{\underset{|}{C}}}}-\overset{F}{\underset{F}{\overset{|}{\underset{|}{C}}}}-$$

Rule. Although this is a valuable route to fluoro aliphatics, few significant advances and little use have been made of it since the development of better reagents and methods for monofluorination.

New techniques devised for the use of hydrogen fluoride moderate the powerful Lewis acid character of the anhydrous acid for additions to oxiranes (epoxides). Polar basic solvents such as diethyl ether, tetrahydrofuran, and dioxane have been used. The acid has also been complexed with a weak base such as potassium fluoride[162] or diisopropylamine.[160] In reactions where a powerful proton donor is required, boron trifluoride

$$KF{\cdot}HF\ (K^+HF_2^-) \qquad (i\text{-}C_3H_7)_2NH{\cdot}n\,HF \quad n = 2 \text{ or } 3$$

etherate is used with hydrogen fluoride to give *in situ* tetrafluoroboric acid.[164] Perchloric acid has also been used as a proton donor in 48% aqueous hydrofluoric acid.[165]

For the purpose of this review, all of these techniques for moderating the reactivity and nature of hydrogen fluoride are considered under the

[159] W. Bockemüller, "Organic Fluorine Compounds," in *Newer Methods of Preparative Organic Chemistry*, translated and revised from German by C. J. Kibler, Interscience, New York, 1948, pp. 229–248.

[160] K. Wiechert, "Use of Hydrogen Fluoride in Organic Reactions" in *Newer Methods of Preparative Organic Chemistry*, translated and revised from German by J. E. Jones, Interscience, New York, 1948.

[161] M. Kilpatrick and J. G. Jones, "Anhydrous Hydrogen Fluoride as a Solvent and a Medium for Chemical Reactions," in *The Chemistry of Non-aqueous Solvents*, Vol. 2, J. J. Lagowski, Ed., Academic Press, New York, 1967, Chapter 2.

[162] (a) J. A. Wright and N. F. Taylor, *Carbohyd. Res.*, **3**, 333 (1967); (b) **6**, 347 (1968).

[163] (a) G. Aranda, J. Jullien, and J. A. Martin, *Bull. Soc. Chim. Fr.*, **1965**, 1890; (b) **1966**, 2850.

[164] J. Cantacuzène and D. Ricard, *Bull. Soc. Chim. Fr.*, **1967**, 1587.

[165] F. H. Lincoln, W. P. Schneider, and G. B. Spero, U.S. Pats. 2,867,633 to 2,867,637 (1959) [*C.A.*, **53**, 11447h to 11452a (1959)].

general heading of addition reactions of hydrogen fluoride. Since these reactions have been reviewed earlier,[16] only selected recent work will be reviewed here.

An alternative method for adding hydrogen fluoride to oxiranes is use of boron trifluoride etherate without added hydrogen fluoride.

$$\underset{\text{(oxirane, O bridging)}}{R_2C\!-\!CR_2} + BF_3\cdot\text{etherate} \longrightarrow \left[R_2\!-\!\underset{+}{\overset{OBF_3^-}{\overset{|}{C}}}\!-\!CR_2\right] + \text{ether}$$

The exact nature and degree of charge separation in the intermediate is not important to us now. The intermediate undergoes an intra- or intermolecular fluorine transfer to give a fluorinated species which is converted to fluorohydrin product by a hydrolytic workup.

$$\left[R_2\!-\!\underset{+}{\overset{OBF_3^-}{\overset{|}{C}}}\!-\!CR_2\right] \xrightarrow{F^-} R_2\overset{OBF_2}{\overset{|}{C}}\!-\!\underset{F}{\underset{|}{C}}\!-\!R_2 \xrightarrow[\text{or } H_2O]{OH^-} R_2\overset{OH}{\overset{|}{C}}\!-\!\underset{F}{\underset{|}{C}}R_2$$

The remainder of this section consists of an expansion and updating of earlier reviews.[9, 10, 13, 16, 101]

The replacement of aliphatically bonded halogen by fluorine using hydrogen fluoride is extremely important industrially and has already been reviewed extensively.[166] The only substitution reactions reviewed here are the replacement of oxygen functions (alcohol, carboxylate, or sulfonate) by fluoride.

Hydrogen fluoride is particularly valuable for introducing fluorine into steroids, particularly by addition to oxiranes. Methods of preparing fluorosteroids have been recently reviewed.[4, 10, 167]

In summary, this review covers some selected recent additions of hydrogen fluoride to carbon-carbon double bonds, some selected recent substitutions of oxygen functions (*e.g.*, alcohol, carboxylate, sulfonate) by hydrogen fluoride, the addition of hydrogen fluoride or modified hydrogen fluoride (*e.g.*, $(CH_3)_2NH\cdot3\,HF$, $KF\cdot HF$) to oxiranes, the use of boron trifluoride etherate to add hydrogen fluoride to oxiranes, and, finally, some miscellaneous fluorination reactions of hydrogen fluoride.

Availability and Preparation of Hydrogen Fluoride and Other Reagents

Anhydrous hydrogen fluoride is commercially available and may be purchased in cylinders of laboratory size from the common vendors of

[166] Ref. 15, Sheppard and Sharts, p. 71.

[167] A. A. Akhrem, I. G. Reshetova, and Y. A. Titov, *Russ. Chem. Rev.*, **34**, 926 (1965).

laboratory gases. Alternatively it can be prepared by distilling commercially available 70% aqueous hydrofluoric acid through a copper or mild-steel distilling column. Extremely pure hydrogen fluoride can be obtained

$$KF{\cdot}HF \xrightarrow{500-600^\circ} KF + HF\uparrow$$

by thermal decomposition of potassium bifluoride. Details are provided in reviews.[168, 169]

Potassium hydrogen fluoride (KHF_2), commonly called potassium bifluoride, is commercially available and can be prepared from potassium carbonate and hydrofluoric acid.[109]

Boron trifluoride and boron trifluoride etherate are commercially available. Redistillation of boron trifluoride etherate before use is normally necessary. Directions for the preparation of boron trifluoride and its conversion into the etherate are available.[109] Fluoroboric acid is commerically available and can be easily prepared from hydrofluoric acid and boron trifluoride.

Diisopropylamine trihydrofluoride is prepared *in situ* from commercially available diisopropylamine and hydrogen fluoride.[163]

Scope and Limitations

Hydrogen fluoride is used as a laboratory fluorinating agent today only when superior alternative methods are not available. It is important in halogen fluoride additions from positive halogen reagents (p. 137). Currently hydrogen fluoride finds other principal uses in addition to epoxides and substitutions on sugars. In this review the emphasis is on modern synthetic fluorination methods; the addition reactions of hydrogen fluoride to alkenes and alkynes, formerly very important laboratory reactions, are discussed briefly. In contrast, boron trifluoride etherate, which has been developed more recently as a reagent for effective addition of hydrogen fluoride to oxiranes, is covered more extensively.

Addition of Hydrogen Fluoride to Alkenes and Alkynes*

This classic method for introducing fluorine into organic molecules has been reviewed thoroughly.[5, 9, 16, 155, 168] Additions of hydrogen fluoride to carbon-carbon unsaturation are normally carried out at as low a temperature as possible (−78 to 0°) to minimize the side reactions of rearrangement and polymerization. Because anhydrous hydrogen fluoride is a

[168] Ref. 3, Simons, Vol. I, p. 225.

[169] M. Hudlicky, *Chemistry of Organic Fluorine Compounds*, Macmillan, New York, 1962, pp. 38–44, 65–67.

* Note added in proof: G. A. Olah, M. Nojima, and I. Kerekes, *Synthesis*, **1973**, 779, recently report that 70% hydrogen fluoride/pyridine (or other tertiary amines) is a useful reagent for addition of hydrogen fluoride to olefins and alkynes in tetrahydrofuran solvent. The reaction is run at 0° to 50° at atmospheric pressure without significant polymerization or isomerization of the olefin.

powerful Lewis acid, intermediate carbonium ions or incipient carbonium ions are formed readily when hydrogen fluoride and alkenes are mixed.

The addition of hydrogen fluoride to 1-alkenes (mono-substituted ethylenes) normally occurs in high yield. The lifetime or stability of any intermediate secondary carbonium ion is too small to lead to side reactions.

$$RCH{=}CH_2 + HF \rightarrow \underset{\text{(Good yield)}}{RCHFCH_3}$$

The point is, however, why add hydrogen fluoride to an alkene, for example 1-pentene, to synthesize 2-fluoropentane? Substitution of 2-chloropentane by potassium fluoride is unquestionably more convenient in most laboratories.

Internal alkenes (1,2-disubstituted ethylenes) and cycloalkenes not substituted at the double bond give good yields of products. The possible intermediate secondary carbonium ion reacts mostly with sources of fluoride. Unsymmetrical internal alkenes normally give two isomers, as do unsymmetrically substituted cycloalkenes.

$$R_1CH{=}CHR_2 + HF \rightarrow R_1CHFCH_2R_2 + R_1CH_2CHFR_2$$

[R-substituted cyclohexene] + HF ⟶ [R-substituted fluorocyclohexane] + [isomeric R-substituted fluorocyclohexane]

Alkenes and cycloalkenes that can form intermediate tertiary carbonium ions normally give more side-reaction products than the expected addition product. The ability of hydrogen fluoride to form an intermediate

$$(CH_3)_2C{=}CH_2 + HF \xrightarrow{0^\circ} \underset{\text{Main product}}{\left[-C(CH_3)_2-CH_2- \right]_x} + \underset{\text{Low yield}}{t\text{-}C_4H_9F}$$

carbonium ion with isobutylene makes hydrogen fluoride a good catalyst for the alkylation of isobutylene by isobutane, an industrial alkylation of great importance.

$$CH_3-CH(CH_3)-CH_3 + CH_2{=}C(CH_3)_2 \xrightarrow{HF} \underset{\text{2,2,4-Trimethylpentane}}{CH_3-C(CH_3)_2-CH_2-CH(CH_3)-CH_3}$$

Although the preceding general comments apply to nearly all additions of hydrogen fluoride to alkenes, additions to specific structures are also governed by stereochemistry and by the nature of substituent groups. The results of addition of hydrogen fluoride to pregnene (**18**) and to cholestene (**19**) derivatives illustrate this. The expected addition products **20** and **21**

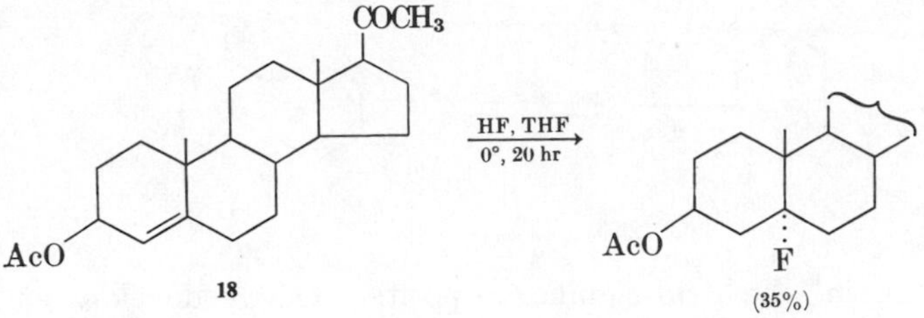

$COCH_3$

AcO

18

HF, THF
0°, 20 hr

AcO F

(35%)

(Ref. 170)

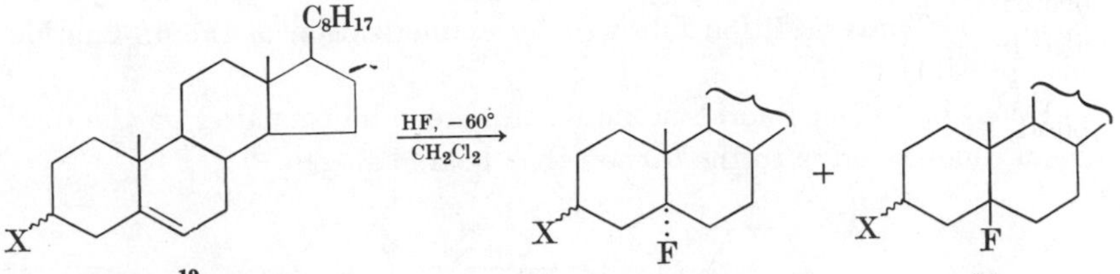

C_8H_{17}

HF, −60°
CH_2Cl_2

X X F + X F

19 **20** **21**

(a) X = β-OH
(b) X = β-OAc
(c) X = α-OH
(d) X = α-OAc

(a) (48%)
(b) (60%)
(c) (79%)
(d) (40%)

(a) (16%)
(b) (3%)
(c) (2%)
(d) (6%)

(Ref. 171)

were obtained from the 3-substituted Δ^5-cholestene derivatives, **19a–d**. In contrast, a rearrangement occurred (methyl migration) to give **24**, as well as the normal adduct (compound **23**), with the 7-hydroxy-Δ^4-cholestenes, **22a** and **22b**.[172] Rearrangements of steroid structures where C-5 is a carbonium ion center are also observed in additions of hydrogen fluoride to 4,5- and 5,6-epoxycholestane derivatives (see p. 201).

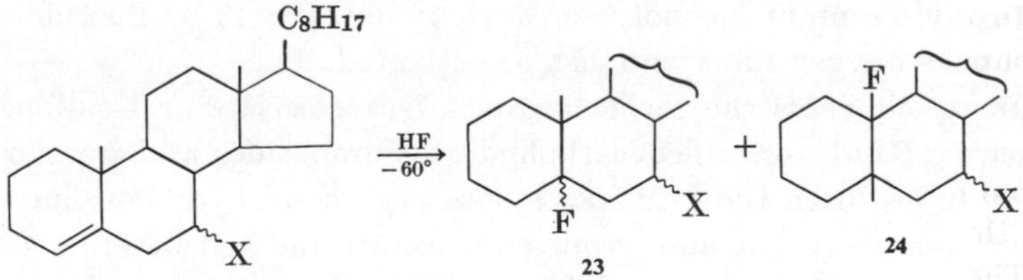

C_8H_{17}

HF
−60°

X F X + F X

22 **23** **24**

(a) X = α-OH
(b) X = β-OH

(a) α-F (97%), β-F (<0.5%)
(b) α-F (40%), β-F (<5%)

(a) (1.5%)
(b) (30%)

[170] A. Bowers, P. G. Holton, E. Denot, M. C. Loza, and R. Urquiza, *J. Amer. Chem. Soc.*, **84**, 1050 (1962).

[171] J.-C. Jacquesy, R. Jacquesy and J. Levisalles, *Bull. Soc. Chim. Fr.*, **1967**, 1649.

[172] J.-C. Jacquesy, R. Jacquesy, and M. Petit, *Tetrahedron Lett.*, **1970**, 2595.

The addition of hydrogen fluoride to 6-deutero-3-acetoxycholestene was shown to be *cis* for the α-fluoro product.[171] This may be a four-center

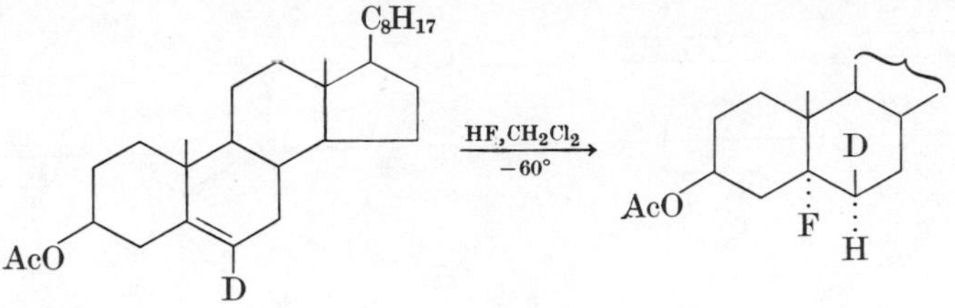

$$\xrightarrow[-60^\circ]{HF, CH_2Cl_2}$$

concerted reaction in which no significant positive charge develops at C_5; this would explain why no rearrangement of the 19-methyl group occurred. *Trans* addition followed by epimerization of the 5β-fluoride is also possible.

When hydrogen fluoride adds to an alkene halogenated on the double bond, fluorine adds to the carbon that bears halogen.[157]

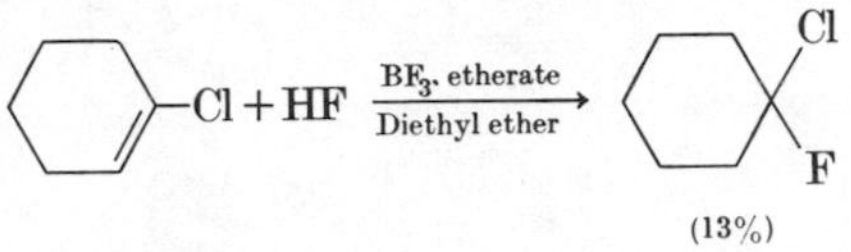

$$\text{-Cl} + HF \xrightarrow[\text{Diethyl ether}]{BF_3\text{ etherate}} \quad \text{(13\%)} \qquad \text{(Ref. 173)}$$

The addition of hydrogen fluoride to 1-alkynes or to symmetrical alkynes gives *gem*-difluorides. Unsymmetrical alkynes give two *gem*-difluorides.[157] The reaction no longer has synthetic value. Preparation of

$$R_1C{\equiv}CR_2 + 2\,HF \rightarrow R_1CF_2CH_2R_2 + R_1CH_2CF_2R_2$$

gem-difluorides is more conveniently achieved by the action of sulfur tetrafluoride on carbonyl compounds.[6]

Substitution of Oxygen Functions by Hydrogen Fluoride

Hydrogen fluoride is not a good general reagent for substituting the hydroxyl group of alcohols* or acyloxy derivatives by fluoride. In most cases the oxygen function must be activated.

In recent years the replacement of 1-carboxylate or 1-sulfonate ester groups of fully esterified carbohydrate furanosides and pyranosides by fluoride has been the principal laboratory use of hydrogen fluoride. The most commonly displaced groups are acetate and benzoate. In most cases these functions are displaced from the 1-carbon atom which is activated by the alpha ring oxygen as shown in the accompanying examples.

The primary side reactions in the reaction of hydrogen fluoride with esterified sugar derivatives are isomerization and hydrolysis of protecting

[173] J. Cantacuzene, R. Jantzen, and D. Ricard, *Tetrahedron*, **28**, 717 (1972).

* Note added in proof: G. A. Olah, M. Nojima and I. Kerekes, *Synthesis*, **1973**, 786, have recently developed the reaction of 70% hydrogen fluoride/pyridine with alcohols at room temperature as a convenient high yield route to aliphatic fluorides.

CH₂OAc, OAc, OAc, AcO —Anhyd. HF, −70°, 5 min→ CH₂OAc, OAc, AcO, F (77%) (Ref. 174)

25 —Anhyd. HF, Room t, 10 min→ 26 (69%) (Ref. 175)

ester groups. Non-sugar derivatives under similar conditions would be expected to give elimination products. To minimize side reactions, the reaction times in hydrogen fluoride are kept as short as possible. For example, in the reaction of hydrogen fluoride with tetra-O-benzoyl-β-D-ribofuranose (**25**), if the reaction time at room temperature is 150 hours rather than 10 minutes, the yield of fluorosugar **26** is 9% rather than 69%; the rest of the initially formed **26** is converted to products **27–29**.[175]

25 —Anhyd. HF, Room t, 150 hr→ 26 + 27 (15%) + 28 + 29 (42%)

The normal conditions for reaction can be used to effect a desired isomerization of a less stable to a more stable isomer.[176] This type of epimer-

1-β-Fluoride —Anhyd. HF, −10 to 0°, 15 min, room t, 10 min→ 1-α-Fluoride (77%)

ization is to be expected where fluorine at the 1-carbon atom is *cis* to the substituent at the 2-carbon atom.

Attempts to replace unactivated hydroxyl by fluoride using hydrogen fluoride usually fail. Rearrangement often occurs and product corresponding to direct replacement is often not obtained. For example,

[174] L. D. Hall and J. F. Manville, *Can. J. Chem.*, **45**, 1299 (1967).

[175] N. Gregersen and C. Pedersen, *Acta Chem. Scand.*, **22**, 1307 (1968).

[176] L. D. Hall, R. N. Johnson, J. Adamson, and A. B. Foster, *Can. J. Chem.*, **49**, 118 (1971).

both 9α- and 11β-hydroxyandrostenedione give 9α-fluoroandrostenedione in unspecified yields.[177] The reaction of 21-acetoxy-Δ^4-pregnene-11β,17α-

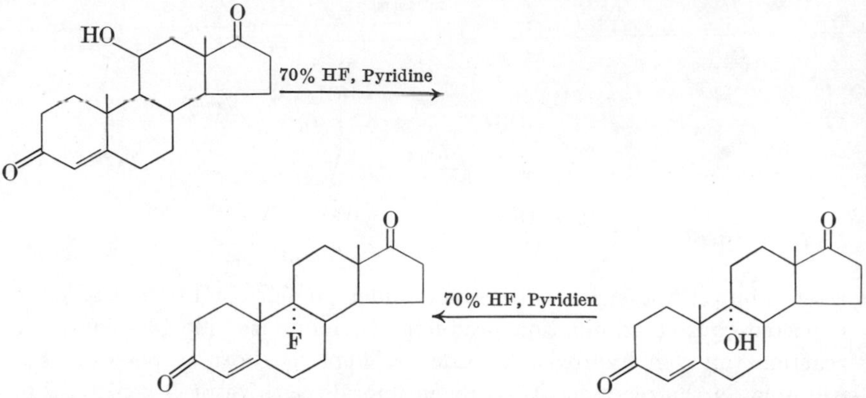

70% HF, Pyridine

70% HF, Pyridien

diol with 70% anhydrous hydrogen fluoride in pyridine gave the 9α-fluoro compound and an elimination product in unspecified yields. The 17α-hydroxyl group did not react.[177]

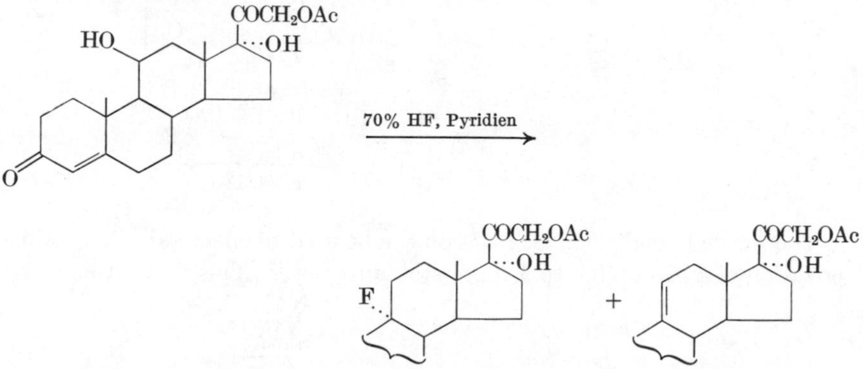

70% HF, Pyridien

Reaction of pregnenolone with hydrogen fluoride in tetrahydrofuran gave a low yield of unrearranged 3β-fluoro-Δ^5-pregnen-20-one.[170]

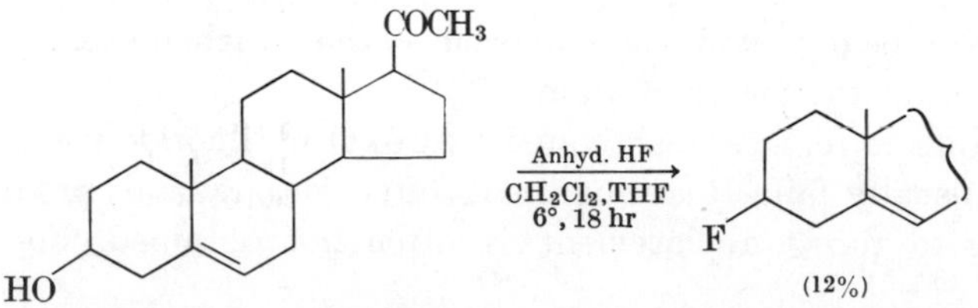

Anhyd. HF
CH_2Cl_2,THF
6°, 18 hr

(12%)

[177] C. G. Bergstrom and R. M. Dodson, *J. Amer. Chem. Soc.*, **82**, 3479(1960).

Addition of Hydrogen Fluoride to Oxiranes

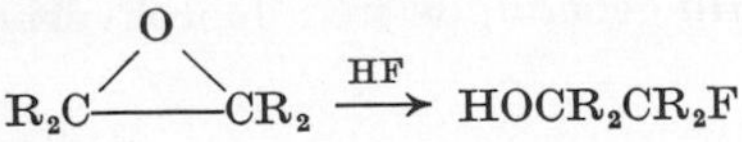
O
R₂C
CR₂
HF
HOCR₂CR₂F

(See Table VI, p. 206.)

Rarely is pure anhydrous hydrogen fluoride used to open an oxirane ring. Under these vigorous conditions, polymerization becomes an important side reaction. In addition many compounds suffer rearrangement in anhydrous hydrogen fluoride or add hydrogen fluoride to available carbon-carbon unsaturation.

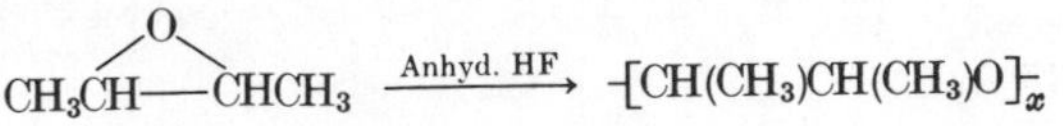
O
CH₃CH—CHCH₃
Anhyd. HF
[CH(CH₃)CH(CH₃)O]ₓ

(Ref. 178)

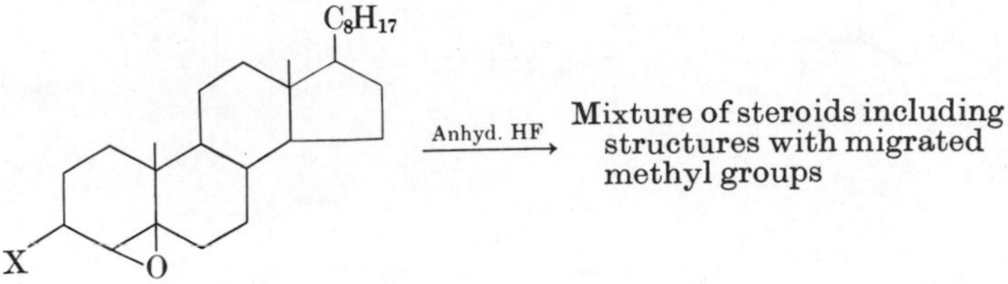
C₈H₁₇
X
O
Anhyd. HF
Mixture of steroids including structures with migrated methyl groups

(Ref. 179)

For addition to the oxirane ring, hydrogen fluoride is normally diluted by a solvent or solvent pair or is modified by reaction with some Lewis base or Lewis acid. The commonly used solvents and modifiers follow.

Solvents.

Nonpolar: benzene, methylene chloride, or chloroform. Polar (with a basic site): diethyl ether, dioxane, tetrahydrofuran, or glycol.

Pairs (nonpolar with polar): chloroform-tetrahydrofuran or benzene-diethyl ether.

Lewis Bases to Reduce Acidity of Medium and Provide a Better Source of Fluoride Ion.

(a) $KF + HF = KHF_2$ (often written KF·HF)

(b) $[(CH_3)_2CH]_2NH + 3\,HF = [(CH_3)_2CH]_2NH \cdot 3\,HF$

Lewis Acids to Increase the Availability of Protons.

$$HF + BF_3\cdot\text{ether} = H^+BF_4^- + \text{ether}$$

$$48\%\ HF\ (aq) + HClO_4 = H^+ + ClO_4^- + 48\%\ HF\ (aq)$$

In studies through 1966 a large number of steroid epoxides had been treated with hydrogen fluoride in a variety of solvents or solvent pairs,

[178] I. Shahak, S. Manor, and E. D. Bergmann, *J. Chem. Soc., C*, **1968**, 2129.

[179] M. Neeman and J. S. O'Grodnick, *Tetrahedron Lett.*, **1971**, 4847.

usually chloroform, chloroform-tetrahydrofuran, methylene chloride, chloroform-ethanol. In general, desired fluorohydrins were obtained from steroids.[4, 9, 10]

$COCH_2OAc$ ···OH O O $\xrightarrow[-30°,\ 4\ hr]{HF,\ CHCl_3—THF}$ HO $COCH_2OAc$ ···OH F O (Ref. 180)

(74%)

$CHCO_2CH_3$ O O··· O O··· $\xrightarrow[CHCl_3]{48\%\ aq\ HF}$ O··· O HO F (Ref. 181)

(−)

Early attempts to prepare 16-fluorosteroids by addition of hydrogen fluoride to 16,17α-epoxy steroids met with repeated failure. The 16,17α-epoxy steroids underwent rearrangements on reaction with hydrogen fluoride.[182, 183] The addition of hydrogen fluoride to 16,17α-epoxypregnan-3α-ol-11,20-dione acetate was recently reinvestigated and most of the products identified.[184] (See the equation on the top of p. 203].

An unusual oxirane opening of 4β,5β-epoxy-3-keto steroids with hydrogen fluoride gives 2α-fluoro-3-keto-4-ene product.[185] (See p. 217.) Although rearrangements are encountered in additions of hydrogen fluoride to oxiranes, rearrangements are much more common when boron trifluoride etherate alone reacts with steroid oxiranes. This topic is discussed on p. 205 and following.

[180] R. F. Hirshmann, R. Miller, J. Wood, and R. E. Jones, *J. Amer. Chem. Soc.*, **78**, 4956 (1956).

[181] J. A. Hogg, G. B. Spero, J. L. Thompson, B. J. Magerlein, W. P. Schenider, D. H. Peterson, O. K. Sebek, H. C. Murray, J. C. Babcock, R. L. Pederson, and J. A. Campbell, *Chem. Ind.* (London), **1958**, 1002.

[182] R. E. Beyler and F. Hoffman, *J. Org. Chem.*, **21**, 572 (1956).

[183] E. L. Shapiro, M. Steinberg, D. Gould, M. J. Gentles, H. L. Herzog, M. Gilmore, W. Charney, E. B. Hershberg, and L. Mandell, *J. Amer. Chem. Soc.*, **81**, 6483 (1959).

[184] D. R. Hoff, *J. Org. Chem.*, **35**, 2263 (1970).

[185] M. Neeman, T. Mukai, J. S. O'Grodnick, and A. L. Rendall, *J. Chem. Soc., Perkin I*, 2300 (1972).

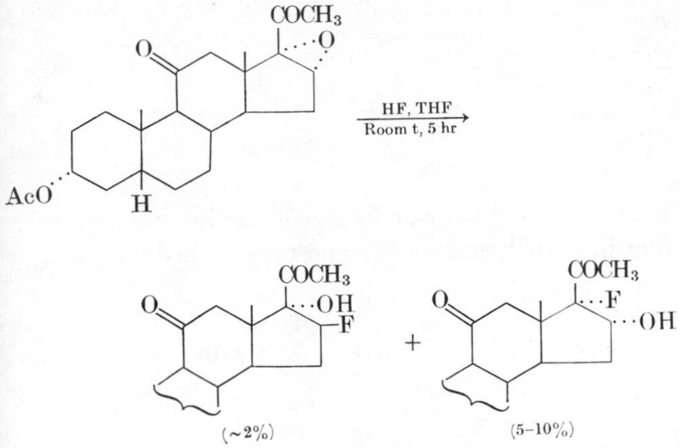

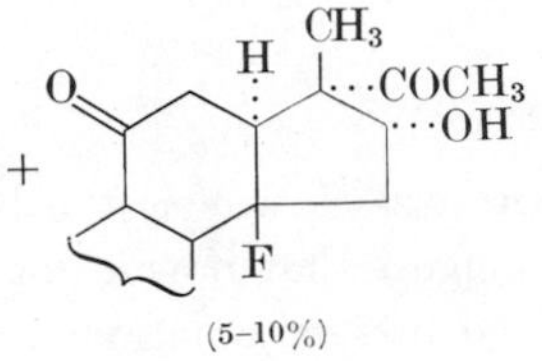

$\xrightarrow[\text{Room t, 5 hr}]{\text{HF, THF}}$

(~2%) + (5–10%)

+ (5–10%) + Three isomeric olefins and starting material

In Table XVI, examples are given of successful addition of hydrogen fluoride to steroid oxiranes with the epoxide function located at 4α,5α; 4β,5β; 5α,6α; 5β,6β; 6α,7α; 9β,11β; 11β,12β; 17β,20; or 20,21.

Before 1965 most of the additions of hydrogen fluoride to the oxirane ring were with steroids. In 1965, amine hydrofluorides, principally diisopropylamine trihydrofluoride, were shown to add hydrogen fluoride to a series of substituted styrene epoxides and alicyclic epoxides. The reagent was formed *in situ* by mixing the appropriate quantities of acid and amine.[163] As can be seen from two examples, hydrogen fluoride adds stereospecifically *trans* and in good yield.

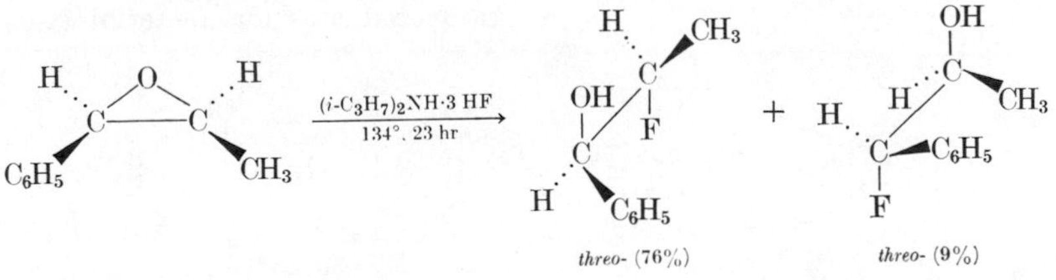

$\xrightarrow[\text{134°, 23 hr}]{(i\text{-}C_3H_7)_2NH\cdot 3\,HF}$

threo- (76%) + *threo*- (9%)

(Ref. 163a)

Addition to *trans*-1-phenyl-2-methyloxirane gives only the *erythro* isomers $C_6H_5CHOHCHFCH_3$ (9%) and $C_6H_5CHFCHOHCH_3$ (74%). The reverse orientation for addition of the *cis* and *trans* epoxides is not expected nor explained by the authors.

$(i\text{-}C_3H_7)_2NH \cdot 3\,HF$, 100°, 9 hr (80%) (Ref. 163b)

In 1967 the use of potassium acid fluoride in glycol was introduced for opening the oxirane ring in carbohydrates.[162a] The reagent has not been

KF·HF, Glycol, reflux, 40 min (66%) (Ref. 162a)

tried for opening other classes of oxiranes. However, the high temperatures required (glycol boils at 198°) suggest that it will not have general use since fluorohydrins usually undergo further reactions and rearrangements under these vigorous conditions.

Several simple dialkyl aliphatic and alicyclic oxiranes were investigated using hydrogen fluoride in mixed chloroform-tetrahydrofuran. Rigid systems gave fluorohydrins more regularly than nonrigid systems. Side reactions often predominated with the nonrigid compounds.[178]

$CH_3CH\text{—}CHCH_3$ (epoxide) $\xrightarrow[0°,\ 3\ hr]{HF,\ CHCl_3,\ THF}$ Polymer (25%), 2-butanone (7%), recovered starting material (61%)

$\xrightarrow[0°,\ 3\ hr]{HF,\ CHCl_3,\ THF}$ Polymer (60%), cyclohexanone (19%), recovered starting material (10%)

$\xrightarrow[0°,\ 3\ hr]{HF,\ CHCl_3,\ THF}$ OH, F + OH, F (80%)

Fluoroboric acid is a strong acid which ionizes much more easily than hydrogen fluoride. Addition of boron trifluoride etherate to hydrogen fluoride gives fluoroboric acid *in situ*. For addition to epoxides, this combination of reagents provides a powerful proton donor (HBF_4 *in situ*) and a source of fluoride ion (HF). This combination has been used in recent years to add hydrogen fluoride to oxiranes.[173, 186, 187]

$$\xrightarrow[(C_2H_5)_2O,\ 7\text{hr.}]{HF,\ BF_3\cdot\text{ether}}$$ (56%) (Ref. 187)

$$C_2H_5CF\overset{O}{—}CHCH_3 \xrightarrow[\text{Room t}]{HF,\ BF_3\cdot\text{ether}} C_2H_5CF_2CHOHCH_3$$ (Ref. 186)

cis-trans mixture (>90%)

The use of fluoroboric acid with hydrogen fluoride is similar in principle to fluorinations of oxiranes carried out with 48% aqueous hydrofluoric acid to which 71% perchloric acid has been added. In this system, perchloric acid serves as a powerful proton source while hydrogen fluoride provides fluoride ion.[165]

In Table VI the preceding reagents for adding hydrogen fluoride to oxiranes are compared with each other and with boron trifluoride etherate.

Boron Trifluoride Etherate as a Reagent for Adding Hydrogen Fluoride to Oxiranes

Boron trifluoride etherate dissolved in an organic solvent, usually benzene, diethyl ether, or a mixture of the two, is not a good general reagent for adding hydrogen fluoride to an oxirane ring. Indeed, the initial studies of the reaction of boron trifluoride etherate with steroid oxiranes were not directed towards fluorohydrins, but rather to rearrange the oxiranes to ketones.[188] The synthesis of steroid fluorohydrins was a

$$\xrightarrow[\text{Benzene}]{BF_3\cdot\text{ether}}$$

(Or α-epoxide)

$$\xrightarrow[\text{Benzene}]{BF_3\cdot\text{ether}}$$

(Or α-epoxide)

[186] J. Cantacuzene and J. Leroy, *Tetrahedron Lett.*, **1970**, 3277.

[187] J. Cantacuzene and M. Atlani, *Tetrahedron*, **26**, 2447 (1970).

[188] (a) H. B. Henbest and T. I. Wrigley, *J. Chem. Soc.*. **1957**, 4596; (b) **1957, 4765.**

TABLE VI. COMPARISON OF REAGENTS USED TO ADD HYDROGEN FLUORIDE TO OXIRANES

Reagent	Strength	Yield of Fluorohydrin To Be Expected	Comments
Anhyd. HF	Very powerful	Poor	Polymerization and rearrangement expected to predominate
BF_3·ether in benzene	Very powerful	Poor	As above
BF_3·ether in diethyl ether	Very powerful	Poor to good	Extensively used with steroids; rearrangements are probable
Anhyd. HF, BF_3·ether catalyst, solvent	Strong	Fair to excellent	Rearrangements do not usually occur
$[(CH_3)_2CH]_2NH\cdot 3\,HF$	Moderate	Good to excellent	No rearrangements reported; only *trans* addition reported; the reagent has not been extensively evaluated
KF·HF in ethylene glycol at 180°	Weak	Fair to good	Use has been limited to carbohydrates; other types of fluorohydrins are not expected to survive the high temperatures required

fortuitous fluorination side reaction. Hydrogen fluoride, properly catalyzed, is usually a better reagent for forming fluorohydrins.

5α,6α- and 5β,6β-Epoxycholestane Derivatives. The reactions of boron trifluoride etherate with 5α,6α- and 5β,6β-epoxysteroids have been most extensively studied.[188–196]

Both 5α,6α- and 5β,6β-epoxycholestane rearranged when treated with boron trifluoride etherate in benzene.[188] In contrast with these

C_8H_{17}

BF_3·ether / Benzene

5α,6α-epoxy — 5β-H(30%) (Ref. (189*a*))
5β,6β-epoxy — 5β-H(98%) (Ref. (188*b*))

rearrangements, the 3β-acetoxy-5,6-epoxycholestanes **30** formed the fluorohydrins **31** in good yield.

C_8H_{17}

BF_3·ether (Ref. 189*a*)

AcO, O, AcO, HO, F

30a 5α,6α-epoxy —Benzene→ **31a** 5α-OH, 6β-F (80%)
30b 5β,6β-epoxy —Benzene-ether→ **31b** 5β-OH, 6α-F (76%)

These initial studies prompted additional investigations of the effect of substitution at the 3β position on the course of the reaction and on the effect of solvent. In general, electron-withdrawing groups at the 3β position were found to increase yields of fluorohydrin by reducing the

[189] (a) A. Bowers and H. J. Ringold, *Tetrahedron*, **3**, 14 (1958); (b) *J. Amer. Chem. Soc.*, **80**, 4423 (1958).

[190] A. Bowers, L. C. Ibáñez, and H. J. Ringold, *J. Amer. Chem. Soc.*, **81**, 5991 (1959).

[191] C. Djerassi and H. J. Ringold, U.S. Pat. 2,983,737 (1961) [*C.A.*, **55**, 21171g (1961)].

[192] H. J. Ringold, A. Bowers, O. Mancera, and G. Rosenkranz, Ger. Pat. 1,096,357 (1961) [*C.A.*, **55**, 27429h (1961)].

[193] J. W. Blunt, M. P. Hartshorn, and D. N. Kirk, *Tetrahedron*, **22**, 3195 (1966).

[194] J. M. Coxon, M. P. Hartshorn, C. N. Muir, and K. E. Richards, *Tetrahedron Lett.*, **1967**, 3725.

[195] B. N. Blackett, J. M. Coxon, M. P. Hartshorn, and K. E. Richards, *Tetrahedron*, **25**, 4999 (1969).

[196] (a) I. G. Guest and B. A. Marples, *Tetrahedron Lett.*, **1969**, 1947; (b) *J. Chem. Soc., C*, **1970**, 1626.

potential for carbonium ion character at C_5, thereby reducing the migratory ability of the hydrogen at C_6 and the methyl group at C_{10}. Electron-donating groups at the 3 position increased the extent of rearranged products. Benzene favored rearrangement while diethyl ether greatly increased the amount of fluorohydrin formation.[188, 189a, 194–196]

The potential complexity of the reaction of boron trifluoride etherate with epoxy-steroids is illustrated by the reaction with 3β-hydroxy-5α,6α-epoxycholestane in which six products were isolated and identified, only one of them a fluorohydrin (yield 2%).[196]

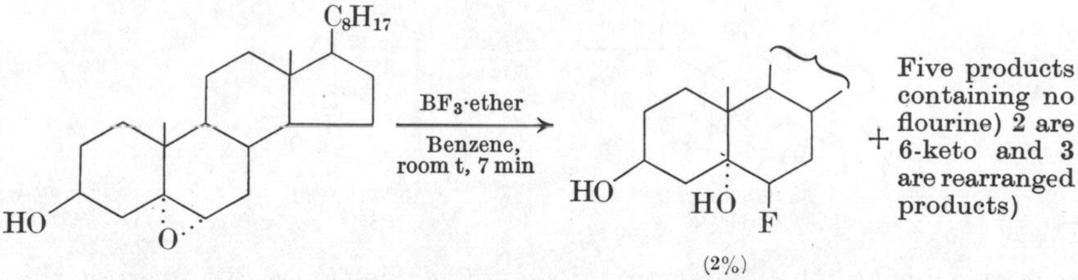

C_8H_{17}

HO O

BF_3·ether

Benzene, room t, 7 min

HO HO F

(2%)

+ Five products containing no flourine) 2 are 6-keto and 3 are rearranged products)

Another factor in the reactions of boron trifluoride etherate with steroid epoxides is the amount of hydrogen fluoride in the etherate. The course of reaction of 3β-acetoxy-5α,6α-epoxycholestane (**30a**) with boron trifluoride etherate is reported to have been totally altered when a saturated solution of the steroid in benzene was treated with hydrogen fluoride-free boron trifluoride etherate.[193]

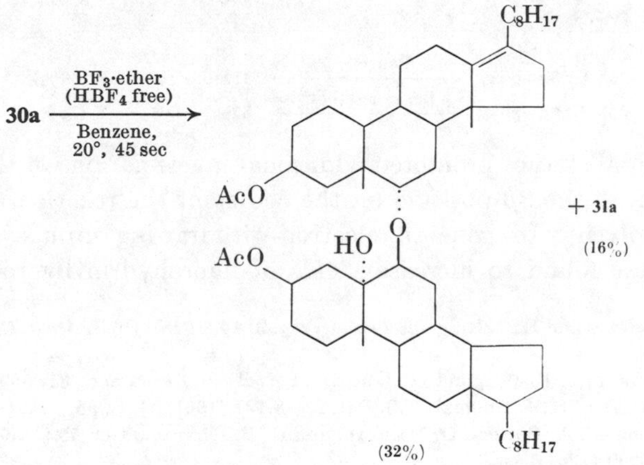

30a $\xrightarrow[\text{Benzene, 20°, 45 sec}]{BF_3\text{·ether } (HBF_4 \text{ free})}$

C_8H_{17}

AcO

O

AcO HO

C_8H_{17}

(32%)

+ **31a**

(16%)

Stereochemistry plays an important role in the reaction of steroids with boron trifluoride etherate. Often the stereochemical effects are not easily separated from the electronic effects of substituent groups. An example in which stereochemistry is decisive is the reaction of 3α-acetoxy-5α,6α-epoxycholestane. Whereas the 3β-acetoxy derivative gave a high yield of fluorohydrin (80%),[189a] the 3α-acetoxy isomer gave fluorohydrin only in 8% yield plus several rearrangement products.[194] In this example the axial 3α-acetoxy group is not effective in removing charge from C_5.

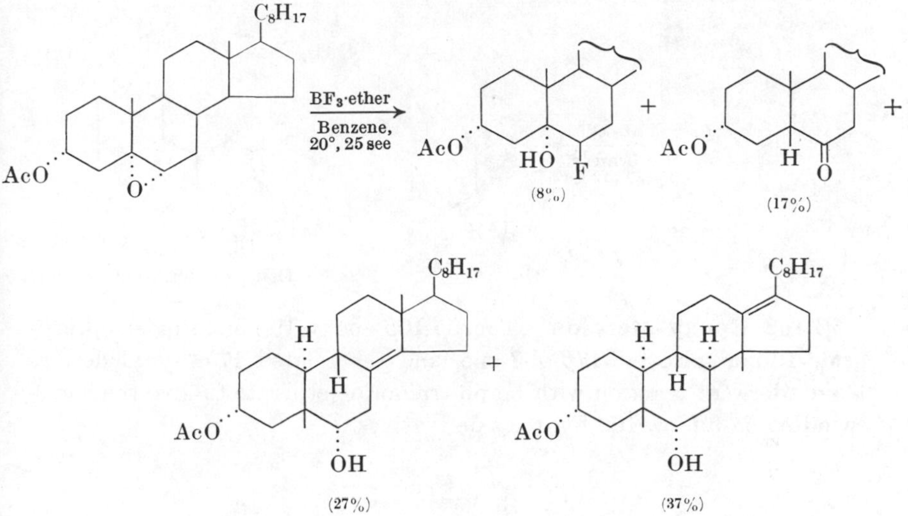

C_8H_{17} AcO O $\xrightarrow[\text{Benzene, 20°, 25 sec}]{BF_3\text{·ether}}$ AcO HO F (8%) + AcO H O (17%) +

C_8H_{17} H H AcO OH (27%) + C_8H_{17} H H H AcO OH (37%)

The reactions of 5,6-epoxycholestane derivatives have been presented in some detail to illustrate the possible complexities in the reaction of boron trifluoride etherate with steroid oxiranes. Other examples of these reactions are given in Table XVII. The opening of the oxirane ring in other than the 5,6 position is not considered in the detail given to the 5,6-epoxycholestanes.

4α,5α- and 4β,5β-Epoxy-steroids. Reaction of 4α,5α-epoxycholestane with boron trifluoride etherate gave primarily the 4-keto derivative, coprostanone.[188a, 195] No fluorohydrin was obtained. 4β,5β-Epoxycholestane under similar conditions gave the fluorohydrin, 4β-hydroxy-

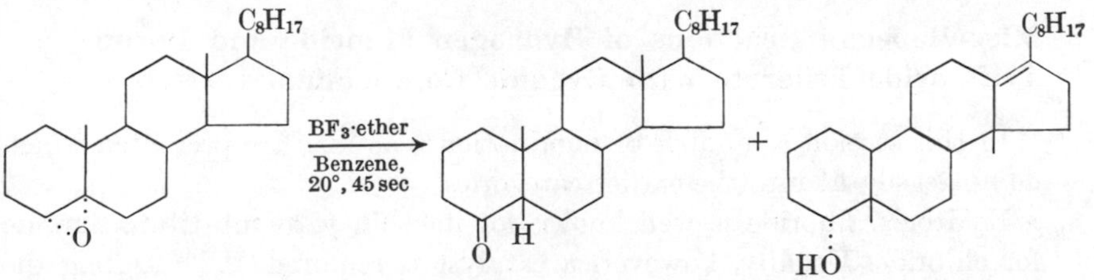

C_8H_{17} O $\xrightarrow[\text{Benzene, 20°, 45 sec}]{BF_3\text{·ether}}$ C_8H_{17} O H (70%) + C_8H_{17} HO (11%)

(Ref. 195)

5α-fluorocholestane (**32**), in only 3% yield. Six rearrangement products were identified (see Table XVII).[195] The expected 4β-hydroxy-5α-fluorocholestane (**32**) underwent rearrangement under reaction conditions to give products similar to those from the starting material.[195]

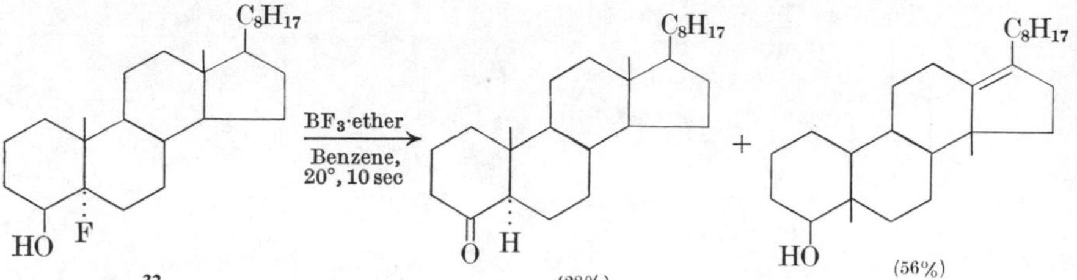

C_8H_{17} HO F **32** $\xrightarrow[\text{Benzene, } 20°, 10 \text{ sec}]{BF_3\text{·ether}}$ C_8H_{17} O H (28%) + C_8H_{17} HO (56%)

Other unidentified derivatives

5β,10β-Epoxy-steroids. The 5β,10β-epoxy-19-norsteroids, 5β,10β-oxido-19-norandrostan-17β-ol-3-one, and the related 17-ethynyl derivative underwent reaction with boron trifluoride etherate to give the corresponding 5α-fluoro, 10β-hydroxy derivatives.[197]

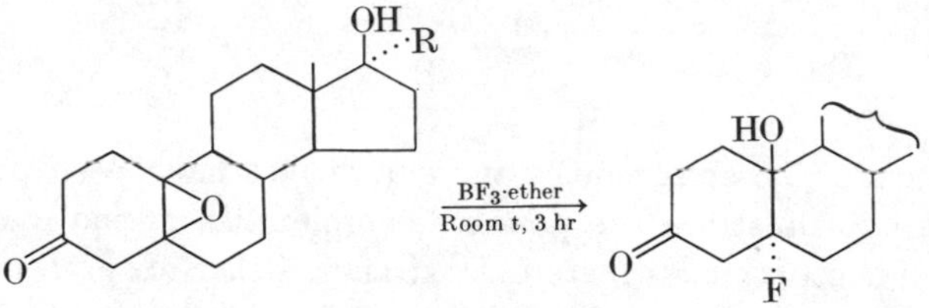

OH R O O $\xrightarrow[\text{Room t, 3 hr}]{BF_3\text{·ether}}$ HO O F

R = H (88%)
R = —C≡CH (78%)

Miscellaneous Steroid Oxiranes. Steroid oxiranes having the epoxide group in positions other than those previously discussed gave rearrangement products with boron trifluoride etherate. Examples can be found in Table XVII. Examples included have the epoxy substitution in the following positions: 7α,8α; 8β,9β; 9α,11α; 9β,11β; 8α,14α.

Miscellaneous Reactions of Hydrogen Fluoride and Boron Trifluoride Etherate with Organic Compounds

In this section a number of fluorination reactions are presented which do not easily fit into the earlier categories.

Hydrogen fluoride is well known for its ability to substitute fluoride for chloride. Usually, however, a catalyst is required.[156, 166] During the

[197] J. P. Ruelas, J. Iriarte, F. Kincl, and C. Djerassi, *J. Org. Chem.*, **23**, 1744 (1958).

addition of hydrogen fluoride to the double bond of ethyl γ-chlorocrotylacetamidomalonate, the γ-chlorine atom was replaced by a fluorine atom.[198]

$$CH_3CCl{=}CHCH_2C(NHCOCH_3)(CO_2C_2H_5)_2 \xrightarrow[\text{Room t, 3 hr; } 30°, \text{ 2 hr}]{HF} CH_3CF_2CH_2CH_2C(NHCOCH_3)(CO_2C_2H_5)_2 \quad (78.5\%)$$

Hydroxyl groups can be replaced using hydrogen fluoride with an ynamine.[199] The intermediate was suggested to be an amide fluoride, or

$$C_6H_5CH_2CH_2OH + HF \xrightarrow[CH_2Cl_2,\ -80°]{C_6H_5C{\equiv}CN(CH_3)_2} C_6H_5CH_2CH_2F \quad (60\text{–}90\%)$$

more probably α-fluoroenamine or its ketene iminium fluoride.[199] This novel replacement has not been explored for synthetic utility.

The reaction of a diazoacetate with hydrogen fluoride to give a fluoroacetate[48] has been extended to preparation of 21-fluorosteroids.[200a,b]

$$XC(=O)CHN_2 \xrightarrow[CH_2Cl_2,\ (i\text{-}C_3H_7)_2O]{HF} XC(=O)CH_2F$$

Boron trifluoride etherate and cyclodecyn-6-ol gave a transannular addition of hydrogen fluoride.[201]

$$\text{cyclodecyn-6-ol (R, OH)} \xrightarrow[CH_2Cl_2]{BF_3\cdot\text{ether}} \text{fluoro-octahydronaphthalene (F, R)}$$

R = H (60–65%), CH_3 (87%)

Conjugated hypofluorination has been reviewed.[202] Although the reaction is not strictly an addition of hydrogen fluoride, it is sufficiently similar to be worth mentioning. A powerful oxidizing agent is used with hydrogen fluoride to generate in effect OH^+ for addition to halogenated alkenes.

$$CX_2{=}CH_2 \xrightarrow[2.\ H_2O]{1.\ CrO_3 + HF} CX_2FCH_2OH \qquad X = F, Cl$$

$$CCl_2{=}CH_2 \xrightarrow[(CH_3CO)_2O]{H_2O_2 + HF} CCl_2FCH_2OH \quad (25\text{–}30\%)$$

$$CF_2{=}CH_2 \xrightarrow{KMnO_4 + HF} CF_3CH_2OH$$

[198] M. Hudlicky, *Collect. Czech. Chem. Commun.*, **32**, 453 (1967).

[199] H. G. Viehe, R. Fuks, and M. Reinstein, *Angew. Chem., Int. Ed. Engl.*, **3**, 581 (1964); H. G. Viehe, *Chemistry of Acetylenes*, Marcel Dekker, New York, 1969.

[200] (a) Ciba-Geigy AG, Ger. Pat. 2,100,324 (1971) [*C.A.*, **75**, 130024v (1971)]; (b) G. Anner and P. Wieland, U.S. Patent 3,758,524 (1973).

[201] B. Rao and L. Weiler, *Tetrahedron Lett.*, **1971**, 927.

[202] L. S. German and I. L. Knunyantz, *Angew. Chem., Int. Ed. Engl.*, **8**, 349 (1969).

An interesting source of hydrogen fluoride was found in the reaction of O-benzyl-2,3-anhydro-β-D-ribopyranoside with potassium fluoride in molten acetamide; the solvent acted as a source of protons. The same product was formed by using potassium acid fluoride in refluxing glycol.[203]

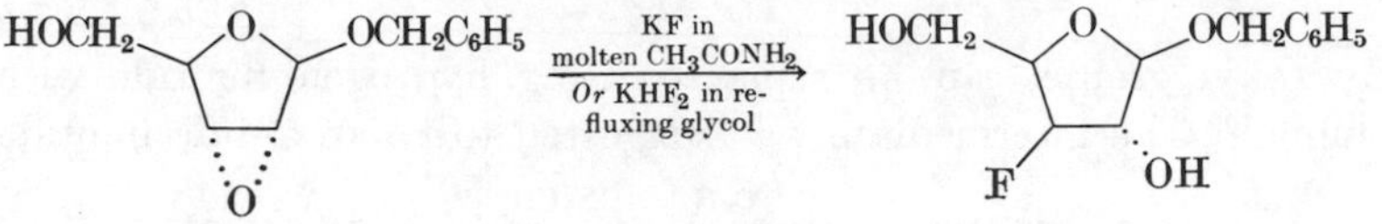

$$\text{HOCH}_2\text{–(2,3-epoxy furanoside)–OCH}_2\text{C}_6\text{H}_5 \xrightarrow[\text{Or KHF}_2\text{ in refluxing glycol}]{\text{KF in molten CH}_3\text{CONH}_2} \text{HOCH}_2\text{–(3-F, 2-OH furanoside)–OCH}_2\text{C}_6\text{H}_5$$

Mechanism

No extensive mechanism studies have been made of the addition of hydrogen fluoride to epoxides or the substitution of carboxylate groups on sugars by hydrogen fluoride. Product analysis permits reasonable mechanisms to be postulated.

Addition of Hydrogen Fluoride to Carbon-Carbon Unsaturation

Most additions of hydrogen fluoride to alkenes and alkynes have been carried out in anhydrous liquid hydrogen fluoride. Under these conditions, hydrogen fluoride is polymeric; evidence indicates the presence of significant hexamer, H_6F_6, in the liquid phase.[168, 204] Although liquid anhydrous hydrogen fluoride is a powerful Lewis acid and is exceeded in proton-donor strength by only a few acids, the concentration of free protons present is small. The mechanism of addition of hydrogen fluoride to

$$H_6F_6 + B{:} \rightarrow B{:}H^+ + H_5F_6^-$$
$$H_6F_6 \rightleftarrows H^+ + H_5F_6^-$$

alkenes has been discussed.[16] It is not clear whether *trans* addition is a two-step process or a one-step concerted process.

Two-Step *trans* Addition.

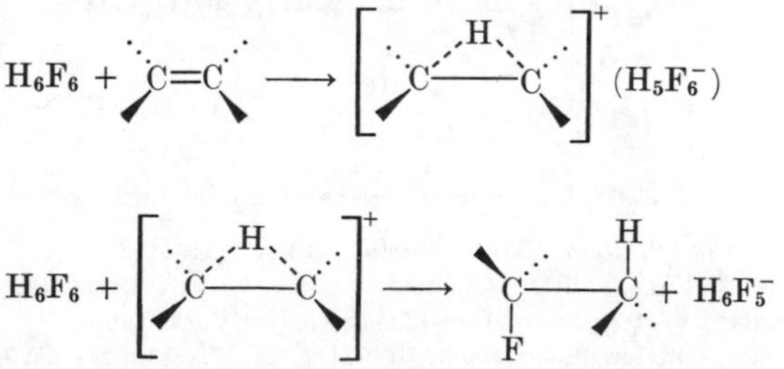

$$H_6F_6 + \text{C=C} \longrightarrow \left[\text{C}\overset{H}{\text{—}}\text{C}\right]^+ (H_5F_6^-)$$

$$H_6F_6 + \left[\text{C}\overset{H}{\text{—}}\text{C}\right]^+ \longrightarrow \text{F–C–C–H} + H_6F_5^-$$

[203] S. Cohen, D. Levy, and E. D. Bergmann, *Chem. Ind.*, (London), **1964**, 1802.

[204] J. H. Simons, *Fluorine Chemistry*, Vol. 5, Academic Press, New York, 1965, pp. 2–14.

One-Step Concerted *trans* Addition. The polymerization and rearrangement of some alkenes during addition of hydrogen fluoride argues for the two-step process for many alkenes.

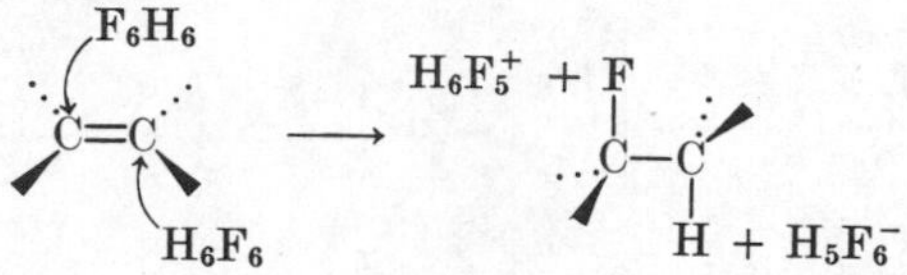

$$F_6H_6 \quad \mathrm{C{=}C} \quad H_6F_6 \longrightarrow H_6F_5^+ + \mathrm{F{-}C{-}C{-}H} + H_5F_6^-$$

Hydrogen fluoride adds *cis* to 6-deutero-3-acetoxycholestane. This apparent abnormal *cis* addition could be the result of a normal *trans* addition followed by epimerization,[172] since 6β-fluoro-steroids are well

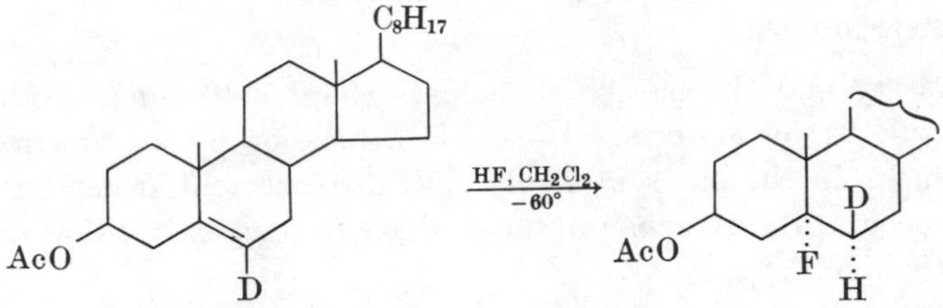

C_8H_{17} ; AcO ; D $\xrightarrow[-60°]{HF,\ CH_2Cl_2}$ AcO ; D ; F ; H (Ref. 171)

known to epimerize easily with hydrogen fluoride to the more stable 6α-fluoro-steroids.[181, 189, 191, 205a] However, some abnormal steric effects could be responsible, as observed in certain electrophilic additions to 5,6-unsaturated steroids.[32] Recently deuterium fluoride (DF) was used to show that isomerization and backbone rearrangements proceed through olefinic intermediates but that *cis* addition and elimination does occur with a specific steroid.[205b]

A four-center addition would also explain a *cis* orientation of the product and has been proposed for gas-phase addition of hydrogen fluoride to carbon-carbon unsaturation. However, theoretical arguments are not confirmed by experiment.[206–208]

Although not discussed earlier in this review, the addition of hydrogen fluoride to unreactive alkenes, halogenated alkenes in particular, is catalyzed by Lewis acids such as boron trifluoride and antimony(V) fluoride. These reactions have been reviewed.[16] Presumably catalysis occurs through formation of the very strong fluoroboric acid or similar

[205] (a) J. A. Edwards, H. J. Ringold, and C. Djerassi, *J. Amer. Chem. Soc.*, **82,** 2318 (1960): (b) J. Barbier, C. Berrier, J.-C. Jacquesy, and R. Jacquesy, *Tetrahedron*, **29,** 1047 (1973).

[206] S. W. Benson and G. R. Haugen, *J. Phys. Chem.*, **70,** 3336 (1966).

[207] J. M. Simmie, W. J. Quiring, and E. Tschuikow-Roux, *J. Phys. Chem.*, **74,** 992 (1970).

[208] E. Tschuikow-Roux, W. J. Quiring, and J. M. Simmie, *J. Phys. Chem.*, **74,** 2449 (1970).

fluoro acid which is a much more powerful proton donor than hydrogen fluoride alone. Mixtures of boron trifluoride and hydrogen fluoride are

$$H_6F_6 + BF_3 \rightarrow H^+BF_4^- + H_5F_5$$

$$H^+BF_4^- + \rangle C{=}C\langle \longrightarrow \left[\left(\rangle C{-}CH\right)^+ \rightleftharpoons \left(\rangle C\overset{H}{\text{---}}C\langle\right)^+\right] + BF_4^-$$

therefore more powerful reagents for adding hydrogen fluoride than either reagent alone. A tabular comparison of acid strengths of various fluoro acids in liquid hydrogen fluoride is available.[204]

Fluoride Substitution of Fully Esterified Sugars by Hydrogen Fluoride

Stereochemistry and the nature of vicinal groups determine whether substitution will be by a normal back-side attack or by an apparent front-side attack. In the cases in Table XV, displacement from the 1-carbon atom occurs; the 1-carbon atom is alpha to oxygen, making this position most active.

When the sugar is a 2-deoxy sugar, normal back-side attack is expected with inversion of configuration at the reaction site (see p. 199).

When a *trans* substituent is present in the 2 position normal, back-side attack can occur or a product from an apparent front-side attack may be isolated. For example, tetra-O-benzoylribofuranose reacted with hydrogen fluoride to give tri-O-benzoyl-β-D-ribofuranosyl fluoride in 70% yield, the apparent product from a front-side fluoride attack. The reaction was studied by nmr and shown to proceed through an intermediate benzoxonium ion.[175] Thus the apparent front-side attack was the result

ROCOCH$_2$ — O — OCOR; RCO$_2$, OCOR $\xrightarrow[\text{10 min}]{\text{HF}}$ [ROCOCH$_2$ — O — O; RCO$_2$, O—C$^+$—R] $\longrightarrow$ ROCOCH$_2$ — O — F; RCO$_2$, OCOR

$R = C_6H_5$

of an intramolecular back-side attack followed by the intermolecular back-side attack by fluoride.

Further evidence for the benzoxonium ion intermediate was found in an nmr investigation of a ring contraction observed during substitution of carbohydrate esters by anhydrous hydrogen fluoride.[209]

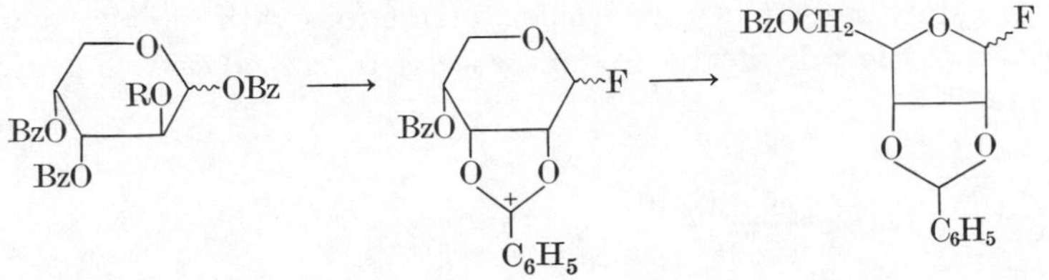

An initial protonation of the leaving group is certainly important. When hydrogen fluoride is used, the rate of displacement is orders of magnitude faster than the rate of substitution when potassium acid fluoride is used. Substitution with the latter reagent requires a solvent (*e.g.*, glycol) at high temperature (150–200°).

In all likelihood most of the substitutions in Table XV proceed by conventional back-side nucleophilic attacks or by some mechanism with

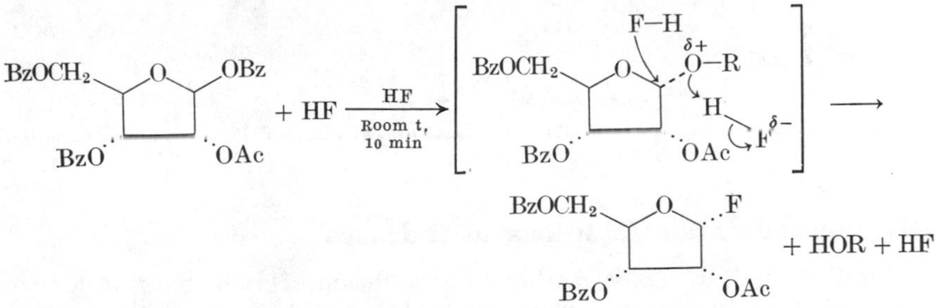

+ HF —HF, Room t, 10 min→

+ HOR + HF

intermediates such as that shown for fluoride substitution of tetra-O-benzoyl-β-D-ribofuranose. Nevertheless, in the absence of additional evidence, a four-center mechanism cannot be ruled out for all substitutions by hydrogen fluoride.[175] The small size of hydrogen fluoride favors four-center reactions, as do the short hydrogen fluoride bond length and its low degree of dissociation. The accompanying example is suggested as a reaction in which a four-center front-side reaction may occur.[175]

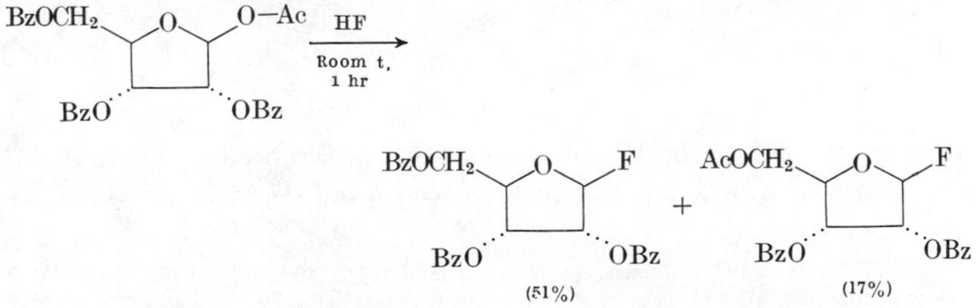

—HF, Room t, 1 hr→

(51%) (17%)

[209] C. Pedersen, *Angew. Chem., Int. Ed. Engl.*, **11**, 241 (1972).

In mechanistic studies of fluoride substitution on sugars with hydrogen fluoride, a major problem is subsequent rearrangement or reaction of the product formed initially. Rearranged products from carbonium ion species that are stabilized by intramolecular interactions with ester groups were proposed in the 2-deoxyglycopyranose series from studies with deuterium fluoride.[210]

BzO, OBz, OBz; $\xrightarrow[0^\circ]{HF,\ C_6H_6}$; F; BzO; OBz; BzO; O; ^+C; O; C_6H_5; F +; F +; OR; BzO; Bz, H; BzO

Hydrogen Fluoride Additions to Oxiranes

Product analysis indicates that *trans* addition of hydrogen fluoride to oxiranes normally occurs. The strong Lewis-acid nature of hydrogen fluoride makes an initial protonation of the oxirane probable. A back-side attack by hydrogen fluoride follows. When one carbon atom is more highly substituted by alkyl groups, fluorine will add to that carbon. Where

R, O, R, H, R + 2 HF ⟶ H- - -$F^{\delta-}$, δ^+, O, R, C, C, R, H, δ^+, R, HF ⟶ R, OH, R, C–C, H, F, R + solvated H^+

steric effects make a back-side attack difficult, the product of front-side attack can form or some side reaction can occur.[211] In cases where *cis*-

[210] (a) I. Lundt and C. Pedersen, *Acta Chem. Scand.*, **25**, 2749 (1971); (b) K. Bock and C. Pedersen, *ibid.*, **25**, 2757 (1971).

[211] J.-C. Lanet and M. Mousseron-Canet, *Bull. Soc. Chim. Fr.*, **1969**, 1751.

C_8H_{17} HO F O HF, $CHCl_3$, 0°, 1 hr HO F F OH

C_8H_{17} HO F O HF, $CHCl_3$, 0°, 1 hr O F HO

fluorohydrins are formed the product probably forms by a two-step process rather than by a one-step front-side addition. The first step is normal addition to give *trans*-fluorohydrin. In a subsequent second step, epimerization occurs. This two-step pathway has been established for the addition of hydrogen fluoride to 4β,5β-epoxyandrostan-17β,3-one.[179]

OAc O O HF, $CHCl_3$, Room t, 1 hr O HO F HF $CHCl_3$ O HO F

The sensitivity of the hydrogen fluoride addition to solvent and substituent groups is illustrated by the reaction of a similar steroid, 4β,5β-epoxy-10β-hydroxyestra-3,17-dione, in ethanol-chloroform.[179, 185]

O HO O O HF, $CHCl_3$—C_2H_5OH 0° HO F O

In general, the addition of hydrogen fluoride to steroid oxiranes in a suitable solvent is not accompanied by extensive skeletal rearrangements. Rearrangements such as that presented earlier (pp. 202–203)[184] are exceptions rather than the rule. In most cases, if a carbonium intermediate forms, it is intercepted by fluoride before rearrangement occurs.

The mechanistic importance of fluoride ion in stereospecific *trans* addition of hydrogen fluoride to oxiranes is emphasized by the reactions of diisopropylamine trihydrofluoride with oxiranes. In this work, yields of fluorohydrin were high and success was achieved with oxiranes that underwent polymerization and rearrangement with hydrogen fluoride alone.[163]

$(i\text{-}C_3H_7)_2NH \cdot 3\,HF$, 110°, 18 hr (65%) (Ref. 163b)

In this system, attack of fluoride ion on the oxirane could occur first and be independent of initial protonation of oxygen. Fluoride ion should be donated by diisopropylammonium fluoride and the proton by hydrogen fluoride.

Similarly, potassium acid fluoride (KF·HF or KHF_2) is effective because of increased availability of fluoride ion.[162a] The conditions for this

KHF_2; Glycol, Reflux, 40 min

reaction are similar to those for alkali fluoride displacement of halide and sulfonate esters. The reaction can certainly occur by an initial fluoride ion attack followed by proton abstraction from solvent or potassium acid fluoride.

F^- → Solvent or KHF_2 → product

The importance of protonation of the oxirane is emphasized by the greater yields obtained when boron trifluoride etherate is used to catalyze the addition of hydrogen fluoride. In this system, fluoroboric acid must act as the protonating species and hydrogen fluoride as the fluoride ion donor.[187]

(HBF_4) HF, BF_3·ether, ether → HF → (60%)

In summary, depending on the reaction system, five mechanisms appear possible for adding hydrogen fluoride to oxiranes.

System 1. Anhydrous hydrogen fluoride in a solvent gives either (a) two-step addition in which hydrogen fluoride donates a proton and hydrogen fluoride subsequently donates a fluoride ion, or (b) one-step concerted *trans* addition with 2 molecules of hydrogen fluoride donating a proton and a fluoride, or (c) concerted *cis* addition of one hydrogen fluoride molecule.

System 2. Hydrogen fluoride is complexed with a Lewis base such as isopropylamine or potassium fluoride forming $(i\text{-}C_3H_7)_2NH\cdot 3\,HF$ or KHF_2, which donates a fluoride ion, and hydrogen fluoride donates a proton.

System 3. Hydrogen fluoride is complexed with a Lewis acid such as boron trifluoride to give fluoroboric acid, $HF\cdot BF_3 \leftrightharpoons HBF_4$, where fluoroboric acid donates a proton to the oxirane and hydrogen fluoride donates the fluoride ion.

Reaction of Boron Trifluoride Etherate and Oxiranes

The reaction of boron trifluoride etherate with oxiranes is highly sensitive to the stereochemistry of and the substitution on the oxirane and to the solvent. In general, nonpolar solvents lead to rearranged products and polar solvents to fluorohydrins. Several examples have been given earlier (see pp. 205–210).

The initial step in the reaction must be formation of an oxonium borate complex **33**. Depending on solvent and substituents, the intermediate will

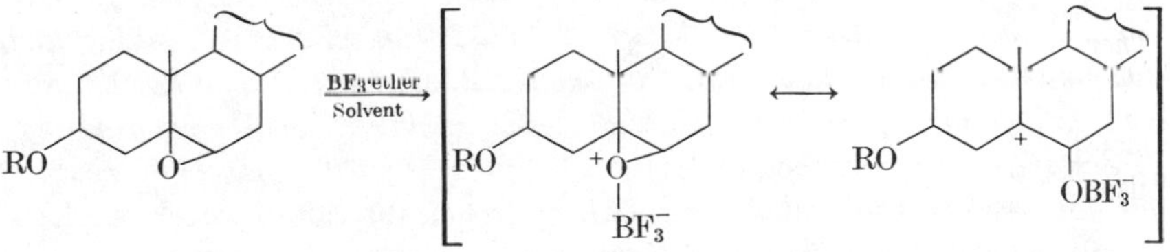

RO O $\xrightarrow[\text{Solvent}]{BF_3\cdot\text{ether}}$ [RO $^+$O BF_3^- ⟷ RO + OBF_3^-]

33a **33b**

behave as if **33a** or **33b** were present. In polar solvents, fluoride ion can be transferred intermolecularly or intramolecularly and fluorohydrins

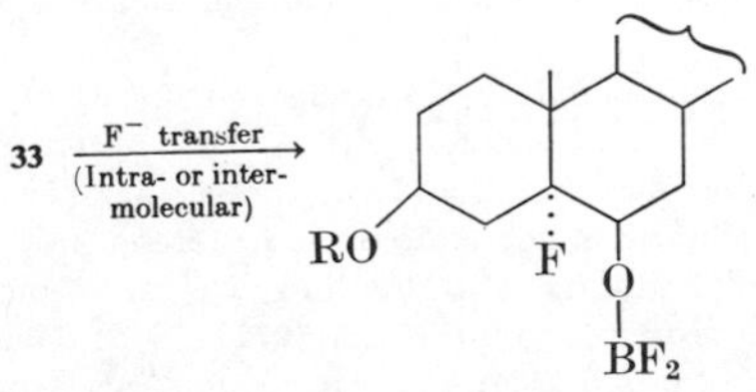

33 $\xrightarrow[\text{(Intra- or inter-molecular)}]{F^-\ \text{transfer}}$ RO F O BF_2

result. Concerted rearrangements can also occur as shown for form **33b** to give rearranged products plus fluorohydrin.

In a nonpolar solvent, rearrangements predominate. Other possibilities

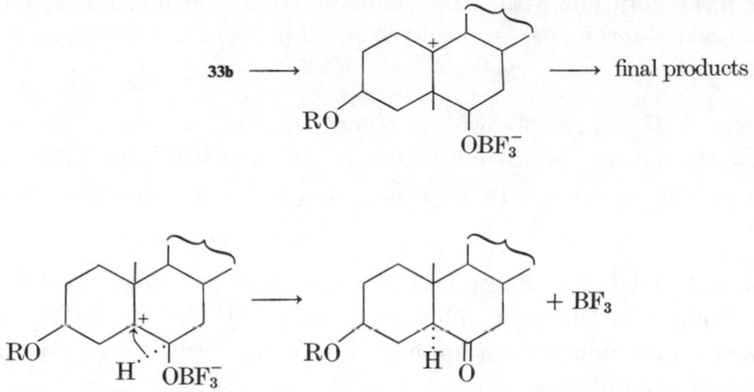

33b ⟶ [structure: RO, +, OBF_3^-] ⟶ final products

[structure: RO, +, H, OBF_3^-] ⟶ [structure: RO, H, O] + BF_3

also exist. Because the reaction of boron trifluoride etherate with oxiranes is not a good general synthetic method for monofluorination, no additional discussion is warranted here, but extensive discussions are available.[4, 179, 194–196]

Experimental Conditions

Reactions with hydrogen fluoride at temperatures below 19° may be carried out at atmospheric pressure in open systems. Above the boiling point (19°), reactions are carried out in autoclaves constructed of mild steel. Because hydrogen fluoride rapidly attacks glass, polyethylene or polypropylene apparatus is normally used. *Hydrogen fluoride rapidly attacks human tissues. All investigators must wear protective clothing and face masks. Rubber gloves must be inspected carefully for pinholes before use. Inhalation of hydrogen fluoride leads rapidly to pulmonary edema. Therefore, work must be conducted in well ventilated hoods with positive air flow.* Before beginning work with hydrogen fluoride research workers should consult manufacturers' data sheets or other reliable sources that discuss the handling and properties of hydrogen fluoride.[74, 169, 212–214, *]

* "Treatment of Hydrogen Fluoride Injuries," A. J. Finkel, *Advan. Fluorine Chem.*, **7**, 199 (1973).

[212] J. F. Gall, "Hydrogen Fluoride," under "The Chemistry and Technology of Fluorine," in Kirk-Othmer, *Encyclopedia of Chemical Technology*, 2nd ed., Vol. 9, Interscience, New York, 1966, pp. 610–625.

[213] *Hydrofluoric Acid*, Chemical Safety Data Sheet SD-25, 1970, Manufacturing Chemists' Association, Inc., 1825 Connecticut Avenue, N.W., Washington, D.C. 20009.

[214] E. I. du Pont de Nemours and Co., Inc., Industrial Chemicals Dept., Wilmington, Del. Bulletins: (a) Hydrofluoric Acid, Anhydrous; Storage and Handling, No. A 72708; (b) Hydrofluoric Acid, 70%; Storage and Handling, No. A 72709; (c) Basic Facts for Safe Handling of Anhydrous Hydrofluoric Acid, No. A 72711-1; (d) Basic Facts for Safe Handling of Hydrofluoric Acid, 70%, No. A 72724; (e) First Aid for Anhydrous Hydrofluoric Acid, No. A 72712-1; (f) First Aid for Hydrofluoric Acid, 70%, No. A 72725.

In addition, first-aid procedures for hydrogen fluoride must be learned by all persons who may contact hydrogen fluoride and by coworkers who are in the area and may be inadvertently exposed or may be required to assist in case of an accident.

A local first-aid station should be equipped with the following items for treating hydrogen fluoride burns.[214e,f]

An aqueous solution of benzalkonium chloride* in a concentration of 0.10–0.13%. An alternative is a solution of 70% denatured ethyl alcohol, or a saturated solution of magnesium sulfate (epsom salt).

Magnesia paste,† or HF ointment,† or "A and D"‡ ointment 0.5%, or "Pontocaine"§ Hydrochloride (available on prescription).

99.5% Pure USP medical oxygen.

Ice cubes (not crushed ice).

Gauze, compression dressings, eye patches.

Towels.

Basins of assorted sizes, shower facilities.

Emergency Treatment for Hydrogen Fluoride Burns

Skin Contact with Hydrogen Fluoride.[214e,f] The following procedures have been recommended for first-aid treatment of external skin contact with hydrogen fluoride.

1. Flush the area immediately with large quantities of cool water. Remove clothing as quickly as possible while flushing. If possible, immerse the affected part(s) in water cooled by ice cubes on the way to the first-aid station.

2. Take the victim to the first-aid station. Keep the affected part immersed in ice water until treatment can begin.

3. Immerse the burned part in an iced solution of benzalkonium chloride for 1–4 hours depending on the appearance and extent of the burn. (An iced solution of 70% ethyl alcohol or epsom salt could also be used.)

* *Benzalkonium Solution.* Solutions of benzalkonium chloride for use as skin treatments should be in concentrations of 0.10 to 0.13%. Benzalkonium chloride is available under the trade name "Zephiran"§ chloride either as an aqueous solution (1:750) or concentrate (17%) which must be diluted. To prepare one gallon of a "Zephiran" solution from the concentrate, mix 1 fluid ounce of "Zephiran" (17%) and 127 fluid ounces of water.

† *Ointment.* Magnesia paste is made by the addition of USP glycerine to USP magnesium oxide powder to form a thick paste.

HF ointment is prepared by mixing 3 ounces of magnesium oxide powder, 4 ounces of heavy mineral oil, and 11 ounces of white petrolatum.

Although these ointments are effective, they tend to harden and become difficult to remove. The use of "A and D" ointment avoids this difficulty.

‡ U.S. Pat., White Laboratories, Inc.

§ U.S. Pat., Winthrop Laboratories, Inc.

The solution may need to be removed periodically from the area being soaked to relieve discomfort caused by the cold.

If immersion is impractical, a compress made by inserting ice cubes between layers of gauze may be applied to the burned part for 1–4 hours, keeping it continually soaked with the benzalkonium solution.

Observe the injured. Be prepared to treat the injured for shock.

4. Consult a physician before undertaking additional treatment.

5. If no physician is available after the injured part has been soaked for up to 4 hours in benzalkonium chloride solution, dry the part gently. Apply magnesia paste, HF ointment, or "A and D" ointment to the burned area, covering it with a compression dressing. Oils and greases should not be applied except under instructions from a physician.

6. If blisters form while the affected part is being soaked in benzalkonium chloride, treatment by a physician is needed immediately. If no physician is available, a medical attendant should cut away all white raised tissue and free the blistered area of debris.

7. Burns around fingernails are extremely painful; the nails may have to be split by a physician. The nail should be split from the distal end to relieve pain, facilitate drainage, and prevent hydrogen fluoride from infiltrating into the deeper structures.

8. After first-aid treatment, care should be directed by a physician.

Anyone who knows or even suspects that he has come in contact with hydrofluoric acid should immediately seek first aid.

Eye Contact with Hydrogen Fluoride. 1. Flush the eyes immediately with large quantities of clean water while holding the eyelids apart. Continue flushing for 15 minutes, then repeat once or twice, as required, at 15-minute intervals.

2. When not flushing the eyes, apply ice compresses intermittently for at least 1 hour.

3. Call an eye physician as soon as possible.

4. To alleviate the pain until the physician arrives, instill 2 drops of 0.5% "Pontocaine"§ hydrochloride solution into the eye. No oils or oily ointments should be instilled unless directed by a physician.

Vapor Inhalation. 1. Remove victim immediately to an uncontaminated atmosphere and get medical help. Keep him warm. *Note:* To prevent the development of severe lung congestion (pulmonary edema) a medical attendant should administer oxygen as soon as possible. As a rule, patients respond satisfactorily to unpressurized inhalation with a respirator-type mask. The oxygen inhalation lasts 1/2 hour, then continues at half-hour intervals for at least 3–4 hours. Oxygen inhalation may then be discontinued if there are no signs of lung congestion, and if the person's breathing is easy and his color good.

2. Keep the person under medical observation for at least 24–48 hours.

Experimental Procedures

The scope of the addition of hydrogen fluoride to alkenes and acetylenes is given in a recent review.[16] Detailed procedures are given for addition of hydrogen fluoride to 1,2-dichloro-2-propene and to 1-octyne[169] and by Henne[5] for addition to ethylene, propylene, and cyclohexene.

The reactions discussed earlier in this section were carried out at atmospheric pressure in plastic or metal (Cu, mild steel) containers. Reactants were simply mixed and stirred from $-70°$ to room temperature. The only difficulty in the procedures is that extensive precautions are required for handling hydrogen fluoride.

2-Fluoro-3-hydroxy-*trans*-decalin. A general method has been developed for the addition of hydrogen fluoride to epoxides.[215] In a polyethylene vessel, a solution of 2,3-epoxy-*trans*-decalin (33 g, 0.22 mol) in a mixture of dry chloroform (50 ml) and dry tetrahydrofuran (60 ml) was cooled to 0°. A cold solution of anhydrous hydrogen fluoride (42 g, 2.2 mol) in dry chloroform (200 ml) was added in small portions, with stirring, during 90 minutes. The solution, which had turned red, was stirred at 0° for an additional 90 minutes and then poured into excess aqueous potassium carbonate. The organic phase was separated, washed with water, and dried, and the solvent was removed through a short Vigreux column. The residue was distilled under reduced pressure. The fraction boiling at 72 100° (0.2 mm) was redistilled, giving a mixture of two isomeric fluorohydrins, bp 75° (0.2 mm). The mixture could be separated by vapor-phase chromatography into two compounds in the proportion 1:3. However, the fluorohydrins are unstable unless kept over dry potassium fluoride at 0° and even then begin to decompose after 1–3 days, so no preparative separation was attempted. The overall yield was 50%.[178]

The same experimental procedure was applied to twelve other epoxides.

9α-Fluorohydrocortisone Acetate (Addition to a Steroid Epoxide).[216] Δ^4-Pregnen-9β,11β-oxido-17α-21-diol-3,20-dione 21-acetate (50 g, 0.080 mol) was dissolved in redistilled chloroform (1 l) in a polyethylene bottle provided with a magnetic stirrer and a polyethylene inlet tube. Anhydrous hydrogen fluoride (approximately 60 g, 3.1 mol) was passed into the solution at 0° with thorough agitation. The addition

[215] H. B. Henbest, M. Smith, and A. Thomas, *J. Chem. Soc.*, **1958**, 3293.
[216] J. Fried and E. F. Sabo, *J. Amer. Chem. Soc.*, **79**, 1130 (1957).

required about 20 minutes. The color of the solution gradually deepened, and at the end of the addition became an intense cherry red. At the same time the solution separated into two layers, with most of the colored material concentrated in the upper, smaller layer.* After $1\frac{1}{2}$ hours at 0° a suspension of sodium bicarbonate in water was added carefully with stirring until the mixture became weakly basic. It was then transferred to glass equipment, and the steroids were isolated from the chloroform extract. The crude crystalline material (55 g) was slurried in ethyl acetate (100 ml) and allowed to crystallize for 45 minutes at 5°. Additional ethyl acetate was added (375 ml), and the mixture was heated to boiling. The bulk of the material went into solution, leaving a sand-like residue consisting mostly of a by-product. The residue (4.3 g., mp 245–250°) was filtered. The filtrate was allowed to crystallize at room temperature for 15 minutes, and the resulting 9α-fluorohydrocortisone acetate was filtered (22 g, mp 228–229°). Concentration of the mother liquors gave in successive crystallizates an additional 3 g of product and 1.5 g of impure by-product. Recrystallization of the combined high-melting fractions (5.8 g) from acetone gave an additional 1.4 g of 9α-fluorohydrocortisone acetate. The total yield of the latter is therefore 26.4 g (50.5%). Analytically pure material was obtained after two recrystallizations from ethyl acetate. This material contained varying amounts of ethyl acetate (5–9%) and was therefore dried at 100° for 2 hours; mp 232–233°; $[\alpha]_D^{23}$ +143° (c 0.50, $CHCl_3$), +127° (c 0.54, acetone).

When the reaction was carried out in chloroform containing 5% ethanol there was only a single phase and the yield of product rose to 60–65%.

The by-product (4.4 g, 9%) was purified by crystallization from acetone and identified as Δ^4-pregnene-11β,17α,21-triol-3,20-dione 21-acetate, the alcohol resulting from hydrolysis of epoxide starting material.

α-D-Glucopyranosyl Fluoride Tetraacetate (Substitution of a Sugar Ester).[149] D-Glucose pentaacetate (5.0 g, 1.25 mol) was dissolved in anhydrous liquid hydrogen fluoride (20 ml) which had been cooled in a polyethylene vessel to −78° (solid carbon dioxide-acetone). The resulting solution was left to warm to room temperature during about 20 minutes and was then poured directly into a mixture of ether and aqueous sodium bicarbonate. The ether layer was washed with water and dried over sodium sulfate. Evaporation of this solution produced a clear syrup which crystallized as needles from ethanol (3.3 g, 74%), mp 107–111°. Recrystallization from aqueous ethanol afforded pure α-D-glucopyranosyl fluoride

* The appropriate color development can be ascertained easily with some experience (light behind polyethylene vessel) and is an excellent criterion for the attainment of the proper concentration of hydrogen fluoride. When too little hydrogen fluoride (color pink to bright red) was used only the epoxide could be isolated.

tetraacetate, mp 110–112°, $[\alpha]_D^{23}$ + 91.0° (c 2.93) (lit mp 108°, $[\alpha]_D^{20}$ +90.1°).

Tri-O-benzoyl-β-D-ribofuranosyl Fluoride.[175] A solution of tetra-O-benzoyl-β-D-ribofuranose (500 mg, 0.885 mmol) was kept in 1 ml of anhydrous hydrogen fluoride for 10 minutes at room temperature. Methylene chloride was then added and the mixture was poured into water; the organic layer was washed with aqueous sodium hydrogen carbonate and dried. Crystallization of the crude syrup (382 mg) from ether-pentane gave 284 mg (69%) of tri-O-benzoyl-β-D-ribofuranosyl fluoride, mp 79–81°. Recrystallization from ether-pentane gave the pure product as colorless crystals, mp 81–83°, $[\alpha]_D^{22}$ +120.3° (c 4, $CHCl_3$).

1-Fluoro-2-indanol.[163b] Indane oxirane (15.4 g, 0.116 mol) was added to 28.2 g (0.175 mol) of diisopropylamine trihydrofluoride, $(i\text{-}C_3H_7)_2NH \cdot 3\,HF$, in a 250-ml flat-bottomed flask connected to a gas bubbler containing a standardized solution of hydrochloric acid. Nitrogen or dry air was passed through the flask while the mixture was magnetically stirred for 5 hours at 78°.

The fluoroalcohols were extracted with several portions of boiling ether until the reaction mixture was decolorized. The mixture was further extracted with 50 ml of boiling benzene. The organic phases were combined, washed twice with distilled water, and dried over potassium carbonate. The product was distilled under vacuum to give 10.4 g (59%) of *cis*- and *trans*-1-fluoro-2-indanol, bp 70–83° (0.1–0.2 mm), as a viscous colorless liquid which was stored at 0°.

PERCHLORYL FLUORIDE

Caution! Fluorinations with perchoryl fluoride are potentially hazardous because explosions are possible. Safety precautions are mandatory when working with this reagent! (See p. 232.)

Availability and Preparation of Reagent

Perchloryl fluoride (ClO_3F) is a gas manufactured by the Pennwalt Corporation, commercially available from Pennwalt and from PCR, Inc. Detailed lists of its physical properties are available.[217, 218] Some of the

[217] *Perchloryl Fluoride*, Booklet DC-1819, Pennsalt Chemicals Corporation, Philadelphia, Pa., 1957, pp. 1–24. Note that the company is now Pennwalt Corp. This booklet was supplemented by Bulletin DS I-1819, "Perchloryl Fluoride PF®, Laboratory Safety."

[218] J. F. Gall, "Perchloryl Fluoride," under the Chemistry and Chemical Technology of Fluorine, in Kirk-Othmer, *Encyclopedia of Chemical Technology*, 2nd ed., Vol. 9, Interscience, New York, 1966, pp. 598–610.

key properties of perchloryl fluoride are bp −46.67°, mp −147.7°, $d^{-190°}$ 2.19, T_c 95.9°.

Perchloryl fluoride was first synthesized from fluorine and a perchlorate salt in 1951,[219] but its structure was not recognized until later.[220] In 1952

$$F_2 + KClO_3 \rightarrow ClO_3F + KF$$

perchloryl fluoride was prepared by the electrolysis of a solution of sodium perchlorate in liquid hydrogen fluoride.[221] This laboratory-scale electroly-

$$Na^+ClO_3^- + 2\,HF \rightarrow ClO_3F + NaF + H_2$$

sis was developed for large-scale commercial production.[222] Alternative syntheses are available.[223]

$$KClO_4 \xrightarrow[\text{or } SbF_5]{FSO_3H} ClO_3F$$

We strongly recommmend that perchloryl fluoride be purchased for laboratory fluorinations, unless the laboratory is specially equipped and experienced personnel are available to handle high-energy fluorine compounds. No work with perchloryl fluoride should be started until pertinent literature is studied, particularly the manufacturer's brochure[217] or a published review,[218] which should be available in most libraries.

Scope and Limitations as a Monofluorination Reagent

Perchloryl fluoride was developed initially as a unique reagent for substituting *two fluorine* atoms for the hydrogen atoms of an activated methylene group.[224]

$$CH_2(CO_2C_2H_5)_2 \xrightarrow[2.\ 2\ FClO_3]{1.\ 2\ Na} CF_2(CO_2C_2H_5)_2$$

$$CH_3COCH_2COCH_3 \xrightarrow[2.\ 2\ FClO_3]{1.\ 2\ Na} CH_3COCF_2COCH_3$$

$$CH_3COCH_2CO_2C_2H_5 \xrightarrow[2.\ 2\ FClO_3]{1.\ 2\ Na} CH_3COCF_2CO_2C_2H_5$$

219 H. Bode and E. Klesper, *Z. Anorg. Allg. Chem.*, **266**, 275 (1951).

220 H. Bode and E. Klesper, *Angew. Chem.*, **66**, 605 (1954).

221 (a) A. Engelbrecht and H. Atzwanger, *Monatsh. Chem.*, **83**, 1087 (1952); (b) *J. Inorg. Nucl. Chem.*, **2**, 348 (1956).

222 G. Barth-Wehrenalp, *J. Inorg. Nucl. Chem.*, **2**, 266 (1956).

223 (a) G. Barth-Wehrenalp, U.S. Pat. 2,942,948 (1960) [*C.A.*, **54**, 21681c (1960)]; (b) G. Barth-Wehrenalp and H. C. Mandell, Jr., U.S. Pat. 2,942,949 (1960) [*C.A.*, **54**, 21681d (1960)]; (c) A. F. Englebrecht, U.S. Pat. 2,942,947 (1960) [*C.A.*, **54**, 21681b (1960)]; (d) W. A. Lalande, Jr., Ger. Pat. 1,026,285 (1958) [*C.A.*, **54**, 20117i (1960)]; (e) H. R. Leech, *Chem. Ind.* (London), **1960**, 242; (f) W. A. Lalande, Jr., U.S. Pat. 2,982,617 (1961) [*C.A.*, **55**, 16926i (1961)].

224 C. E. Inman, R. E. Oesterling, and E. A. Tyczkowski, *J. Amer. Chem. Soc.*, **80**, 6533 (1958).

An excellent reinvestigation in 1966 of the fluorination of diethyl malonate[225] disclosed the true potential of perchloryl fluoride as a *mono*-fluorination reagent. By choice of solvent and conditions, perchloryl fluoride could be controlled to act as a mono- or di-fluorination reagent, as shown in Table VII and Chart 1.[225, 226] Most organic compounds that form

BLE VII. REACTIONS OF PERCHLORYL FLUORIDE: COMPOSITION OF FLUORINATED MIXTURE FROM MALONATE ESTERS

Ester	Base[a] (Solvent)	Products, % Yield: $F_2C(CO_2C_2H_5)_2$	$C_2H_5CH(CO_2C_2H_5)_2$	$CH_2(CO_2C_2H_5)_2$
$_2(CO_2C_2H_5)_2$	$Na(C_2H_5OH)$	36.1	25.2	14.4
	$Na(C_2H_5OH)$	26.2	5.2	40.2
	Na(toluene)	29.2	0	29.3
$_5CH(CO_2C_2H_5)_2$	$Na(C_2H_5OH)$	—	33.3	—
	Na(toluene)	—	4.0	—
$H(CO_2C_2H_5)_2$	$Na(C_2H_5OH)$	85.6	—	—
	Na(toluene)	95.5[b]	—	—

Ester	Base[a] (Solvent)	Products, % Yield (*Continued*): $C_2H_5CF(CO_2C_2H_5)_2$	$FCH(CO_2C_2H_5)_2$	$(C_2H_5)_2C(CO_2C_2H_5)_2$
$_2(CO_2C_2H_5)_2$	$Na(C_2H_5OH)$	15.3	8.7	—
	$Na(C_2H_5OH)$	0.5	28.2	—
	Na(toluene)	0	41.6	—
$_5CH(CO_2C_2H_5)_2$	$Na(C_2H_5OH)$	51.5	—	15.2
	Na(toluene)	96.0	—	0
$H(CO_2C_2H_5)_2$	$Na(C_2H_5OH)$	6.7	7.6	—
	Na(toluene)	—	—	—

The ester, base, and ClO_3F are in the molecular proportion 1:1:1, except in the first reaction where the portion is 1:2:2. Sodium is added to the solvent, malonate added, and then $FClO_3$ bubbled in at 10–15°.
The remaining 4.5% consisted of two unidentified materials.

stable carbanions can be monofluorinated by perchloryl fluoride under

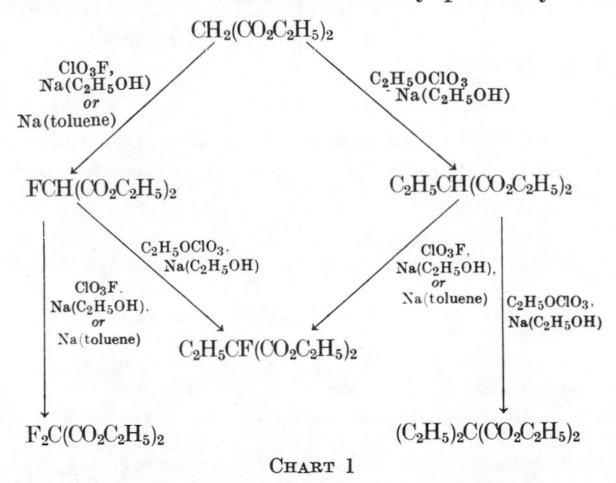

CHART 1

[225] H. Gershon, J. A. A. Renwick, W. K. Wynn, and R. D'Ascoli, *J. Org. Chem.*, **31,** 916 (1966).

[226] Ref. 15, Sheppard and Sharts, pp. 136–148.

appropriate experimental conditions. The competing side reaction of chlorate substitution is discussed under "Mechanism and Side-Reaction Products." Difluorination by perchloryl fluoride has been reviewed.[226, 227]

Monofluorination of Aliphatic Compounds Capable of Difluorination. Compounds such as diethyl malonate, malononitrile, ethyl acetoacetate, and 2,4-pentanedione can be *gem*-difluorinated. However, if one equivalent of base is used and the reaction with perchloryl fluoride is carried out in an inert aprotic solvent, monofluorination is readily accomplished. (See Table VII.)

$$CH_2(CO_2C_2H_5)_2 \xrightarrow{\text{Na, Toluene}} Na^+[:CH(CO_2C_2H_5)_2]^-$$

$$Na^+[:CH(CO_2C_2H_5)_2]^- \xrightarrow{FClO_3} FCH(CO_2C_2H_5)_2 + NaClO_3$$

Monofluorination of Monosubstituted Malonic Esters, Acetoacetic Esters, and Cyanoacetic Esters. Monosubstituted malonic esters and related compounds have been widely used as precursors to α-fluoroalkanoic acids. The monosubstituted compound is converted to an alkali metal derivative by an appropriate base (*e.g.*, Na, NaH, K).

The carbanion salt is suspended in an appropriate solvent and perchloryl fluoride gas is passed in. The resulting monofluoro malonic ester is then saponified and decarboxylated by standard techniques to give an α-fluoroalkanoic acid. Many examples of this type of reaction are given in

$$RCH(CO_2C_2H_5)_2 \xrightarrow[\text{or Na, Ethanol}]{\text{Na, Toluene}} Na^+[CR(CO_2C_2H_5)_2]^- \xrightarrow[0^\circ]{ClO_3F}$$

R = alkyl or aryl

$$FCR(CO_2C_2H_5)_2 \xrightarrow[\substack{2.\ H_2O,\ H^+ \\ 3.\ \text{Heat}}]{1.\ OH^-,\ H_2O} RCHFCO_2H$$

Table XVIII.[228–231]

For monofluorination of substituted cyanoacetic esters, aprotic solvents must be used to prevent side reactions involving the cyano group. For example, sodium ethoxide in ethanol as a base results in imidate derivatives.[228]

$$RCH(CO_2C_2H_5)CN \xrightarrow[2.\ ClO_3F]{1.\ NaOC_2H_5,\ C_2H_5OH} RCF(CO_2C_2H_5)C(OC_2H_5){=}NH$$

[227] V. M. Khutoretskii, L. V. Okhlobystina, and A. A. Fainzil'berg, *Russ. Chem. Rev*, **36,** 145 (1967).

[228] H. Gershon, S. G. Schulman, and A. D. Spevack, *J. Med. Chem.*, **10,** 536 (1967).

[229] H. Gershon and R. Parmegiani, *J. Med. Chem.*, **10,** 186 (1967).

[230] B. Cavalleri, E. Bellasio, A. Vigevani, and E. Testa, *Farm., Ed. Sci.*, **24,** 451 (1969) [*C.A.*, **71,** 112570h (1969)].

[231] B. Cavalleri, E. Bellasio, and E. Testa, S. Afr. Pat. 68 02,632 (1968) [*C.A.*, **71,** 12845h (1969)].

Monofluorination of Organometallic Compounds. Organometallic compounds are sources of carbanions. As a result, a great variety of organometallic compounds react with perchloryl fluoride. When proper solvent and reaction conditions are used, an organometallic compound is readily converted to a monofluorinated organic derivative. In experimental work so far reported, alkyl and aryl lithium compounds and Grignard reagents have been successfully monofluorinated. In the following examples the term "inverse addition" indicates that the organometallic compound was added to a solution of perchloryl fluoride.[232]

$$n\text{-}C_{12}H_{25}Li + ClO_3F \xrightarrow[-78°,\ \text{inverse addition}]{\text{THF/pet. ether (2:1)}} \underset{(39\%)}{n\text{-}C_{12}H_{25}F} + \underset{(20\%)}{C_{12}H_{26}}$$

$$C_6H_{11}\text{-}MgBr + ClO_3F \xrightarrow[-78°,\ \text{inverse addition}]{\text{HMPT/THF/(C}_2\text{H}_5)_2\text{O (1:1:1)}} \underset{(68\%)}{C_6H_{11}\text{-}F} + \underset{(25\%)}{C_6H_{12}}$$

$$C_6H_5MgBr + ClO_3F \xrightarrow[-78°,\ \text{inverse addition}]{(C_2H_5)_2O} \underset{(36\%)}{C_6H_5F} + \underset{(50\%)}{C_6H_6}$$

$$2,2'\text{-dilithiobiphenyl} + ClO_3F \xrightarrow[-78°]{\text{THF/pet. ether (2:1)}} \underset{(94\%)}{2,2'\text{-difluorobiphenyl}} + 2\text{-X-}C_6H_4\text{-}C_6H_5$$

X = H or F (trace)

Monofluorination of Nitroaliphatic Compounds. Mono- and dinitro aliphatic compounds are readily converted by strong bases to metallic derivatives (carbanions) which react with perchloryl fluoride to give monofluoronitro aliphatic compounds. Nitroaliphatic compounds are

$$K^+[:C(NO_2)_2CH_2OCH_3]^- \xrightarrow[20°,\ 15\ \text{min}]{\text{ClO}_3\text{F, Acetone}} FC(NO_2)_2CH_2OCH_3$$

sufficiently strong acids to be monofluorinated by perchloryl fluoride in an aprotic media without first forming a salt (the solvent acts as base).

[232] M. Schlosser and G. Heinz, *Chem. Ber.*, **102**, 1944 (1969).

Numerous examples appear in the United States[20, 233, 234] and Russian literature.[235]

Monofluorination of Ethers and Related Compounds.* A variety of enol derivatives react with perchloryl fluoride to form monofluorinated products.

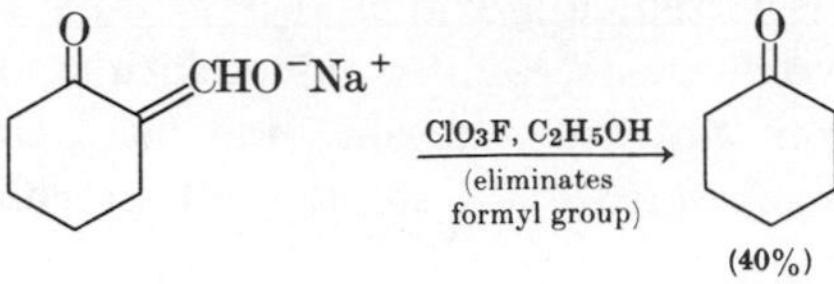

O CHO⁻Na⁺ → (ClO_3F, C_2H_5OH; eliminates formyl group) O F (40%) (Refs. 236, 237)

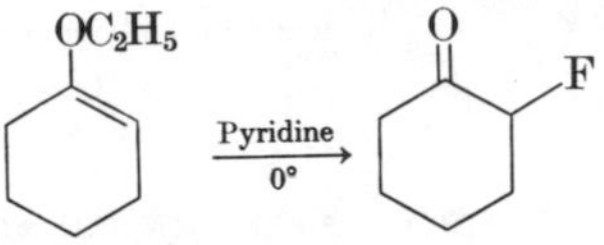

OC_2H_5 → (Pyridine, 0°) O F (Ref. 238)

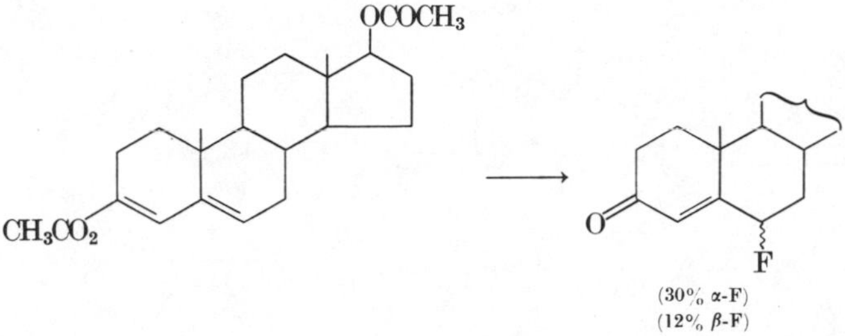

OCOCH$_3$ CH_3CO_2 → O F (30% α-F) (12% β-F) (Ref. 239)

The reactions of perchloryl fluoride with steroid derivatives, particularly enol ethers and enamines, to give monofluorinated steroids have been reviewed.[4, 9, 10] Many of the examples cited in these reviews are not repeated here.

Mechanism and Side-Reaction Products

No studies have been made specifically to determine the mechanism of reaction of perchloryl fluoride with organic compounds. Several mechanisms have been proposed which include perchloryl fluoride as a source of

[233] M. J. Kamlet and H. G. Adolph, *J. Org. Chem.*, **33**, 3073 (1968).

[234] J. P. Lorand, J. Urban, J. Overs, and Q. A. Ahmed, *J. Org. Chem.*, **34**, 4176 (1969).

[235] (a) L. V. Okhlobystina and V. M. Khutoretskii, *Bull. Acad. Sci. USSR, Div. Chem. Sci., Engl. Transl.*, **1969**, 1095; (b) V. M. Khutoretskii, L. V. Okhlobystina, and A. A. Fanzil'berg, *ibid.*, **1970**, 333; (c) L. V. Okhlobystina, V. M. Khutoretskii, and A. A. Fainzil'berg, *ibid.*, **1971**, 1387.

* Note added in proof: The oxofluorination of estra-1,3,5(10),6-tetraene-3,17β-diol diacetate with perchloryl fluoride gave 7α-fluoro-6-oxo-17β-estradiol diacetate. M. Neeman, Y. Osawa and T. Mukai, *J. Chem. Soc., Perkin I*, **1973**, 1462.

[236] (a) N. L. Allinger and H. M. Blatter, *J. Org. Chem.*, **27**, 1523 (1962); (b) A. S. Kende, *Tetrahedron Lett.*, No. 14, 13 (1959).

[237] B. L. Shapiro and M. M. Chrysam III, *J. Org. Chem.*, **38**, 880 (1973).

[238] S. Nakanishi, K. Morita, and E. V. Jensen, *J. Amer. Chem. Soc.*, **81**, 5259 (1959).

[239] Y. Osawa and M. Neeman, *J. Org. Chem.*, **32**, 3055 (1967).

"positive fluorine."[226] For the reaction of perchloryl fluoride with carbanions or other nucleophiles, we favor an initial attack by an electron pair on the chlorine of perchloryl fluoride.[240] Localized nucleophiles such

$$RO^- + FClO_3 \longrightarrow [RO\cdots ClO_3\cdots F]^- \longrightarrow ROClO_3 + F^-$$

as alkoxides give direct fluoride ion displacement. Mesomeric nucleophiles form an intermediate that can undergo intramolecular transfer of fluoride ion and form the thermodynamically very stable carbon-fluorine bond.

$$CH_3C(CO_2C_2H_5)=C(O^-)OC_2H_5 + FClO_3 \longrightarrow [H_3C(CO_2C_2H_5)C(C(O)OC_2H_5)\text{–}O\text{–}ClF(O)_2O^-] \longrightarrow H_3C(CO_2C_2H_5)C(F)C(O)OC_2H_5 + ClO_3^-$$

Recently organometallic reagents (RM) that have no mesomeric anions were also found to give good yields of monofluoro compounds.[232] This can also be explained by an attack on chlorine. The substitution that normally leads to the perchloryl product must be considerably slower at low temperature, permitting the intermediate adduct to be formed reversibly. This intermediate may have sufficient lifetime to transfer fluoride intramolecularly.

Normal temperature

$$RM + FClO_3 \longrightarrow [R^-\cdots M^+\cdots ClO_3\cdots F] \xrightarrow{\text{Fast}} RClO_3 + MF$$

Transition state

Low temperature

$$RM + FClO_3 \rightleftharpoons [R\text{–}ClF(O)_2O^-]M^+ \longrightarrow RF + MClO_3$$

Intermediate

[240] W. A. Sheppard, *Tetrahedron Lett.*, **1969**, 83.

The formation of an N—F product in reaction of 2,2,6,6-tetramethylpiperidine may be one example of attack on the fluorine of perchloryl fluoride, probably because the steric hindrance of four methyl groups prevents attack on the central chlorine.[241] Note that piperidine gives the expected perchlorate.[241]

$$\text{2,2,6,6-tetramethylpiperidine (NH)} + FClO_3 \longrightarrow \text{2,2,6,6-tetramethylpiperidine (N—F)}$$

$$\text{piperidine (NH)} + FClO_3 \longrightarrow \text{piperidine (N—ClO}_3\text{)}$$

Experimental Procedures and Hazards

Procedures for monofluorination with perchloryl fluoride are quite simple for most compounds. A carbanion salt is formed *in situ* by an appropriate base-solvent system (see Table VIII). The salt may be dissolved or suspended. Gaseous perchloryl fluoride is bubbled into the solution or suspension. For some compounds inverse addition is desirable.

Adequate cooling is required during the exothermic addition of perchloryl fluoride. The temperature required for reaction must be determined experimentally.

The procedure for reaction of an organometallic reagent with perchloryl fluoride is similar. Lower temperatures (to $-78°$) are required and special solvent systems are necessary.[232]

The possibility of explosions is always present when perchloryl fluoride is added to an organic solvent or other organic compounds. Perchloryl fluoride is a powerful oxidizing agent and must be accorded great respect. Before any experimental work is started, a full knowledge of the potential hazards must be understood. Reference 218 is recommended as a source of such knowledge. Unfortunately, the Pennwalt New Products Booklet on perchloryl fluoride that emphasizes laboratory safety is no longer readily available.[217] Some quotations taken from the literature are cited below in order to emphasize the hazards.

"Gaseous perchloryl fluoride (ClO_3F) reacts strongly with most aliphatic and nonaromatic heterocyclic amines and contacting the undiluted amines with perchloryl fluoride often results in violent uncontrolled oxidations and/or explosions."[241]

[241] D. M. Gardner, R. Helitzer, and D. H. Rosenblatt, *J. Org. Chem.*, **32**, 1115 (1967).

TABLE VIII. SUMMARY OF REACTION CONDITIONS FOR MONOFLUORINATION BY PERCHLORYL FLUORIDE

Substrate	Base Used	Solvent Systems
Substituted malonic esters	Na, NaH	Toluene, C_2H_5OH, DMF/C_2H_5OH
Substituted cyanoacetic esters	Na, K	C_2H_5OH, DMF, toluene
Nitroaliphatic compounds	Na, NaH, K (in some cases no base is used)	THF, HMPT, DMF, CH_3OH, dimethoxyethane, acetone, CH_3CN
Enol derivatives	Na, $Li^+[N(i\text{-}C_3H_7)_2]^-$; no base in some reactions	C_2H_5OH, pyridine, dioxane, toluene
Grignard reagents		Mixed solvents: some components are ether, THF, HMPT, petroleum ether
Lithium reagents		Mixed solvents: some components are ether, THF, HMPT, petroleum ether, toluene

"Although this reaction [preparation of 1-(*m*-bromophenyl)-3-(*p*-bromophenyl)-2,2-difluoro-1,3-propanedione] has been carried out successfully ten times in the course of this work, it is considered very dangerous; all due precautions should be taken. The same procedure used in these laboratories for the fluorination of a similar compound, ethyl *p*-fluorobenzoylacetate (author: by perchloryl fluoride), resulted in a violent explosion that completely demolished a 0.25-in. Plexiglas 'safety shield.' "[242]

"Sicherheitsmassnahmen: Im Verlauf von rund 100 Ansätzen mit Perchlorylfluorid kam es dreimal zu heftigen Explosionen. In zwei dieser Fälle war ein anomaler Reaktionsverlauf vorauszusehen gewesen. Auch bei Ansätzen mit kleinsten Substanzmengen wird dringend geraten, einen Schutzschild aus Sicherheitsglas zu verwenden oder zumindest das Reaktionsgefäss mit einer dicken Plastikfolie als Splitterschutz abzudecken."[232]

". . . the perchloryl fluoride gas and methyl alcohol vapor mixture exploded, when he added a final portion of sodium methylate."[243]

"Also, Dr. H. A. DeWalt of Wyeth Institute for Medical Research was killed in a lab explosion last year, and Wyeth believes nitroperchlorylbenzene may have been the agent."[244]

"Perchloryl fluoride is a valuable reagent but suffers from the disadvantage that it reacts only with strongly nucleophilic centers and that the by-product of its reaction, chloric acid, is dangerously explosive in admixture with organic compounds."[245]

From these statements it should be clear that sound and complete safety precautions must be observed when reactions involving perchloryl

[242] S. A. Fuqua and R. M. Silverstein, *J. Org. Chem.*, **29**, 395 (1964); see footnote 2.
[243] *Chem. Eng. News*, **37**, 60 (July 13, 1959).
[244] *Chem. Eng. News*, **38**, 62 (Jan. 25, 1960).
[245] D. H. R. Barton, *Pure Appl. Chem.*, **21**, 285 (1970).

fluoride are taking place. Remote handling is required and personnel must be shielded. The use of minimum quantities of chemicals for the purpose required is absolutely necessary. For normal work five 1-g experiments are preferred to one 5-g experiment. The simplicity of reaction techniques makes repeated small runs highly desirable. Also, perchloryl fluoride should not be condensed into a reaction mixture or solvent for later reaction. The experiment should be arranged so that perchloryl fluoride is consumed as added.

2-Fluoropropanoic Acid.[246] Dry dimethylformamide (500 ml) and 56.5% sodium hydride dispersion in mineral oil (12.8 g, 299 mmol) were placed in a three-necked flask fitted with a dropping funnel, a gas inlet tube, and a condenser. The flask was flushed with nitrogen, and diethyl methylmalonate (50 g, 295 mmol) was added slowly over 1.5 hours with stirring. Stirring was continued until effervescence ceased. The mixture was cooled in ice to 7°. Perchloryl fluoride was introduced intermittently until a temperature increase was no longer observed and excess perchloryl fluoride had been added (31.2 g, 305 mmol). The contents of the flask were flushed with nitrogen, and 600 ml of water was slowly added. The mixture was extracted with ether, and the extracts were washed with water, dried, and evaporated. The residue on fractional distillation gave crude diethyl methylfluoromalonate (30.1 g, 55%), bp 80–88° (13 mm), $n^{25}D$ 1.4013.

The impure ester (27.4 g, 143 mmol) was heated under reflux with 6 *N* hydrochloric acid (500 ml) for 36 hours, by which time a homogeneous solution was obtained. The solution was cooled, diluted with water, and extracted with ether. The extracts were washed with water, dried, and evaporated. On fractional distillation the residue gave impure 2-fluoropropanoic acid (6.74 g, 51%), bp 63–66° (13 mm), $n^{25}D$ 1.3837. Further fractionation gave the pure acid.

Diethyl Dodecylfluoromalonate.[229] Sodium dispersion (9.3 g, 0.41 g-at.) was suspended in 500 ml of dry toluene and 135 g (0.41 mol) of diethyl dodecylmalonate added. Perchloryl fluoride (41.5 g, 0.405 mol) was added to the well-stirred suspension in a rapid stream, keeping the temperature at 10–15° by means of an ice bath. Upon completion of addition of the gas, the inorganic materials were removed by filtration, and the toluene was flash-evaporated under vacuum. The residue was filtered through a fine sintered-glass filter and distilled. The yield of product was 122 g (87%), bp 200–207° (10 mm), $n^{25}D$ 1.4316.

[246] F. L. M. Pattison, R. L. Buchanan, and F. H. Dean, *Can. J. Chem.*, **43**, 1700 (1965).

Ethyl 2-Cyano-2-fluorobutyrate.[228] Sodium dispersion (23 g, 1 g-at.) was suspended in 1 liter of dry toluene and heated to 50°. To the mixture was added 145 g (1.03 mol) of ethyl 2-cyanobutyrate sufficiently slowly to keep the temperature of the reaction mixture below 90°. The excess ester was necessary to assure complete consumption of the sodium. The system was purged with dry nitrogen and kept at 10–15° by external cooling. Perchloryl fluoride was added in a rapid stream. Upon completion of the reaction, as evidenced by cessation of heat evolution, the system was again purged with dry nitrogen. The inorganic salts were removed by filtration, dissolved in water, and extracted with toluene. The combined toluene layers were washed with water, flash-evaporated, and the residue was distilled. Ethyl 2-cyano-2-fluorobutyrate was collected at 67–68° (10 mm), n^{25}_D 1.3956, in 80% yield.

6α- and 6β-Fluorotestosterone Acetate.[239] A slow stream of perchloryl fluoride was passed through a solution of 6.14 g of testosterone enol diacetate in 90 ml of 95% aqueous dioxane at 22° for 4 hours. During this time 52.5 ml of water was added to the reaction mixture in portions of 5.1, 5.4, 6.3, 15.3, and 20.4 ml at half-hour intervals; thus a final concentration of 60% aqueous dioxane was produced. Nitrogen was bubbled through the reaction mixture for 10 minutes and the solvent was flash-evaporated at 22° to afford 8.64 g of a yellow oil. After careful chromatography, which is described in the original paper, 6β-fluorotestosterone acetate (1.8 g, 30%) and 6α-fluorotestosterone acetate (0.73 g, 12%) were obtained.

Fluoroferrocene.[247] A solution of lithioferrocene (0.01 mol) in 125 ml of ether, prepared under a helium atmosphere from bromoferrocene (2.65 g, 0.0100 mol) and *n*-butyllithium (0.011 mol) was cooled to −70° and diluted with 250 ml of dry tetrahydrofuran. When a mixture of perchloryl fluoride (0.040 mol) and helium was bubbled in, the reaction mixture changed from an orange suspension to a deep-red solution. While the solution was warmed to 23°, the excess perchloryl fluoride was removed by vigorously bubbling helium through the reaction mixture. The workup consisted of reduction by aqueous sodium thiosulfate (to remove perchlorylferrocene), separation of the organic layer, removal of the solvent, and column chromatography on alumina. Elution of the column with hexane afforded 0.20 g (10%) of crude fluoroferrocene, mp 113–116°, whose gas-liquid chromatogram showed a single peak of slightly longer retention time than ferrocene. Low-temperature recrystallization of the crude compound from methanol gave orange crystals, mp 116–118°.

[247] F. L. Hedberg and H. Rosenberg, *J. Organomet. Chem.*, **28**, C14 (1971).

Fluorobenzene.[232] To a three-necked flask equipped with a stirrer, thermometer, and a mercury-sealed pressure-relief vent was added 40 ml of tetrahydrofuran. The flask was cooled to −78°. Perchloryl fluoride (5 mmol) was condensed into the flask as measured by a rotameter. An ethereal solution containing 5 mmol of phenyllithium and some lithium bromide was added dropwise to the flask over a 10-minute period. The solution, which had changed to a golden color during addition, was allowed to warm to room temperature at which time a Gilman test was negative. Toluene (0.14 g) was added as an internal analytical standard. The reaction mixture was washed three times with 50-ml portions of water. Analysis of the resulting solution by gas chromatography [2 m Carbowax 20M (15%), 90°; 1.5 m di-*n*-decyl phthalate and Bentonite (5%)), 80°] established yields of 42% of fluorobenzene and 12% of benzene.

When phenylmagnesium bromide was substituted for phenyllithium under analogous conditions, the yields were 36% of fluorobenzene and 50% of benzene.

NITROSYL FLUORIDE

Nitrosyl fluoride is a new reagent that provides a controlled method of monofluorination. The reagent is particularly good for reaction with sensitive steroids through addition to a double bond.[4]

$$\rangle C{=}C\langle_{H} + 2\,NOF \longrightarrow HF + F{-}\overset{|}{\underset{|}{C}}{-}\overset{|}{C}{=}NNO_2 \xrightarrow{\text{Hydrolysis}} F{-}\overset{|}{\underset{|}{C}}{-}\overset{O}{\overset{\|}{C}}{-}$$

Other Hetero-Fluorine Reagents

Other hetero-fluorine reagents have been used to form C–F bonds, usually through addition to olefinic bonds. However, few general synthetic routes have developed. Halogen fluoride addition (pp. 137–157) is really

$$Y{-}F + \rangle C{=}C\langle \longrightarrow Y{-}\overset{|}{\underset{|}{C}}{-}\overset{|}{\underset{|}{C}}{-}F$$

a fluoride ion addition to a halonium ion intermediate. Most of these addition reagents are not readily available or are limited to use with highly selective substrates, such as fluoroolefins. Thus reagents such as NO_2–F,* FCO–F, RCO–F, FSO–F, F_3S–F, F_5S–SF_5 (source of F_5S–F), and FSO_2–F add to unsaturated centers to form new C–F bonds but only if the center is highly electron deficient, as in fluoroolefins.[248] Other reagents such as R_2NF suffer from inherent problems in handling and control of reactions. Phenyliododifluoride ($C_6H_5IF_2$) has been reported to

* Note added in proof: G. A. Olah and M. Nojima, *Synthesis*, **1973**, 785 report a general method for nitrofluorination of alkenes. Nitronium tetrafluoroborate in 70% hydrogen fluoride/pyridine reacts with alkenes at −78 to 0° to give the *vicinal*-fluoronitroalkane in good yield.

[248] Ref. 15, Sheppard and Sharts, pp. 110–127.

add to a selected olefinic bond in a steroid,[249] but usually it is a source of fluorine for *cis* addition of F—F to olefins.[249]

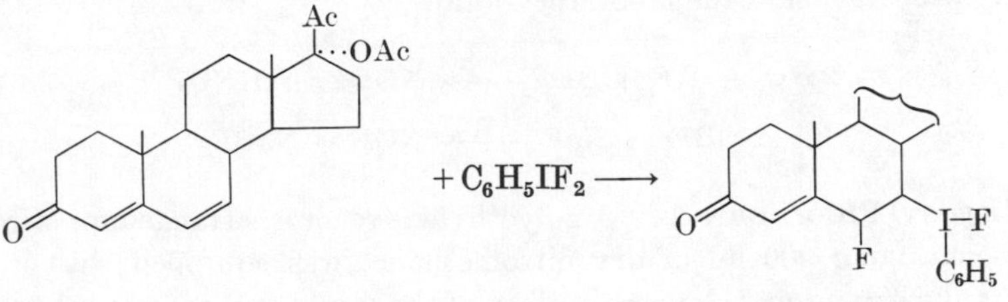

$$+ C_6H_5IF_2 \longrightarrow$$

(Or possibly the 7-fluoro-6-iodo isomer)

Nitrofluorination of olefins and acetylenes can be accomplished by a mixture of anhydrous hydrogen fluoride in concentrated nitric acid at low temperature. Effectively F and NO_2 are added 1,2 (from F—NO_2 generated *in situ*?). The addition has been described only for olefins substituted

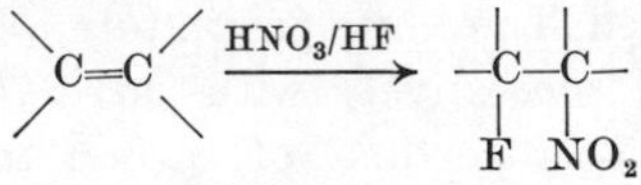

$$\rangle C{=}C\langle \xrightarrow{HNO_3/HF} -\underset{F}{\overset{|}{C}}-\underset{NO_2}{\overset{|}{C}}-$$

with electronegative groups.[250, 251] Three examples have been reported for addition to simple, nonfluorinated olefins.[250]

$$CHCl{=}CH_2 \xrightarrow{-60^\circ} \underset{(59)\%}{CHFClCH_2NO_2}$$

$$CH_2ClCH{=}CH_2 \xrightarrow{-10^\circ} CH_2ClCHFCH_2NO_2 + CH_2ClCH(NO_2)CH_2F \quad \text{(19\% total)}$$

$$CH_2{=}CHCO_2CH_3 \xrightarrow{0^\circ} \underset{(\sim 10\%)}{CH_2FCH(NO_2)CO_2CH_3}$$

Acetylenes activated by an electron-donating group are also nitrofluorinated, but a mixture of isomeric olefins is formed.[252]

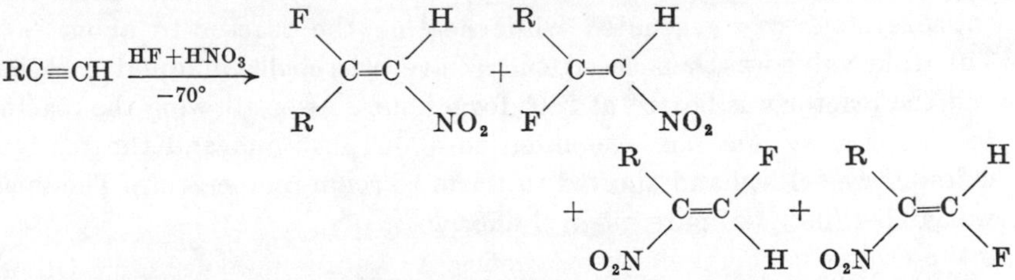

$$RC{\equiv}CH \xrightarrow[-70^\circ]{HF+HNO_3} (F)(R)C{=}C(H)(NO_2) + (R)(F)C{=}C(H)(NO_2) + (R)(O_2N)C{=}C(F)(H) + (R)(O_2N)C{=}C(H)(F)$$

R = C_6H_5, CH_3 (37%), n-C_4H_9 (35%), CH_2OCH

[249] P. G. Holton, A. D. Cross, and A. Bowers, *Steroids*, **2**, 71 (1963).
[250] I. L. Knunyants, L. S. German, and I. N. Rozhkov, *Bull. Acad. Sci. USSR, Div. Chem. Sci., Engl. Transl.*, **1963**, 1794.
[251] B. L. Dyatkin, E. P. Mochalina, and I. L. Knunyants, *Russ. Chem. Rev.*, **35**, 417 (1966);
[252] A. Ostaszynski and J. Wielgat, *Rocz. Chem.*, **45**, 1345 (1971) [*C.A.*, **76**, 33725n (1972)].

Availability and Preparation of Reagent

Nitrosyl fluoride is available commercially.[253] Convenient procedures for its preparation in the laboratory follow.[254]

$$2\,NO_2 + HF + BF_3 \longrightarrow NOBF_4 + HNO_3$$

$$NOBF_4 + NaF \xrightarrow{\text{Heat}} NOF + NaBF_4$$

Nitrosyl Fluoroborate. A polyethylene reactor, arranged for stirring and containing 900 ml of dry nitromethane, was equipped with a gas inlet tube extending below the surface of the liquid and an exit tube leading to a T-tube through which was passed a slow stream of dry nitrogen. The reactor was cooled to −20° with stirring; 60 g (3.0 mol) of anhydrous hydrogen fluoride, 204 g (3.9 mol) of boron trifluoride, and 276 g (6.0 mol) of nitrogen dioxide were each added slowly in that order. The rate of addition of each reagent was controlled so that the temperature was kept below 0°. During the addition of nitrogen dioxide, precipitation of nitrosyl fluoroborate occurred. The white precipitate was collected by filtration in a dry box, washed successively with three 75-ml portions of nitromethane and three 75-ml portions of carbon tetrachloride, and dried overnight in a heated (50–100°) vacuum desiccator. The yield was 341 g (97%). An analytical sample was purified further by sublimation.

Nitronium tetrafluoroborate was not a contaminant since the product was inert to benzene at 20°. (NO_2BF_4 reacts vigorously with benzene to give nitrobenzene.[255]) Moreover, pure nitrosyl fluoride was obtained directly in the next step.

Nitrosyl Fluoride. In a dry box, 117 g (1.0 mol) of dry nitrosyl fluoroborate and 84 g (2.0 mol) of dry sodium fluoride were mixed intimately and placed in a 300-ml prefluorinated Monel reactor. Nickel tubing (with a T-joint and a Monel Hoke valve leading to a vacuum pump) joined the reactor to a 300-ml Monel cylinder which served as a receiver. Both reactor and receiver were equipped with safety rupture disks. The entire system was evacuated while heating the reactor to about 80°. The Hoke valve was then closed, the receiver was cooled in liquid nitrogen, and the reactor was heated at 250° for 6 hours. After allowing the reactor to cool, the system was evacuated to about 0.15 mm and the receiver cylinder was closed and allowed to warm to room temperature. The yield was 48.5 g (99%) of pure nitrosyl fluoride.

[253] Ozark-Mahoning Company, Special Chemicals Division, 1870 South Boulder, Tulsa, Oklahoma 74119.

[254] (a) S. Andreades, *J. Org. Chem.*, **27**, 4157 (1962); (b) B. L. Dyatkin, E. P. Mochalina, R. A. Bekker, S. R. Sterlin, and I. L. Knunyants, *Tetrahedron*, **23**, 4291 (1967).

[255] G. Olah, S. Kuhn, and A. Mlinko, *J. Chem. Soc.*, **1956**, 4257.

A simplified procedure uses nitrogen dioxide and potassium or cesium fluoride.[256] In a typical run, 5 g (33 mmol) of 99% cesium fluoride was

$$2\,NO_2 + MF \rightarrow NOF + MNO_3$$
$$M = K \text{ or } Cs$$

dried at 300° for 2 hours, powdered under vacuum, and placed in a 150-ml prefluorinated Monel vessel. Nitrogen dioxide (2.85 mmol) was condensed into the vessel at −78° and allowed to warm to ambient temperature. The volatile material, pure nitrosyl fluoride (1.37 mmol), was removed after 5 days. The reaction time varied from 1 to 5 days in several runs and is dependent on the particle size and dryness of the salt.

The rate of preparation of nitrosyl fluoride can be increased by allowing the reaction to occur in a prefluorinated metal vessel above 90°. Nitrogen dioxide (2.47 mmol) and excess potassium fluoride were heated to 90° for 2.5 hours. Complete reaction occurred, giving 1.19 mmol (48%) of nitrosyl fluoride. At 300° the reaction occurred within 15 minutes with a slightly lower yield. Cesium and potassium fluoride give solely nitrosyl fluoride in about the same yield.

Scope and Limitations

Nitrogen oxide fluorides are a well-known class of reagents which have been studied extensively.[257, 258] However, until recently, reactions with organic substrates were mostly limited to highly fluorinated olefins and other fluorinated unsaturated compounds.[254, 257] Boswell found that by proper choice of conditions nitrosyl fluoride could be used to introduce a single fluorine atom into steroids. Initial addition of F—NO to a double bond is followed by reaction of the adduct with a second molecule of nitrosyl fluoride to give a nitrimine; hydrolysis leads to the α-fluoroketone (see equation on p. 236). Nitrosyl fluoride usually adds only to electron-rich double bonds.

Nitrosyl fluoride has been added to tetramethylethylene to give a fluoronitroso dimer (like nitrosyl chloride addition) and an α-fluoroketone. The dimer readily lost fluoride by hydrolysis during chromatography and formed 3-methyl-3-hydroxy-2-butanone oxime.[259]

[256] C. T. Ratcliffe and J. M. Shreeve, *Chem. Commun.*, **1966**, 674.

[257] R. Schmutzler, *Angew. Chem., Int. Ed. Engl.*, **7**, 440 (1968).

[258] C. Woolf, *Advan. Fluorine Chem.*, **5**, 1 (1965).

[259] (a) G. A. Boswell, Jr., *Chem. Ind.* (London), **1965**, 1929; (b) *J. Org. Chem.*, **33**, 3699 (1968).

$$(CH_3)_2C{=}C(CH_3)_2 \xrightarrow{NOF} [(CH_3)_2CFCH(CH_3)\overset{O}{\overset{\uparrow}{N}}]_2 + (CH_3)_2CFCOCH_3$$

$$\downarrow$$

$$(CH_3)_2C(OH)C(CH_3){=}NOH$$

Although some unsaturated steroids give the dimer of the F—NO adduct, most steroids give the α-fluoronitrimine which is often not isolated but converted to the α-fluoroketone during workup. Examples of addition to steroid unsaturation at positions 2–3, 4–5, 5–6, 9–11, or 16–17 have been reported, but the major use has been with steroids unsaturated at 5–6 and 9–11. A 9-fluoro substituent potentiates the biological activity of a steroid, and the 5-fluoro-6-keto system provides the functionality needed for conversion to the active 6-fluoro steroids via sulfur tetrafluoride.[260]

A Markownikoff orientation of addition, assuming polarization $\overset{\delta-}{F}{-}\overset{\delta+}{NO}$, is usually found. In general, the fluorine adds to the more highly substituted carbon. For 5–6 olefinic steroids, fluorine always adds at carbon atom 5. Keto and acetate groups and unactivated olefinic or acetylenic bonds are not attacked, but hydroxyl or amino groups are rapidly oxidized.

A recent review provides a more extensive discussion of the reactions of nitrosyl fluoride with steroids and the conversion of products to other monofluoro steroids.[4]

Mechanism

The addition of nitrosyl fluoride to steroids appears similar to nitrosyl chloride addition in some respects, although products of different types are often formed. Both nitrosyl chloride and fluoride add to an olefinic bond to form initially a nitrosyl halide adduct. If the adduct is not sterically hindered, the typical nitroso dimer forms in both cases. For example, the absence of the 10β-methyl in the 19-nor-4- and -5-ene series allows dimerization of the initially formed nitrosyl fluoride adduct to compete with tautomerization and nitrosation; a nitroso fluoride dimer and an α-fluoronitrimine therefore are formed in comparable amounts.

However, in most steroids, such as 10β-methyl-5-enes, formation of the nitroso dimer is sterically prevented. The nitroso chloride from nitrosyl chloride addition is oxidized to nitro chloride before it tautomerizes to the oxime. But because nitrosyl fluoride does not seem to be as effective an oxidizing agent as nitrosyl chloride, the nitroso fluoride adduct tautomerizes to fluoro oxime. The transient fluoro oxime is then nitrosated by

[260] G. A. Boswell, Jr., *J. Org. Chem.*, **31**, 991 (1966).

nitrosyl fluoride to give an N-nitrosonitrone that rearranges spontaneously to a nitrimine. Boswell cites a series of experimental observations in support of this mechanism.[259b]

AcO XNO When X = F AcO X H NO AcO F N OH —HF FNO AcO F +N—N=O O− When X = Cl, NOCl AcO Cl H NO_2 AcO F +N—N—O− O AcO F N—NO_2

The mechanism of initial addition of nitrosyl fluoride to the olefinic bond is not known. The addition of nitrosyl chloride to steroid-5-enes has been shown to be *trans* whether carried out in ether, carbon tetrachloride, or methylene chloride. However, some evidence favors a *trans* ionic addition,[261] whereas other evidence favors a *trans* free-radical addition.[262]

The reaction of nitrosyl fluoride with α,β-unsaturated carbonyl compounds proceeds slowly with partial conversion to nitro compounds. Since nitrosyl fluoride does not readily add to electron-deficient olefins,[254] the nitro products are probably formed by a radical process similar to that observed in the reaction of impure nitrosyl chloride and steroid 5-enes.[262] Unactivated olefins such as cholest-2-ene also react slowly with nitrosyl fluoride to give 2-fluoro-3-nitro derivatives (probably by a radical process) and 2α-fluoro-3-keto derivatives. Chromatography of intermediate nitro fluorides results in dehydrofluorination and isolation of vinyl

[261] A. Hassner and C. Heathcock, *J. Org. Chem.*, **29,** 1350 (1964).

[262] W. A. Harrison, R. H. Jones, G. D. Meakins, and P. A. Wilkinson, *J. Chem. Soc.*, **1964,** 3210.

nitro derivatives. Note that some α-fluoroketone is formed directly, analogous to the reaction of nitrosyl fluoride with tetramethylethylene. Presumably the α-fluoroketone is the result of direct reaction of a slightly hindered fluorooxime intermediate with nitrosyl fluoride.

Experimental Procedures

5α-Fluoro-3β-hydroxycholestane-6-nitrimine Acetate and 5α-Fluoro-3β-hydroxycholestan-6-one Acetate.[259b] A dry, 500-ml polyethylene bottle was equipped with a magnetic stirring bar and polyethylene gas inlet and gas exit tubes. In it was placed a solution of cholesteryl acetate (40.0 g, 93.9 mmol) in methylene chloride or carbon tetrachloride (100–250 ml). The exit tube was protected by a drying tube containing calcium chloride, and the system was swept with a stream of nitrogen to remove moisture and air. The reactor was cooled in an ice bath while a slow stream of nitrosyl fluoride (15.0 g, 310 mmol, normally 0.33–0.5 times the moles of steroid olefin) was passed into the stirred solution during 3–7 hours. After an initial induction period of 0.25–2 hours, depending on the rate of flow, the solution became deep blue, gradually turned green, and finally took on a straw color. The reaction mixture was washed with water and saturated salt solution, dried over magnesium sulfate, and evaporated under reduced pressure. The viscous residue of 34.0 g (72%) of 5α-fluoro-3β-hydroxycholestane-6-nitrimine acetate was recrystallized from methylene chloride-hexane to give 31.6 g (66%) of pure fluoronitrimine, mp 158–163, $[\alpha]_D^{23}$ −68°. In other workups the crude product was chromatographed on neutral alumina (activity III, 20–30 g/g of starting olefin) and the eluted solids were crystallized from the appropriate solvent system to furnish the fluoroketone, 5α-fluoro-3β-hydroxycholestan-6-one acetate.

9α-Fluoro-3β-hydroxy-11-nitrimino-5α-pregnan-20-one Acetate.[263] Methylene chloride (800 ml) was dried over phosphorus pentoxide and distilled. 3β-Hydroxy-5α-pregn-9(11)-en-20-one acetate (2.0 g, 5.59 mmol) was added and the reaction mixture was swept with dry nitrogen, while it was cooled in an ice bath. Then the reaction mixture was saturated with nitrosyl fluoride. The vessel was sealed and stored at 3° for 10 days. The solution was then poured into water and washed with saturated sodium bicarbonate and sodium chloride solutions. The aqueous washes were backwashed with methylene chloride. The combined organic solutions were dried over sodium sulfate and evaporated to give 2.5 g of a green foam. The foam was dissolved in 1:1 benzene-hexane, passed

[263] J. P. Gratz and D. Rosenthal, *Steroids*, **14**, 729 (1969).

through 10 g of alumina, concentrated, and passed through 25 g of Florisil in 1:1 benzene-hexane. The eluate was recrystallized from methylene chloride-hexane to give 1.10 g (45%) of the fluoronitrimine.

ELECTROPHILIC FLUORINATIONS BY HYPOFLUORITES

The reactions of bromine and chlorine with double bonds often proceed through a bromonium or chloronium ion intermediate. In contrast, fluorine is too electronegative to form a fluoronium ion required for controlled addition of fluorine to double bonds. Usually elemental fluorine gives nonselective radical attack on double bonds leading to an extensive mixture of products. Recently some control of molecular fluorine has been achieved by use of low temperatures, inert solvents, and metallic halide catalysts.[264] Aqueous fluorination,[18, 19] which undoubtedly involves hypofluorous acid, also gives controlled fluorination but has limited utility (p. 133). An alternative to be discussed in this section is the generation of an incipient fluoronium-like species from a Y—F reagent where Y is a highly electronegative species.

Recently hypofluorites, primarily trifluoromethyl hypofluorite, have been developed as useful reagents for electrophilic fluorination.[245] A strong electronegative group such as perfluoroalkyl is required to stabilize the hypofluorite group and activate the O—F bond for electrophilic reaction. Thus trifluoromethyl hypofluorite adds a new dimension to fluorination; CF_3O—F adds to olefinic or other unsaturated centers, and the CF_3O group often can be hydrolyzed to leave a monofluoro product. Oxygen difluoride also adds to olefinic compounds,[248] but appears too reactive (like elementary fluorine) to be generally useful.

Availability and Preparation of the Reagents*

Trifluoromethyl hypofluorite (fluoroxytrifluoromethane) is available commercially.[265] Its preparation requires reaction of elementary fluorine with various carbonyl derivatives or methanol.[266] Thus it is prepared in high yields by reaction of fluorine in the presence of a silver(II) fluoride catalyst with carbon monoxide,[267] carbonyl fluoride,[267a, 268] or methanol.[267a] Fluorination of potassium cyanate[269] and other oxygen-containing

[264] Ref. 15, Sheppard and Sharts, p. 99.

* Note added in proof: Hypofluorites have been recently reviewed by M. Lustig and J. M. Shreeve, *Advan. Fluorine Chem.*, **7**, 175 (1973). This article discusses primarily synthesis and properties, but not the use of this agent in electrophilic fluorination of organic compounds.

[265] "Technical Bulletin on Fluoroxytrifluoromethane," PCR, Inc., P.O. Box 1466, Gainesville, Florida 32601.

[266] C. J. Hoffman, *Chem. Rev.*, **64**, 91 (1964).

[267] (a) K. B. Kellogg and G. H. Cady, *J. Amer. Chem. Soc.*, **70**, 3986 (1948); (b) J. A. C. Allison and G. H. Cady, *ibid.*, **81**, 1089 (1959); (c) R. S. Porter and G. H. Cady, *ibid.*, **79**, 5625 (1957); (d) W. P. Van Meter and G. H. Cady, *ibid.*, **82**, 6005 (1960).

[268] R. T. Lagemann, E. A. Jones, and P. J. H. Woltz, *J. Chem. Phys.*, **20**, 1768 (1952).

[269] A. Ya. Yakubovich, M. A. Englin, and S. P. Makarov, *J. Gen. Chem. USSR, Engl. Transl.*, **30**, 2356 (1960).

compounds[270] with fluorine gives trifluoromethyl hypofluorite and other hypofluorites.

$$CO, COF_2, CH_3OH \xrightarrow[AgF_2]{F_2} CF_3OF$$

$$COF_2 \xrightarrow[\text{Electrolysis}]{HF} CF_3OF \quad \text{(Ref. 271)}$$

$$KNCO \xrightarrow{F_2} CF_3OF$$

$$CO_2 \xrightarrow[CsF]{F_2} CF_2(OF)_2 \quad \text{(Ref. 272)}$$

$$(CF_3)_3COK \xrightarrow[-78^\circ]{F_2} (CF_3)_3COF \quad \text{(Ref. 273)}$$

$$(CF_3)_2(C_2F_5)COCs \xrightarrow[-78^\circ]{F_2} (CF_3)_2(C_2F_5)COF \quad \text{(Ref. 273)}$$

$$SOF_4 \xrightarrow{F_2} SF_5OF \quad \text{(Ref. 274)}$$

Because handling elementary fluorine is extremely hazardous and requires elaborate safety precautions and equipment, we do not recommend that hypofluorites be prepared in the normal laboratory equipped for organic synthesis. However, if synthesis of a hypofluorite is planned, the references above on hypofluorites and *reviews on handling fluorine should be studied carefully and all safety precautions should be taken. The hypofluorites are strong oxidizing agents, similar to fluorine, and should be handled only behind adequate shields;* fires and explosions do occur on contact with reactive organic reagents. The toxicity of trifluoromethyl hypofluorite and other fluoroalkyl hypofluorites has not been determined, but *these hypofluorites are expected to be highly toxic like fluorine and oxygen difluoride.*[275]

Scope and Limitations

The use of trifluoromethyl hypofluorite and other hypofluorites in fluorination of organic compounds has been reviewed by Barton.[245] In his words: "Hypofluorites are powerful but selective fluorinating agents with unusual electrophilic character." They have been used principally

[270] G. H. Cady and K. B. Kellogg, *J. Amer. Chem. Soc.*, **75,** 2501 (1953).

[271] S. Nagase, T. Abe, H. Baba, and K. Kodaira, U.S. Pat. 3,687,825 (1972) [*C.A.*, **77,** 164024*b* (1972)].

[272] (a) F. A. Hohorst and J. M. Shreeve, *J. Amer. Chem. Soc.*, **89,** 1809 (1967); (b) P. G. Thompson, *ibid.*, **89,** 1811 (1967); (c) R. L. Cauble and G. H. Cady, *ibid.*, **89,** 1962 (1967).

[273] D. H. R. Barton, R. H. Hesse, M. M. Pechet, G. Tarzia, H. T. Toh, and N. D. Westcott, *Chem. Commun.*, **1972,** 122.

[274] J. K. Ruff and M. Lustig, *Inorg. Chem.*, **3,** 1422 (1964).

[275] Ref. 15, Sheppard and Sharts, p. 453.

on steroids, but recently have been shown to provide a convenient synthesis of fluoronucleic bases and nucleosides, as well as a method to introduce fluorine into simple organic molecules.* The earlier work on reactions of hypofluorites has been reviewed.[266]

In general, hypofluorites are used only at low temperatures (—78°), preferably with the reactant dissolved in an inert solvent such as fluorotrichloromethane. Cosolvents are often used to increase solubility, but oxidizable cosolvents such as methanol, acetone, and tetrahydrofuran must be used with *caution and adequate safety protection against explosions and fires since the hypofluorites are powerful oxidizing agents.*

Olefins

Generally the hypofluorites (R_fO–F) add to electron-rich olefins by stereospecific *cis* addition. Usually a *cis*-1,2-difluoro adduct is the main by-product. Little has been done with unactivated olefins other than a

$$R_fO\text{–}F + YC(=C)X \longrightarrow Y,X\text{-}C(F)\text{–}C(OR_f) + Y,X\text{-}C(F)\text{–}C(F)$$

gas-phase, light-induced addition of trifluoromethyl hypofluorite to ethylene, which is quantitative.[267b] Stilbene gives the CF_3O—F and F–F adducts, both *cis*. Most examples of addition to olefins have been obtained with steroids.

Aromatic Derivatives

Aromatic rings that are activated to electrophilic substitution are also susceptible to fluorination or oxidation.[276a] For example, N-acetyl-β-naphthylamine gives α-fluorination as the main reaction. However, N-acetyl-α-naphthylamine gives a product that results from both fluorine substitution and CF_3O–F addition. 2,6-Dimethylphenol gives chiefly the dimer of 6-fluoro-2,6-dimethylcyclohexa-2,4-dienone.[276a] The natural product khellin (**34**) was cleanly oxidized by trifluoromethyl hypofluorite to the quinone. Presumably electrophilic addition of trifluoromethyl hypofluorite was followed by elimination, like the mechanism proposed for other electrophilic reagents such as nitric acid.[276]

The naturally occurring bacteriocide and fungicide griesiofulvin is fluorinated at an activated aromatic position and an activated vinyl

* Note added in proof: Schiff bases react with trifluoromethyl hypofluorite to give products from fluorination on carbon and nitrogen (ArCF=NR and $ArCF_2NFR$). J. Leroy, F. Dudragne, J. C. Adenis, and C. Michaud, *Tetrahedron Lett.*, **1973**, 2771.

[276] (a) D. H. R. Barton, A. K. Ganguly, R. H. Hesse, S. N. Loo, and M. M. Pechet, *Chem. Commun.*, **1968**, 806.

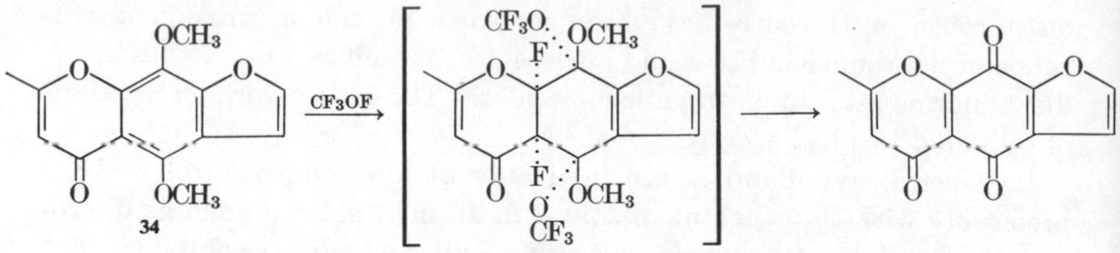

position when it reacts with trifluoromethyl hypofluorite (see Table p. 396).[276b]

Benzene reacts with trifluoromethyl hypofluorite at low temperatures to give a complex mixture but, if the mixture is irradiated, fluorobenzene in 65% yield and trifluoromethyl phenyl ether in 10% yield are obtained.[277] Low-temperature irradiation of a toluene-trifluoromethyl hypofluorite mixture gives fluorination of both the ring and methyl group.[277] Electrophilic fluorination of aromatic compounds by hypofluorites appears to have limited utility but can be expected as a complicating side reaction in reactants containing an aromatic ring.

Aliphatic Derivatives

Under the normal low-temperature reaction conditions, saturated carbon-hydrogen bonds are not attacked by hypofluorites. However, irradiation promotes monofluorosubstitution. Thus cyclohexane gives fluorocyclohexane in 44% yield by irradiation with trifluoromethyl hypofluorite at −78° in fluorotrichloromethane.[277] Isobutyric acid gives almost equal substitution of the tertiary and primary hydrogen.[277] In the two examples of reactions with amino acids, the carbon-hydrogen bonds substituted were as remote as possible from the amine function, presumably because the reaction was run in hydrogen fluoride and the amine group was completely protonated.[277] When irradiated in hydrogen fluoride solution, 1-aminoadamantane reacted with trifluoromethyl hypofluorite to give essentially equal amounts (25–27%) of tertiary and secondary hydrogen monofluoro substitution. Only the secondary hydrogens remote from the NH_2 (NH_3^+) function were substituted.[277]

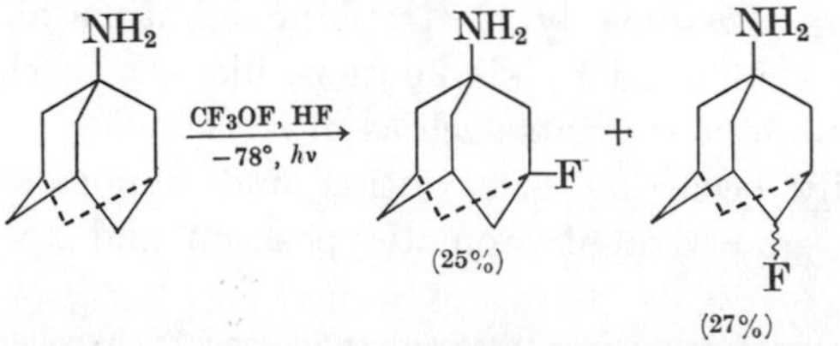

(25%) (27%)

[276] (b) D. H. R. Barton, R. H. Hesse, L. Ogunkoya, N. D. Westcott, and M. M. Pechet, *J. Chem. Soc., Perkin I*, **1972**, 2889.

[277] J. Kollonitsch, L. Barash, and G. Doldouras, *J. Am. Chem. Soc.*, **92**, 7494 (1970).

Steroids*

The major use of hypofluorites has been developed by Barton, but unfortunately the experimental details are mostly limited to communications which contain only partial information on yields and by-products.[245] Most of the reactions are additions to steroid double bonds activated by an oxygen or amine function that is allylic or attached to an olefinic linkage.

Double Bond at C_2-C_3. Vinyl ethers and enamines of cholestanone give 2α-fluorocholestanone with trifluoromethyl hypofluorite or other hypofluorites. The adducts must be intermediates to the fluoroketone.[278]

C_8H_{17} AcO → [F, AcO, X] → F, O

$X = OCF_3$ or F

Adducts can be isolated when the enol acetate reacts with trifluoromethyl hypofluorite, but they are readily converted to the fluoroketone by mild alkaline hydrolysis. A series of hypofluorites, R_fO–F and SF_5O–F, give 2α-fluorocholestanone in about the same yield as trifluoromethyl hypofluorite. In some cases intermediate adducts were noted or isolated.[273, 278]

Double Bonds at C_3-C_4 and C_5-C_6. Only the diene systems with 3-alkoxy, 3-acetoxy, or 3-(1-pyrrolidinyl) activation have been studied. Again the vinyl ether and enamine systems lead to a single α-fluoroketone product.[278] The vinyl acetate, however, gives the 5,6-difluoro adduct along with the 6-fluoro-Δ^4-ketone, presumably formed by elimination of acetyl fluoride from the adduct. (Equations on p. 248).

Double Bond at C_4-C_5. A 4,5-double bond allylic to a keto or acetoxy group leads to an adduct that is converted by hydrolysis to an α-fluoro-α,β-unsaturated ketone. The 4,5-difluoro adduct may also be formed.[279]

The corresponding 4,5-unsaturated steroid with no 3-substituent gave only a complex mixture.[279] Note, however, that the 4,5-double bond allylic

* Note added in proof: Use of trifluoromethyl hypofluorite to prepare 12-fluoro-corticosteroids was recently reported by D. H. R. Barton, R. H. Hesse, M. M. Pechet and T. J. Tewson, *J. Chem. Soc., Perkin I*, **1973**, 2365.

[278] (a) D. H. R. Barton, L. S. Godinho, R. H. Hesse, and M. M. Pechet, *Chem. Commun.*, **1968**, 804; (b) D. H. R. Barton and R. H. Hesse, U.S. Pat. 3,687,943 (1972) [≡S.Afr. Patent 68 05, 836; *C.A.* **73** 109978h (1970)].

[279] D. H. R. Barton, L. J. Danks, A. K. Ganguly, R. H. Hesse, G. Tarzia, and M. M. Pechet, *Chem. Commun.*, **1969**, 227.

R

O

F

R = CH_3O-, N—

AcO

AcO F F +

O F

CF_3O—F adduct Hydrolysis

O

O F

CF_3OF

AcO

AcO F OCF_3 + AcO F F

Hydrolysis

O F

to a keto group is less reactive than an unactivated 9,11-double bond (see below).

Double Bond at C_5–C_6. Pregnenolone acetate, which has an unactivated double bond, gave only modest yields of a *cis* CF_3O—F adduct which was characterized by hydrolysis to 6α-fluoroprogesterone.[279] The major product of the reaction of this unactivated double bond was a mixture of fluorinated by-products that were not readily separated or identified.

Double Bond at C_9–C_{11}. If the 9,11 double bond is activated by an allylic hydroxyl or ester group in the 12-position, *cis* addition of CF_3O—F is the major reaction (*cis* F—F addition is minor). The 12-benzoate ester

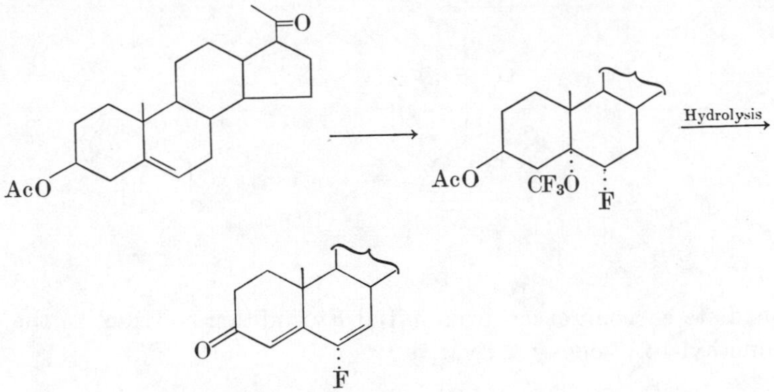

AcO — → AcO CF$_3$O F —Hydrolysis→ O F

gave adducts with a series of hypofluorites; the adducts were not isolated but were hydrolyzed to the 9α-fluoro-11-ketone.

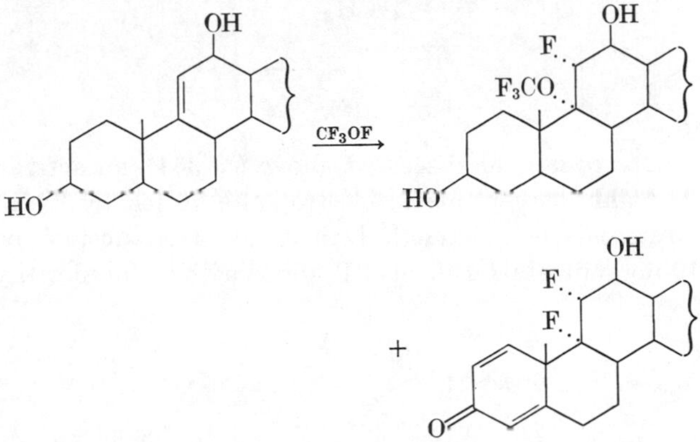

OH HO —CF_3OF→ OH F F$_3$CO HO + OH F F O (Ref. 279)

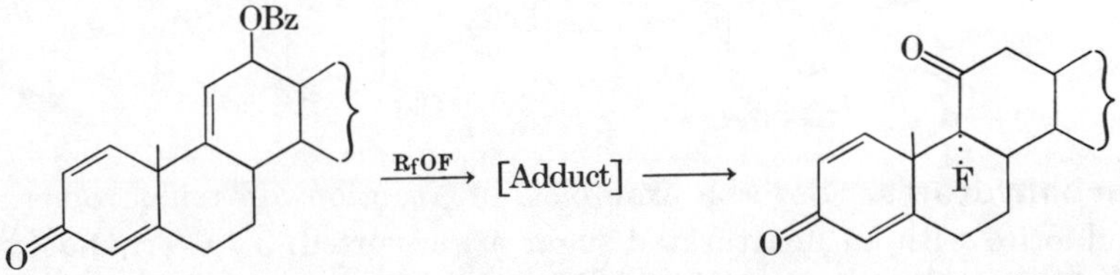

OBz O —R_fOF→ [Adduct] → O F O

(Ref. 273)

With no 12-oxygen substituent to activate the double bond, either a complex mixture formed or the 10-methyl group migrated with aromatization of the A ring.[279] Note that in this substrate the unactivated 9,11 bond is more reactive toward the hypofluorite than the 4,5-double bond activated by a 3-keto function. (Equation on p. 250).

16-Methylene. Trifluoromethyl hypofluorite reacts with a 16-methylene derivative like other electrophilic reagents. The proposed intermediate carbonium ion from initial F^+ addition collapses to the 16-fluoromethyl-16,17-epoxy derivative.[279]

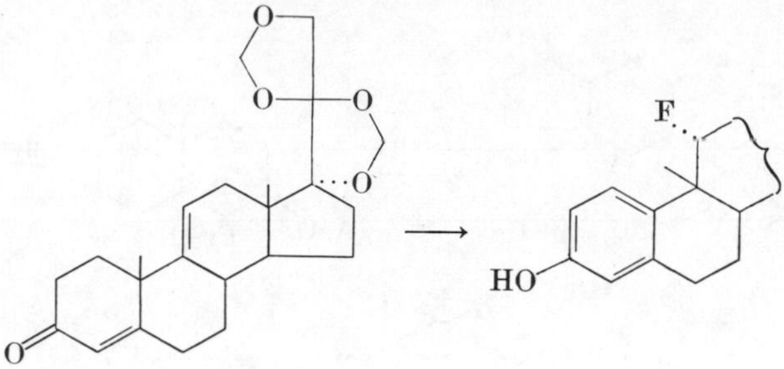

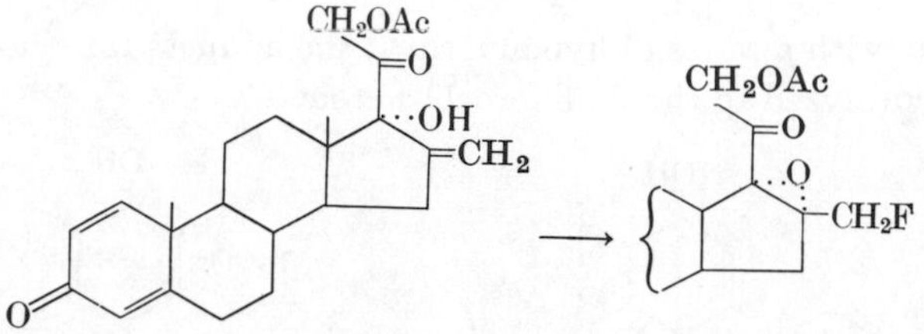

Aromatic Steroids. As discussed above (p. 354), an activated aromatic ring is highly susceptible to electrophilic attack by hypofluorite. The aromatic steroids, estrone methyl ether and estrone acetate, both give 10β-fluoro-19-norandrosta-1,4-diene-3,17-dione with trifluoromethyl hypofluorite.[276]

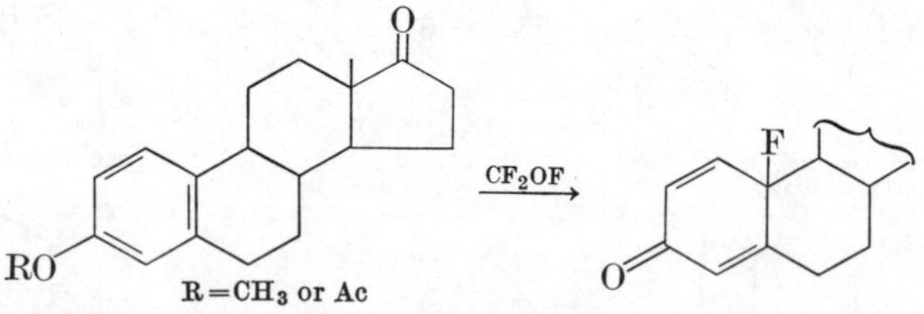

R=CH_3 or Ac

Carbohydrates. Several examples of reaction of trifluoromethyl hypofluorite with an unsaturated sugar are reported. 3,4,6-Tri-O-acetyl-D-glucal gave the four possible fluorinated products resulting solely from *cis* addition of CF_3O–F or F–F.[280] In contrast to the steroid examples, the yield of F–F adduct exceeds that of the CF_3O–F adduct (see "mechanism," p. 251).

Nucleic Bases and Nucleosides. Trifluoromethyl hypofluorite provides a convenient synthetic route to the important fluoronucleic

[280] (a) J. Adamson, A. B. Foster, L. D. Hall, and R. H. Hesse, *Chem. Commun.*, **1969**, 309; (b) J. Adamson, A. B. Foster, L. D. Hall, R. N. Johnson, and R. H. Hesse, *Carbohyd. Res.*, **15**, 351 (1970).

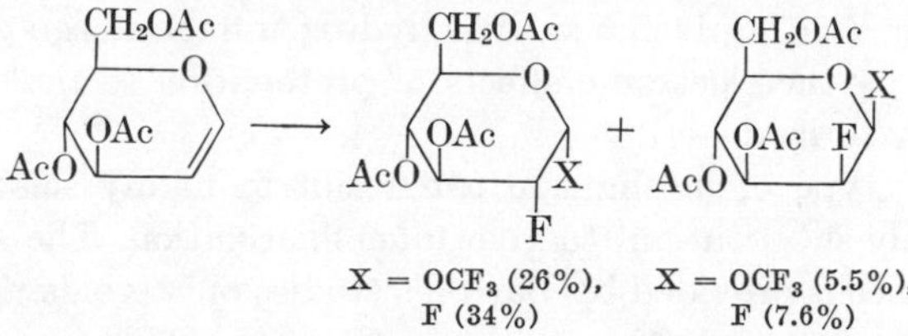

X = OCF_3 (26%), F (34%)   X = OCF_3 (5.5%), F (7.6%)

bases and fluoronucleosides such as 5-fluorouracil derivatives. For example, uracil adds trifluoromethyl hypofluorite at −78°; the crude adduct reacts with triethylamine in aqueous methanol during the workup to give 5-fluorouracil in 84% yield.[281] This reaction is applicable to a series of nucleic bases and even to nucleosides with similarly high yields.

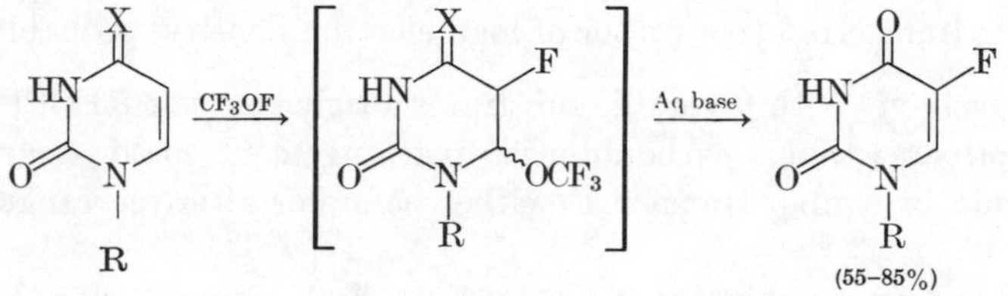

(Refs. 281–283)

(55–85%)

Limitations. In the discussion of the utility of the hypofluorite reagents, a general pattern is apparent. These highly potent electrophilic reagents are also powerful oxidants. Consequently they must be used at low temperatures under carefully controlled conditions of high dilution. Any functionality that is easily oxidized or attacked by an electrophile must be protected, or side reactions become predominant.

Mechanism

Hypofluorites are highly electrophilic reagents which could be considered a source of F^+ as proposed for perchloryl fluoride (see p. 230). Although the reaction products are sometimes similar to those that would be expected for F^+ or a fluoronium ion intermediate (analogous to other positive halogens such as Cl^+), the evidence is against such an intermediate. However, in the transition state of hypofluorite reactions the fluorine appears to have fluoronium ion character, and a β-fluorocarbonium ion could actually form. At temperatures approaching room temperature and above and at low temperatures under irradiation, the hypofluorites, like fluorine, react by a radical mechanism. Usually a mixture of products is formed, although at low temperature a good yield of a discrete product is sometimes found. Photofluorination by trifluoromethyl hypofluorite is considered to be a radical chain reaction with $CF_3O\cdot$ the

[281] M. J. Robins and S. R. Naik, *J. Amer. Chem. Soc.*, **93**, 5277 (1971).

chain carrier[277, 284] Solvents such as hydrogen fluoride can give control of orientation by the electronic effects of protonation so that only selected products are formed.

The ionic type of mechanism can result in highly selective addition (or effectively substitution) to give monofluorination. The major mechanistic evidence is provided by Barton's studies on steroids.[245] The orientation observed in the addition products, the nature of by-products, and the rearrangements that are found provide the basis for the following observations and conclusions about the mechanism.

1. The hypofluorites are polarized so that the fluorine is electrophilic. This does not mean a free F^+ in an ion pair or in any other form.

$$R_f\overset{\delta-}{O}-\overset{\delta+}{F}$$

2. Fluorine is transferred to a center of high electron density, probably in a concerted fashion. With favorable substrates a tight ion pair $\overset{+}{R}F\overset{-}{O}CF_3$ could be generated so that a carbonium ion rearrangement could occur; $^-OCF_3$ could add or simply transfer F^- either before or after rearrangement.

3. The products from addition of CF_3O—F or F—F always have the added groups *cis*. A concerted fluorination is favored, but a tight ion pair that does not allow loss of geometry in an intermediate is also possible. Adducts may be unstable. Often products resulting from decomposition or hydrolysis of initially formed adduct are isolated.

$$-\overset{|}{C}=\overset{|}{C}- \quad F-OCF_3 \qquad \text{or} \qquad -\overset{|}{C}=\overset{|}{C}- \quad F\diagdown O-CF_2\diagup F$$

In the addition of hypofluorites to vinyl ethers, acetates, and enamines, fluorine is always transferred to the 2-carbon atom of the double bond corresponding to the center of highest electron density; for example, in a Δ^2-steroid, depending on the vinyl substituent, one product can predominate or all three products shown can form. The α-fluoroketone did not form from the adducts under the reaction conditions but resulted from basic hydrolysis.

When the incipient carbonium ion is more highly stabilized by allylic interaction as in 3,4,6-tri-O-acetyl-D-glucal, the F—F adduct predominates. The argument is that the CF_3O^- species has more time to transfer

[282] M. J. Robins and S. R. Naik, *Chem. Commun.*, **1972**, 18.

[283] D. H. R. Barton, R. H. Hesse, H. T. Toh, and M. M. Pechet, *J. Org. Chem.*, **37**, 329 (1972).

[284] J. Kollonitsch, L. Barash, and G. A. Doldouras, Abstracts, 162nd National ACS Meeting, Washington, D.C., 1971, FLUO 19.

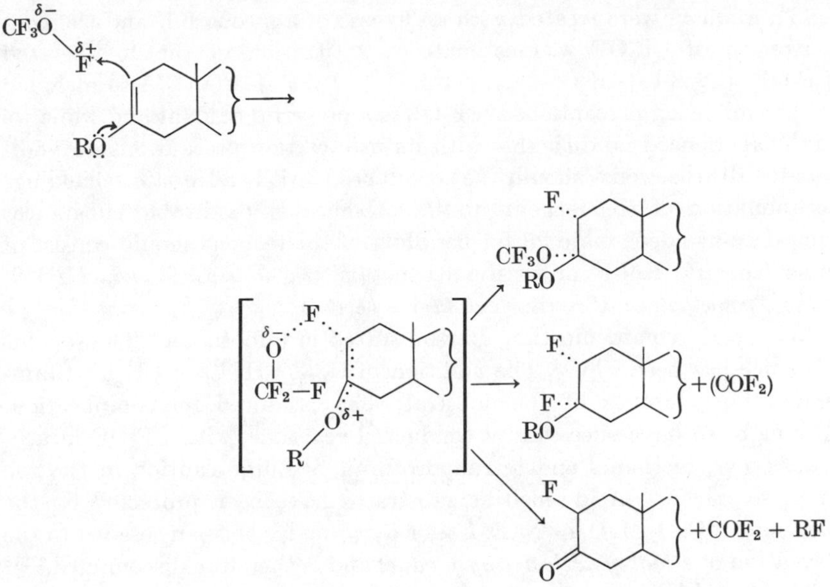

fluoride and eliminate carbonyl fluoride. However, the geometry of this system may be more favorable for the six-membered-ring transition state to eliminate carbonyl fluoride.

Two examples of trapping of the incipient carbonium ions have been given, a 19-methyl group migration (p. 250) and epoxide formation (p. 250).[279] Each is analogous to examples with electrophilic reagents known to involve carbonium ion intermediates.

Extensive studies are needed to elucidate fully the mechanism of hypofluorite addition, but restrictions on reaction conditions and the hazardous nature of the reagent make such studies difficult.

Hazards and Experimental Procedures

Explicit experimental details for the hypofluorite reactions are meager since reactions with steroids, nucleic bases, nucleosides, and sugars are described mostly in a series of preliminary communications. The lack of detail is unfortunate because hypofluorites are extremely reactive reagents—as reactive as elemental fluorine—that *can cause fires and explosions* if handled incorrectly. The hazards of these reagents have been experienced by colleagues in our laboratories, and we stress that *the potential danger in using hypofluorites must be recognized and proper safety precautions taken before they are used.*

Barton recently published the following general advice.[283] "Solutions of CF_3OF were prepared by passing the gaseous reagent into $CFCl_3$ at

$-78°$; aliquots were treated with an excess of aqueous KI, and the concentration of CF_3OF was estimated by titration of the I_2 liberated ($CF_3OF + 2\ KI + H_2O \rightarrow I_2 + 2\ KF + 2\ HF + CO_2$)." Also included are the following precautions. "CF_3OF is a powerful oxidant and while we have experienced no difficulty with its use, certain precautions are indicated: all reactions should be conducted with adequate shielding; accumulation of the reagent in the presence of oxidizable substances should be avoided; material for handling of the reagent should consist of glass, Teflon®, Kel-F, or passivated metals. *On no account should PVC, rubber, polyethylene or similar substances be used.*"

In an early communication, Barton stated in a footnote: "The solvent of choice has been $CFCl_3$. The addition of CCl_4, $CHCl_3$, or CH_2Cl_2 to improve the solubility of the substrate has produced no complication. Although we have successfully conducted reactions with CF_3OF in acetone, ether, methanol and tetrahydrofuran, we urge caution in the use of these solvents. Acid-sensitive substrates have been protected by the inclusion of CaO, MgO, or NaF. Use of pyridine for this purpose led to the formation of a *highly explosive by-product* and is therefore discouraged."[278]

The following procedures are derived primarily from the communications. The synthesis of a steroid is taken from a patent example; the only other experimental information is the general statement in the initial publication that the steroids were allowed to react with hypofluorite in fluorotrichloromethane at $-78°$.[278] Subsequent reports refer to this work.

Fluorobenzene.[277] A solution of benzene (25 mmol) in 80 ml of trichlorofluoromethane in a special Kel-F reactor equipped with a transparent Kel-F window[285] was irradiated at $-78°$ (acetone-dry ice bath) under stirring while trifluoromethyl hypofluorite gas (20 mmol) was passed in during 1 hour. After 30 minutes of further irradiation the solvent was distilled. Analysis of the residue by quantitative gas-liquid chromatography indicated fluorobenzene, 65%, and trifluoromethoxybenzene ($C_6H_5OCF_3$), 10%. A similar experiment, without irradiation, gave a complex mixture of products with only 17% of fluorobenzene; addition of CF_3OF in the dark followed by uv irradiation also gave fluorobenzene in low yield. Irradiation with a very high luminous flux density is of key importance for the success of this method.

2α-Fluorocholestanone.[278b] A solution of 3-methoxy-$\Delta^{2,3}$-cholestene (300 mg, 0.75 mmol) in fluorotrichloromethane (200 ml) containing calcium oxide (200 mg) was treated at $-78°$ with a slow stream of trifluoromethyl hypofluorite. Excess oxidant was removed by sweeping with

[285] J. Kollonitsch, G. A. Doldouras, and V. F. Verdi, *J. Chem. Soc., B*, **1967**, 1093.

nitrogen. The solution was then filtered and the precipitate washed with methylene chloride. The filtrate was washed with aqueous sodium bicarbonate, and concentrated to small bulk. Crystallization from hexane gave 2α-fluorocholestanone in 70% yield.

Photofluorination of L-Isoleucine.[277] L-Isoleucine (10 mmol) dissolved in 50 ml of hydrogen fluoride was photofluorinated at −78° for 1 hour with trifluoromethyl hypofluorite present. The solvent was evaporated and the residue was dissolved in water and made basic. Amino acid analysis (Spinco-Beckman) indicated the presence of 3.9 mmol of *trans*-3-methyl-L-proline (*ca.* 39%) besides small amounts of other amino acids. The authors proposed that δ-fluoro-L-isoleucine was formed initially, but was cyclized to proline by the aqueous base.

5-Fluorouracil. (See equation, p. 251) A.[281] Reaction of 0.336 g (0.003 mol) of uracil with a twofold excess of trifluoromethyl hypofluorite in methanol-fluorotrichloromethane at −78° resulted in complete loss of the uracil chromophore at 260 nm. Solvent and excess reagent were removed at 20°. The resulting, somewhat unstable adduct mixture (presumably 5-fluoro-6-trifluoromethoxy-5,6-dihydrouracils) was dissolved in a 10% solution of triethylamine in 50% aqueous methanol and allowed to stand at room temperature for 24 hours. The solution was evaporated to dryness. Recrystallization of the resulting solid from water gave 0.33 g (85%) of 5-fluorouracil.

B.[283] Uracil (0.336 g, 3.00 mmol) in a mixture of trifluoroacetic acid (6 ml) and water (20 ml) was added to a solution of trifluoromethyl hypofluorite (4.5 mmol) in trichlorofluoromethane (50 ml) at −78° in a pressure bottle. The precipitated uracil redissolved in the aqueous layer when the mixture was warmed to room temperature. The mixture was vigorously stirred for 15 hours. The excess trifluoromethyl hypofluorite was removed with nitrogen, and the solvent was removed under reduced pressure. The solid residue was sublimed at 210–230° (0.5 mm) to give crude 5-fluorouracil (0.365 g, 94%), mp 260–270°. Recrystallization from methanol-ether gave pure 5-fluorouracil (0.33 g, 85%).

3,4,6-Tri-O-acetyl-D-glucal and Trifluoromethyl Hypofluorite.[280] (See equation p. 251). 3,4,6-Tri-O-acetyl-D-glucal in trichlorofluoromethane at −78° and trifluoromethyl hypofluorite gave a mixture of four products which could be resolved by thin-layer chromatography [Kieselgel (Merck) 7731, ether-light petroleum (1:1), detection with concd. H_2SO_4] and gas-liquid chromatography (Pye 104 Chromatograph, SE 30, 170°, flame ionization detection). Elution of the mixture from Kieselgel (Merck, 7734) gave [in order of elution and after crystallization, in each case, from ether-light petroleum (bp 100–120°)]: trifluoromethyl 3,4,6-tri-O-acetyl-2-deoxy-2-fluoro-α-D-glucopyranoside, 26%, mp 84–85°, [α]D +158° ($CHCl_3$); 3,4,6-tri-O-acetyl-2-deoxy-2-fluoro-α-D-glucopyranosyl fluoride, 34%, mp 91–92°, [α]D +138° ($CHCl_3$); trifluoromethyl 3,4,6-tri-O-acetyl-2-deoxy-2-fluoro-β-D-mannopyranoside, 5.5%, mp 96–97°, [α]D −24° ($CHCl_3$); and 3,4,6-tri-O-acetyl-2-deoxy-2-fluoro-β-D-mannopyranosyl fluoride, 7.6%, mp 113–114°, $[\alpha]_D$ −3.5° ($CHCl_3$).

CONDENSATION REACTIONS TO INTRODUCE MONOFLUOROCARBON UNITS

All the preceding sections are devoted to methods in which a single fluorine is introduced selectively by forming a new carbon-fluorine bond, often in a highly complex and sensitive molecule such as a steroid or a sugar. However, the classical method for preparing a complex monofluoro aliphatic organic compound, often for biological studies, is by a sequence starting with a simple monofluoro reactant. It usually involves condensation at an active methylene group by reagents such as fluoroacetic acid or ester, fluoroketones, or fluoroaldehydes. More recently, reactive intermediates such as fluorinated carbenes or radicals have been shown to be capable of introducing a fluorinated unit through substitution or addition.[286a] Chlorofluorocarbene is one of the few such intermediates that has been studied to any extent, and not primarily in synthetic schemes. Another is monofluorocarbene which adds stereospecificially in low yield to alkenes and cycloalkenes.[286b] Chlorofluorocarbene and bromofluorocarbene have been used to synthesize a number of cyclopropane derivatives.[286c,d]

In this section the classical condensation reactions for preparing monofluoro aliphatic derivatives are reviewed briefly, although this approach is now more of historical interest than practical value. However, on occasions it could be the best or only synthetic route and should always be held in reserve. It was the method first used to prepare certain important biological chemicals such as fluorouracil[287] and fluorocarboxylic

$$CHBr_2F \xrightarrow{2\ n\text{-}C_4H_9Li} LiCHBrF \xrightarrow{-LiBr} :CHF$$

$$C_3H_7CH{=}CHC_3H_7 \xrightarrow{:CHF} \text{1-fluoro-2,3-dipropylcyclopropane} + \text{cyclohexane (18\%), 2,3-dimethylbutene (10\%), 1-heptane (3\%), } trans\text{-4-octene (3\%)}$$

syn and *anti* (2%)

[286] (a) Ref. 15, Sheppard and Sharts, pp. 370–382; (b) M. Schlosser and G. Heinz, *Chem. Ber.*, **104**, 1934 (1971) (c) D. Seyferth and L. J. Murphy, *J. Organomet. Chem.*, **49**, 117 (1973), (d) D. Seyferth and S. P. Hopper, *ibid.*, **51**, 77 (1973).

[287] C. Heidelberger, *Progr. Nucl. Acid Res. Mol. Biol.*, **4**, 1 (1965).

acid derivatives.[12] A major contribution to this field comes from Bergmann's laboratory in Israel. Unfortunately, recent reviews that cover condensation reactions as routes to monofluoro aliphatic compounds are not exhaustive.[12, 286]

Many simple monofluoro aliphatic acids and esters are analogs of chemicals important in biological studies of metabolism and for pharmacological use. Pattison has pioneered research in this area, including fluoro compounds that occur in nature, and has written excellent reviews.[7, 12] The key reagent is fluoroacetic acid or ester. As examples we cite the synthesis of fluorocitric acid, which is the highly toxic reagent that blocks citrate metabolism and which is synthesized *in vivo* from fluoroacetic acid. Fluoropyruvic acid can be synthesized from the same intermediate.

$$C_2H_5OCOCH_2F + C_2H_5OCOCO_2C_2H_5 \xrightarrow{KOC_2H_5} \underset{\text{Diethyl fluorooxalacetate}}{C_2H_5OCOCHFCOCO_2C_2H_5} + C_2H_5OH \quad \text{(Ref. 288)}$$

$$\text{Diethyl fluorooxalacetate} \xrightarrow[\text{Pyridine}]{CH_3CO_2H} \xrightarrow{H^+} \underset{\text{Fluorocitric acid } (>50\%)}{\begin{array}{c} FCHCO_2H \\ | \\ HOCCO_2H \\ | \\ CH_2CO_2H \end{array}} \quad \text{(Ref. 289)}$$

$$\text{Diethyl fluorooxalacetate} \xrightarrow[-CO_2]{\text{Concd. HCl}} \underset{\text{Fluoropyruvic acid}}{FCH_2COCO_2H} \quad \text{(Ref. 288)}$$

For the synthesis of a series of α-monofluoro alkanoic acids a condensation route was compared to the halogen fluoride addition and perchloryl fluoride substitution methods.[34] Condensation was inferior because of a low overall yield, a lengthy procedure, and need for large quantities of ethyl fluoroacetate (expensive and highly toxic).

$$FCH_2CO_2C_2H_5 + ClCO_2C_2H_5 \rightarrow \underset{(28\%)}{CHF(CO_2C_2H_5)_2}$$

$$RBr + CHF(CO_2C_2H_5)_2 \xrightarrow[DMF]{NaH} \underset{(51\text{–}81\%)}{RCF(CO_2C_2H_5)_2} \xrightarrow{\text{Hydrolysis}} \underset{(42\text{–}80\%)}{RCHFCO_2H}$$

$R = C_2H_5$, n- and i-C_3H_7, n-C_4H_9, $F(CH_2)_4$, $C_2H_5O_2C(CH_2)_4$, n-$C_{16}H_{33}$, $C_6H_5CH_2$

5-Fluorouracil **35**, which is an important anticancer drug, was first prepared by a condensation of fluoroacetate.[287, 290]

[288] I. Blank, J. Mager, and E. D. Bergmann, *J. Chem. Soc.*, **1955**, 2190.

[289] P. J. Brown and B. C. Saunders, *Chem. Ind.* (London), **1962**, 307.

[290] R. Duschinsky, E. Pleven, and C. Heidelberger, *J. Amer. Chem. Soc.*, **79**, 4559 (1957).

$$FCH_2CO_2C_2H_5 + HCO_2C_2H_5 \xrightarrow{K} KOCH{=}CFCO_2C_2H_5 \xrightarrow{C_2H_5SC(=NH)NH_2-HBr}$$

[2-ethylthio-4-hydroxy-5-fluoropyrimidine: OH, F, N, C_2H_5S, N] $\xrightarrow[H_2O]{H^+}$ [5-fluorouracil: O, F, H, N, O, N, H] **35**

A modification was recently reported by Bergmann.[291]

$$FCH_2CO_2C_2H_5 + HCO_2C_2H_5 \xrightarrow[(n\text{-}C_4H_9)_2O,\ 0°]{NaH} [HCOCHFCO_2C_2H_5] \xrightarrow[20°]{CH_3COCl}$$

$$\underset{(-)}{CH_3CO_2CH{=}CFCO_2C_2H_5} \xrightarrow[\text{Benzyl isothiouronium chloride}]{NaOCH_3,\ CH_3OH,}$$ [2-benzylthio-4-hydroxy-5-fluoropyrimidine: OH, F, N, $C_6H_5CH_2S$, N] (30%) $\xrightarrow{\text{Hydrolysis}}$

5-Fluorouracil (**35**)
(40%)

The first synthesis of a fluorosugar was from diethyl fluorooxalacetate, prepared from fluoroacetate.[101] More recently the fluoropentitols (see below) were prepared by a Claisen-type condensation. Total synthesis of fluorosugars is of limited utility because complications arise from the creation of multiple asymmetric centers.[101]

[CO_2CH_3–CH–O, CH_2–O, $C(CH_3)_2$ acetonide] $+ CH_2FCO_2C_2H_5 \xrightarrow[(C_2H_5)_2O]{Na}$ [$COCHFCO_2C_2H_5$–CH–O, CH_2–O, $C(CH_3)_2$] $\xrightarrow{KBH_4}$

[$CHOHCHFCH_2OH$–CH–O, CH_2–O, $C(CH_3)_2$] $\xrightarrow{H_2SO_4}$ H–C(–OH)($CHOHCHFCH_2OH$)(CH_2OH) (Ref. 292)

[(±)-2-deoxy-2-fluororibitol separated from mixture of fluoropentitols]

[291] E. D. Bergmann, I. Shahak, and I. Gruenwald, *J. Chem. Soc., C*, **1967**, 2206.
[292] P. W. Kent and J. E. G. Barnett, *J. Chem. Soc.*, **1964**, 2497.

Five other recent examples of syntheses of monofluoro aliphatic compounds by condensation reactions follow.

A malonic ester approach to fluorinated amino acids was unsuccessful because lactone formation (loss of hydrogen fluoride) occurred too readily.[293]

$$CH_2FCH_2CHBrCH_3 + CH_2(CO_2C_2H_5)_2 \xrightarrow{Na,\ C_2H_5OH} \underset{(64\%)}{CH_2FCH_2CH(CH_3)CH(CO_2C_2H_5)_2} \xrightarrow[H_2O/C_2H_5OH]{NaOH} \underset{(47\%)}{CH_2FCH_2CH(CH_3)CH(CO_2H)_2} \xrightarrow{Heat} CH_2FCH_2CH(CH_3)CH_2CO_2H + \text{4-methyl-}\delta\text{-valerolactone } (CH_3)$$

α-Fluoroalkylphosphonates were used for preparation of fluoroolefins. In the example given, the starting diethyl α-fluorobenzylphosphonate was prepared using fluoroalkylamine reagent (**FAR**, see p. 299).[294]

$$C_6H_5CHFP(O)(OC_2H_5)_2 \xrightarrow{NaH,\ DMF} [C_6H_5CFP(O)(OC_2H_5)_2] \xrightarrow{ArCHO} C_6H_5CF(ArCHO^-)\text{—}P(O)(OC_2H_5)_2 \xrightarrow{H_2O} \underset{(40\%)}{C_6H_5CF{=}CHAr}$$

A series of γ-fluoropentamethine cyanine dyes was prepared to study the effect of fluorine substitution on absorption spectra. Again fluoroacetate was the initial starting material for preparation of the essential intermediate, fluoromalonaldehyde.

$$FCH_2CO_2Na + 2\,[(CH_3)_2NCHCl]^+Cl^- \longrightarrow \left[(CH_3)_2N\text{—}CH{=}CF\text{—}CH{=}N(CH_3)_2\right]^+ Cl^- \xrightarrow{H_2O(K_2CO_3)}$$

$$(CH_3)_2N\text{—}CH{=}CF\text{—}CH{=}O \xrightarrow{NaOH} \left[O{\cdots}CH{\cdots}CF{\cdots}CH{\cdots}O\right]^- Na^+ \quad \text{(Ref. 295)}$$

[293] M. Hudlický, B. Kakáč, and I. Lejhancová, *Collect. Czech. Chem. Commun.*, **32,** 183 (1967).

[294] E. D. Bergmann, I. Shahak, and J. Appelbaum, *Isr. J. Chem.*, **6,** 73 (1968).

[295] C. Reichardt and K. Halbritter, *Ann. Chem.*, **737,** 99 (1970).

$$\text{1-ethyl-2-methylquinolinium } BF_4^- + [\text{O}\cdots\text{CH}\cdots\text{CF}\cdots\text{CH}\cdots\text{O}]^- Na^+ \xrightarrow{Ac_2O,\ AcONa}$$

$$\text{bis(1-ethylquinolin-2-yl)-fluoropentamethine cyanine (cation)} \quad \text{(Ref. 296)}$$

Organometallic reagents condense with fluorocarboxylic acids, acid halides, and esters to give the corresponding ketones and alcohols.

$$RCHFCO_2C_2H_5 + R'MgX \rightarrow \underset{(40\text{–}55\%)}{RCHFCOR'} + \underset{(5\text{–}35\%)}{RCHFC(OH)R'_2} \quad \text{(Ref. 297)}$$

$$FCH_2COCl + CH_3MgBr \rightarrow \underset{(41\%)}{FCH_2C(OH)(CH_3)_2} \quad \text{(Ref. 298)}$$

Diazoalkanes condense with acyl fluorides to give α-fluoro acids and α-fluoroketones.[299]

$$RCHN_2 + FCO_2C_2H_5 \xrightarrow{-N_2} \underset{(62\text{–}85\%)}{RCHFCO_2C_2H_5}$$

$$RCHN_2 + FCOCH_3 \xrightarrow{-N_2} \underset{(38\text{–}62\%)}{RCHFCOCH_3}$$

A number of carbonyl additions of fluorodinitromethane have recently been reported.[300]

$$RCHO + FC(NO_2)_2H \longrightarrow RCH(OH)CF(NO_2)_2$$

SELECTION OF METHODS FOR MONOFLUORINATION

The best methods for preparing monofluoroaliphatic compounds are concisely outlined in Table IX, in which the reagents are summarized and evaluated by type of fluorine substitution. To aid in choosing the most applicable method, comments dealing with limitations and interfering functionality are included. "Activated" primary or secondary fluoride refers to a fluoride that is allylic, benzylic, or alpha to other groups such as carbonyl, ester, or nitrile.

[296] C. Reichardt and K. Halbritter, *Chem. Ber.*, **104**, 822 (1971).
[297] E. Elkik and H. Assadi-Far, *Bull. Soc. Chim. Fr.*, **1970**, 991.
[298] G. A. Olah and J. M. Bollinger, *J. Amer. Chem. Soc.*, **90**, 947 (1968).
[299] E. D. Bergmann and I. Shahak, *Isr. J. Chem.*, **3**, 73 (1965).
[300] H. G. Adolph, *J. Org. Chem.*, **35**, 3188 (1970).

Aliphatic Primary Fluorides. The most practical synthesis is the reaction of an anhydrous alkali metal fluoride with an alkyl halide or sulfonate ester in an appropriate reaction medium. When alkali metal fluorides fail, silver fluoride or silver tetrafluoroborate should be used with the appropriate bromide or iodide. The reaction of **FAR** or R_2PF_3/R_3PF_2 with a primary alcohol should be attempted if the preceding reactions fail. Because of the possibility of an explosion, the reaction of perchloryl fluoride with an organolithium or Grignard reagent should be given the lowest priority.

Benzylic or Allylic Fluorides. The most practical synthesis is the reaction of an alkali metal fluoride with a benzylic or allylic halide or sulfonate ester. Silver fluoride or silver tetrafluoroborate on a bromide or iodide is the method of second choice. **FAR** on an allylic or benzylic alcohol may give favorable results. Again the last reagent to be tried should be perchloryl fluoride on the appropriate organolithium or Grignard reagent.

Secondary Fluorides. No method is clearly preferred. If a halogen or sulfonate ester group is activated by an alpha substituent (*e.g.*, cyano, keto, or carbalkoxy), substitution using an alkali metal fluoride is preferred. The options of the synthetic routes may indicate that addition of halogen fluoride or hydrogen fluoride to a double bond is best. Sometimes the reaction of **FAR** with a secondary alcohol is preferred. Where other methods fail, the reaction of perchloryl fluoride with an organolithium or Grignard reagent should be cautiously attempted.

Tertiary Fluorides. The best method is the addition of halogen fluoride to a substituted alkene followed by reduction to remove the halogen. When rearrangement or the polymerization side reaction does not prevail, addition of hydrogen fluoride is appropriate. If an anion of the desired tertiary skeleton can be formed, the use of perchloryl fluoride may be attempted.

Steroidal Fluorides. The methods appropriate for primary aliphatic fluorides may be used to synthesize fluoromethyl steroids.

Ring-substituted steroids are best prepared by addition of hydrogen fluoride to an epoxide or by addition of halogen fluoride or hydrogen fluoride to a double bond. The use of **FAR** to replace a hydroxyl group depends upon the ring position and stereochemistry of the hydroxyl group. Although, in the past, silver fluoride and silver tetrafluoroborate have been used frequently to substitute iodide and bromide, the method is not expected to give good yields of the desired fluoride.

Fluoroketo steroids are prepared by addition of nitrosyl fluoride or trifluoromethyl hypofluorite to a double bond. For nitrosyl fluoride, the

TABLE IX. SUMMARY OF METHODS

Reagent and Reaction Type		Primary RCH_2F Unactivated	Primary RCH_2F Activated
Halogen fluoride addition (*in situ* generation)	$R_2C{=}CR_2 \xrightarrow{\text{“X—F”}} R_2CXCFR_2$	Limited because F usually oriented to m highly substituted carbon	
Fluoroalkylamine reagent (**FAR**)	$ROH \rightarrow RF$	Good	Fair, carbonium rearrangement often a proble
Inorganic fluorides Alkali metal fluoride, including AgF, $AgBF_4$	$R{-}X \rightarrow R{-}F$ X = halides, esters	Halides or sulfonate substitution usually g to excellent, KF preferred	
R_4NF		Not reported	Good on esters
Phosphorus fluorides	X = OH or OR	Good for substitution of OH or silyl et limited examples	
Hydrogen fluoride addition: Direct; Moderated by complexing; Indirect from BF_3	$>C{=}C< \rightarrow >CH{-}CF<$; epoxide $>C{-}O{-}C< \rightarrow >COH{-}CF<$	Limited because additions usually give carbon with most substitution	
Substitution	$RX \rightarrow RF$ X = OH or ester	Rare	Primarily polyl groups

EPARE ALIPHATIC FLUORIDES

econdary R_2CHF		Tertiary R_3CF	Sugars	Steroid	Comments, Limitations, and Interfering Functionality
ctivated	Activated				
d to sub- tituted lefins; roduct as α-halo ıbstitution or further eaction	Good with F; usually oriented *gem* to group that stabilized carbonium ion	Good yield addition to trisubstituted olefin	Used extensively	Used extensively	Used with most functional groups; except groups easily oxidized by positive halogen reagent
r; elimination or earrangement often redominates		Only useful on bridgehead alcohols	No success	Used extensively; yields highly variable depending on types of OH	Unreactive to most functional groups but reacts with free acid and amine groups
lide or sulfonate substitu- ion 'air, often liminated	Good	Halides usually eliminate	Used extensively	Used extensively	Generally good with few limitations
r to good, mited xamples	Good on sulfonate esters	—	Primary use	Not reported	Reagent difficult to get anhydrous; water impurities cause major variations in results
od for substitution of OH r silyl ether, limited xamples		Good, but limited examples	Not reported	Good on one example	Reagent difficult to prepare and handle; other active hydrogen groups will also react
ed extensively, excellent n epoxides		Excellent for both double bonds and epoxides	Used occasionally	Used extensively on epoxides and activated double bonds	Major disadvantage: HF is highly corrosive and toxic; can give substitution of halides as side reaction
ce	Primarily polyhalo groups	Not usual	Used frequently for selective substitution of ether or ester groups	Occasionally	Classical method of commercial importance, usually antimony halide catalyzed; will also add to unsaturated functions

TABLE IX. Summary of Methods

Reagent and Reaction Type	Primary RCH_2F Unactivated	Activated
Perchloryl fluoride $R_3CH \rightarrow R_3C^- \rightarrow R_3CF$	Poor to good on RCH_2Li or RCH_2MgX	
Nitrosyl fluoride $(H)\text{—}C{=}C\!\! < \;\longrightarrow\; (H)\text{—}C(NO)\text{—}C(F)\text{—} \;\longrightarrow\; O_2NN{=}C\text{—}C(F)\text{—} \;\longrightarrow\; \text{—}C({=}O)\text{—}C(F)\text{—}$	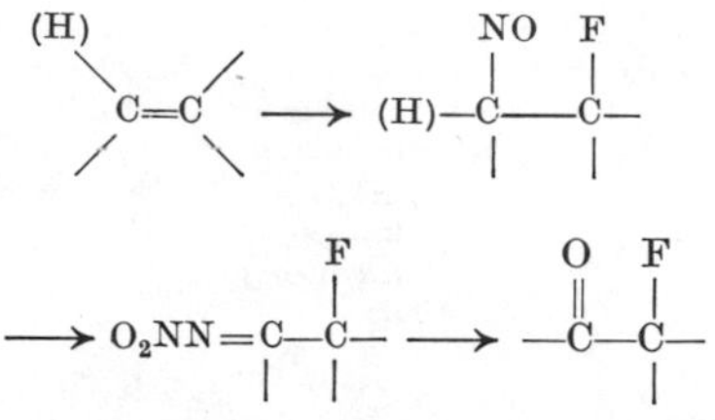	Not practical
Hypofluorites $>C{=}C< \;\longrightarrow\; \text{—}C(F)\text{—}C(F)\text{—} + \text{—}C(F)\text{—}C(OCF_3)\text{—}$; $>CH\text{—} \;\longrightarrow\; >CF\text{—}$	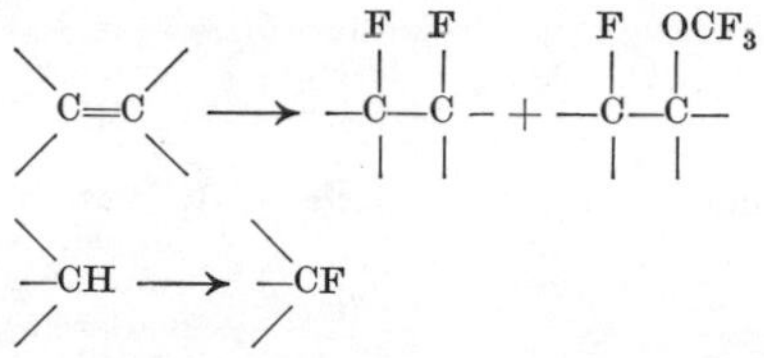 A few examples of C—H substitution, usua by irradiation	

bond should be activated by an appropriate function, e.g., enol ether or enol acetate.

Fluorosugars. This is a highly specialized area and no single reagent works well for all stereochemistry. The best methods use: silver fluoride or silver tetrafluoroborate to substitute fluoride for bromide, iodide, or sulfonate ester groups; tetra-*n*-butylammonium fluoride to substitute fluoride for acetoxy or benzoyloxy groups; and halogen fluoride or hydrogen fluoride to add to double bonds.

TABULAR SURVEY

Tables I–IX appear in the text. Tables X–XX follow and are arranged in the order in which the various fluorinating agents are discussed in the text.

Tables X, XI, XIX, and XX provide all examples known through June, 1973. Tables XII–XVII include all significant examples since earlier reviews, and selected examples from earlier literature that illustrate the methods.

REPARE ALIPHATIC FLUORIDES (*Continued*)

Secondary R_2CHF					
nactivated	Activated	Tertiary R_3CF	Sugars	Steroid	Comments, Limitations, and Interfering Functionality
	Good with α-carbonyl or nitro to form anion; disubstitution must be controlled	Good with α-carbonyl or nitro to stabilize anion	Not reported	Used extensively	Potent oxidizing agent that can explode violently if not used correctly; active hydrogen site, particularly amine and phenols are attacked
oor unless sterically induced	Excellent for addition to electron-rich double bonds	Orients to carbon with substitution to stabilize carbonium ion	Not reported, and probably very limited potential	Used primarily only on steroids to introduce α-fluoroketone function	Primarily of value for steroids but could have other selected applications; hydrogen or anion groups rapidly oxidized
rimarily by C—H substitution through irradiation	Enol ethers or acetates; excellent to give α fluoro ketones	Can be useful for substitution or addition	Gives expected mixtures in limited studies	Numerous examples of addition	Powerful oxidizing agent that must be used with great caution

Within each table, compounds are arranged in the order of increasing number of carbon atoms. Compounds with the same number of carbon atoms are arranged in order of increasing complexity.

Abbreviations for solvents are: ether, diethyl ether; diglycol, diethyleneglycol, DMF; N,N-dimethylformamide; DMS, dimethyl sulfone; DMSO, dimethyl sulfoxide; glyme, dimethoxyethane; diglyme, 2,2′-dimethoxyethyl ether; glycol, ethyleneglycol; HMPT, hexamethylphosphortriamide; NMA, N-methylacetamide; NMP, N-methylpyrrolidone; TMS, tetramethylene sulfone.

Other abbreviations used in formulas are Ac for acetyl ($COCH_3$), OAc for acetate ($OCOCH_3$), Bz for benzoyl (COC_6H_5), OBz for benzoate ($OCOC_6H_5$), OTs for $OSO_2C_6H_4CH_3$-*p* (tosylate), Ms for OSO_2CH_3 (mesyl), and Bzy for $CH_2C_6H_5$ (benzyl). Positive halogen reagents are listed on p. 138 and in footnote *a* of Table X. The fluoroalkylamine reagent, $(C_2H_5)_2NCF_2$-CFClH, is abbreviated as **FAR**. Unless stated otherwise HF means anhydrous hydrogen fluoride.

Yields are reported whenever available.

TABLE X. ADDITION OF HALOGEN FLUORIDE

$$\rangle C{=}C\langle + XF \longrightarrow -\underset{F}{\overset{|}{C}}-\overset{X}{\underset{|}{C}}-$$

	Reactant	Reagent(s)[a]	Condition(s)[b]	Product(s) and Yield(s) (%)	Refs.
C_1	Cyanogen chloride	ClF	25°	CF_2ClNCl_2 (>90)	53
C_2	Ethylene	Br_2, KF, H_2O	—	$BrCH_2CH_3$ (mostly), $BrCH_2CH_2F$ (trace?)	27
		Cl_2, $NaHF_2$, H_2O	Room t, 1:—:1	$ClCH_2CH_2F$ (58)[c]	301
		Cl_2, CaF_2, H_2O	Room t, 1:—:1	$ClCH_2CH_2F$ (—)	301
		NBA, HF, CH_2Cl_2	−40°, 5 hr, 1°:5:1	FCH_2CH_2Br (23)	35
		Cl_2, HF	−20° to −30°, $C_2H_4 + Cl_2$ passed through HF for 3.5 hr[d]	$ClCH_2CH_2F$ (5), $ClCH_2CH_2Cl$ (17)	61
		HCM, HF	0°, C_2H_4 added over 2 hr, 1.2:13:1	$ClCH_2CH_2F$ (63)	61
		HCM, HF, C_6H_6	3-hr addition	$C_6H_5CH_2CH_2Cl$ (8), $ClCH_2CH_2F$ (—)	61
	1,1-Difluoroethylene	HCM, HF	−10°, $C_2H_2F_2$ added during 90 min; 1.6:17:1	CF_3CH_2Cl (71)	61
	1,1-Dichloroethylene	Cl_2, HF	−20°, Cl_2 added during 2 hr, —:15:1	CCl_2FCH_2Cl (21)	61
C_3	Propylene	HCM, HF	0°, C_3H_6 added over 2 hr; 1.2:13:1	CH_3CHFCH_2Cl (62)	61
		DBH, HF, CH_2Cl_2	−80°, 2 hr; room t, overnight; 1:5:1	CH_3CHFCH_2Br (38)	35
		NBS, HF, ether	−70°, 2 hr; 0°, 2 hr; 1.2:30:1	CH_3CHFCH_2Br (30)	73c

		NIS, HF, ether	−70°, 2 hr; 0°, 2 hr; 1.2:30:1	CH_3CHFCH_2I (38)	73c
	Allyl bromide	NBA, HF, ether	−80°, 1.5 hr; 8°, overnight; 1:7.8:1	$BrCH_2CHFCH_2Br$ (22), $FCH_2CHBrCH_2Br$ (32)	35
C_4	Vinyl acetate	Br_2, AgF, CH_3CN or C_6H_6	[e]	$BrCH_2CHFOAc$ (—)	40
	Ethyl diazoacetate	NBS, HF, ether	−70°, 15 hr; 1:3.3:1	$BrFCHCO_2C_2H_5$ (59)	48
		Br_2, HF, ether	−70°, 1.5 hr; 1:20:2	$BrFCHCO_2C_2H_5$ (—), $Br_2CHCO_2C_2H_5$ (—), $C_2H_5OCH_2CO_2C_2H_5$ (—)	48
	Methyl acrylate	NBA, HF, CH_2Cl_2/THF	−80°, 1 hr; 0°, 18 hr; 1.3:50:1	$FCH_2CHBrCO_2CH_3$ (50)	36
		NBA, HF, ether/pyridine	−20°	$FCH_2CHBrCO_2CH_3$ (low), $BrCH_2CHFCO_2CH_3$ (6), $BrCH_2CH(OC_2H_5)CO_2CH_3$ (trace)	33
		DBH, HF, ether/CCl_4 or $CHCl_3$	−40°, 30 min; −30°, 4 hr; 1:14:2	$FCH_2CHBrCO_2CH_3$ (45)	33
		HCM, HF	0°, ester added over 1 hr; 1:15:1	$FCH_2CHClCO_2CH_3$ (31)	61
	Methyl propiolate	NBA, HF, ether	−78°, 3 hr; −20°, overnight; 1:10:1	$BrC{\equiv}CCO_2CH_3$ (36)	46
	1-Butene	NBA, HF, $CHCl_3$/ether	−60°, 3.5 hr; 1:5.6:1	$C_2H_5CHFCH_2Br$ (55)	35

Note: References 301–386 are on pp. 404–406.

[a] The abbreviations used for positive halogen reagents are: NBA, N-bromoacetamide; NIS, N-iodosuccinimide; NBS, N-bromosuccinimide; DBH, 1,3-dibromo-5,5-dimethylhydantoin; NCS, N-chlorosuccinimide; HCM, hexachloromelamine.

[b] The ratios represent the molar ratios for the reacting species: positive halogen source : fluoride source (HF or MF) : alkene.

[c] The product was isolated as vinyl fluoride after dehydrochlorination at 400° over soda lime.

[d] An improved yield was claimed with urea catalysis but no details were given.

[e] Conditions were not stated but were probably room t, 30 min, 1:6:1.

TABLE X. ADDITION OF HALOGEN FLUORIDE (*Continued*)

	Reactant	Reagent(s)[a]	Condition(s)[b]	Product(s) and Yield(s) (%)	Refs.
C_4 (*contd.*)	2-Butene (*trans*)	NCS, HF, ether	−78°, 1 hr; 0°, 1 hr; 1:25:1	*erythro*-$CH_3CHFCHClCH_3$ (25), *erythro*-N-succinimido: *erythro*-$NCH(CH_3)CHClCH_3$ succinimide (major)	73c
		NBS, HF, ether	−70°, 2 hr; 0°, 2 hr; 1.2:30:1	*erythro*-$CH_3CHBrCHFCH_3$ (45) (100% stereospecific)	73c
		NIS, HF, ether	−70°, 2 hr; 0°, 2 hr; 1.2:30:1	*erythro*-$CH_3CHICHFCH_3$ (42) (98% stereospecific)	73c
	2-Butene (*cis*)	NCS, HF, ether	−78°, 1 hr; 0°, 1 hr; 1:25:1	*threo*-$CH_3CHFCHClCH_3$ (21), *threo*-succinimide $N{-}CH(CH_3)CHClCH_3$ (major)	73c
		NBS, HF, ether	−70°, 2 hr; 0°, 2 hr; 1.2:30:1	*threo*-$CH_3CHBrCHFCH_3$ (51) (100% stereospecific)	73c
		NIS, HF, ether	−70°, 2 hr; 0°, 2 hr; 1.2:30:1	*threo*-$CH_3CHICHFCH_3$ (47) (98% stereospecific)	73c
	Isobutylene	$(CH_3)_3COCl$, HF, ether	Low t	$(CH_3)_2CFCH_2Cl$ (low)	73b
		NBS, HF, ether	−70°, 2 hr; 0°, 2 hr; 1:13:1	$(CH_3)_2CFCH_2Br$ (41)	73b
		NIS, HF, ether	−70°, 2 hr; 0°, 2 hr; 1:13:1	$(CH_3)_2CFCH_2I$ (—)	73b
	1,4-Dichloro-2-butyne	NBA, HF, ether	−78°, 3 hr; −20°, overnight; 1:10:1	*trans*-$ClCH_2CF{=}CBrCH_2Cl$ (—)	46

C_5	Methyl methacrylate	DBH, HF, $CHCl_3$	−50°, 10 min; −20°, 4 hr; 1:7:2	$BrCH_2CF(CH_3)CO_2CH_3$ (59)	33
	Allyl acetate	NBA, HF, ether	−80°, 75 min; 1:10:1	$BrCH_2CHFCH_2OAc$ (low), $BrCH_2CHFCH_2OH$ (low)	33
	Allyl trichloroacetate	NBA, HF, pyridine-ether	−80°, 1 hr; 1.2:26:1	$BrCH_2CHFCH_2OCOCCl_3$ (69)	33
	2-Methyl-2-butene	NIS, HF, ether	1:9:1[f]	$CH_3CHICF(CH_3)_2$ (68)	73b
		$(CH_3)_3COCl$, HF, ether	−78°, 1 hr; 0°, 1 hr; 1:9:1	$CH_3CHClCF(CH_3)_2$ (11)	73b
	3-Methyl-1-butene	NBS, HF, CCl_4/THF	−65 to −70°, 1 hr; 1.1:36:1	i-$C_3H_7CHFCH_2Br$ (—)	68
	3,4-Dihydro-2H-pyran	NBS, HF, ether	−78°, 0.5–2 hr; 0°, 0.5–2 hr; 1.2:29:1	(A) *trans* (55–60), *cis* (16)	40
		Br_2, AgF, CH_3CN or C_6H_6	[e]	**A**, *trans* (84), *cis* (16)	40
		I_2, AgF, CH_3CN or C_6H_6	[e]	*trans* (95), *cis* (5)	40
		NIS, HF, ether	[e]	Black tar fluorine-free	40

Note: References 301–386 are on pp. 404–406.

[a] The abbreviations used for positive halogen reagents are: NBA, N-bromoacetamide; NIS, N-iodosuccinimide; NBS, N-bromosuccinimide; DBH, 1,3-dibromo-5,5-dimethylhydantoin; NCS, N-chlorosuccinimide; HCM, hexachloromelamine.

[b] The ratios represent the molar ratios for the reacting species; positive halogen source : fluoride source (HF or MF) : alkene.

[e] Conditions were not stated but were probably room t, 30 min, 1:6:1.

[f] Conditions were not stated but were probably −78°, 2 hr; 0°, 2 hr.

TABLE X. ADDITION OF HALOGEN FLUORIDE (*Continued*)

	Reactant	Reagent(s)[a]	Condition(s)[b]	Product(s) and Yield(s) (%)	Refs.
C_5 (*contd.*)	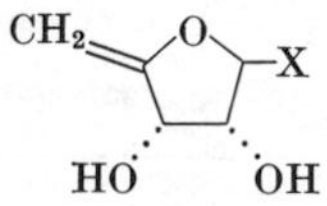 CH_2, O, X, HO, OH X = purine or pyrimidine base protected by Ac or Bz.	I_2, AgF, CH_3CN	Room t, 2 hr; 5:6:1	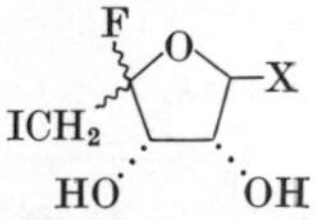 F, O, X, ICH_2, HO, OH (—)	45
	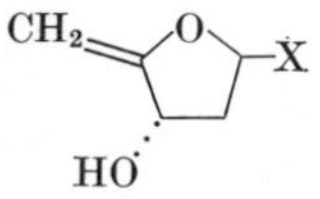 CH_2, O, X, HO X = cytosine, 5-halocytosine, uracil, 5-halouracil, or 5-alkyluracil protected by Ac or Bz.	I_2, AgF, CH_3CN	Room t, 2 hr; 5:6:1	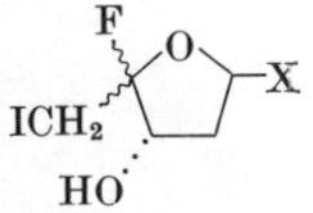 F, O, X, ICH_2, HO (—)	45
C_6	Benzene	I_2, AgF	Room t, addition; heat; 1:2:20	C_6H_5I (10)	28
	1-Hexene	NBS, HF, CCl_4/THF	-65 to $-70°$, 1 hr; 1.1:36:1	n-$C_4H_9CHFCH_2Br$ (—)	68
		NBA, HF, ether	$-80°$, 2 hr; room t, overnight; 1:12.8:1	" (79)	35
		DBH, HF, CCl_4	$-30°$, 0.5–4 hr;[g] 1:26:2	" (64–66)	35
	Tetramethylethylene	NBS, HF, CCl_4/THF	-65 to $-70°$, 1 hr; 1.1:36:1	$(CH_3)_2CFCBr(CH_3)_2$ (—)	68

	NBS, HF, ether	−78°, 2 hr; 1:10:1	" (60)	73a
	$(CH_3)_3COCl$, HF, ether	−78°, 1 hr; 0°, 1 hr; 1:25:1	$(CH_3)_2CFCCl(CH_3)_2$ (46)	73a
	NIS, HF, ether	−70°, 2 hr; 0°, 2 hr; 1:10:1	$(CH_3)_2CFCI(CH_3)_2$ (55)	73a
2-Ethyl-1-butene	NBA, HF, ether	−80°, 2 hr; room t, overnight; 1:10.6:1	$(C_2H_5)_2CFCH_2Br$ (56)	35
Cyclohexene	HCM, HF, ether	−20°, 3 hr; (1:5:1)	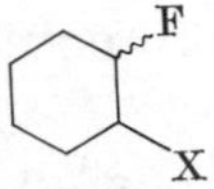 **B**, X = Cl (51)	61
	N-Chlorophthalimide, HF, $CHCl_3$/di-*n*-hexyl ether[h]	−65°; 1.1:24:1	**B**, X = Cl (—)	68
	NBA, HF, $CHCl_3$	−65°, 90 min; 0°, 1 hr	**B**, X = Br (40)	68
	NBA, HF, ether or THF/CH_2Cl_2[i]	−80°, 2 hr; 0°, 2 hr; 1.3:17:1	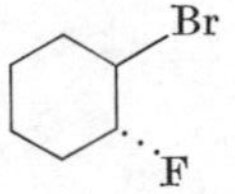 (48)	32, 68

Note: References 301–386 are on pp. 404–406.

[a] The abbreviations used for positive halogen reagents are: NBA, N-bromoacetamide; NIS, N-iodosuccinimide; NBS, N-bromosuccinimide; DBH, 1,3-dibromo-5,5-dimethylhydantoin; NCS, N-chlorosuccinimide; HCM, hexachloromelamine.

[b] The ratios represent the molar ratios for the reacting species; positive halogen source : fluoride source (HF or MF) : alkene.

[g] Varied conditions: the alkene could be added dropwise, the DBH portionwise, or all reagents added together without significant effect on yield.

[h] Also reported in the patent is the use of N-chloropalmitamide, N-chloropropionamide, and N-chlorotoluamide with methyl *n*-butyl ether, benzophenone, and anthraquinone, respectively, as proton acceptors in place of di-*n*-hexyl ether.

[i] The use of N-bromophthalimide or N-bromodecenoylamide with chloroform as solvent and diethyl or diisoamyl ether as proton acceptors was also reported without complete experimental details or yields.

TABLE X. ADDITION OF HALOGEN FLUORIDE (*Continued*)

	Reactant	Reagent(s)[a]	Condition(s)[b]	Product(s) and Yield(s) (%)	Refs.
C_6 (*contd.*)	Cyclohexene (*contd.*)	Br_2, AgF, C_6H_6	[j]	trans-1-bromo-2-fluorocyclohexane (80)	31
		I_2, AgF, cyclohexene/pet. ether/pyridine	Reflux, 2 hr; 1:2:20	1,3-cyclohexadiene Isolated as adduct with maleic anhydride (10)	28
		NIS, HF, ether or dioxane/ CH_2Cl_2[k]	−80°, 2 hr; 0°, 1 hr	trans-1-fluoro-2-iodocyclohexane (73)	32, 68
		AgF, I_2	−8°, CH_3CN	" (60)[l]	302
		I_2, AgF, C_6H_6		" (63)	31
	Ethyl crotonate	I_2, AgF, ethyl crotonate	Reflux; 1:2:8	$ICH_2CH{=}CHCO_2C_2H_5$ (20)	28
	Dimethyl maleate	NBA, HF, CH_2Cl_2/THF	−80°, 1 hr; 0°, 18 hr; 1.3:50:1	*threo*-$CH_3OCOCHFCHBrCO_2CH_3$ (55)	36b, 303a
	Dimethyl fumarate	NBA, HF, CH_2Cl_2/THF	−80°, 1 hr; 0°, 18 hr; 1.3:50:1	*erythro*-$CH_3OCOCHFCHBrCO_2CH_3$ (58)	36b, 303a
	5-Hexenoic acid	NBA, HF, CH_2Cl_2/ fluorenone	−80°, 1 hr; 0°, 1 hr	$BrCH_2CHFCH_2CH_2CH_2CO_2H$ (—)	68

	Substrate	Reagents	Conditions	Product (Yield, %)	Refs.
	1-Hexyne	NBA, HF, ether	$-78°$, 3 hr; $-20°$, overnight; 1:10:1	n-$C_4H_9CF{=}CHBr$ (48) (*trans*:*cis* 95:5)	46
		Br_2, AgF, CH_3CN	Room t; 1.1:1:1	n-$C_4H_9C{\equiv}CBr$ (—) (trace of BrF addition product)	46
	3-Hexyne	NBA, HF, THF/pyridine	$-78°$, 2 hr; 0°, 3 hr; 1.3:10:1	$C_2H_5CF{=}CBrC_2H_5$ (28) (*trans*:*cis* 78:22)	46
		Br_2, AgF, CH_3CN	Room t; 1.1:1:1	$C_2H_5CBr{=}CBrC_2H_5$ (—)	46
	$CH_3OC(CF_3)_2C{\equiv}CH$	NBA, HF	$-78°$, 2 hr; 0°, 3 hr; 1.3:10:1	No reaction	46
	$CH_3OC(CF_3)_2C{\equiv}CCl$	NBA, HF	$-78°$, 2 hr; 0°, 3 hr; 1.3:10:1	No reaction	46
	$CH_3O_2CC{\equiv}CCO_2CH_3$	NBA, HF	$-78°$, 2 hr; 0°, 3 hr; 1.3:10:1	Intractable product	46
C_7	Toluene	I_2, AgF	Add at room t, then heat; 1:2:20	p-$IC_6H_4CH_3$ (15)	28
	1-Heptene	NBA, HF, ether	$-80°$, 2 hr; room t, overnight; 1:9.4:1	n-$C_5H_{11}CHFCH_2Br$ (60–77)	35, 74
	Methylenecyclohexane	Br_2, AgF, C_6H_6		F, CH_2Br (59)	31
		I_2, AgF, C_6H_6		F, CH_2I (70)	31

Note: References 301–386 are on pp. 404–406

[a] The abbreviations used for positive halogen reagents are: NBA, N-bromoacetamide; NIS, N-iodosuccinimide; NBS, N-bromosuccinimide; DBH, 1,3-dibromo-5,5-dimethylhydantoin; NCS, N-chlorosuccinimide; HCM, hexachloromelamine.

[b] The ratios represent the molar ratios for the reacting species; positive halogen source : fluoride source (HF or MF) : alkene.

[j] No temperature was stated. Finely divided silver(I) fluoride, vigorous stirring, and slow initial addition were critical. Addition of "ClF" using $Cl_2/AgF/C_6H_6$ was reported "less successful than either 'BrF' or 'IF'."

[k] The use of N-iodobenzamide or N-iodoisobutyramide with chloroform as solvent and acetone and cyclohexanone as proton acceptors was also reported as a patent example without complete experimental details or yields.

[l] Andreatta and Robertson[60] could not obtain this product, but only N-acetyl-2-iodocyclohexylamine.

TABLE X. Addition of Halogen Fluoride (*Continued*)

	Reactant	Reagent(s)[a]	Condition(s)[b]	Product(s) and Yield(s) (%)	Refs.
C_7 (*contd.*)	Norbornene	NBA, HF, ether	−80°, 14 hr; 1.1:14:1	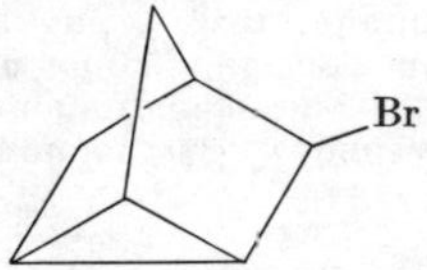 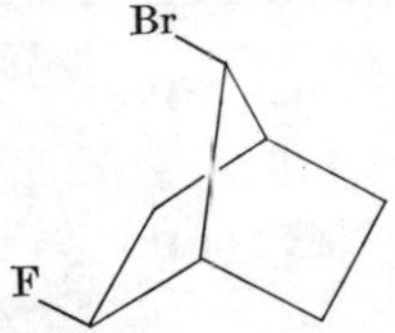 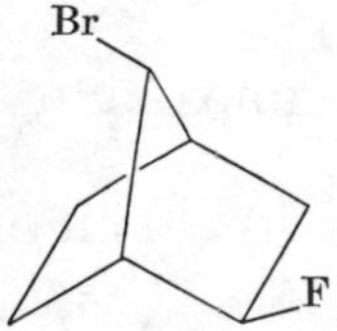 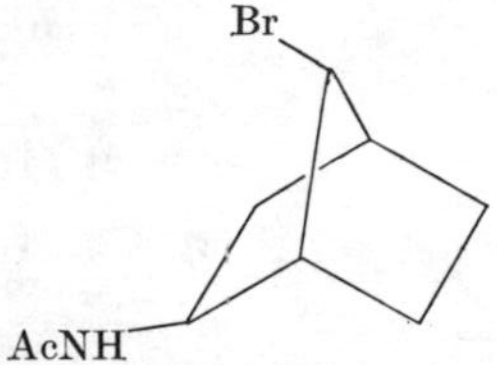 Br (33), F, Br (20), Br, F (34), Br, AcNH (small?)	62
	Anisole	I_2, AgF	Add at room t; heat; 1:2:20	p-$IC_6H_4OCH_3$ (57)	28

	Methyl 5-hexenoate	NBS, HF, octane/THF	$-80°$, 1 hr; 0°, 1 hr	$BrCH_2CHF(CH_2)_3CO_2CH_3$ (—)	68
	3,4-Dihydro-2H-pyran-2-methyl acetate	Br_2, AgF, CH_3CN or C_6H_6	[e]	X, F, O, CH_2OAc **C**, X = Br *trans*-diaxial (57), *trans*-diequatorial (20), *cis* (18)	40
		I_2, AgF, CH_3CN or C_6H_6	[e]	**C**, X = I *trans*-diaxial (78), *trans*-diequatorial (17), *cis* (*ca.* 5)	40
C_8	1-Octene	NBA, HF, ether	$-80°$, 2 hr; room t, overnight; 1:9.3:1	$CH_3(CH_2)_5CHFCH_2Br$ (70)	35
		NBS, HF, CCl_4/THF	-65 to $-70°$, 1 hr; 1.1:36:1	" (—)	68
	Styrene	NBA, HF, ether	$-80°$, 2 hr; room t, overnight; 1:22.7:1	$C_6H_5CHFCH_2Br$ (25)	35
		NBA, HF, ether (79)/ pyridine (21)	$-80°$, 1.5 hr; 8°, overnight; 1:13.4:1	" (68)	35
		Br_2, AgF, C_6H_6	[j]	" (71)	31
		I_2, AgF, C_6H_6	[j]	$C_6H_5CHFCH_2I$ (78)	31

Note: References 301–386 are on pp. 404–406.

[a] The abbreviations used for positive halogen reagents are: NBA, N-bromoacetamide; NIS, N-iodosuccinimide; NBS, N-bromosuccinimide; DBH, 1,3-dibromo-5,5-dimethylhydantoin; NCS, N-chlorosuccinimide; HCM, hexachloromelamine.

[b] The ratios represent the molar ratios for the reacting species; positive halogen source : fluoride source (HF or MF) : alkene.

[e] Conditions were not stated, but were probably room t, 30 min, 1:6:1.

[j] No temperature was stated. Finely divided silver(I) fluoride, vigorous stirring, and slow initial addition were critical. Addition of "ClF" using Cl_2/AgF/C_6H_6 was reported "less successful than either 'BrF' or 'IF'."

TABLE X. ADDITION OF HALOGEN FLUORIDE (*Continued*)

	Reactant	Reagent(s)[a]	Condition(s)[b]	Product(s) and Yield(s) (%)	Refs.
C_8 (*contd.*)	Diethyl maleate	NBA, HF, CH_2Cl_2/THF	−80°, 1 hr; 0°, 18 hr; 1.3:50:1	*threo*-$C_2H_5O_2CCHFCHBrCO_2C_2H_5$ (51)	36b, 303a
	Diethyl fumarate	NBA, HF, CH_2Cl_2/THF	−80°, 1 hr; 0°, 18 hr; 1.3:50:1	*erythro*-$C_2H_5O_2CCHFCHBrCO_2C_2H_5$ (55)	36b, 303a
	Phenylacetylene	NBA, HF, ether	−78°, 3 hr; 0°, 3 hr; 1.3:10:1	*trans*-$C_6H_5CF{=}CHBr$ (—)	46
C_9	Indene	Br_2, AgF, C_6H_6	[j]	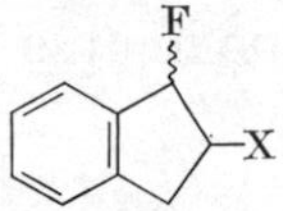 **D**, X = Br (—)	31
		I_2, AgF, C_6H_6	[j]	**D**, X = I (—)	31
	3,4-O-Acetyl-D-arabinal	NBS, HF, ether	−70°, 2 hr; 0°, 2 hr; 1.2:26:1	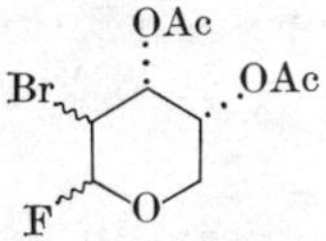 (90)[m]	303b
		Br_2, AgF, CH_3CN		Syrup (93) consisting of 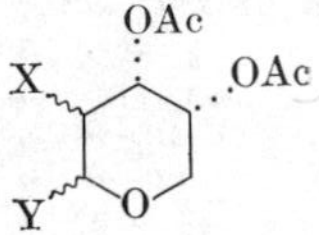 **A**, X = Br—, Y = F··· (62); X = Br···, Y = F— (20); X = Br···, Y = F··· (18)	40
		I_2, AgF, CH_3CN	[e]	Syrup (82) consisting of **A** X = I—, Y = F··· (49); X = I···, Y = F— (38); X = I···, Y = F··· (13)	40

Substrate	Reagents	Conditions	Product(s) and Yield(s) (%)	Refs.
3,4-Di-O-acetyl-L-arabinal	NBS, HF, ether	−70°, 2 hr; 0°, 2 hr; 1.2:26:1	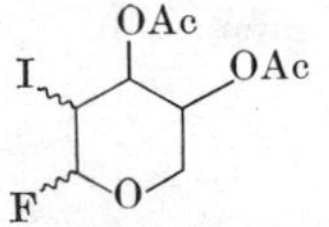 2-iodo-3,4-di-O-acetyl pyranosyl fluoride (I, OAc, OAc, F) (90)[n]	303b
D-Xylal diacetate	Br_2, AgF, CH_3CN	[e]	Syrup (91) consisting of 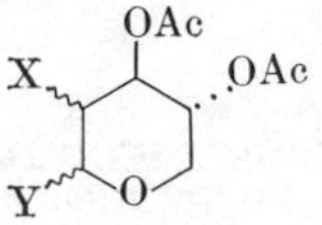 X, Y-substituted 3,4-di-O-acetyl pyranose (**B**): **B**, X = Br—, Y = F··· (55); X = Br···, Y = F— (32); X = Br···, Y = F··· (5)[o]; X = Br—, Y = F— (8)	40
	I_2, AgF, CH_3CN	[e]	Syrup (—) consisting of **B**: X = I—, Y = F··· (46); X = I···, Y = F— (43); X = I···, Y = F··· (8)[o]; X = I—, Y = F— (3)[o]	40

Note: References 301–386 are on pp. 404–406.

[a] The abbreviations used for positive halogen reagents are: NBA, N-bromoacetamide; NIS, N-iodosuccinimide; NBS, N-bromosuccinimide; DBH, 1,3-dibromo-5,5-dimethylhydantoin; NCS, N-chlorosuccinimide; HCM, hexachloromelamine.

[b] The ratios represent the molar ratios for the reacting species; positive halogen source : fluoride source (HF or MF) : alkene.

[e] Conditions were not stated but were probably room t, 30 min, 1:6:1.

[j] No temperature was stated. Finely divided silver(I) fluoride, vigorous stirring, and slow initial addition were critical. Addition of "ClF" using $Cl_2/AgF/C_6H_6$ was reported "less successful than either 'BrF' or 'IF'."

[m] This is the crude yield of a mixture of isomers from which 2-bromo-2-deoxy-β-D-arabinopyranosyl fluoride was isolated in 24% yield.

[n] This is the crude yield of a mixture of isomers from which 2-bromo-2-deoxy-β-L-arabinopyranosyl fluoride was isolated in 19% yield.

[o] This assignment of structure is tentative.

TABLE X. ADDITION OF HALOGEN FLUORIDE (*Continued*)

	Reactant	Reagent(s)[a]	Condition(s)[b]	Product(s) and Yield(s) (%)	Refs.
C_{10}	4-Phenyl-1-butene	I_2, AgF	Room t; 1:2.5:2.6	$CH_2{=}CHCHICH_2C_6H_5$ (—) (isolated as 3-benzylamino-4-phenyl-1-butene hydriodide)	28
	α-Pinene	NBS, HF, CCl_4/THF	−65 to −70°, 1 hr; 1.1:36:1	F, CH_3, Br, CH_3, CH_3 (—)	68
	1-Decene	NBA, HF, ether	−80°, 2 hr; room t, overnight; 1:10.1:1	n-$C_8H_{17}CHFCH_2Br$ (78)	35
	2,6-Dimethyl-7-octen-4-one	NBS, HF, CCl_4/THF	−65 to −70°, 1 hr; 1.1:36:1	i-$C_4H_9COCH_2CH(CH_3)CHFCH_2Br$ (2)	68
	Citronellol	NBS, HF, CCl_4/THF	−65 to −70°, 1 hr; 1.1:36:1	$(CH_3)_2CFCHBrCH_2CH_2CH(CH_3)CH_2CH_2OH$ (—)	68
C_{11}	2-Methoxynaphthalene	I_2, AgF	80°, 30 min; 1:2:20	1-$IC_{10}H_6OCH_3$-2 (47), 2,2′-dimethoxy-1,1′-binaphthyl (30)	28
	Ethyl cinnamate	NBA, HF, CH_2Cl_2/THF	−80°, 1 hr; 0°, 18 hr; 1.3:50:1	$C_6H_5CHFCHBrCO_2C_2H_5$ (60)	36b
C_{12}	Acenaphthylene	I_2, AgF	[j]	Y, H, X, Z; X = H, Y = F, Z = F (—) (*cis*- and *trans*-); X = H, Y = I, Z = F (—); X = F, Y = F, Z = I (—)	31

Methyl 10-undecenoate	NBA, HF, $CHCl_3$	Added at −55°, then −40°, 3 hr; 1:7:1	$BrCH_2CHF(CH_2)_8CO_2CH_3$ (72)	35
	DBH, HF, $CHCl_3$	−80°, 1.5 hr; 8°, overnight; 1:7.2:1	,, (72)	35
3,4,6-Tri-O-acetyl-D-glucal	Cl_2, AgF, CH_3CN/C_6H_6	Room t, 30 min; 1:6:1	Syrup (91) consisting of 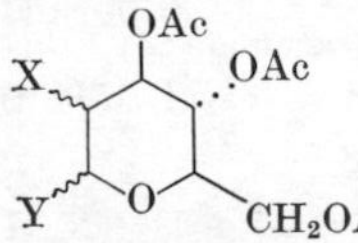 X = Cl—, Y = F··· (16); X = Cl···, Y = F— (62); X = Cl···, Y = F··· (6); X = Cl—, Y = F— (16)	44a
	NCS, HF, ether	[e]	No fluorine-containing product	44a
	NBS, HF, ether	−78°, 0.5–2 hr; 0°, 0.5–2 hr; 1.2:29:1	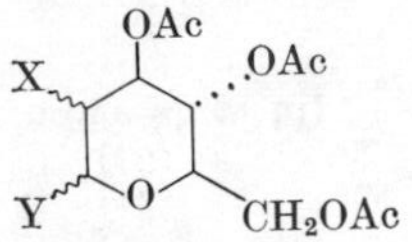 **A,** X = Br—, Y = F··· (55); **B,** X = Br···, Y = F— (30); **C,** X = Br···, Y = F··· (9); **D,** X = H, Y = F··· (8)	40, 47

Note: References 301–386 are on pp. 404–406.

[a] The abbreviations used for positive halogen reagents are: NBA, N-bromoacetamide; NIS, N-iodosuccinimide; NBS, N-bromosuccinimide; DBH, 1,3-dibromo-5,5-dimethylhydantoin; NCS, N-chlorosuccinimide; HCM, hexachloromelamine.

[b] The ratios represent the molar ratios for the reacting species; positive halogen source : fluoride source (HF or MF) : alkene.

[e] Conditions were not stated but were probably room t, 30 min, 1:6:1.

[j] No temperature was stated. Finely divided silver(I) fluoride, vigorous stirring, and slow initial addition were critical. Addition of "ClF" using $Cl_2/AgF/C_6H_6$ was reported "less successful than either 'BrF' or 'IF'."

TABLE X. ADDITION OF HALOGEN FLUORIDE (*Continued*)

	Reactant	Reagent(s)[a]	Condition(s)[b]	Product(s) and Yield(s) (%)	Refs.
C_{12} (*contd.*)	3,4,6,-Tri-O-acetyl-D-glucal (*contd.*)	NBS, HF, ether	−60°, 2 hr; 0°, 2 hr; 1.2:28:1	**A** (26),[p] **C** (13)	39, 41
		Br_2, AgF, CH_3CN	Room t, 30 min; 1:6:1	Crude syrup (97) consisting of **A** (70) **B** (21) **C** (9) Recrystallized **A** (56) **B** (6)	40, 47
		Br_2, AgF, C_6H_6	Room t, 30 min; 1:6:1	Overall yield (83) consisting of **A** (42) **B** (42) **C** (16)	40
		NIS, HF, ether	−70°, 2 hr; 0°, 2 hr; 1.2:30:1	(No overall yield) 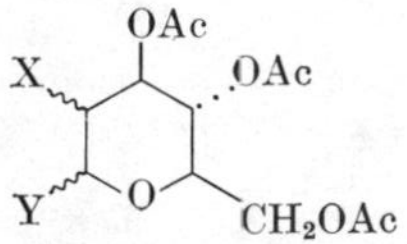 [structure: pyranose ring with X, OAc, OAc, Y, CH_2OAc] **E**, X = I—, Y = F··· (71) **F**, X = I···, Y = F— (3) **G**, X = I···, Y = F··· (23)	40
		NIS, HF, ether	−70°, 2 hr; 0°, 2 hr; 1.2:30:1	**E** (22) **G** (2)	304
		I_2, AgF, CH_3CN or C_6H_6	Room t, 30 min; 1:3:1	Syrup (98) consisting of **E** (60) **F** (34) **G** (6)	40
	Tri-O-acetyl-D-galactal	NBS, HF, ether	−70°, 2 hr; 0°, 2 hr; 1.2:29:1	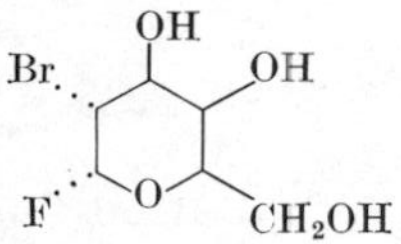 [structure: pyranose ring with Br, OH, OH, F, CH_2OH] (51)[q]	305

		Br_2, AgF, CH_3CN	[e]	X, Y-substituted 3,4-di-OAc-6-CH_2OAc tetrahydropyran; **H**, X = Br—, Y = F··· (34); X = Br···, Y = F— (43); X = Br···, Y = F··· (24)	40
		I_2, AgF, CH_3CN or C_6H_6	[e]	Syrup (96) consisting of **H**, X = I—, Y = F··· (25); X = I···, Y = F— (41); X = I···, Y = F··· (34)	40
C_{13}	3,4-Dihydro-2H-pyran-2-methyl *p*-toluenesulfonate	Br_2, AgF, CH_3CN or C_6H_6		X, Y-substituted 6-CH_2OTs tetrahydropyran; **A**, X = Br—, Y = F··· (66); X = Br···, Y = F— (25); X = Br···, Y = F··· (6)	40
		NBS, HF, ether	[f]	Mixture (20) of **A**, X = Br—, Y = F··· (55); X = Br···, Y = F··· (45)	40
		I_2, AgF, CH_3CN or C_6H_6	[e]	**A**, X = I—, Y = F··· (76); X = I···, Y = F— (18); X = I···, Y = F··· (5)	40

Note: References 301–386 are on pp. 404–406.

[a] The abbreviations used for positive halogen reagents are: NBA, N-bromoacetamide; NIS, N-iodosuccinimide; NBS, N-bromosuccinimide; DBH, 1,3-dibromo-5,5-dimethylhydantoin; NCS, N-chlorosuccinimide; HCM, hexachloromelamine.

[b] The ratios represent the molar ratios for the reacting species; positive halogen source : fluoride source (HF or MF) : alkene.

[e] Conditions were not stated but were probably room t, 30 min, 1 : 6 : 1.

[f] Conditions were not stated but were probably −78°, 2 hr; 0°, 2 hr.

[p] The structure was incorrectly reported initially as the β-D-manno derivative. The correct α-D-manno structure was established by X-ray analysis[41] and agrees with reported results.[40]

[q] The product was isolated only after treatment with sodium methoxide.

TABLE X. Addition of Halogen Fluoride (*Continued*)

	Reactant	Reagent(s)[a]	Condition(s)[b]	Product(s) and Yield(s) (%)	Refs.
C_{14}	1,2-Diphenylethylene	NBA, HF, THF	−80°, 1.5 hr; 8°, overnight; 1:45:1	$C_6H_5CHFCHBrC_6H_5$ (25)	35
	Diphenylacetylene	NBA, HF, ether	−78°, 3 hr; −20°, overnight; 1:40:1	Intractable product	46
C_{18}	1-Octadecene	NBA, HF, ether/dioxane	−80°, 2 hr; 0°, 7 hr; 1:33.6:1	$CH_3(CH_2)_{15}CHFCH_2Br$ (98 crude, not separated from allene)	35
C_{19}	1,4,9(11)-Androsta-triene-3,17-dione	NBA, HF, $(C_2H_5)_2CHCO_2H$/$CHCl_3$/THF	Room t, 17 hr; 1.5:27:1	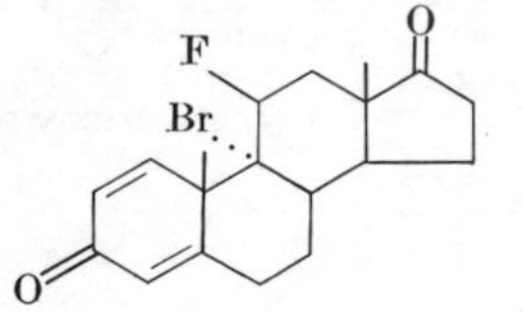 F, Br, O (41)	30
	Δ^5-Androstene-3β,17β-diol	NBS, HF, CH_2Cl_2/THF	−70°, 1.5 hr; 1.1:—:1	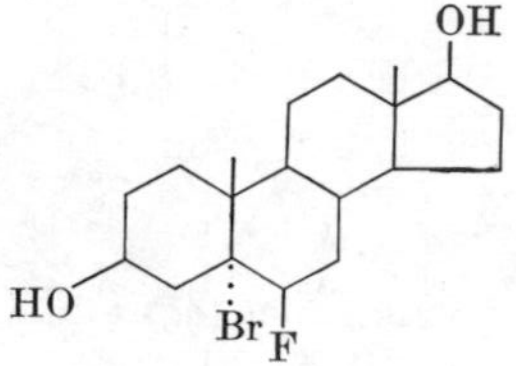 OH, HO, Br, F (80)	68[r]
		NBS, HF, CH_2Cl_2	−70°, 1.5 hr	" (35)	
		NBS, HF, toluene/THF	−65°, 1.5 hr	" (78)	
		NBS, HF, toluene/THF	0°, 1 hr	" (48)	
		NBS, HF, toluene	0°, 1 hr	" (25)	
		NBS, HF, toluene	+15°, 15 min	" (15)	

C_{20}	17α-Ethynyl-$\Delta^{5(10)}$-estrene-3β,17β-diol	NBA, HF, CH_2Cl_2/THF	−80°, 1 hr; 0°, 1 hr; 1.1:75:1	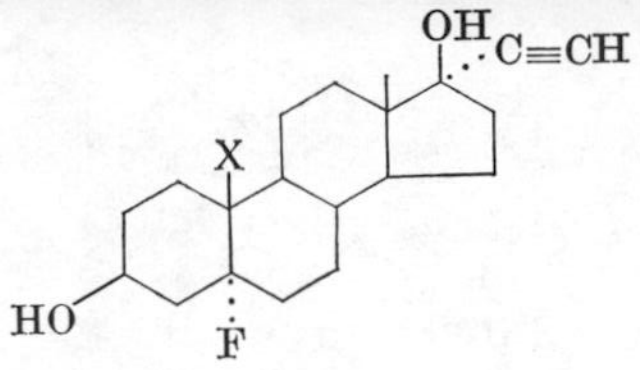 **B**, X = Br (—)	306
		NIS, HF, CH_2Cl_2/THF	−80°, 2 hr; 0°, 12 hr; 1.1:75:1	**B**, X = I (—)	306
	17β-Acetoxy-$\Delta^{5(10)}$-estren-3β-ol	NBA, HF, CH_2Cl_2/THF	−80°, 1 hr; 0°, 1 hr; 1.1:75:1	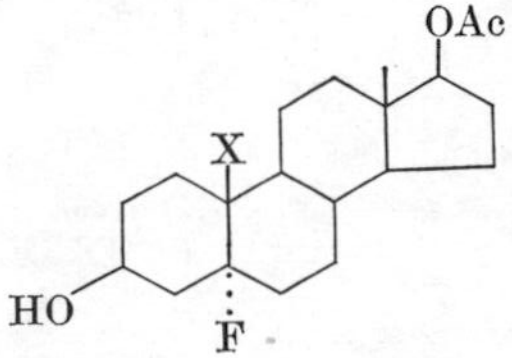 **C**, X = Br (—)	306
		NIS, HF, CH_2Cl_2/THF	−80°, 2 hr; 0°, 12 hr; 1.1:75:1	**C**, X = I (—)	306
	17α-Ethyl-$\Delta^{5(10)}$-estrene-3β,17β-diol	NBA, HF, CH_2Cl_2/THF	−80°, 1 hr; 0°, 1 hr	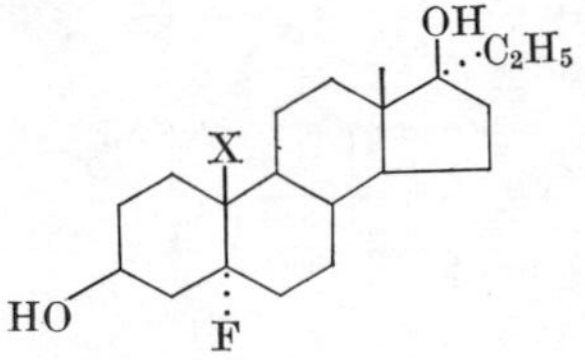 **D**, X = Br (—)	306
		NIS, HF, CH_2Cl_2/THF	−80°, 2 hr; 0°, 12 hr; 1.1:75:1	**D**, X = I (—)	306

Note: References 301–386 are on pp. 404–406.

[a] The abbreviations used for positive halogen reagents are: NBA, N-bromoacetamide; NIS, N-iodosuccinimide; NBS, N-bromosuccinimide; DBH, 1,3-dibromo-5,5-dimethylhydantoin; NCS, N-chlorosuccinimide; HCM, hexachloromelamine.

[b] The ratios represent the molar ratios for the reacting species; positive halogen source : fluoride source (HF or MF) : alkene.

[r] This patent tabulates a series of steroids (pregnenes, pregnadienes, androstenes, androstadienes, etc.) that are claimed to add BrF to unsaturation in the positions 1–2, 5–6, 7–8, 9–11, 14–15, and 16–17, but since no experimental details of physical properties of the products are reported they are not included in this table.

TABLE X. ADDITION OF HALOGEN FLUORIDE (*Continued*)

	Reactant	Reagent(s)[a]	Condition(s)[b]	Product(s) and Yield(s) (%)	Refs.
C_{21}	$\Delta^{4,6}$-Androstadien-17β-ol-3-one acetate	NBA, HF, CH_2Cl_2/THF	−80°, 1 hr	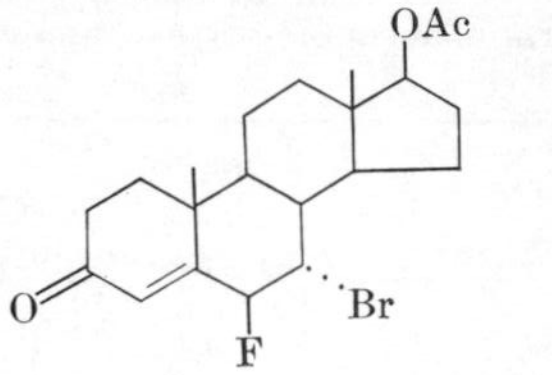 OAc, Br, F, O (—)	68
	21-Fluoro-1,4,9(11)-pregnatriene-3,20-dione	NCS, HF, $(C_2H_5)_2CHCO_2H$/$CHCl_3$/THF	Room t, 48 hr; 1.3:24:1	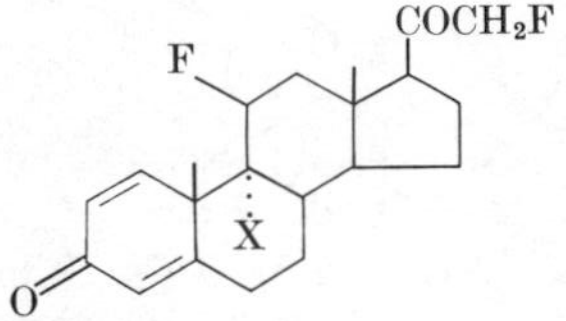 $COCH_2F$, F, X, O; **A**, X = Cl (—)	307
		NBA, HF, $(C_2H_5)_2CHCO_2H$/$CHCl_3$/THF	Room t, 1.5 hr; 1.2:24:1	**A**, X = Br (—)	307
		NIS, KF, DMSO	Room t, 16 hr; 1.2:40:1	**A**, X = I (—)	307
	$\Delta^{4,9(11)}$-Pregnadiene-3,20-dione	NBS, HF, CCl_4/THF	−65 to −70°, 1 hr; 1.1:36:1	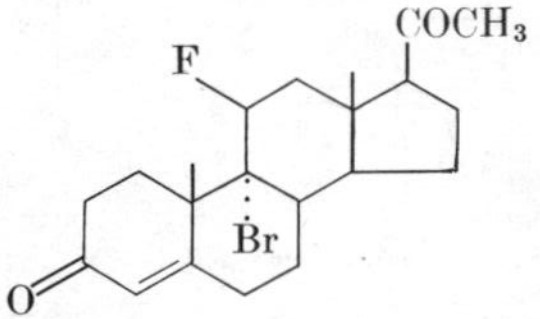 $COCH_3$, F, Br, O (—)	68
	17α-Bromo-21-fluoro-4,9(11)-pregnadiene-3,20-dione	NBA, HF, $(C_2H_5)_2CHCO_2H$/$CHCl_3$/THF	Room t, 1.5 hr; 1:24:1	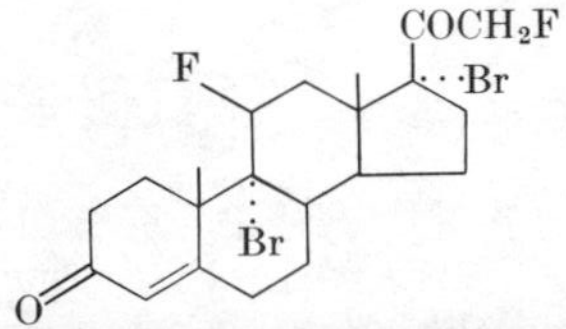 $COCH_2F$, Br, F, Br, O (—)	308

6α,21-Difluoro-17α-hydroxy-4,9(11)-pregnadiene-3,20-dione	NBA, HF, $(C_2H_5)_2CHCO_2H$/ $CHCl_3$/THF	Room t, 1.5 hr; 1:24:1	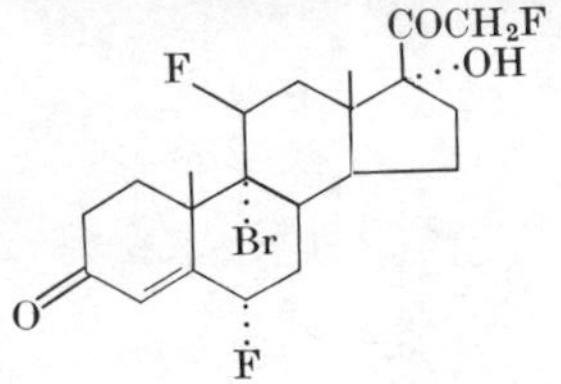 $COCH_2F$, ···OH, F, Br, F, O (—)	309
21-Fluoro-4,9(11)-pregnadiene-3,20-dione	NCS, HF, $(C_2H_5)_2CHCO_2H$/ $CHCl_3$/THF	Room t, 48 hr; 1.3:24:1	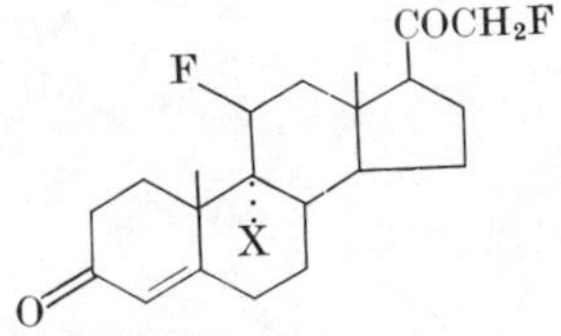 $COCH_2F$, F, X, O **B**, X = Cl (—)	307
	NBA, HF, $(C_2H_5)_2CHCO_2H$/ $CHCl_3$/THF	Room t, 1.5 hr; 1.2:24:1	**B**, X = Br (—)	307
	NIS, KF, DMSO	Room t, 16 hr; 1.2:40:1	**B**, X = I (—)	307
17α-Hydroxy-21-fluoro-4,9(11)-pregnadiene-3,20-dione	NCS, HF, $(C_2H_5)_2CHCO_2H$/ $CHCl_3$/THF	Room t, 40 hr; 1.1:30:1	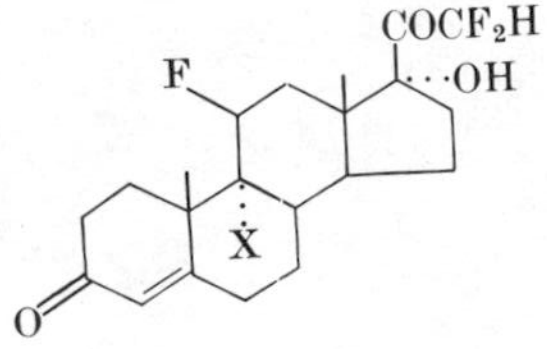 $COCF_2H$, ···OH, F, X, O **C**, X = Cl (—)	308
	NBA, HF, $(C_2H_5)_2CHCO_2H$/ $CHCl_3$/THF	Room t, 1.5 hr; 1:24:1	**C**, X = Br (—)	308

Note: References 301–386 are on pp. 404–406.

[a] The abbreviations used for positive halogen reagents are: NBA, N-bromoacetamide; NIS, N-iodosuccinimide; NBS, N-bromosuccinimide; DBH, 1,3-dibromo-5,5-dimethylhydantoin; NCS, N-chlorosuccinimide; HCM, hexachloromelamine.

[b] The ratios represent the molar ratios for the reacting species; positive halogen source : fluoride source (HF or MF) : alkene.

TABLE X. ADDITION OF HALOGEN FLUORIDE (*Continued*)

	Reactant	Reagent(s)[a]	Condition(s)[b]	Product(s) and Yield(s) (%)		Refs.
C_{21} (*contd.*)	16α,17α-Oxido-$\Delta^{5,9(11)}$-pregnadien-3β-ol-20-one	NBA, HF, CH_2Cl_2/THF	−80°, 1 hr; 1:25–100:1.1	COCH$_3$, O, HO, Br, F	(—)	29
		NBS, HF, CCl_4/THF	−65 to −70°, 1 hr; 1.1:36:1	"	(—)	68
	Δ^2-Pregnen-20-one	NBS, HF, CCl_4/THF	−65 to −70°; 1.1:36:1	COCH$_3$, F, Br	(—)	68
	Δ^4-Pregnen-3β-ol-20-one	NBA, HF, CH_2Cl_2/THF	−80°, 1 hr; 0°, 1 hr; 1.05:20:1	COCH$_3$, HO, F, Br	(—)	68
	Δ^5-Pregnen-3β-ol-20-one	NBA, HF, CH_2Cl_2/THF	−80°, 1 hr; 0°, 1 hr; 1.05:20:1	COCH$_3$, HO, Br, F	(75)	68

	Reactant	Reagents	Conditions	Product(s) (% Yield)		Refs.
		NIS, HF, CH_2Cl_2/THF	−80°, 2 hr; 0°, 12 hr; 1.2:68:1	$COCH_3$; HO; F; I	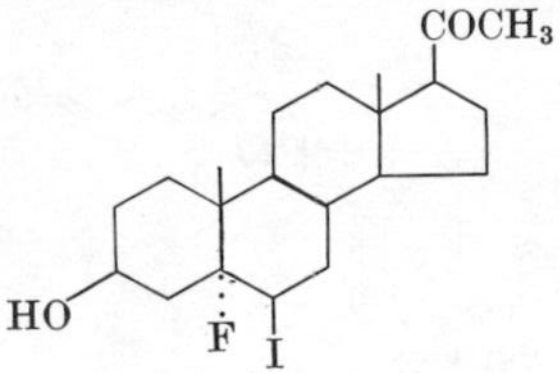 (55)	37
	Δ^5-Androstene-3β,17β-diol-17-acetate	NIS, HF, CH_2Cl_2/THF	−80°, 3 hr; 0°, 16 hr; 1.7:220:1	OAc; HO; F; I	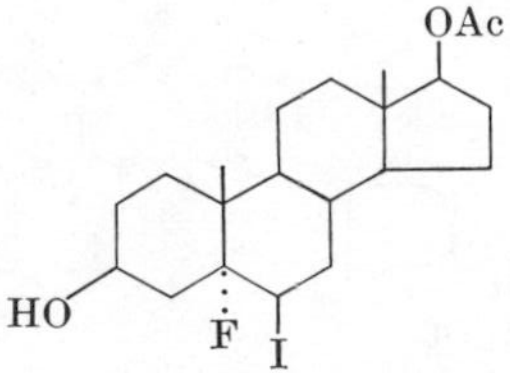 (49)	37
C_{22}	6α-Methyl-21-fluoro-4,9(11)-pregnadiene-3,20-dione	NCS, HF, $(C_2H_5)_2CHCO_2H$/$CHCl_3$/THF	Room t, 48 hr; 1.3:24:1	$COCH_2F$; F; Cl; O; CH_3	(−)	307
	6α-Methyl-17α-hydroxy-21-fluoro-4,9(11)-pregnadiene	NBA, HF, $(C_2H_5)_2CHCO_2H$/$CHCl_3$/THF	Room t, 1–5 hr; 1:24:1	$COCH_2F$; OH; F; Br; O; CH_3	(−)	309

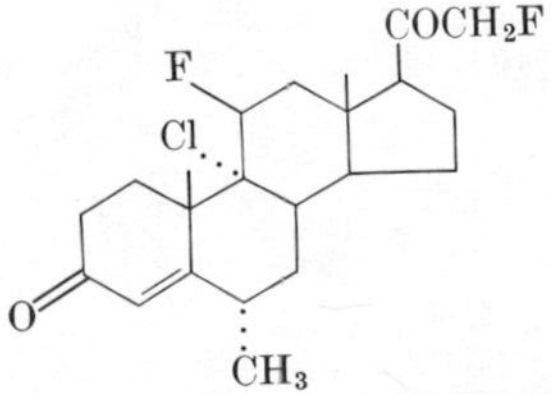

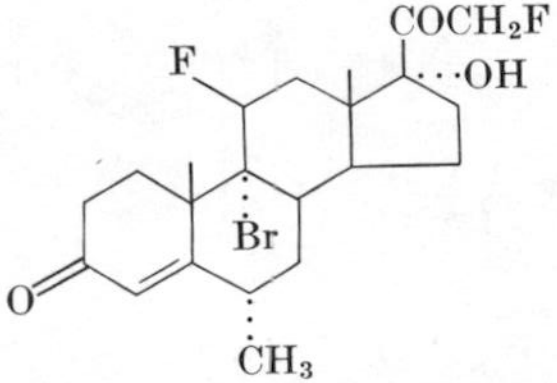

Note: References 301–386 are on pp. 404–406.

[a] The abbreviations used for positive halogen reagents are: NBA, N-bromoacetamide; NIS, N-iodosuccinimide; NBS, N-bromosuccinimide; DBH, 1,3-dibromo-5,5-dimethylhydantoin; NCS, N-chlorosuccinimide; HCM, hexachloromelamine.

[b] The ratios represent the molar ratios for the reacting species; positive halogen source : fluoride source (HF or MF) : alkene.

TABLE X. ADDITION OF HALOGEN FLUORIDE (*Continued*)

	Reactant	Reagent(s)[a]	Condition(s)[b]	Product(s) and Yield(s) (%)	Refs.
C_{22} (*contd.*)	21-Methyl-Δ^5-pregnene-3β,17α-diol-20-one	NBA, HF, CH_2Cl_2/THF	−80°, 16 hr; 1.05:25:1	HO, Br, F, COC_2H_5, OH (−)	69
C_{23}	6α,21-Difluoro-17α-hydroxy-4,9(11)-pregnadiene-3,20-dione 17-acetate	NBA, HF, $(C_2H_5)_2CHCO_2H$/$CHCl_3$/THF	Room t, 1.5 hr; 1:24:1	O, F, Br, F, $COCH_2F$, OAc (−)	309
	1,4,9(11)-Pregnatriene-17α,21-diol-3,20-dione 21-acetate	NBA, HF, $(C_2H_5)_2CHCO_2H$/$CHCl_3$/THF	Room t, 17 hr	O, F, Br, $COCH_2OAc$, OH (51)	30
	17α-Hydroxy-21-fluoro-4,9(11)-pregnadiene-3,20-dione 17-acetate	NCS, HF, $(C_2H_5)_2CHCO_2H$/$CHCl_3$/THF	Room t, 48 hr; 1.1:30:1	O, F, Cl, $COCH_2F$, OAc (−)	308

	NBA, HF, $(C_2H_5)_2CHCO_2H$/ $CHCl_3$/THF	Room t, 1.5 hr; 1:24:1	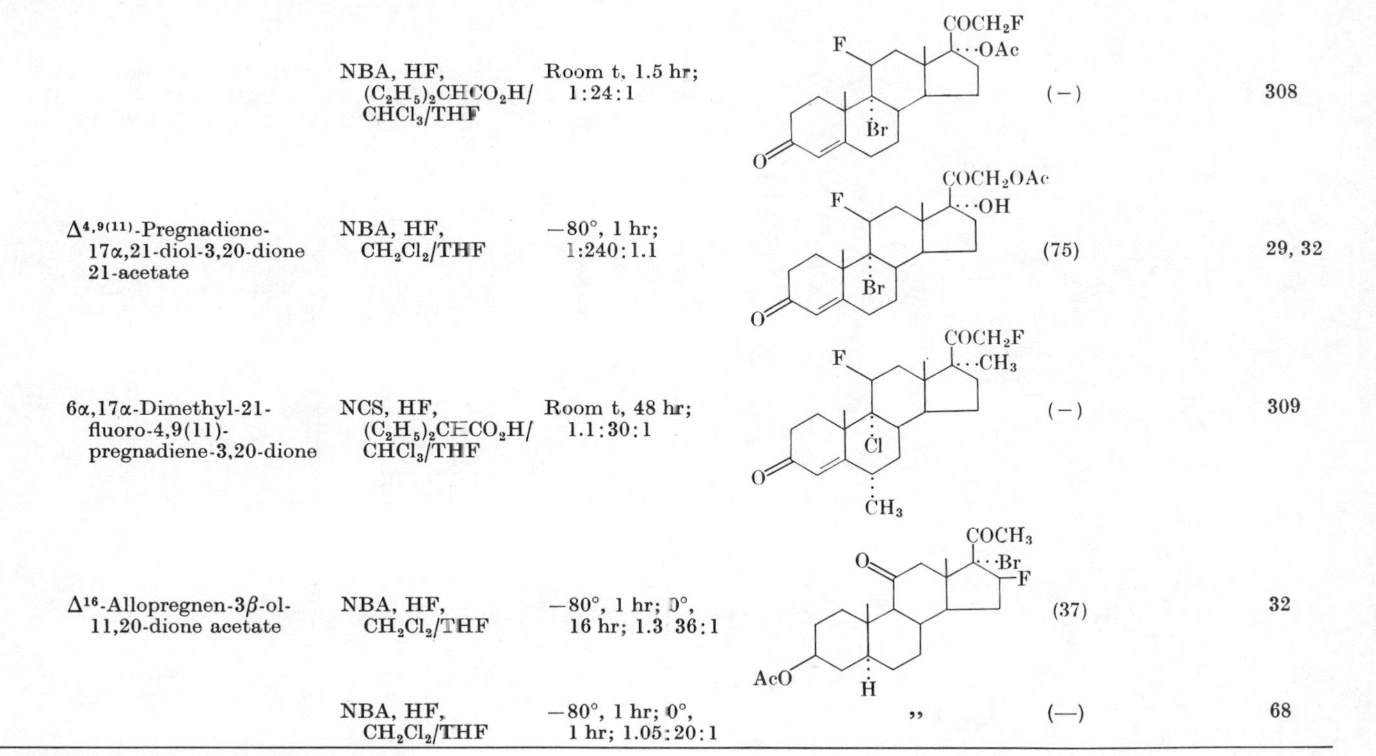	(–)	308
$\Delta^{4,9(11)}$-Pregnadiene-17α,21-diol-3,20-dione 21-acetate	NBA, HF, CH_2Cl_2/THF	–80°, 1 hr; 1:240:1.1		(75)	29, 32
6α,17α-Dimethyl-21-fluoro-4,9(11)-pregnadiene-3,20-dione	NCS, HF, $(C_2H_5)_2CHCO_2H$/ $CHCl_3$/THF	Room t, 48 hr; 1.1:30:1		(–)	309
Δ^{16}-Allopregnen-3β-ol-11,20-dione acetate	NBA, HF, CH_2Cl_2/THF	–80°, 1 hr; 0°, 16 hr; 1.3 36:1		(37)	32
	NBA, HF, CH_2Cl_2/THF	–80°, 1 hr; 0°, 1 hr; 1.05:20:1	,,	(—)	68

Note: References 301–386 are on pp. 404–406.

[a] The abbreviations used for positive halogen reagents are: NBA, N-bromoacetamide; NIS, N-iodosuccinimide; NBS, N-bromosuccinimide; DBH, 1,3-dibromo-dimethylhydantoin; NCS, N-chlorosuccinimide; HCM, hexachloromelamine.

[b] The ratios represent the molar ratios for the reacting species; positive halogen source : fluoride source (HF or MF) : alkene.

TABLE X. ADDITION OF HALOGEN FLUORIDE (*Continued*)

	Reactant	Reagent(s)[a]	Condition(s)[b]	Product(s) and Yield(s) (%)	Refs.
C_{23} (*contd.*)	Δ^{2}-Androstene-3,17β-diol diacetate	NBA, HF, CH_2Cl_2/THF	−80°, 1 hr; 0°, 1 hr	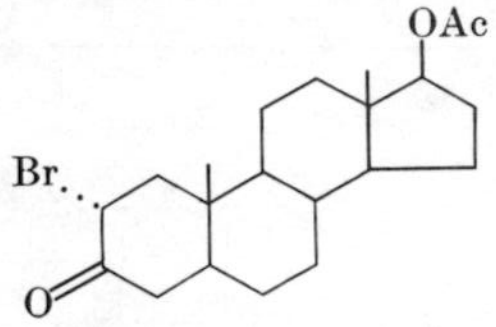 (72)[g]	32
	Δ^{5}-Pregnen-3β-ol-20-one acetate	NIS, HF, CH_2Cl_2/THF	−80°, 2 hr; 0°, 5 hr; 1.1:73:1	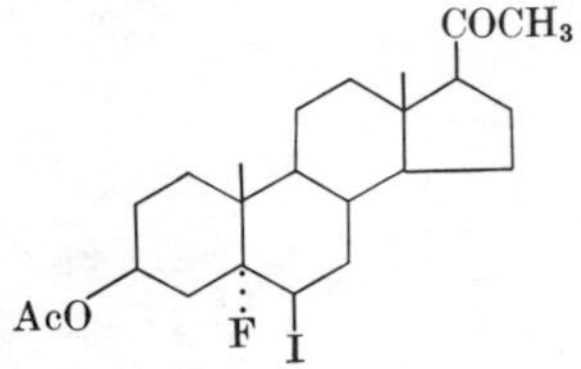 (45)	37
	Δ^{5}-Pregnene-3β,17α-diol-20-one 17-acetate	NBA, HF, CH_2Cl_2/THF	−80°, 1 hr; 0°, 2–16 hr; 1:25–100:1.1	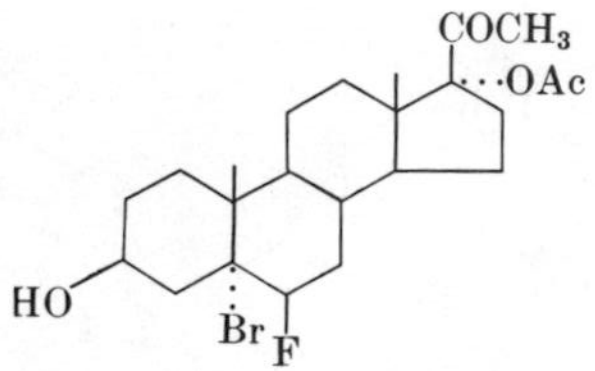 (−)	29
	6-Fluoro-$\Delta^{4,17(20)}$-pregnadiene-11β,21-diol-3-one 21-acetate	NBS, HF, CCl_4/THF	−65 to −70°, 1 hr; 1.1:36:1	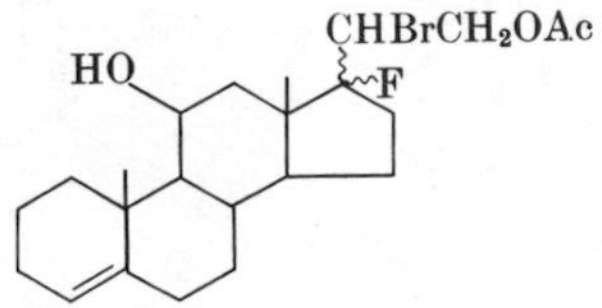 (−)	68

C_{24}	6α-Methyl-17α-hydroxy-21-fluoro-4,9(11)-pregnadiene-3,20-dione 17-acetate	NBA, HF, $(C_2H_5)_2CHCO_2H$/$CHCl_3$/THF	Room t, 1.5 hr; 1:24:1	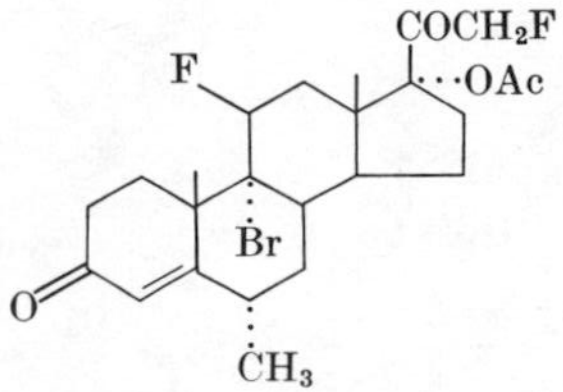 (structure: F; Br; CH_3; $COCH_2F$; ···OAc) (−)	309
	6α-Fluoro-17-hydroxy-21-acetoxy-16α-methyl-1,4,9(11)-pregnatriene-3,20-dione	NCS, HF, N-methyl-pyrrolidone	−30°, 24 hr; 6:60:1	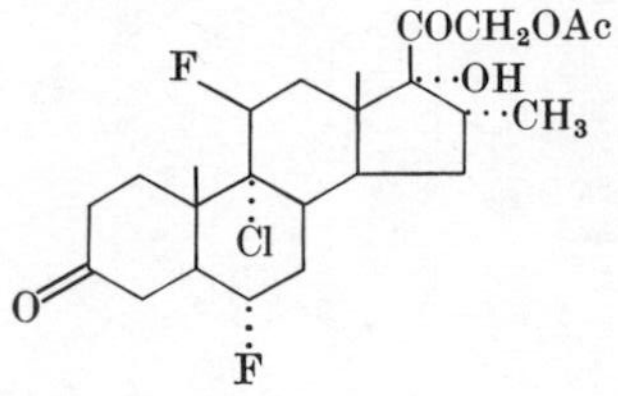 (structure: F; Cl; F; $COCH_2OAc$; ···OH; ···CH_3) (52)	310
C_{25}	$\Delta^{4,6}$-Pregnadiene-17α-21-diol-3,20-dione diacetate	NBA, HF, CH_2Cl_2/THF	−80°, 1 hr; 1:85:1.2	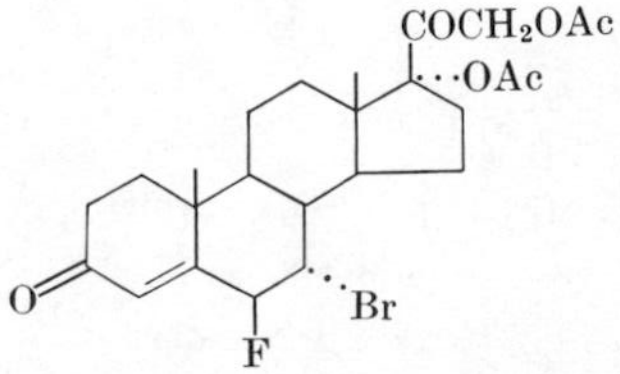 (structure: F; ···Br; $COCH_2OAc$; ···OAc) (35)	29, 32

Note: References 301–386 are on pp. 404–406.

[a] The abbreviations used for positive halogen reagents are: NBA, N-bromoacetamide; NIS, N-iodosuccinimide; NBS, N-bromosuccinimide; DBH, 1,3-dibromo-5,5-dimethylhydantoin; NCS, N-chlorosuccinimide; HCM, hexachloromelamine.

[b] The ratios represent the molar ratios for the reacting species; positive halogen source: fluoride source (HF or MF): alkene.

[s] Additional examples are also reported of the conversion of enol acetates to α-bromoketones instead of adding bromine fluoride.

TABLE X. ADDITION OF HALOGEN FLUORIDE (*Continued*)

	Reactant	Reagent(s)[a]	Condition(s)[b]	Product(s) and Yield(s) (%)	Refs.
C_{25} (*contd.*)	3β-Hydroxy-21-acetoxy-16α,18-dimethyl-5-pregnen-20-one	NBS, HF, CH_2Cl_2/THF	−70°, 1 hr; 1:7:1	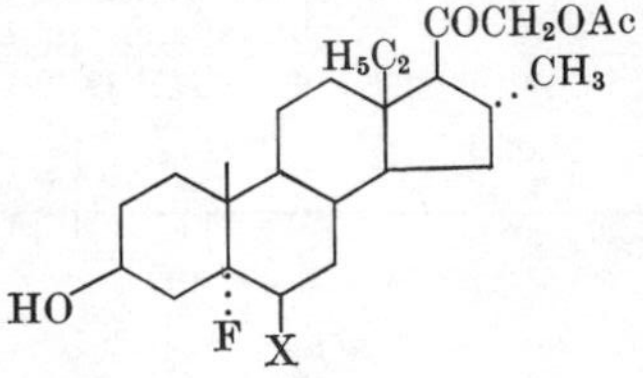 HO, F, X, H_5C_2, $COCH_2OAc$, CH_3; X = Br (55), H (13)	38
		NBS, HF, DMF/CH_2Cl_2	Inverse addition at −40°; −12°, 5 min; 1.1:10:1	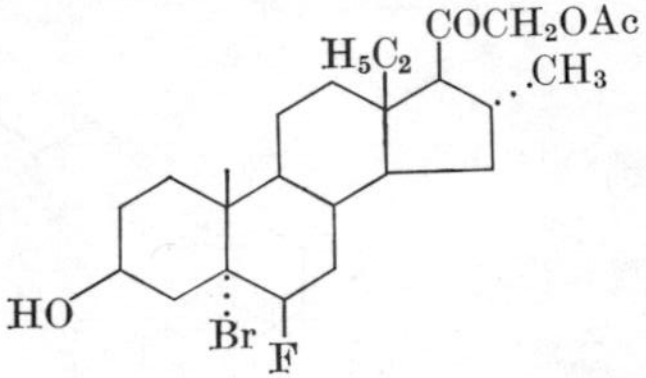 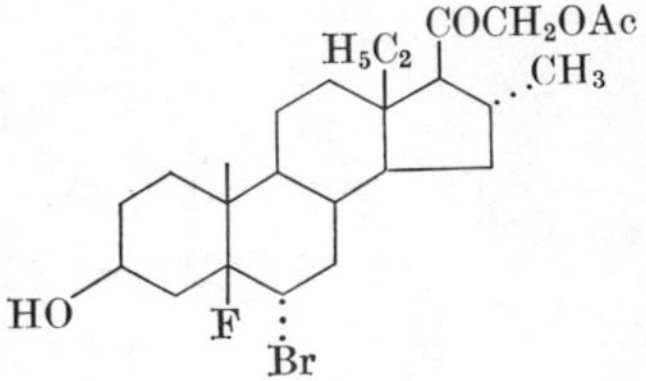 HO, Br, F, H_5C_2, $COCH_2OAc$, CH_3 (70),	38
				HO, F, Br, H_5C_2, $COCH_2OAc$, CH_3 (10)	
C_{27}	17α-Hydroxy-21-fluoro-4,9(11)-pregnadiene-3,20-dione 17-caproate	NBA, HF, $(C_2H_5)_2CHCO_2H$/$CHCl_3$/THF	Room t, 1.5 hr; 1:22:1	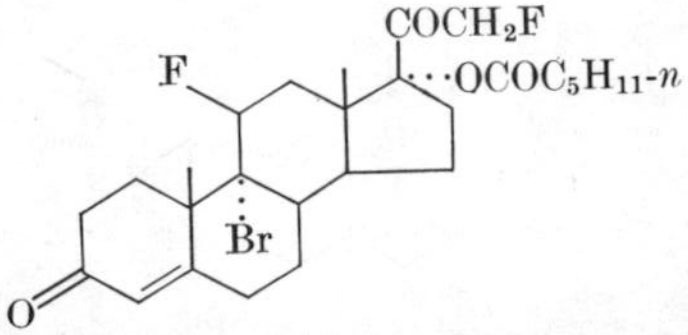 O, F, Br, $COCH_2F$, $OCOC_5H_{11}$-*n* (—)	308

	$\Delta^{8(14)}$-Cholestenol	NBS, HF, CCl_4/THF	−65 to −70°, 1 hr; 1.1:36:1	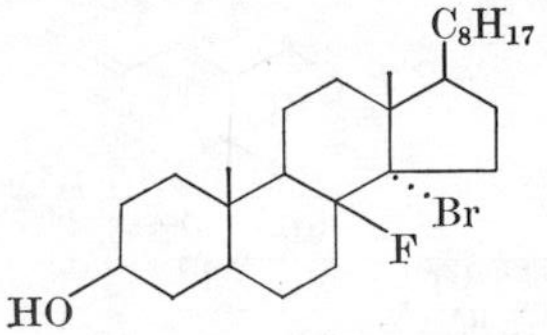 (−)	68
	$\Delta^{8(9)}$-Cholestenol	NBS, HF, CCl_4/THF	−65 to −70°, 1 hr; 1.1:36:1	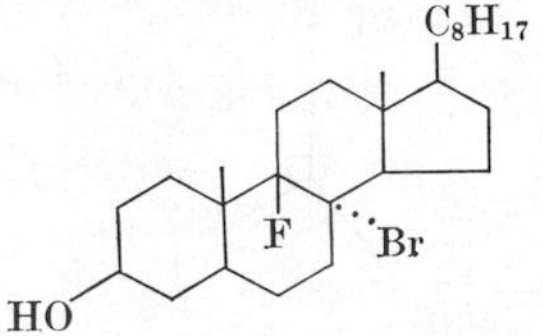 (−)	68
C_{28}	6α-Methyl-17α-hydroxy-21-fluoro-4,9(11)-pregnadiene-3,20-dione 17-caproate	NBA, HF, $(C_2H_5)_2CHCO_2H$/$CHCl_3$/THF	Room t, 1.5 hr; 1:24:1	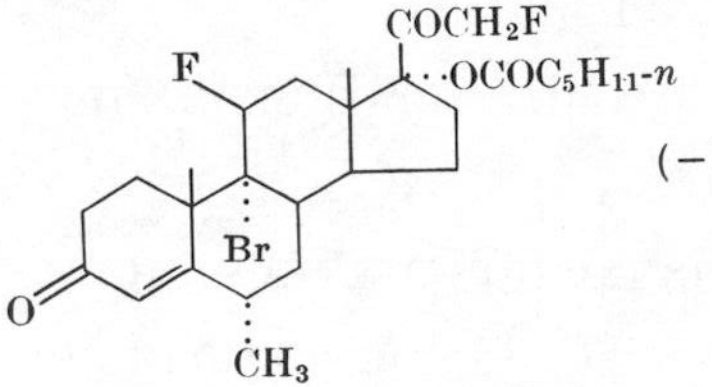 (−)	309
C_{30}	Lupeol	NBS, HF, CCl_4/THF	−65 to −70°, 1 hr; 1.1:36:1	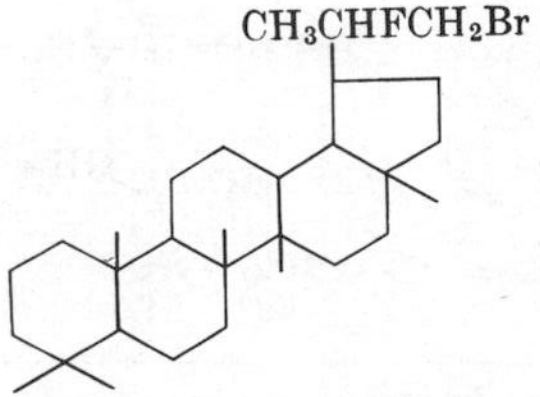 (−)	68

Note: References 301–386 are on pp. 404–406.

[a] The abbreviations used for positive halogen reagents are: NBA, N-bromoacetamide; NIS, N-iodosuccinimide; NBS, N-bromosuccinimide; DBH, 1,3-dibromo-5,5-dimethylhydantoin; NCS, N-chlorosuccinimide; HCM, hexachloromelamine.

[b] The ratios represent the molar ratios for the reacting species; positive halogen source: fluoride source (HF or MF): alkene.

TABLE XI. REPLACEMENT OF THE HYDROXYL GROUP BY FLUORINE USING THE α-FLUOROALKYLAMINE (YAROVENKO) REAGENT[a] (**FAR**)

$$\geq C{-}OH + R_2NCF_2R' \longrightarrow \geq C{-}F + R_2NCOR' + HF$$

	Reactant	Condition(s)[b]	Product(s) and Yield(s) (%)	Refs.
C_2	CH_3CO_2H	[c]	CH_3COF (59)	76
	CH_3COSH	[c]	CH_3COF (50)	76
C_3	$C_2H_5CO_2H$	[c]	CH_3CH_2COF (44)	76
	$C_2H_5CO_2Na$	[c]	CH_3CH_2COF (87)	76
C_4	n-$C_3H_7CO_2H$	[c]	n-C_3H_7COF (46)	76
	n-$C_3H_7CH_2OH$	[c]	n-C_4H_9F (67)	76
	3-Bromo-1-butanol	1:1, ether, room t, 15 hr	$CH_3CHBrCH_2CH_2F$ (73)	85, 84
	4-Bromo-2-butanol	1:1, ether, room t, 15 hr	$CH_3CHFCH_2CH_2Br$ (68)	85
	3-Iodo-1-butanol and 4-iodo-2-butanol (35:65)	1:1, ether, room t, 15 hr	$CH_3CHICH_2CH_2F$, $CH_3CHFCH_2CH_2I$ (total 62)	85
C_5	i-$C_3H_7CH_2CH_2OH$	[c]	i-$C_3H_7CH_2CH_2F$ (35)	76
	$(CH_3)_2C(OH)CO_2CH_3$	1:1.5, CH_2Cl_2, 0°, 24 hr	$(CH_3)_2CFCO_2CH_3$ (13), $CH_2{=}C(CH_3)CO_2CH_3$ (10)	81
C_6	1,2-*O*-Isopropylidene-α-D-xylofuranose	—	ClCHF–C[$(C_2H_5)_2N$] (cyclic orthoamide, structure) (—)	103
	2-(N,N-Diethylamino)ethanol	1:1.5, CH_2Cl_2, 0°, 24 hr	$(C_2H_5)_2NCH_2CH_2F$ (45)	81
	N,N-Dimethyl-DL-serine methyl ester	1:2, CH_2Cl_2, 0°, 36 hr	$(CH_3)_2NCH(CH_2F)CO_2CH_3$ (25)	311
	$HOCH_2CH_2CH(CH_3)CH(NH_2)CO_2H$	1:2.5, ether, room t, 20 hr	Some reaction, but only starting material isolated (70)	84
C_7	Benzoic acid	$(CH_3)_2NCF_2H{\cdot}3\,HF$, 1:2, mixed at 0°, then room t, 1 hr	C_6H_5COF (90)	312

	Benzyl alcohol	1:1.5, CH_2Cl_2, 0°, 24 hr	$C_6H_5CH_2F$ (40)	81
	N-Methylhydroxyproline methyl ester	1:2, CH_2Cl_2, 0°, 36 hr	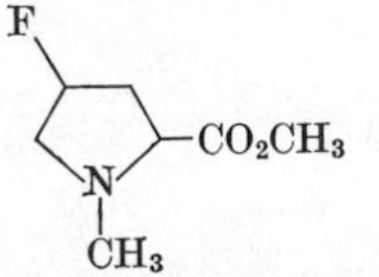 (33)	311
	N,N-Dimethyl-DL-threonine methyl ester	1:2, CH_2Cl_2, 0°, 36 hr	$CH_3CHFCH[N(CH_3)_2]CO_2CH_3$ (60)	311
C_8	1-Phenyl-2,2,2-trichloroethanol	1:1.5, CH_2Cl_2, 0°, 24 hr	$C_6H_5CH(CCl_3)OCOCHClF$ (90)	81
	1,2,3,5-di-*O*-Methylene-α-D-glucofuranose	CH_2Cl_2, room t, 3 hr	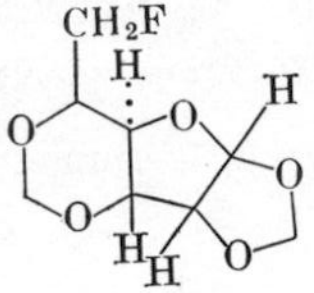 (58)	103
	1-Hydroxybicyclo[2.2.2]octane	1:2, 60–65°	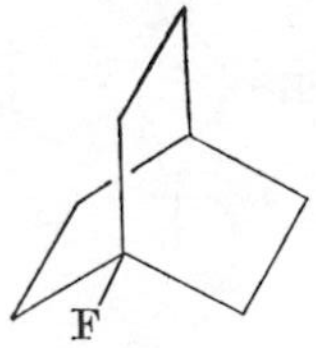 (64)	88
	1-Hydroxy-4-fluorobicyclo[2.2.2]octane	1:2.2, 100°	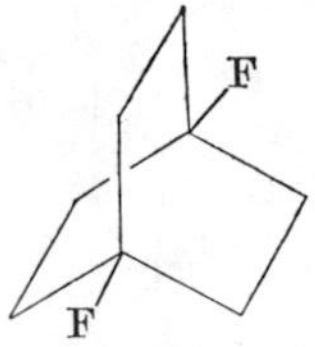 (49)	88

Note: References 301–386 are in pp. 404–406.

[a] The fluoroalkylamine reagent used is 2-chloro-1,1,2-trifluorotriethylamine, $(C_2H_5)_2NCF_2CHFCl$, unless stated otherwise under conditions.

[b] The ratio of numbers is the ratio of moles of reactant to moles of fluoroalkylamine reagent.

[c] Exact conditions not specified. From the discussion in ref. 76 we presume that reagents were added together dropwise at room temperature. In general, equimolar amounts of reactant and reagent were used.

TABLE XI. Replacement of the Hydroxyl Group by Fluorine Using the α-Fluoroalkylamine (Yarovenko) Reagent[a] (**FAR**) (*Continued*)

	Reactant	Condition(s)[b]	Product(s) and Yield(s) (%)	Refs.
C_9	1-Hydroxy-4-methylbicyclo[2.2.2]octane	1:1.5, 120°	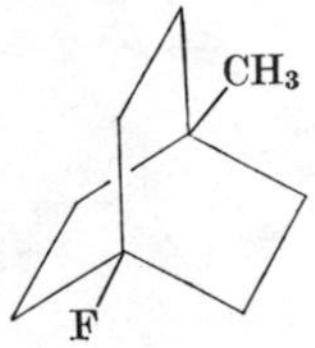 1-Fluoro-4-methylbicyclo[2.2.2]octane (50)	88
	Phthalimidomethanol	1:1.5, CH_2Cl_2, 0°, 24 hr	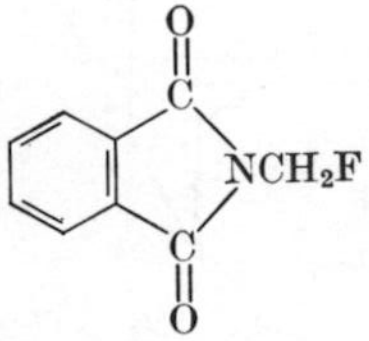 N-(fluoromethyl)phthalimide (NCH_2F) (78)	81
	Methyl mandelate	1:1.5, CH_2Cl_2, 0°, 24 hr	Product isolated as $C_6H_5CHFCONH_2$ (80)	81
	2′-Deoxyuridine	$(CH_3)_2CO$ or DMF, 70°, 5 min	Cyclization only; 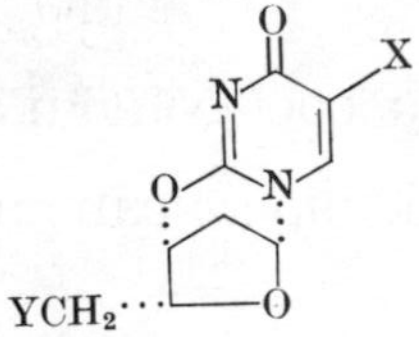 **A**, X = H, Y = OH (71)[d]	313
	2′-Deoxy-5-fluorouridine	$(CH_3)_2CO$ or DMF 70°, 30 min	**A**, X = F, Y = OH (48)[d]	313

C_{10}	5′-Deoxy-5′-chlorothymidine	$(CH_3)_2CO$ or DMF, 70°, 30 min	**A**, X = CH_3, Y = Cl (87)[a]	313
	Thymidine	$(CH_3)_2CO$ or DMF, 70°, 30 min	**A**, X = CH_3, Y = OH (75)[d]	313
	β-Phthalimidoethanol	1:1.5, CH_2Cl_2, 0°, 24 hr	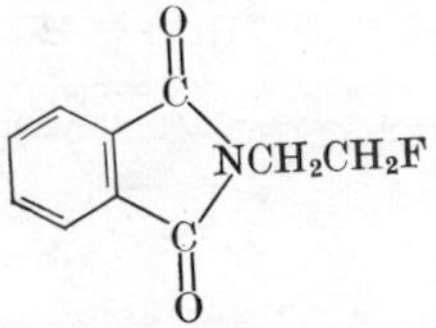 NCH_2CH_2F (−)	81
	1-Phenyl-2,2-dibromoethanol	1:1.5, CH_2Cl_2, 0°, 24 hr	$C_6H_5CHFCHBr_2$ (95)	81
	3,4-$Cl_2C_6H_3CHOHCO_2C_2H_5$	1:1.4, CH_2Cl_2, 0–5°, 15–18 hr	3,4-$Cl_2C_6H_3CHFCO_2C_2H_5$ (55)	314
	2-Hydroxyadamantane	1 hr, 98°	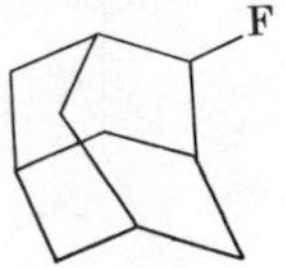 F (−)	315
	3-*t*-Butylcyclohexanol (70:30/*trans*:*cis*)	1:1.8, CH_2Cl_2, −10°, 49 hr	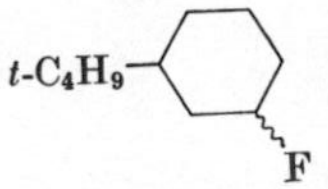 *t*-C_4H_9 F *cis* (10), *trans* (3), *t*-butylcyclohexenes (33)	86
	4-*t*-Butylcyclohexanol (70:30/*trans*:*cis*)	1:1, ether, 0°, 14 hr[e]	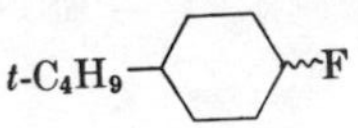 *t*-C_4H_9 F (9; 90:10/*cis*:*trans*), **B**; 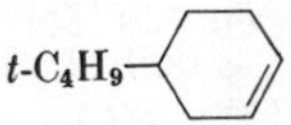 *t*-C_4H_9 (68) **C**	86

Note: References 301–386 are in pp. 404–406.

[a] The fluoroalkylamine reagent used is 2-chloro-1,1,2-trifluorotriethylamine, $(C_2H_5)_2NCF_2CHFCl$, unless stated otherwise under conditions.

[b] The ratio of numbers is the ratio of moles of reactant to moles of fluoroalkylamine reagent.

[d] The reagent was used as dehydrating agent only; no fluorination occurred.

[e] Minor changes in yield and isomer ratio resulted from variation in reaction time (and, in some cases, temperature).

TABLE XI. REPLACEMENT OF THE HYDROXYL GROUP BY FLUORINE USING THE α-FLUOROALKYLAMINE (YAROVENKO) REAGENT[a] (**FAR**) (*Continued*)

	Reactant	Condition(s)[b]	Product(s) and Yield(s) (%)	Refs.
C_{10} (*contd.*)	4-*t*-Butylcyclohexanol (70:30/*trans:cis*)	1:1, ether/$(C_2H_5)_3N$, 35°, 132 hr[e]	**B** (24; 82:18/*cis:trans*), **C** (39)	86
		1:1, CH_3CN/$(C_2H_5)_3N$, 82°, 20 hr[e]	**B** (17; 80:20/*cis:trans*), **C** (—)	
	4-*t*-Butylcyclohexanol-1-d	1:1.1, ether/$(C_2H_5)_3N$, reflux, 5 days	1-fluoro-1-deuterio-4-*t*-C_4H_9-cyclohexane: *cis* (12.5), *trans* (1.1), *t*-C_4H_9-cyclohexane-D (37)	86
	1-Hydroxy-4-acetoxybicyclo[2.2.2]octane	1:1.1, 120–130°	1-fluoro-4-$OCOCH_3$-bicyclo[2.2.2]octane (50)	88
	2-(N,N-Dibutylamino)ethanol	1:1.5, CH_2Cl_2, 0°, 24 hr	$(n\text{-}C_4H_9)_2NCH_2CH_2F$ (30)	81
C_{11}	3-R′-4-R-C_6H_3-$CHOHCO_2C_2H_5$ R = OCH_3, R′ = H R = R′ = OCH_2O	1:1.4, CH_2Cl_2, 0–5°, 15–18 hr	3-R′-4-R-C_6H_3-$CHFCO_2C_2H_5$ R = OCH_3, R′ = H (74) R + R′ = —OCH_2O— (64)	314

	threo-DL-Serine ethyl ester	1:1.5, CH_2Cl_2, 0°, several hr	C_6H_5—CH—CH—$CO_2C_2H_5$ (O—C(CHClF)=N ring) (81)	81
	$C_6H_5CHOHP(O)(OC_2H_5)_2$	1:1, CH_2Cl_2 room t, 3 days	$C_6H_5CHFP(O)(OC_2H_5)_2$ (25)	294
C_{12}	1,2:5,6-Di-*O*-isopropylidene-α-D-glucofuranose	1:1.3, CH_2Cl_2 25°, 22 hr	3-*O*-acyl-1,2:5,6-di-*O*-isopropylidene-α-D-glucofuranose (RO; R = —COCHClF) (74)	102
	1,2:3,4-Di-*O*-isopropylidene-α-D-galactopyranose	1:1.5, CH_2Cl_2, 25°, 22 hr	1,2:3,4-di-*O*-isopropylidene-α-D-galactopyranose with $CH_2OCOCHFCl$ (96)	102
	R,R′-C_6H_3-$CHOHCO_2C_2H_5$; R = OCH_3, R′ = OCH_3; R = OC_2H_5, R′ = H	1:1.4, CH_2Cl_2, 0–5°, 15–18 hr	R,R′-C_6H_3-$CHFCO_2C_2H_5$; R = OCH_3, R′ = OCH_3 (59); R = OC_2H_5, R′ = H (63)	314

Note: References 301–386 are in pp. 404–406.

[a] The ratio of numbers is the ratio of moles of reactant to moles of fluoroalkylamine reagent.
[b] The fluoroalkylamine reagent used is 2-chloro-1,1,2-trifluorotriethylamine, $(C_2H_5)_2NCF_2CHFCl$, unless stated otherwise under conditions.
[c] Minor changes in yield and isomer ratio resulted from variation in reaction time (and, in some cases, temperature).

TABLE XI. REPLACEMENT OF THE HYDROXYL GROUP BY FLUORINE USING THE α-FLUOROALKYLAMINE (YAROVENKO) REAGENT[a] (**FAR**) (*Continued*)

	Reactant	Condition(s)[b]	Product(s) and Yield(s) (%)	Refs.
C_{12} (*contd.*)	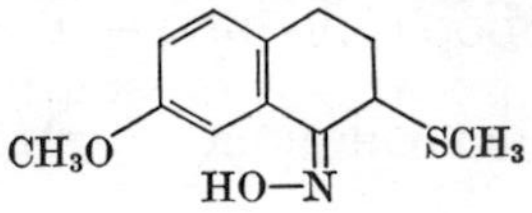 CH_3O, SCH_3, HO—N	1:2, dioxane, 0°, 0.5 hr	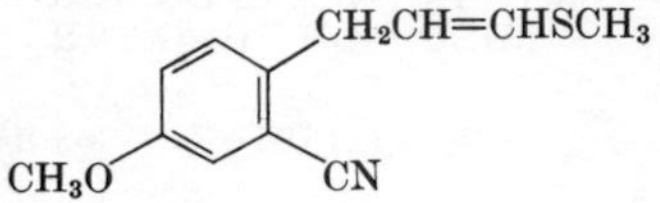 $CH_2CH{=}CHSCH_3$, CH_3O, CN (46, *cis* and *trans*)	316
C_{13}	N,N-Dimethyl-β-phenylserine ethyl ester	1:2, CH_2Cl_2, 0°, 36 hr	$C_6H_5CHFCH(N(CH_3)_2)CO_2C_2H_5$ (80)	311
	Ethyl 2-carbethoxy-2-acetamido-3-methyl-5-hydroxyvalerate	1:1, ether, room t, 20 hr	$FCH_2CH_2CH(CH_3)C(CO_2C_2H_5)_2NHCOCH_3$ (low) (impure)	84
C_{14}	Benzoin	1:1.5, CH_2Cl_2, 0°, 24 hr	$C_6H_5COCHFC_6H_5$ (80)	81
	1-Hydroxy-4-phenylbicyclo[2.2.2]octane	1:2.2, 130–140°	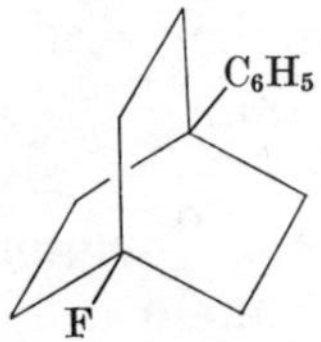 C_6H_5, F (56)	88
C_{15}	3-[2-(3,5-Dimethyl-2-oxocyclohexyl)-2-hydroxyethyl]-glutarimide	1:2, CH_2Cl_2, add at 5°, room t, overnight	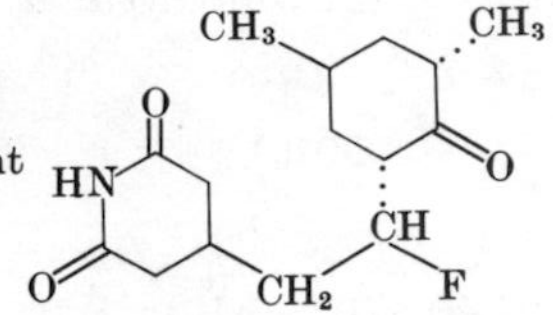 CH_3, CH_3, O, HN, O, O, CH, CH_2, F (13)	317a

C_{16}	5′-*O*-Tosylthymidine	$(CH_3)_2CO$ or DMF 70°, 30 min	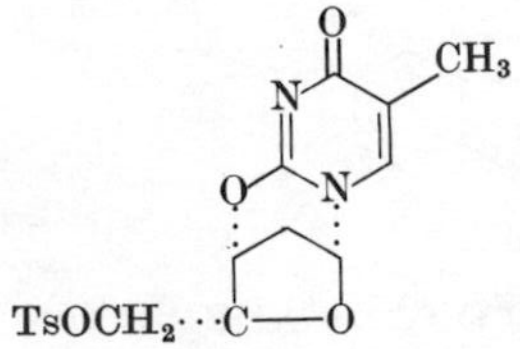 [structure: $TsOCH_2$ anhydro-thymidine] (100)[d]	313
	N,N-Dimethyl-*p*-(O-carbethoxy)-β-phenylserine ethyl ester	1:2, CH_2Cl_2, 0°, 36 hr	p-$C_2H_5OCOC_6H_4CHFCH(N(CH_3)_2)CO_2C_2H_5$ (90) (isolated as hydrochloride)	311
	n-Octyl (+)-mandelate	1:1.5, CH_2Cl_2, 0°, 24 hr	$(-)$-$C_6H_5CHFCO_2C_8H_{17}$-*n* (95)	81
	n-Hexadecyl alcohol	1:1.5, CH_2Cl_2, 0°, 24 hr	*n*-$C_{15}H_{31}CH_2F$ (70)	81
	Morphine	1:2, 25°, 30 min HF catalyst	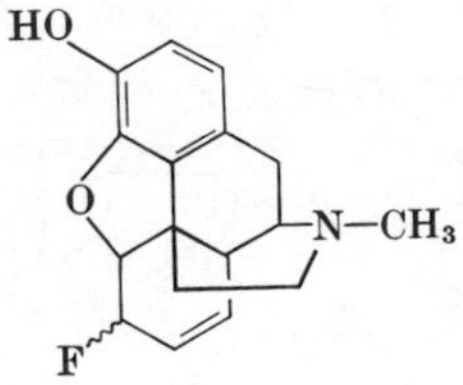 [structure: HO, O, F, N—CH_3 fluoro-deoxymorphine] (−)	100

Note: References 301–386 are in pp. 404–406.

[a] The ratio of numbers is the ratio of moles of reactant to moles of fluoroalkylamine reagent.
[b] The fluoroalkylamine reagent used is 2-chloro-1,1,2-trifluorotriethylamine, $(C_2H_5)_2NCF_2CHFCl$, unless stated otherwise under conditions.
[d] The reagent was used as dehydrating agent only; no fluorination occurred.

TABLE XI. Replacement of the Hydroxyl Group by Fluorine Using the α-Fluoroalkylamine (Yarovenko) Reagent[a] (**FAR**) (*Continued*)

	Reactant	Condition(s)[b]	Product(s) and Yield(s) (%)	Refs.
C_{16} (*contd.*)	R = H, alkyl, alkenyl, alkynyl	1:1–1.5, THF, room t, 30 min	(–), CH_3 (–) X = ···F, —OCOCHClF	318a
	R = H, alkyl, alkenyl, alkynyl	1:1–1.5, THF, room t, 30 min	(–), CH_3 (–) X = ···F, —OCOCHClF	318a
	F, OH	1:1–1.5, THF, room t, 30 min	(–), CH_3 (–) X = ···F, —OCOCHClF	318a
	Methyl 12-hydroxystearolate	1:2, CH_2Cl_2, 0°, 12 hr	$n\text{-}C_6H_{13}CHFCH_2C{\equiv}C(CH_2)_7CO_2CH_3$ (70)	87
	Methyl ricinoleate	1:2, CH_2Cl_2, 0°, 12 hr	$n\text{-}C_6H_{13}CHFCH_2CH{=}CH(CH_2)_7CO_2CH_3$ (90)	87
	Methyl 12-hydroxystearate	1:2, CH_2Cl_2, 0°, 12 hr	$n\text{-}C_6H_{13}CHF(CH_2)_{10}CO_2CH_3$ (75)	87
C_{17}	Estradiol-3-methyl ether	1:1.5, THF, room t, 30 min	CH_3O, R, CH_3 (47) R = ···F (26)	90b

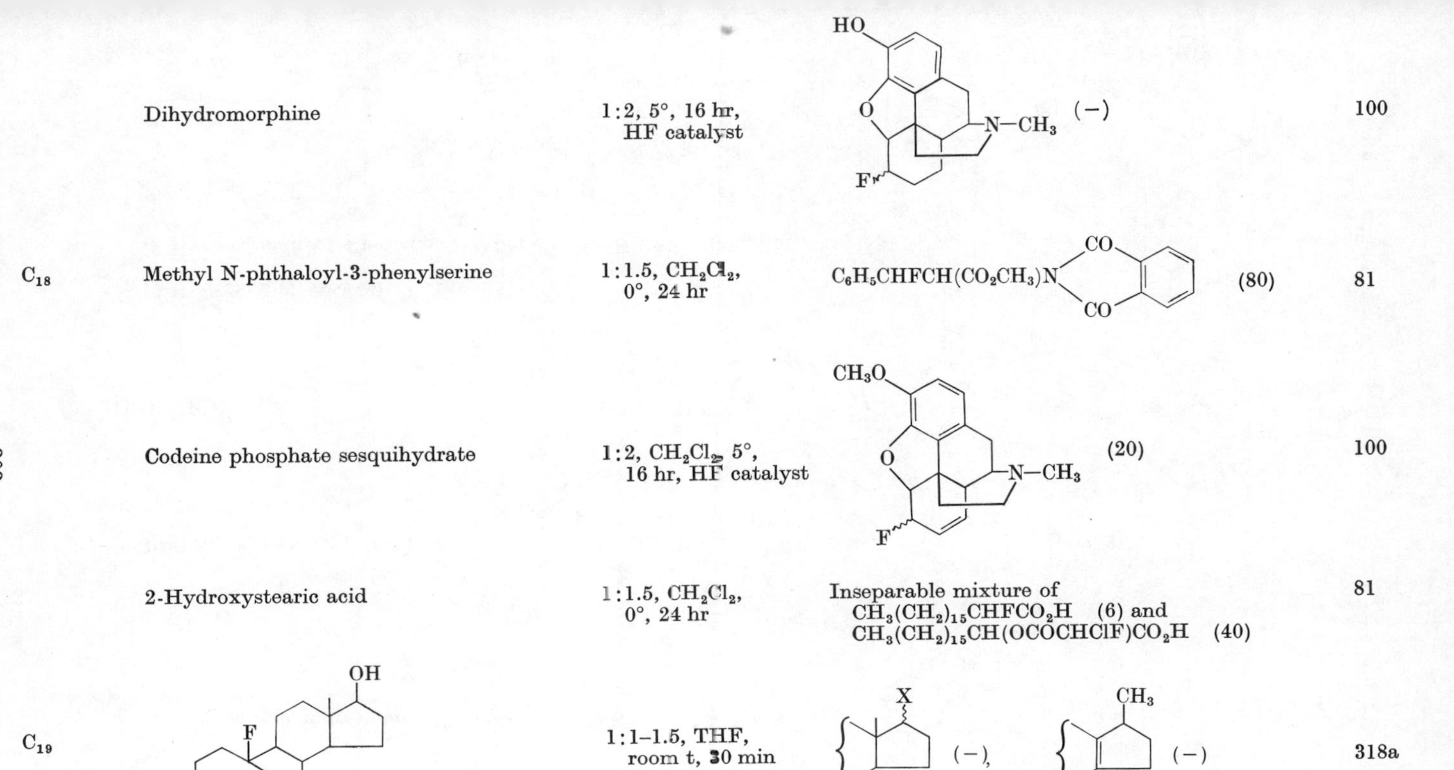

	Dihydromorphine	1:2, 5°, 16 hr, HF catalyst	(structure: HO, O, N—CH_3, F) (–)	100
C_{18}	Methyl N-phthaloyl-3-phenylserine	1:1.5, CH_2Cl_2, 0°, 24 hr	$C_6H_5CHFCH(CO_2CH_3)N$ (phthalimido, CO, CO) (80)	81
	Codeine phosphate sesquihydrate	1:2, CH_2Cl_2, 5°, 16 hr, HF catalyst	(structure: CH_3O, O, N—CH_3, F) (20)	100
	2-Hydroxystearic acid	1:1.5, CH_2Cl_2, 0°, 24 hr	Inseparable mixture of $CH_3(CH_2)_{15}CHFCO_2H$ (6) and $CH_3(CH_2)_{15}CH(OCOCHClF)CO_2H$ (40)	81
C_{19}	(structure: OH, F, O)	1:1–1.5, THF, room t, 30 min	(structure: X) (–), (structure: CH_3) (–); X = ··· F, —OCOCHClF	318a

Note: References 301–386 are in pp. 404–406.

[a] The fluoroalkylamine reagent used is 2-chloro-1,1,2-trifluorotriethylamine, $(C_2H_5)_2NCF_2CHFCl$, unless stated otherwise under conditions.

[b] The ratio of numbers is the ratio of moles of reactant to moles of fluoroalkylamine reagent.

TABLE XI. REPLACEMENT OF THE HYDROXYL GROUP BY FLUORINE USING THE α-FLUOROALKYLAMINE (YAROVENKO) REAGENT[a] (**FAR**) (*Continued*)

	Reactant	Condition(s)[b]	Product(s) and Yield(s) (%)	Refs.
C_{19} (*contd.*)	11α-Hydroxy-4-androstene-3,17-dione	THF	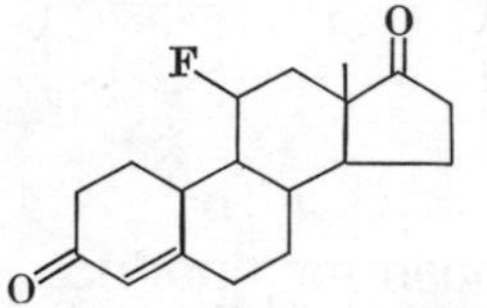 F (<45)	95
	3β-Hydroxyandrost-5-en-17-one	1.5:1, CH_2Cl_2, reflux, 15 min	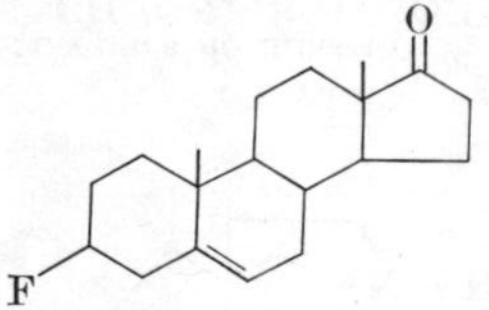 F (58)	90b
	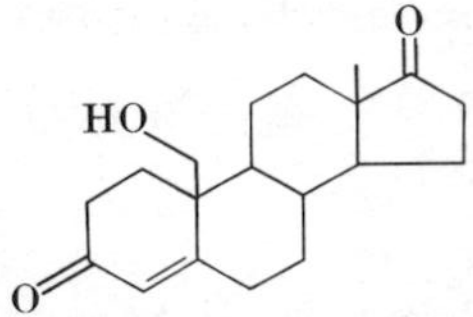 HO O O 19-Hydroxyandrost-4-ene-3,17-dione	1:1.4, CH_3CN or THF, reflux, 45 min	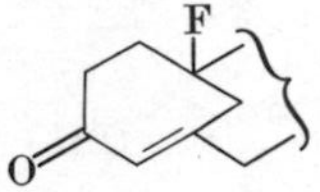 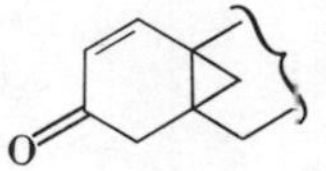 F O (38), O (47)	90a, 98, 318b
	19-Hydroxyandrost-4,6-diene-3,17-dione	CH_3CN, reflux, 1 hr	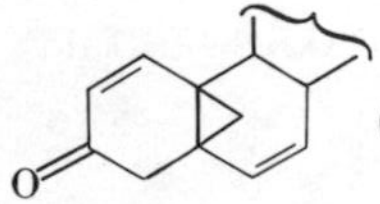 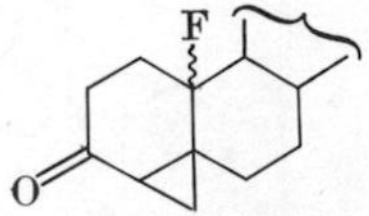 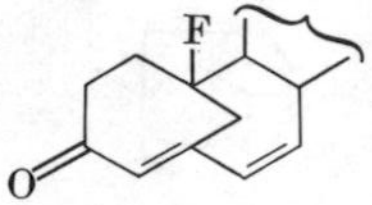 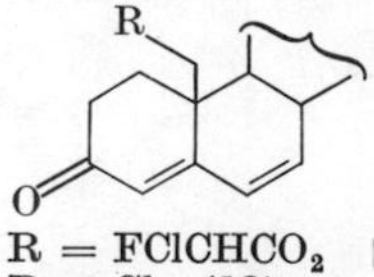 O (9), F O (23), F O (8), R O; R = $FClCHCO_2$ (4); R = Cl (13)	99

Reactant	Conditions	Product(s) and Yield(s) (%)	Refs.
19-Hydroxyandrost-4,7-diene-3,17-dione	CH_3CN, reflux, 1 hr	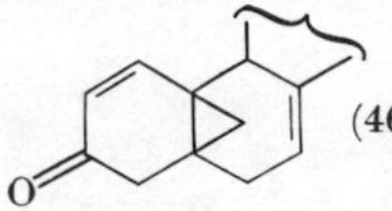 (46), 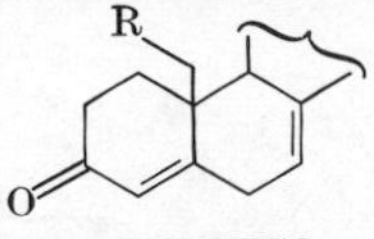 R = $FClCHCO_2$ (—)	99
Testosterone	Excess **FAR**, CH_2Cl_2, 25[c]	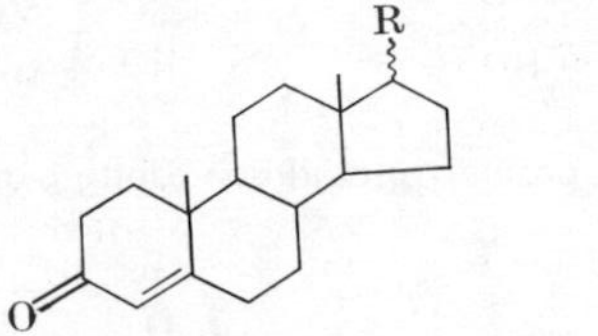 **A**, R = F · · · (—)	89
	1:1.5, CH_3CN, room t, 30 min	**A**, R = F · · · (18) R = $FClCHCO_2$— (27), 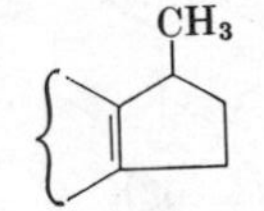 (25)	90
5β-Androstane-3,17-dione-19-carboxylic acid	~1:2, THF, 0–5°, 1 hr	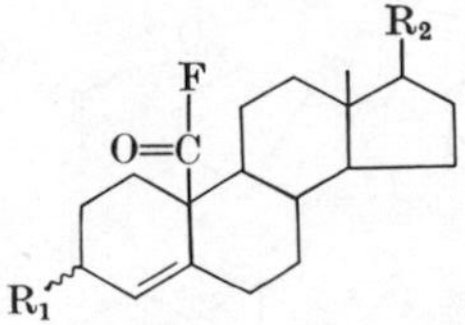 $R_1 = R_2$ = carbonyl (–)[f]	319

Note: References 301–386 are in pp. 404–406.

[a] The fluoroalkylamine reagent used is 2-chloro-1,1,2-trifluorotriethylamine, $(C_2H_5)_2NCF_2CHFCl$, unless stated otherwise under conditions.
[b] The ratio of numbers is the ratio of moles of reactant to moles of fluoroalkylamine reagent.
[f] Many examples, in which R_1 and R_2 are ester and carbonyl groups, were claimed in the patent without experimental details.

TABLE XI. Replacement of the Hydroxyl Group by Fluorine Using the α-Fluoroalkylamine (Yarovenko) Reagent[a] (**FAR**) (*Continued*)

	Reactant	Condition(s)[b]	Product(s) and Yield(s) (%)	Refs.
C_{19} (*contd.*)	3β-Hydroxyandrost-5-en-17-one	1:2,[g] CH_2Cl_2, 0–5°, 3–16 hr	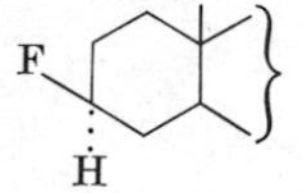 (90)	89, 96, 91, 320
	3β-Hydroxy-5α:6α-epoxy-17-ketoandrostane	~1:2, CH_2Cl_2, 0°, 20 hr	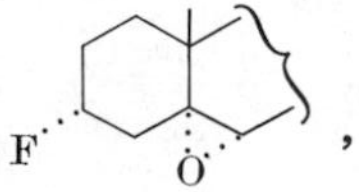 (20), 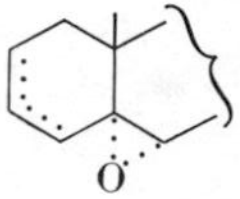 (80)[h]	321
	3β-Hydroxy-5β:6β-epoxy-17-ketoandrostane	~1:2, CH_2Cl_2, 0°, 1.5 hr	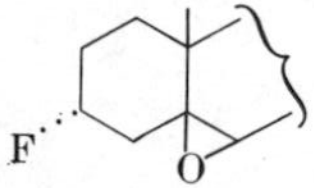 (40–50)	321
	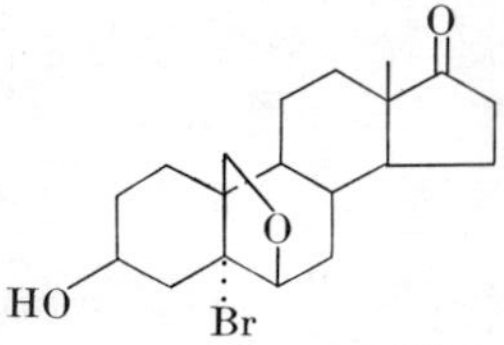	~1:2, CH_2Cl_2, 0°, 30 hr	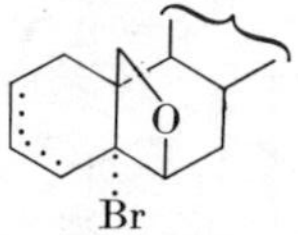 (−)[h] No fluoro compound isolated	321
	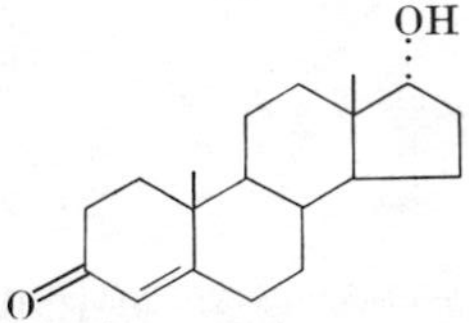 Epitestosterone	1:1.5,[i] THF, room t, 14 hr	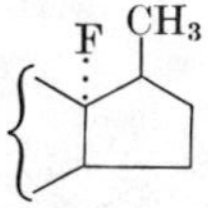 (7), 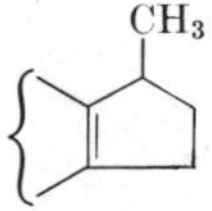 (80)	90, 322

3β-Fluoro-5-methyl-6β-hydroxy-19-nor-(5β)-androst-9-en-17-one	1:2, CH_2Cl_2, 0°, 5 min	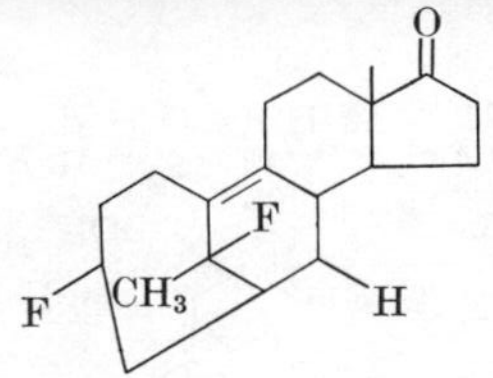 (93)	91
11α,17α-Dihydroxy-4-pregnene-3,20-dione	THF	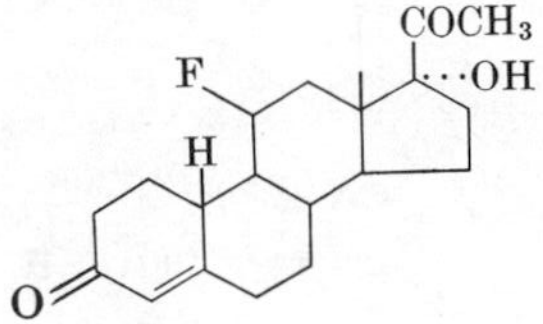 (45)	95
3β-Hydroxy-5α-androstan-17-one	1.5:1, CH_2Cl_2, 0°, 16 hr	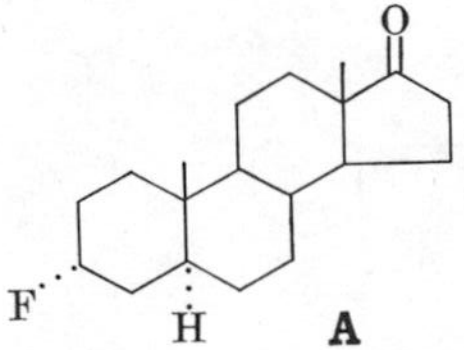 **A** (42), 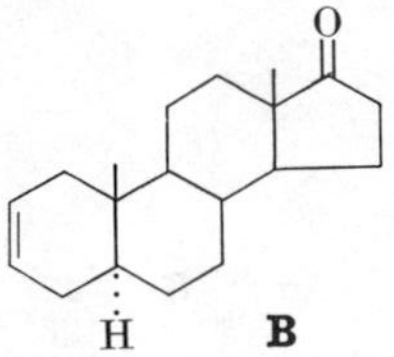 **B** (34)	90b

Note: References 301–386 are in pp. 404–406.

[a] The fluoroalkylamine reagent used is 2-chloro-1,1,2-trifluorotriethylamine, $(C_2H_5)_2NCF_2CHFCl$, unless stated otherwise under conditions.
[b] The ratio of numbers is the ratio of moles of reactant to moles of fluoroalkylamine reagent.
[g] $(i\text{-}C_3H_7)_2NCF_2CHClF$ and $(n\text{-}C_4H_9)_2CHClF$ have also been used. Yields were not reported.
[h] This product is a mixture of the Δ^2 and Δ^3 isomers.
[i] Reference 322 cites 17 patent examples in which 13-fluoro or 17-fluorosteroids were formed. No yields were given. The solvents CH_3CN and ether were also used.

TABLE XI. REPLACEMENT OF THE HYDROXYL GROUP BY FLUORINE USING THE α-FLUOROALKYLAMINE (YAROVENKO) REAGENTS[a] (**FAR**) (*Continued*)

	Reactant	Condition(s)[b]	Product(s) and Yield(s) (%)	Refs.
C_{19} (*contd.*)	3β-Hydroxy-5α-androstan-17-one	1.5:1, THF, 0°, 16 hr	**A** (18), **B** (19)	90b
		1.5:1, CH_3CN, 0°	**B** (23)	90b
		1:1.6, CH_2Cl_2, 5°, 18 hr	**A** (35), **B** (—)	89, 96, 320
		$FClCHCF_2NR_1R_2$ 1:1.6, CH_2Cl_2, 5°, 18 hr	**A** (—)	96, 320
	3α-Hydroxy-5α-androstan-17-one	1:1.6, CH_2Cl_2, 5°, 18 hr	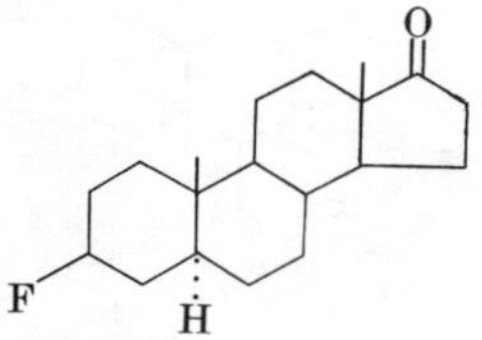 [structure: 3-fluoro-5α-androstan-17-one] (16) + **B** (–)	89, 96, 320
		1:1, THF, room t	**B** (100)	90b
C_{20}	3α-Hydroxy-17β-acetoxyestr-5(10)-ene	1:1.5, CH_2Cl_2, 0°, 20 min	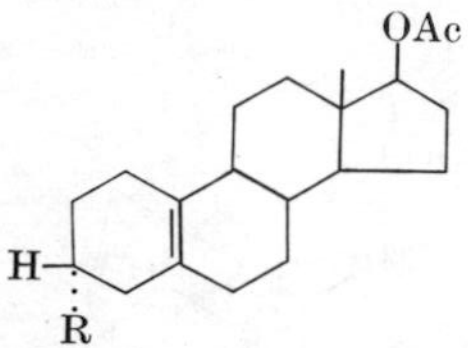 [structure: 3-R-17-OAc-estr-5(10)-ene], [structure: 3-F ring fragment] (17), R = F (4), R = OCOCHClF (15) 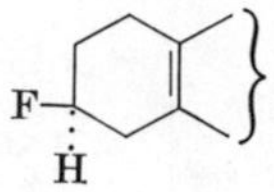	93, 323

			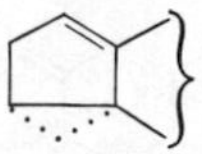 (7), 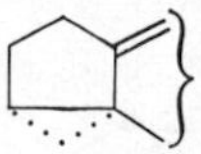 (24), 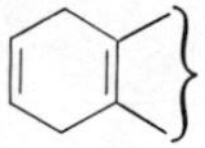 (19), 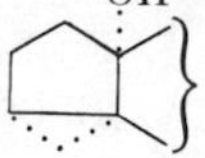 (15)	
	3β-Hydroxy-17β-acetoxyestr-5(10)-ene	CH_2Cl_2, 0°, 20 min	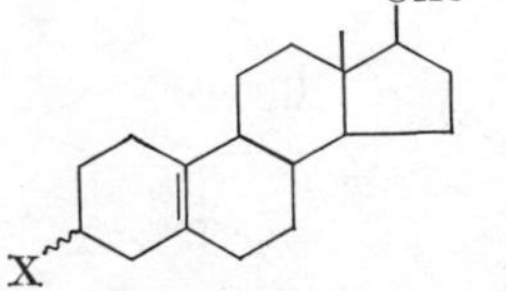 X = α-F (0) X = β-F (90) X = β-ClFCHCO$_2$ (3)	93, 323
C_{21}	15α-Hydroxy-11-ketoprogesterone	1:2, CH_2Cl_2, 5°, 4.5 hr	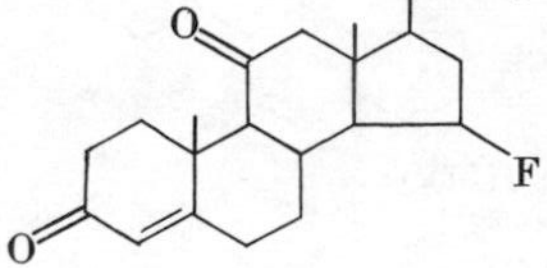 D (43), 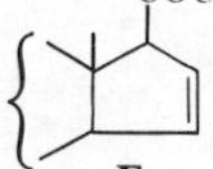 E (20)	89, 96, 317b 320

Note: References 301–386 are in pp. 404–406.

[a] The fluoroalkylamine reagent used is 2-chloro-1,1,2-trifluorotriethylamine, $(C_2H_5)_2NCF_2CHFCl$, unless stated otherwise under conditions.

[b] The ratio of numbers is the ratio of moles of reactant to moles of fluoroalkylamine reagent.

[j] Successful fluorinations were carried out in unreported yields where the group $—NR_1R_2$ was the following: pyrollidino, 2-methylpyrrolidino, 2,2-dimethylpyrrolidino, morpholino, piperidino. The patents should be consulted for a list of other derivatives of **FAR** claimed.

TABLE XI. Replacement of the Hydroxyl Group by Fluorine Using the α-Fluoroalkylamine (Yarovenko) Reagent[a] (**FAR**) (*Continued*)

	Reactant	Condition(s)[b]	Product(s) and Yield(s) (%)	Refs.
C_{21} (*contd.*)	15α-Hydroxy-11-ketoprogesterone (*contd.*)	1:2, CH_2Cl_2, 5°, 16 hr, $(C_2H_5)_3N\cdot HF$ catalyst	**D** (63), **E** (16)	96, 320
	11α-Hydroxy-pregn-4-ene-3,20-dione	1:2, CH_2Cl_2, 5°, 14 hr	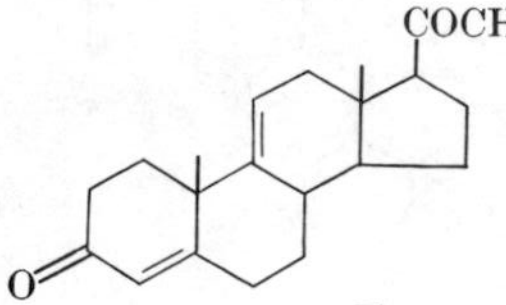 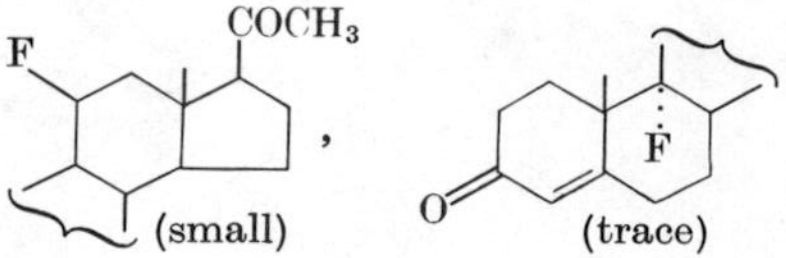 COCH₃ (large) **F**; F, COCH₃ (small), F (trace)	89, 96, 320
		1:1.5, CH_2Cl_2, 0°, 16 hr	(86)	90b
	15α-Hydroxyprogesterone	1:2, CH_2Cl_2, 5°, 14 hr	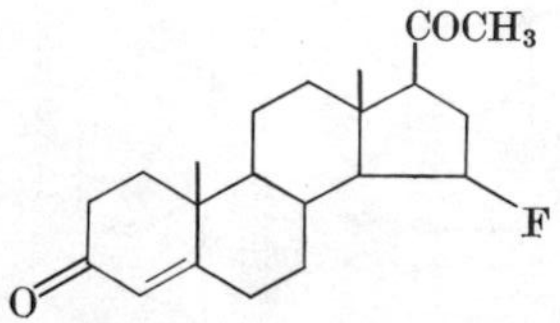 COCH₃, F (—)	96, 320

Reactant	Conditions[a,b]	Products and Yields (%)	Refs.
6β,11α-Dihydroxypregn-4-ene-3,20-dione	1:6, CH_2Cl_2, 5°, 3 hr	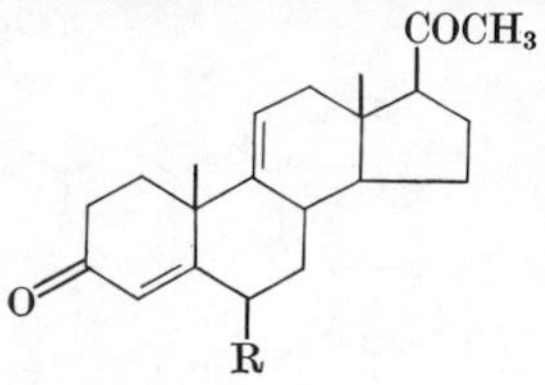 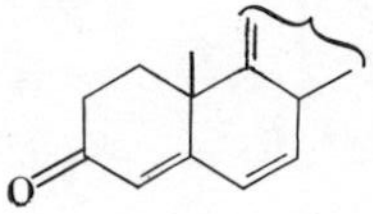 (structure) R = F··· (11); R = HO— (36); R = $FClCHCO_2$— (31), (structure) (22)	89, 96, 320
3β,19-Dihydroxyandrost-5-en-17-one 3β-acetate	1:1.5, CH_2Cl_2, 0°, 17–42 hr, distil solvent, chromatograph over Florisil	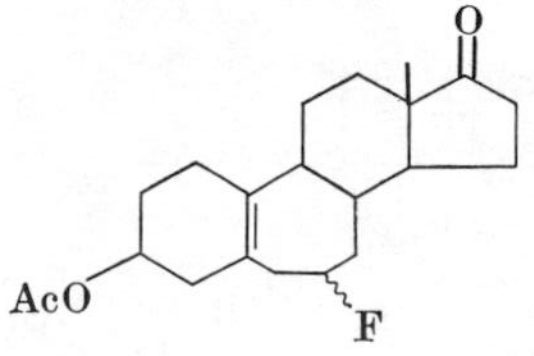 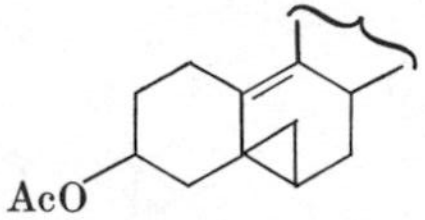 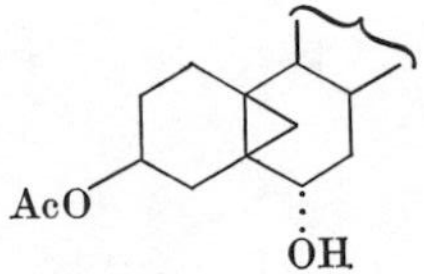 A (31), (structure) (5), (structure) (11) 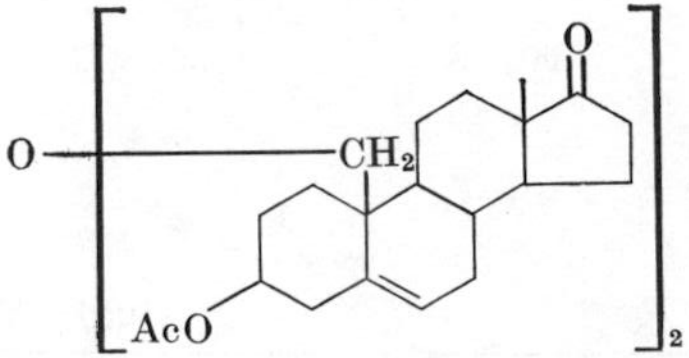 + (structure) (32)	104

Note: References 301–386 are in pp. 404–406.

[a] The fluoroalkylamine reagent used is 2-chloro-1,1,2-trifluorotriethylamine, $(C_2H_5)_2NCF_2CHFCl$, unless stated otherwise under conditions.
[b] The ratio of numbers is the ratio of moles of reactant to moles of fluoroalkylamine reagent.

TABLE XI. Replacement of the Hydroxyl Group by Fluorine Using the α-Fluoroalkylamine (Yarovenko) Reagent[a] (**FAR**) (*Continued*)

	Reactant	Condition(s)[b]	Product(s) and Yield(s) (%)	Refs.
C_{21} (*contd.*)	3β,19-Dihydroxyandros-5-en-17-one 3β-acetate (*contd.*)	1:1.5, CH_2Cl_2, −20°, 48 hr workup over alumina column	**A** (63)	104
		1:1.5, CH_3CN, reflux, 1 hr workup over alumina column	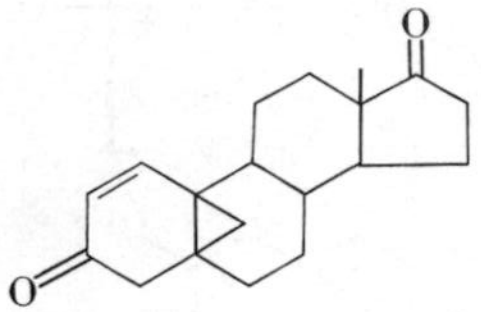 (64)	104
	3β,6β-Dihydroxy-5α-bromoandrostan-17-one 3-acetate	CH_2Cl_2, room t, 24 hr	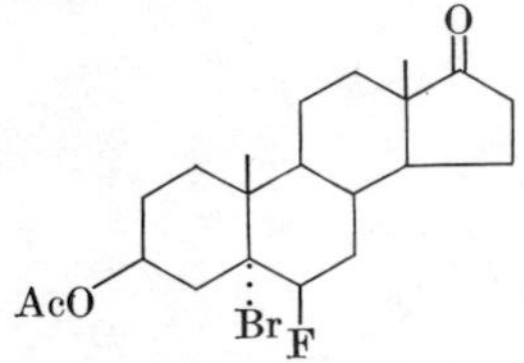 AcO, Br, F (25)	90b
	3α-Hydroxypregn-5-en-20-one	1:1, THF, room t, 10 min	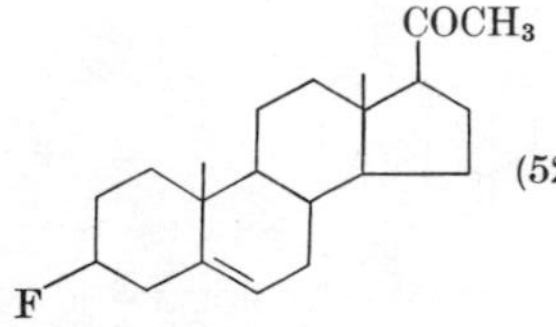 $COCH_3$, F (52)	90b
	6β-Hydroxy-3:5-cyclopregnan-20-one	1:2, CH_2Cl_2, 5°, 14 hr	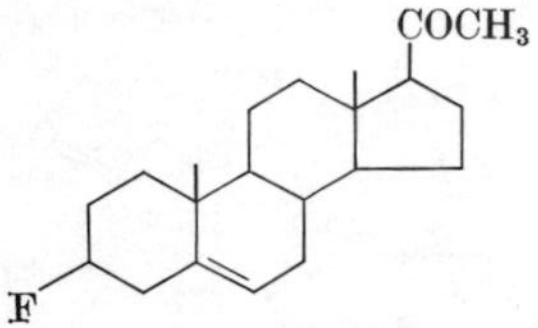 $COCH_3$, F (69)	89, 96, 320

6β-Hydroxytestosterone 17-acetate	1:2, Diglyme, reflux, 1 hr	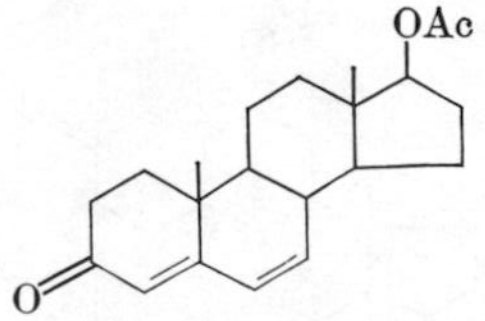 [structure: 17-OAc steroid 3-one, 4,6-diene] (44) No fluoro compound isolated	90b
17β-Acetoxy-5α-androst-2-en-1α-ol	1:1, CH_2Cl_2, room t, 30 min	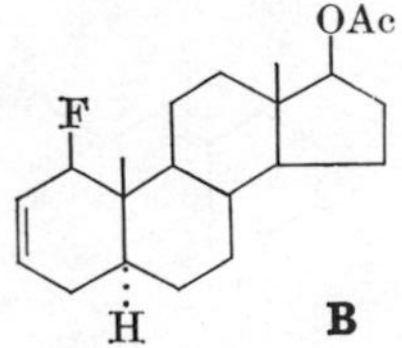 [structure: 1-F, 17-OAc, 5α-H, 2-ene] **B** (53)	92a
	1:2, CH_2Cl_2, room t, 30 min	**B** (main product), 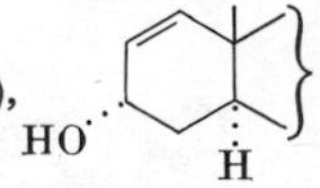 [structure: HO, H] (72 total)	92b
1α-Hydroxy-5α-androstane-3,17-dione	1:1, CH_2Cl_2, reflux, 15 min	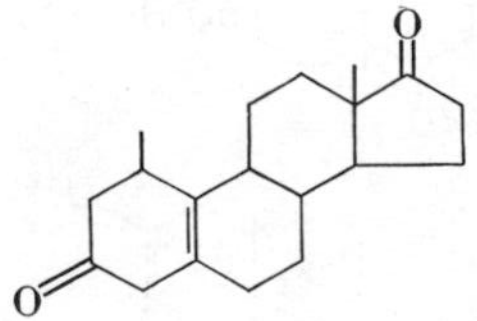 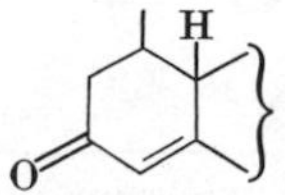 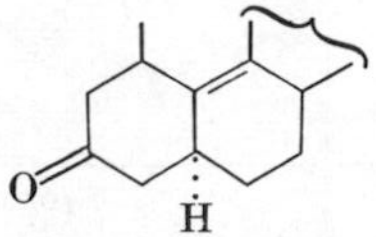 [structure: 3,17-dione] (21), [structure] (Trace), [structure: H] (12)	92b

Note: References 301–386 are in pp. 404–406.

[a] The fluoroalkylamine reagent used is 2-chloro-1,1,2-trifluorotriethylamine, $(C_2H_5)_2NCF_2CHFCl$, unless stated otherwise under conditions.

[b] The ratio of numbers is the ratio of moles of reactant to moles of fluoroalkylamine reagent.

TABLE XI. REPLACEMENT OF THE HYDROXYL GROUP BY FLUORINE USING THE α-FLUOROALKYLAMINE (YAROVENKO) REAGENT[a] (**FAR**) (*Continued*)

	Reactant	Condition(s)[b]	Product(s) and Yield(s) (%)	Refs.
C_{21} (*contd.*)				
	3β,19-Dihydroxyandrost-5-en-17-one 3-acetate	1.5:1, CH_2Cl_2, room t, 12 hr	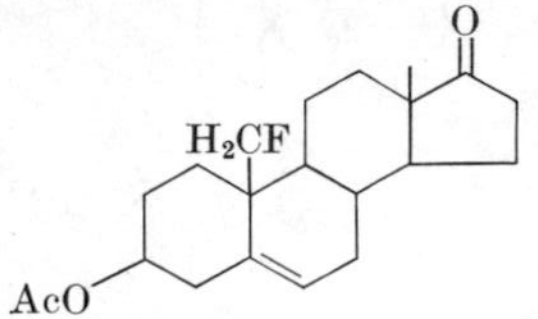 O; H_2CF; AcO (−)	97
	19-Hydroxy-10α-testosterone 17-acetate	1.5:1, CH_2Cl_2, room t, 12 hr	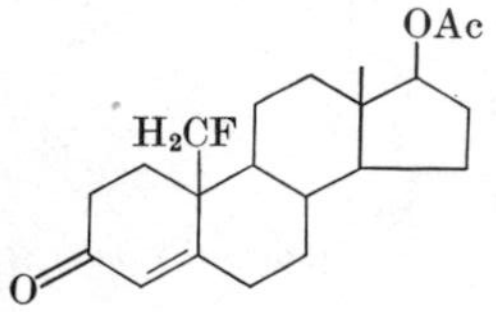 OAc; H_2CF; O (−)	153a
C_{22}	16β-Hydroxymethyl-17α-pregn-4-ene-3,20-dione	1:2.5, CH_2Cl_2, reflux, 7 hr	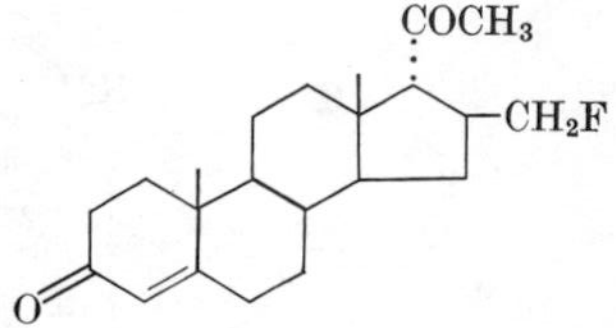 $COCH_3$; CH_2F; O (46)	89, 96, 320
	16α,17α-Dihydroxy-6α-methyl-4-pregnene-3,20-dione	1:2,[k] THF, 25°, 3 hr, HF catalyst	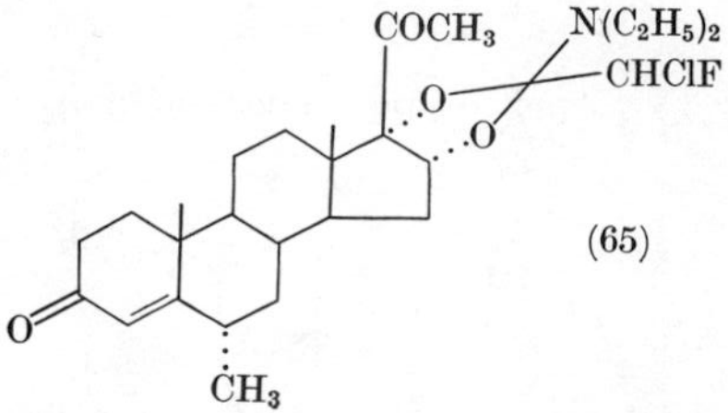 $COCH_3$; $N(C_2H_5)_2$; CHClF; O; O; O; CH_3 (65)	79

Reactant	Conditions	Product(s) and Yield(s) (%)	Refs.
2-Hydroxymethyl-5α-androst-2-en-17β-ol acetate	1:2, THF, reflux, 1 hr	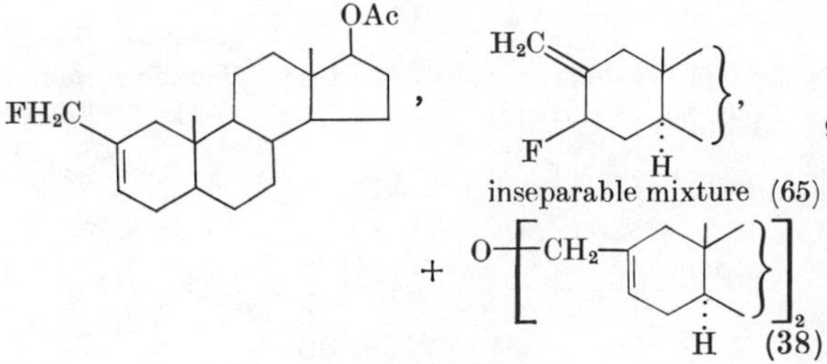 FH_2C (steroid, OAc), H_2C (F, H) inseparable mixture (65) + $O[CH_2$ (H) $]_2$ (38)	90
Methyl 3,11-diketo-16β-hydroxy-*trans*-1,4,17(20)-pregnatrien-21-oate	1:1, CH_2Cl_2, 5°, 3.5 hr	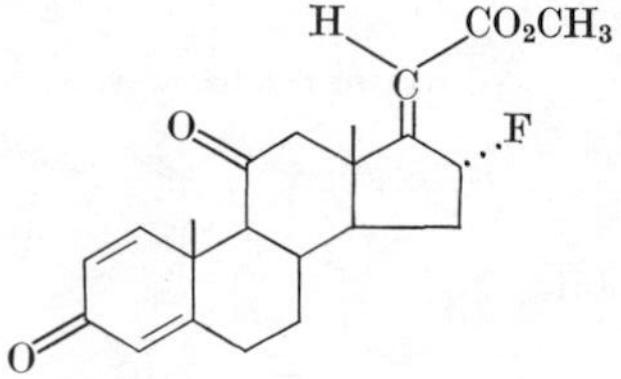 H, CO_2CH_3, F (79)	94
Methyl 3,11-diketo-16α-hydroxy-*trans*-1,4,17(20)-pregnatrien-21-oate	1:1, CH_2Cl_2, 5°, 3.5 hr	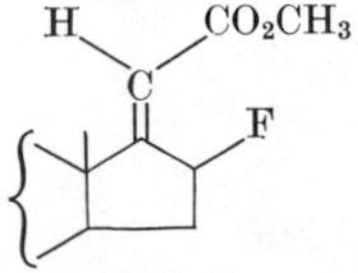 H, CO_2CH_3, F (71)	94
Methyl 3,11-diketo-16α-hydroxy-*cis*-1,4,17(20)-pregnatrien-21-oate	1:1, CH_2Cl_2, 5°, 3.5 hr	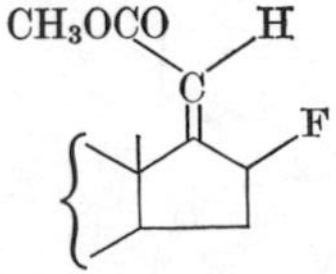 CH_3OCO, H, F (57)	94

Note: References 301–386 are in pp. 404–406.

[a] The fluoroalkylamine reagent used is 2-chloro-1,1,2-trifluorotriethylamine, $(C_2H_5)_2NCF_2CHFCl$, unless stated otherwise under conditions.
[b] The ratio of numbers is the ratio of moles of reactant to moles of fluoroalkylamine reagent.
[k] $(C_2H_5)_2NCF_2CHF_2$ could also be used in place of **FAR**.

TABLE XI. Replacement of the Hydroxyl Group by Fluorine Using the α-Fluoroalkylamine (Yarovenko) Reagent[a] (**FAR**) (*Continued*)

	Reactant	Condition(s)[b]	Product(s) and Yield(s) (%)	Refs.
C_{22} (*contd.*)	6α-Methyl-15α-hydroxy-11-ketoprogesterone	—	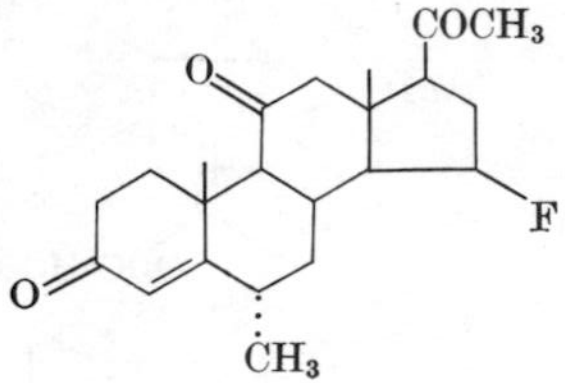 $COCH_3$; F; CH_3; O (high)	96, 320
C_{23}	9α-Fluoro-11β,16α,17α,21-tetrahydroxy-1,4-pregnadiene-3,20-dione 21-acetate	Excess, CH_2Cl_2, 25°, 16 hr	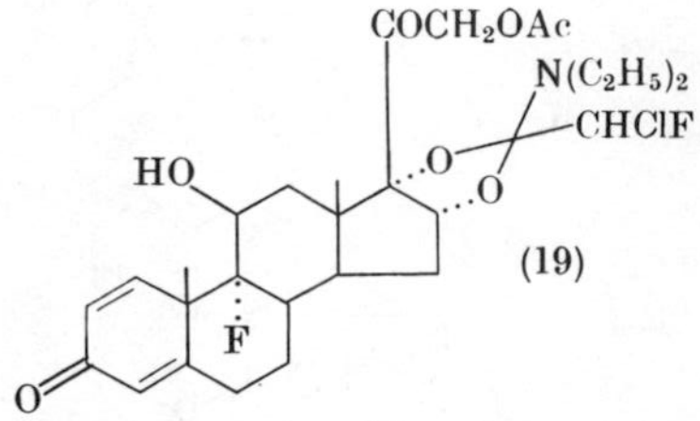 $COCH_2OAc$; $N(C_2H_5)_2$; CHClF; HO; O; F (19) (1:18 isomeric mixture)	79
	11β,16α,21-Trihydroxypregna-1,4,17(20)-trien-3-one 21-acetate	1:2.5, CH_2Cl_2, 5°, 18 hr	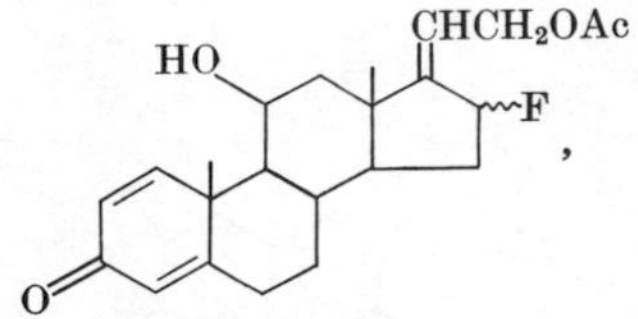 $CHCH_2OAc$; HO; F; O; Mixture of 16α- and β-epimers (19), 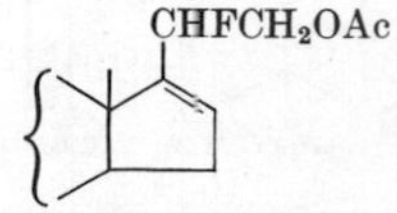 $CHFCH_2OAc$; Mixture of 20α- and β-epimers (38)	96, 320

Reactant	Conditions	Product(s) and Yield(s) (%)	Refs.
11β,20,21-Trihydroxypregna-1,4,16-trien-3-one 21-acetate	1:2.5, CH_2Cl_2, 5°, 4 hr	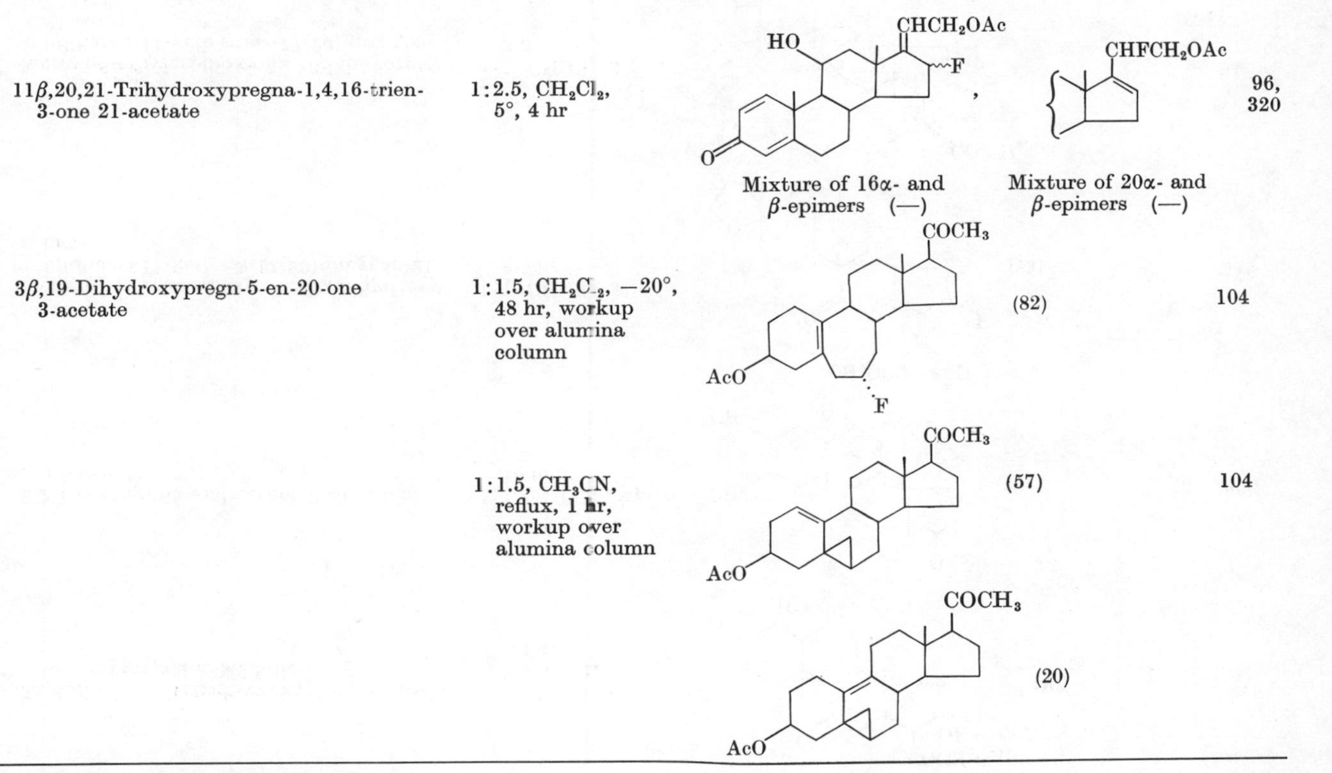 Mixture of 16α- and β-epimers (—), Mixture of 20α- and β-epimers (—)	96, 320
3β,19-Dihydroxypregn-5-en-20-one 3-acetate	1:1.5, CH_2Cl_2, −20°, 48 hr, workup over alumina column	(82)	104
	1:1.5, CH_3CN, reflux, 1 hr, workup over alumina column	(57)	104
		(20)	

Note: References 301–386 are in pp. 404–406.

[a] The fluoroalkylamine reagent used is 2-chloro-1,1,2-trifluorotriethylamine, $(C_2H_5)_2NCF_2CHFCl$, unless stated otherwise under conditions.

[b] The ratio of numbers is the ratio of moles of reactant to moles of fluoroalkylamine reagent.

TABLE XI. REPLACEMENT OF THE HYDROXYL GROUP BY FLUORINE USING THE α-FLUOROALKYLAMINE (YAROVENKO) REAGENTS[a] (**FAR**) (*Continued*)

	Reactant	Condition(s)[b]	Product(s) and Yield(s) (%)	Refs.
C_{23} (*contd.*)				
	5α,6β,11β,17α Tetrahydroxy-16α-chloro-21-acetoxypregnane-3,20-dione	1:1, CH_2Cl_2, 0°, 3.5 hr, HF catalyst	$COCH_2OAc$, OH, Cl, HO, O, HO, F (40)	94
	2,2-Dimethyl-5α-androstane-3β,17β-diol 17-acetate	1:1.25, THF, warm, 10 min	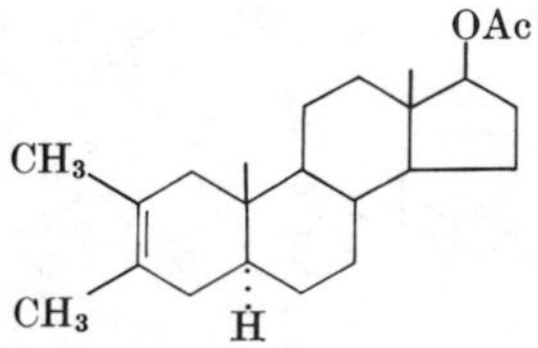 OAc, CH_3, CH_3, H (50)	90b
C_{24}	Methyl 3-ethylenedioxy-5α,16α-dihydroxy-6β-fluoro-11-keto-*cis*-17(20)-pregnen-21-oate	1:1, CH_2Cl_2, 5°, 3.5 hr	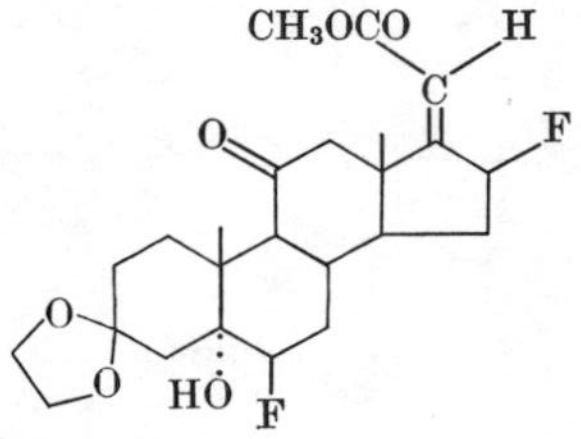 CH_3OCO, H, C, O, F, O, O, HO, F (83)	94
	Methyl 3-ethylenedioxy-5α,16β-dihydroxy-6β-fluoro-11-keto-*trans*-17(20)-pregnen-21-oate	1:1, CH_2Cl_2, 5°, 3.5 hr	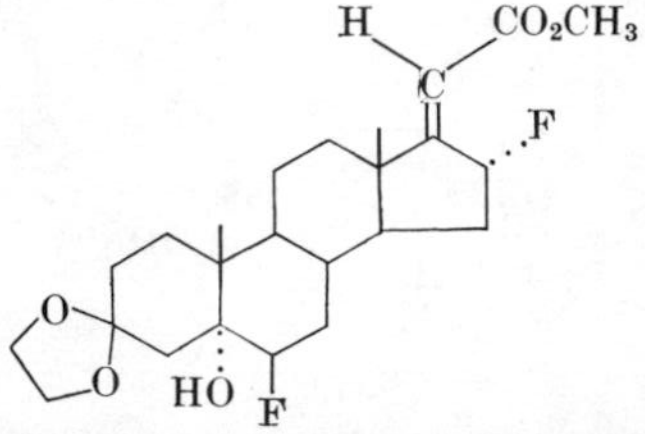 H, CO_2CH_3, C, F, O, O, HO, F (93)	94

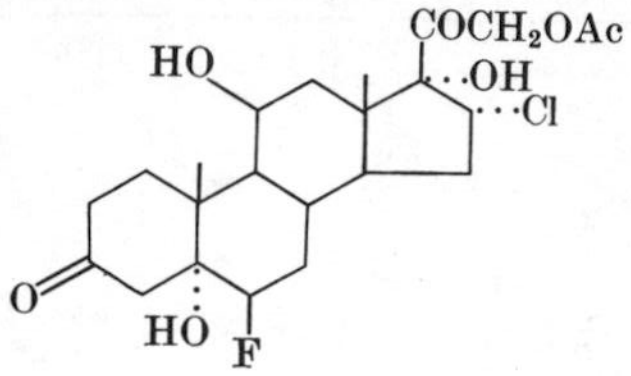

Reactant	Conditions	Product(s) (% yield)	Refs.
Methyl 3-ethylenedioxy-5α,16β-dihydroxy-6β-fluoro-11-keto-*cis*-17(20)-pregnen-21-oate	1:1, CH_2Cl_2, 5°, 3.5 hr	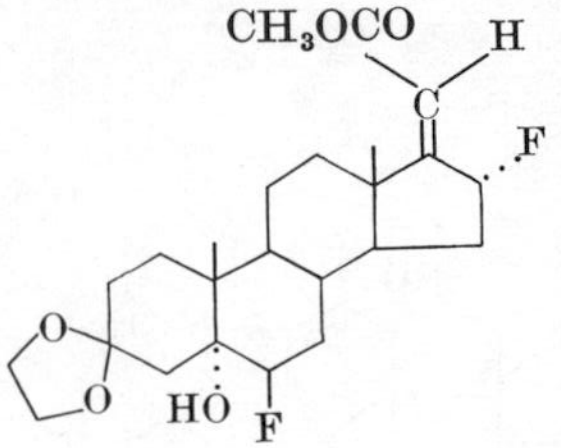 CH_3OCO, H, C, F, HO, F, O, O (78)	94
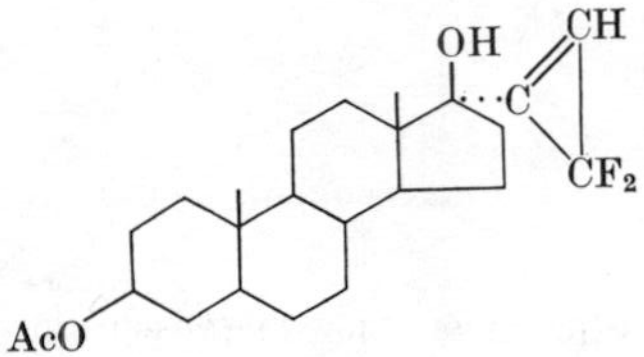 OH, CH, C, CF_2, AcO	CH_2Cl_2	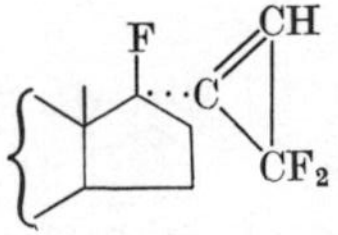 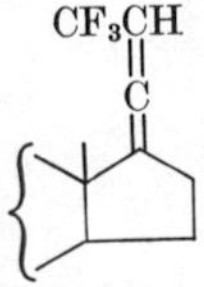 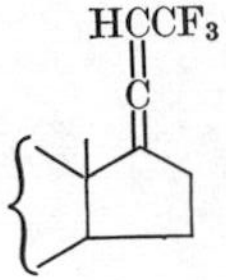 F, CH, C, CF_2 (6), CF_3CH=C= (25), $HCCF_3$=C= (1)	325
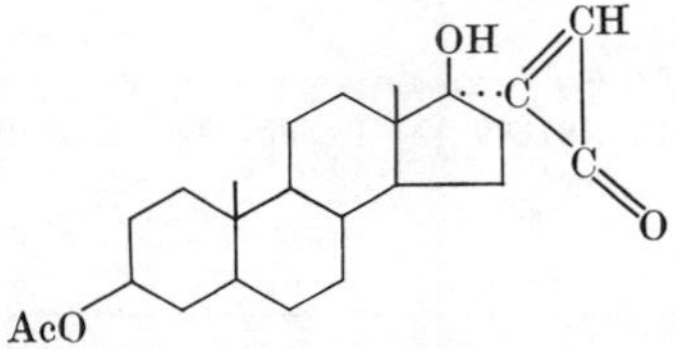 OH, CH, C, C, O, AcO	CH_2Cl_2	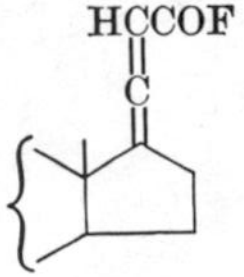 HCCOF=C= (80)	325

Note: References 301–386 are in pp. 404–406.

[a] The fluoroalkylamine reagent used is 2-chloro-1,1,2-trifluorotriethylamine, $(C_2H_5)_2NCF_2CHFCl$, unless stated otherwise under conditions.
[b] The ratio of numbers is the ratio of moles of reactant to moles of fluoroalkylamine reagent.

TABLE XI. Replacement of the Hydroxyl Group by Fluorine Using the α-Fluoroalkylamine (Yarovenko) Reagent[a] (**FAR**) (*Continued*)

	Reactant	Condition(s)[b]	Product(s) and Yield(s) (%)	Refs.
C_{25}	Methyl 3-ethylenedioxy-5α,16β-dihydroxy-6β-methyl-11-keto-*trans*-17(20)-pregnen-21-oate	1:1, CH_2Cl_2, 5°, 3.5 hr	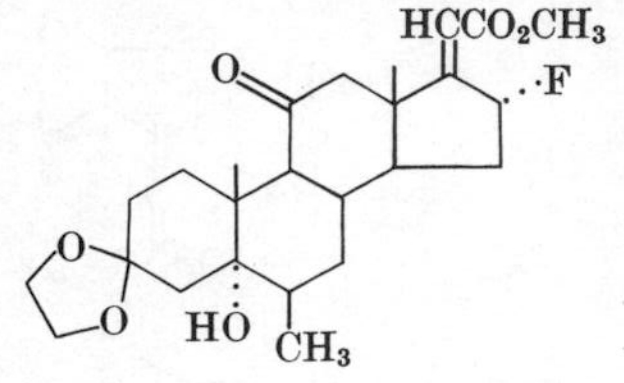 $HCCO_2CH_3$, O, F, O, O, HO, CH_3 (77)	94
	Methyl 3-ethylenedioxy-5α,16α-dihydroxy-6β-methyl-11-keto-*cis*-17(20)-pregnen-21-oate	1:1, CH_2Cl_2, 5°, 3.5 hr	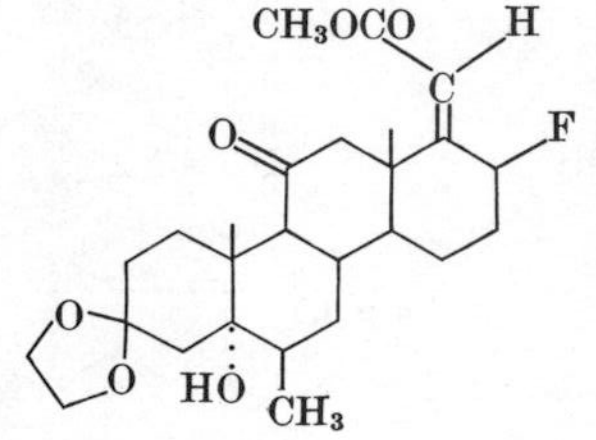 CH_3OCO, H, C, O, F, O, O, HO, CH_3 (64)	94
	3β,14β,19-Trihydroxycard-5-enolide 3-acetate	1:2.1, CH_2Cl_2, −20°, 30 hr	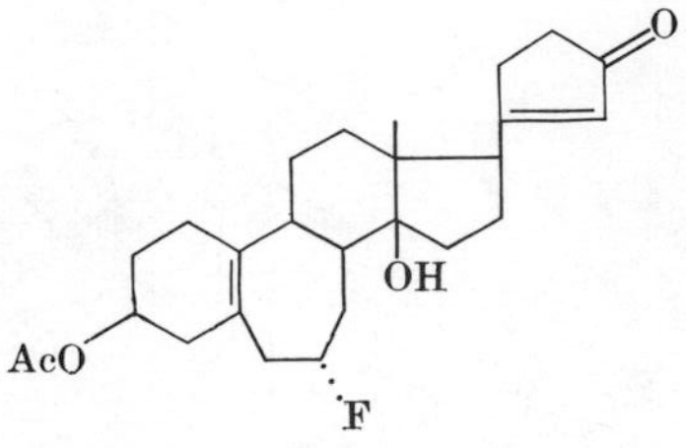 O, OH, AcO, F (25)	326

	17α,20:20,21-Bismethylenedioxypregn-5-ene-3β-19-diol 3β-acetate	1:1.5, CH_2Cl_2, −20°, 48 hr, workup over alumina column	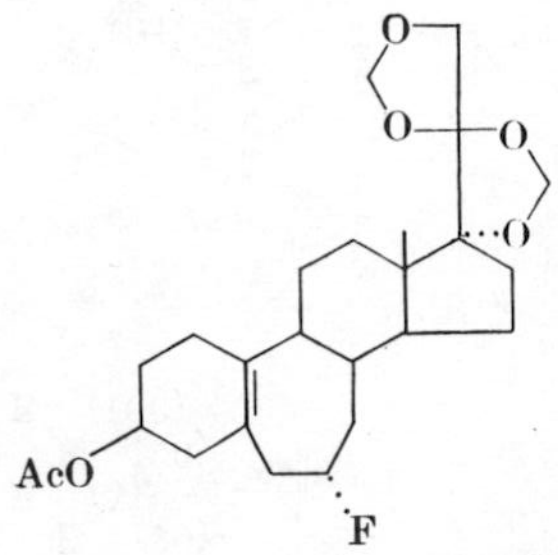 (77)	104
C_{26}	6α-Hydroxymethyl-3β,17α-dihydroxy-5α-pregnan-20-one 3,17-diacetate	1:2.8, CH_2Cl_2, reflux, 7 hr	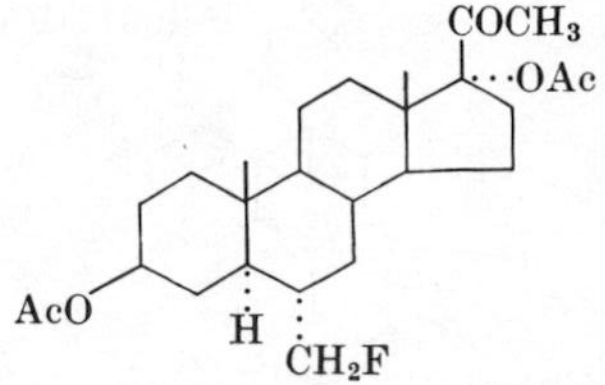 (71)	96, 320
C_{27}	9α-Fluoro-11β,16α,17α,21-tetrahydroxy-1,4-pregnadiene-3,20-dione	Excess **FAE**, THF, 25°, 16 hr, HF catalyst	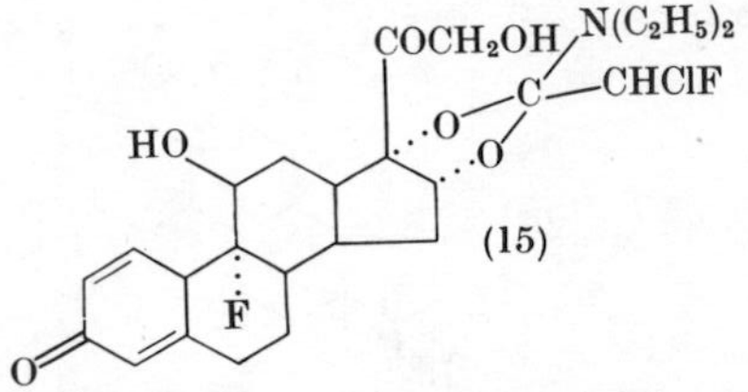 (15)	79

Note: References 301–386 are in pp. 404–406.

[a] The fluoroalkylamine reagent used is 2-chloro-1,1,2-trifluorotriethylamine, $(C_2H_5)_2NCF_2CHFCl$, unless stated otherwise under conditions.

[b] The ratio of numbers is the ratio of moles of reactant to moles of fluoroalkylamine reagent.

TABLE XI. REPLACEMENT OF THE HYDROXYL GROUP BY FLUORINE USING THE α-FLUOROALKYLAMINE (YAROVENKO) REAGENT[a] (**FAR**) (*Continued*)

	Reactant	Condition(s)[b]	Product(s) and Yield(s) (%)	Refs.
C_{27} (*contd.*)				
	3β-Fluoro-19-hydroxy-5-cholestene	1:2, CH_2Cl_2, −10°, 3 hr	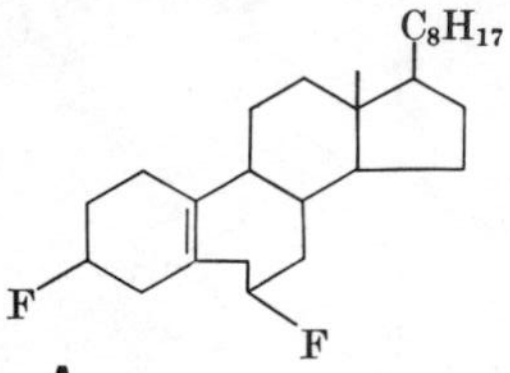 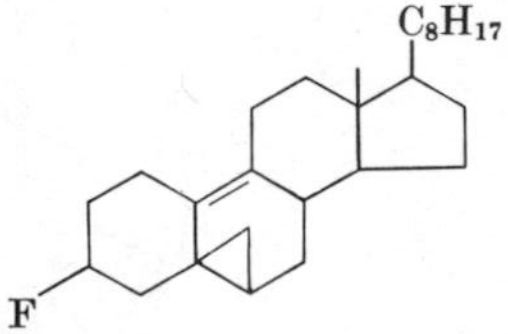 **A** (40), **B** (50)	105
	3β-Fluoro-6β-hydroxy-5β,19(5β)-cyclocholestane	1:2, CH_2Cl_2, −10°, 1.5 hr	**A** (80), **B** (20)	105
	3β-Fluoro-7β-hydroxy-β-homo-5(10)-cholestene	1:2, CH_2Cl_2, −10°, 1.5 hr	**A** (5), **B** (90)	105
	3β-Fluoro-6β-hydroxymethyl-5-cholestene	1:2, CH_2Cl_2, −10°, 1.5 hr	**B** (100)	105
	3β-Fluoro-5-methyl-6β-hydroxy-19-nor-Δ^9(5β)-cholestene	1:2, CH_2Cl_2, 0°, 5 min	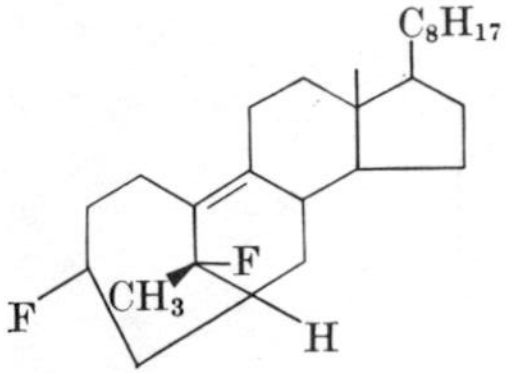 (84)	327

	Reactant	Conditions[b]	Product(s) and Yield(s) (%)	Refs.
	Cholesterol	1:1, CH_2Cl_2, 0°, 3 hr	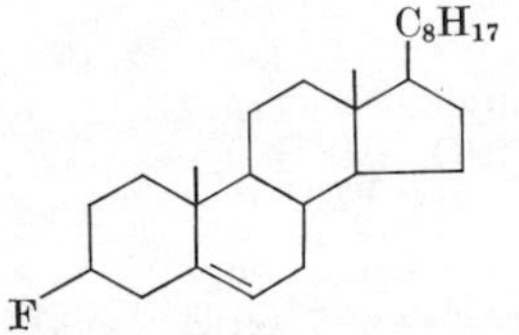 3-Fluoro steroid with C_8H_{17} side chain (90)	91
		1:2, $CHCl_3$, 0°, 5 hr	" (90)	105
	Methyl deserpidate	1:3, CH_2Cl_2, 0°, 1 hr; 25°, 22 hr	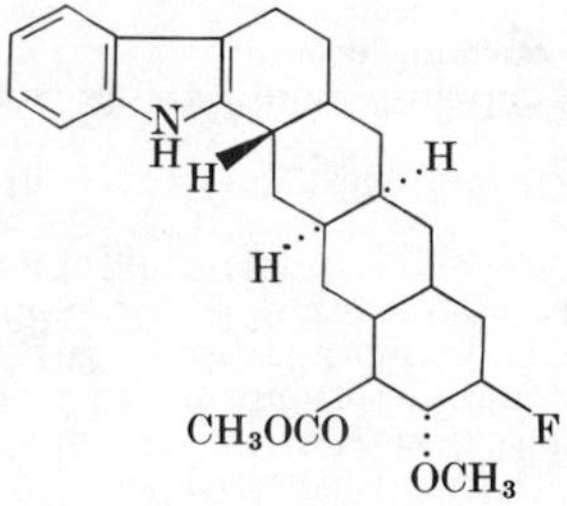 Fluoro product with CH_3OCO, OCH_3, F (—)	325
C_{28}	Methyl reserpate[l]	1:3, CH_2Cl_2, 0°, 1 hr; 25°, 2 hr	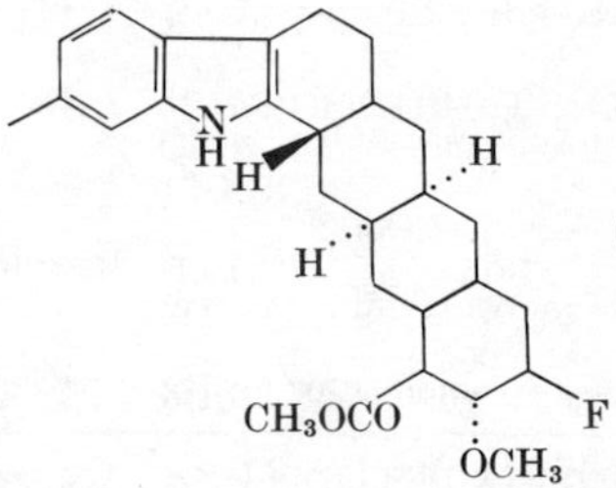 Fluoro product with CH_3OCO, OCH_3, F (36)	324

Note: References 301–386 are in pp. 404–406.

[a] The fluoroalkylamine reagent used is 2-chloro-1,1,2-trifluorotriethylamine, $(C_2H_5)_2NCF_2CHFCl$, unless stated otherwise under conditions.

[b] The ratio of numbers is the ratio of moles of reactant to moles of fluoroalkylamine reagent.

[l] Fluorination was also reported to prepare a series of lower alkyl 18-deoxy-18-fluororeserpates and deserpidates including propyl, *sec*-butyl, *n*-hexyl, and *n*-octyl.

TABLE XII. Monofluorosubstitution of Organic Halides or Sulfonic Esters

$$\geq C{-}X + MF \longrightarrow \geq C{-}F + MX \quad \text{and} \quad \geq C{-}OSO_2R + MF \longrightarrow \geq C{-}F + MOSO_2R$$

	Halide or Ester	Metal Fluoride[a]	Condition(s)	Product(s) and Yield(s) (%)	Ref.
C_3	2-Chloropropene	KF	NMP, 160–180°, 6–8 hr	2-Fluoropropene (—)	328
C_4	Methyl α-bromoisobutyrate	AgF	140–145°, 3 hr	Methyl α-fluoroisobutyrate (20)	329
	Dichloromaleonitrile	KF mixed with other fluorides	250°, vapor phase	FCCN ‖ ClCCN (62)	330
	4-Chlorobutyryl chloride	KF	TMS, 195°	Cyclopropanecarbonylfluoride (70)	120
C_5	2,3-Dibromo-2-methylbutane	AgF	CH_3CN, room t, 1 hr	$(CH_3)_2CFCHBrCH_3$ (34)	73b
	Ethyl 2-bromopropionate	KF	CH_3CONH_2, 132°	Ethyl 2-fluoropropionate (65–70)	331
C_6	2-Bromo-3-fluoro-2,3-dimethylbutane	AgF	CH_3CN, −20°, 4 hr; 0°, 24 hr	$(CH_3)_2CFCF(CH_3)_2$ (5), $(CH_3)_2CFC(CH_3){=}CH_2$ (43)	73a
	6-Bromo-1-hexene[b]	KF	Diglycol, 90°, 2 hr	6-Fluoro-1-hexene (40)	118
	Ethyl 2-bromobutyrate	KF	CH_3CONH_2, 132°	Ethyl 2-fluorobutyrate (—)	331
C_7	1-Chloroheptane[b]	KF	Diglycol, reflux	1-Fluoroheptane (39)	118
	1-Bromoheptane	LiF	HMPT, 160–180°, 9.5 hr	1-Heptene (54)	115
	Benzyl bromide	KF	NMP, heat to distil product	Benzyl fluoride (70)	332
	2-Bromo-3-methoxy-2,3-dimethylbutane	AgF	CH_3CN, mixed at 0°; room t, 24 hr	2-Fluoro-3-methoxy-2,3-dimethylbutane (20), 3-methoxy-2,3-dimethyl-1-butene (35)	73a
	Ethyl 2-bromovalerate	KF	CH_3CONH_2, 132°	Ethyl 2-fluorovalerate (70–75)	331
C_8	Bromocyclooctatetraene	AgF	Pyridine, room t, 6 d	Fluorocyclooctatetraene (67)	333
	$C_6H_5CH_2CH_2Br$	$(C_2H_5)_4NF$	CH_3CN	$C_6H_5CH_2CH_2F$ (4), $C_6H_5CH{=}CH_2$ (96)	117
	α-Bromoacetanilide	KF	Diglycol, 125–130°, 2 hr	α-Fluoroacetanilide (45)	121

	α-Chloroacetanilide	KF	Diglycol, 125–130°, 2 hr	α-Fluoroacetanilide (53)	121
C_9	Ethyl 7-bromoheptanoate[b]	AgF	60–80°	Ethyl 7-fluoroheptanoate (—)	152
	α-Chloropropionanilide	KF	Diglycol, 125–130°, 2 hr	α-Fluoropropionanilide (64)	121
	α-Bromopentanoyl-N,N-diethylamide	KF	Diglycol, 125–130°, 2 hr	α-Fluoropentanoyl-N,N-diethylamide (69)	121
	Methyl 2-bromo-2-phenylacetate	KF	CH_3CN, 120°, 6 hr	Methyl 2-fluoro-2-phenylacetate (98)	334
	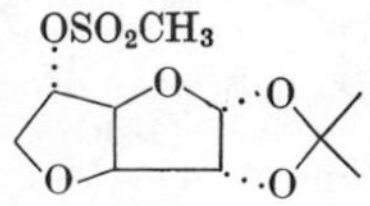 [structure: OSO_2CH_3 bicyclic furanose acetonide]	KF	Glycol, reflux, 1 hr	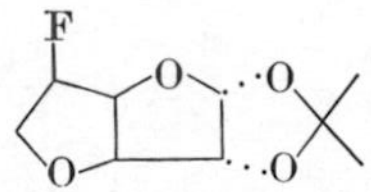 [structure: F bicyclic furanose acetonide] (75)	151
	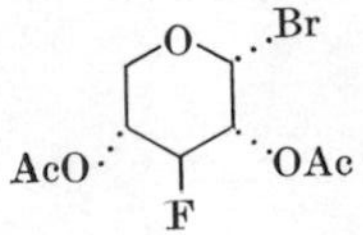 [structure: pyranosyl bromide, Br, AcO, OAc, F]	AgF	CH_3CN, room t, 1 hr	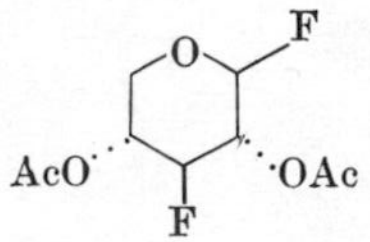 [structure: pyranosyl fluoride, F, AcO, OAc, F] (46)	335a
C_{10}	α-Bromobutyranilide	KF	Diglycol, 125–130°, 2 hr	α-Fluorobutyranilide (66)	121
	Ethyl 2-bromo-2-phenylacetate	KF	DMF, 145°, 8 hr	Ethyl 2-fluoro-2-phenylacetate (53)	336
		KF	Diglycol, temp and time varied	,, (0–40)	336
C_{12}	α-Bromo-γ-phthalimidobutyramide	KF	Diglycol, 130°, 2.5 hr	α-Fluoro-γ-phthalimidobutyramide (58)	121

Note: References 301–386 are on pp. 404–406.

[a] A stoichiometric excess of fluoride was normally used.
[b] This older example is included to permit comparison with newer fluorinations.

TABLE XII. MONOFLUOROSUBSTITUTION OF ORGANIC HALIDES OR SULFONIC ESTERS (*Continued*)

	Halide or Ester	Metal Fluoride[a]	Condition(s)	Product(s) and Yield(s) (%)	Ref.
C_{12} (*contd.*)	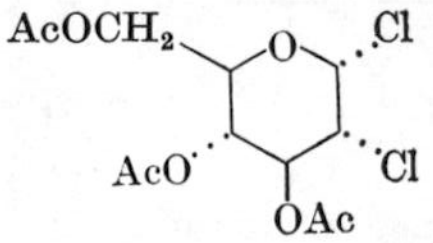 AcOCH$_2$, O, Cl, AcO, Cl, OAc (pyranose structure)	$AgBF_4$	Ether, 0°, various methods of mixing; 10–25 min	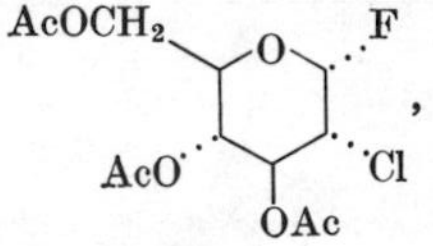 AcOCH$_2$, O, F, AcO, Cl, OAc (pyranose structure), **A** (55–68); 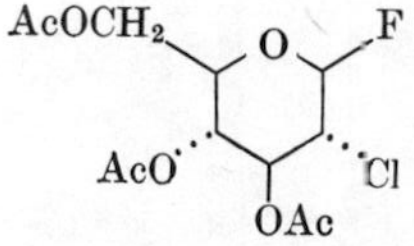 AcOCH$_2$, O, F, AcO, Cl, OAc (pyranose structure) **B** (32–44)	123
		$AgBF_4$	Toluene, 0°, 1 hr	**A** (76), **B** (1)	123
	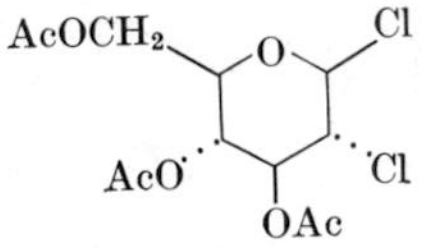 AcOCH$_2$, O, Cl, AcO, Cl, OAc (pyranose structure)	$AgBF_4$	Ether, 0°; various methods of mixing; 10–25 min	**A** (68–75), **B** (25–32)	123
		$AgBF_4$	Toluene, 0°, stir 1 hr	**A** (77), **B** (1)	123
	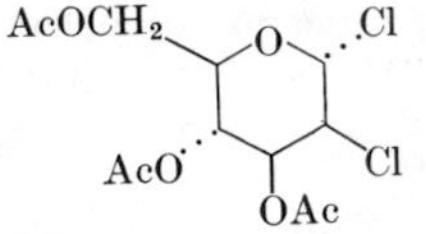 AcOCH$_2$, O, Cl, AcO, Cl, OAc (pyranose structure)	$AgBF_4$	Ether, 0°; various methods of mixing; 10–25 min	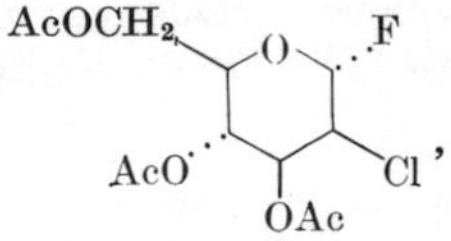 AcOCH$_2$, O, F, AcO, Cl, OAc (pyranose structure), **C** (82–89)	123

Reactant	Fluoride	Conditions	Product(s) and Yield(s) (%)	Refs.
			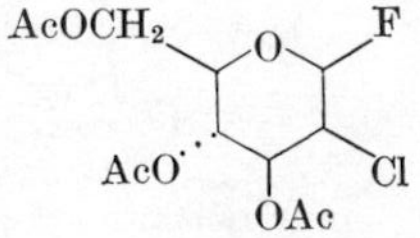 AcOCH$_2$ O F AcO Cl OAc **D** (11–18)	
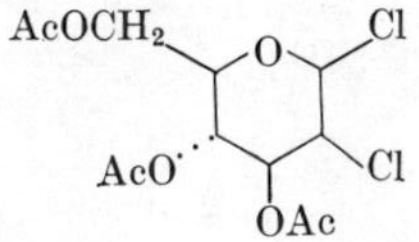	$AgBF_4$	Toluene, 0°, stir 1 hr	**C** (91), **D** (1)	123
AcOCH$_2$ O Cl AcO Cl OAc	$AgBF_4$	Ether, 0°; various methods of mixing; 10–25 min	**C** (75–90), **D** (9.5–25)	123
	$AgBF_4$	Toluene, 0°, stir 1 hr	**C** (91), **D** (1)	123
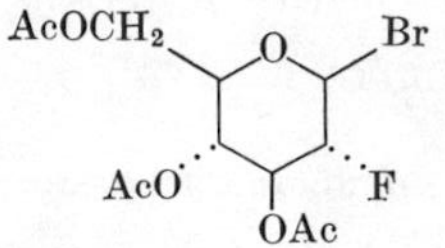 AcOCH$_2$ O Br AcO F OAc	AgF	CH_3CN, room t, stir 17 hr	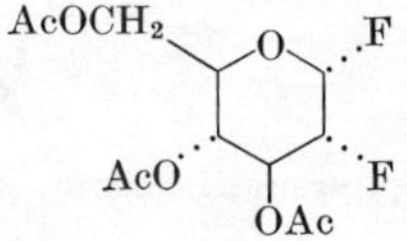 AcOCH$_2$ O F AcO F OAc (41)	176
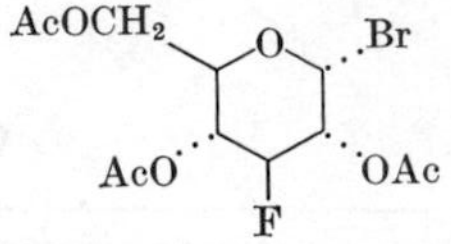 AcOCH$_2$ O Br AcO OAc F	AgF	CH_3CN, room t, 1 hr	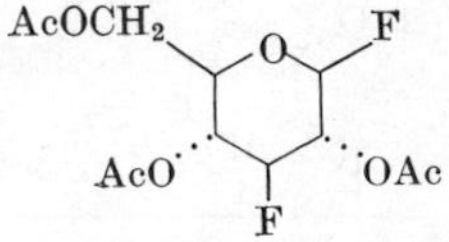 AcOCH$_2$ O F AcO OAc F (77)	335[a]

Note: References 301–386 are on pp. 404–406.

[a] A stoichiometric excess of fluoride was normally used.

TABLE XII. Monofluorosubstitution of Organic Halides or Sulfonic Esters (*Continued*)

	Halide or Ester	Metal Fluoride[a]	Condition(s)	Product(s) and Yield(s) (%)	Ref.
	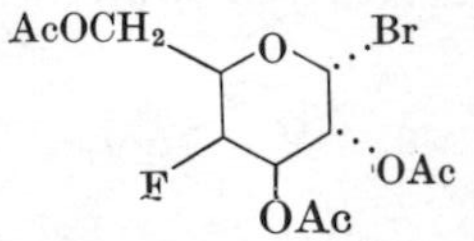 $AcOCH_2$, O, Br, F, OAc, OAc	AgF	CH_3CN, 25°, 1 hr	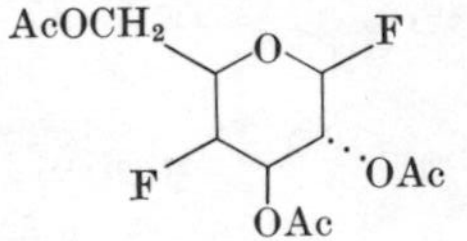 $AcOCH_2$, O, F, F, OAc, OAc (50)	335b
C_{13}	α-Bromoheptananilide	KF	Diglycol, 125–130°, 2 hr	α-Fluoroheptananilide (61)	121
	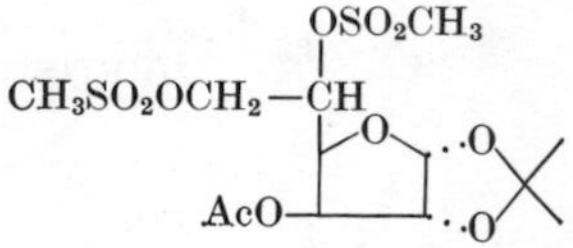 OSO_2CH_3, $CH_3SO_2OCH_2$—CH, O, O, O, AcO	KF	Glycol, reflux, 2 hr	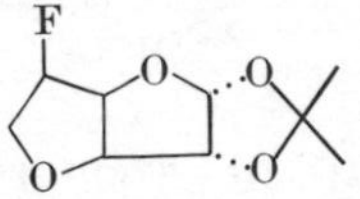 F, O, O, O, O (48)	151
	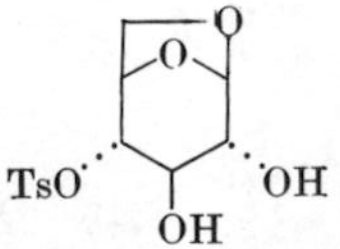 O, O, TsO, OH, OH	KHF_2	Glycol reflux, 1.25 hr	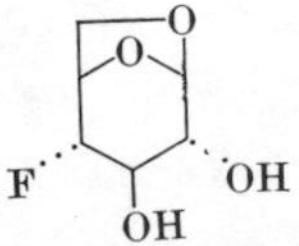 O, O, F, OH, OH (56)	150
C_{14}	α-Bromoheptan-*p*-toluidide	KF	Diglycol, 125–130°, 2 hr	α-Fluoroheptan-*p*-toluidide (62)	121
	2-Bromo-2-phenylacetanilide	KF	Diglycol, 125–130°, 2 hr	2-Fluoro-2-phenylacetanilide (32)	121
	2-Chloro-2-phenylacetanilide	KF	Diglycol, 125–130°, 2 hr	2-Fluoro-2-phenylacetanilide (25)	121
	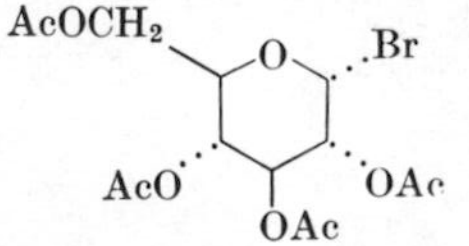 $AcOCH_2$, O, Br, AcO, OAc, OAc	AgF	CH_3CN, room t, shake overnight	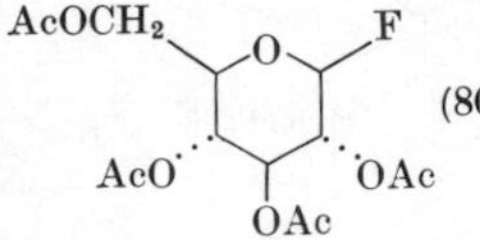 $AcOCH_2$, O, F, AcO, OAc, OAc (80)	149

	Substrate	Fluoride	Conditions	Product(s) (Yield %)	Refs.
	AcOCH$_2$, AcO, OAc, OAc, Br (pyranosyl bromide)	AgF	CH_3CN, room t, shake overnight	AcOCH$_2$, AcO, OAc, OAc, F (pyranosyl fluoride) (—)	149
	AcOCH$_2$, AcO, OAc, OAc, Cl (pyranosyl chloride)	$AgBF_4$	Ether, 0°, 15 min	AcOCH$_2$, AcO, OAc, OAc, F (pyranosyl fluoride), **A** (48); AcOCH$_2$, AcO, OAc, OAc, F (pyranosyl fluoride) **B** (33)	123
		$AgBF_4$	$ClCH_2CH_2Cl$, 0°, 1.5 hr	**A** (30), AcOCH$_2$, AcO, OAc, OAc, OAc (pyranose tetraacetate) (10)	123
C_{15}	α-Tosyloxyacetanilide	KF	Diglycol, 90–100°, 40 min	α-Fluoroacetanilide (54)	121

Note: References 301–386 are on pp. 404–406.

[a] A stoichiometric excess of fluoride was normally used.

TABLE XII. MONOFLUOROSUBSTITUTION OF ORGANIC HALIDES OR SULFONIC ESTERS (*Continued*)

	Halide or Ester	Metal Fluoride[a]	Condition(s)	Product(s) and Yield(s) (%)	Ref.
C_{16}	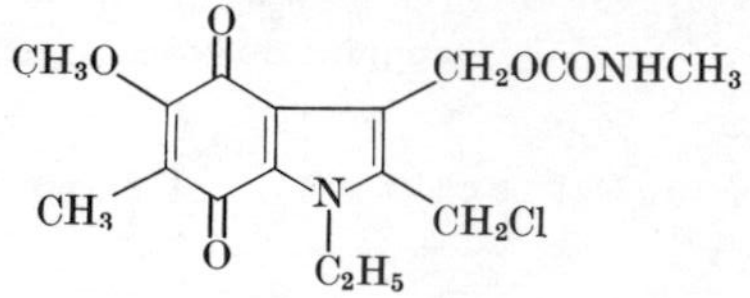 CH_3O, $CH_2OCONHCH_3$, CH_3, CH_2Cl, N–C_2H_5 (O, O)	AgF	CH_3CN, reflux, 2 hr	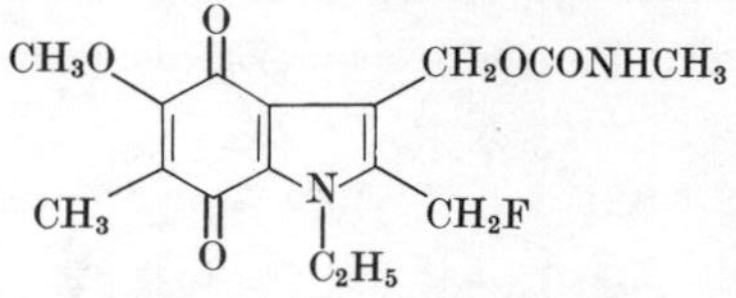 CH_3O, $CH_2OCONHCH_3$, CH_3, CH_2F, N–C_2H_5 (O, O) (30)	119
	α-Tosylpropionanilide	KF	Diglycol, 95–100°, 40 min	α-Fluoropropionanilide (63)	121
C_{18}	α-Bromo-γ-phthalimidobutyramide	KF	Diglycol, 130°, 2.5 hr, 25 mm	α-Fluoro-γ-phthalimidobutyramide (55)	337
C_{19}	N-Benzyl-α-bromo-γ-phthalimido-butyramide	KF	Diglycol, 130°, 2.5 hr, 25 mm	N-Benzyl-α-fluoro-γ-phthalimido-butyramide (40)	337
	Methyl 12-tosyloxystearate	$(n\text{-}C_4H_9)_4NF$	CH_3CN, room t	Methyl 12-fluorostearate (40)	153b
	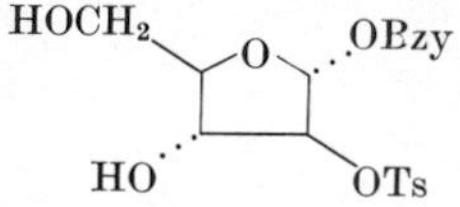 $HOCH_2$, O, OBzy, HO, OTs	KF	CH_3CONH_2, 200°	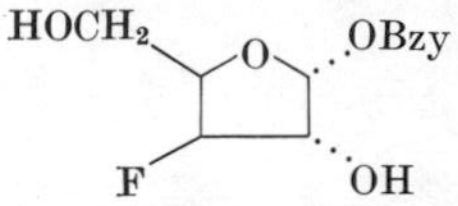 $HOCH_2$, O, OBzy, F, OH (50) (rearrangement product)	203
	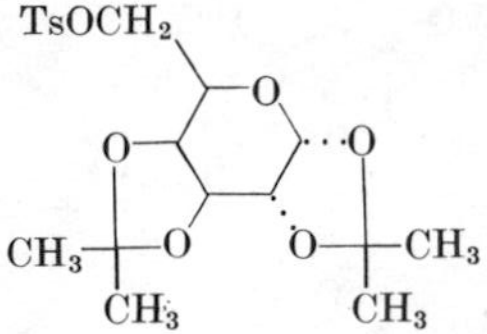 $TsOCH_2$, O, O, O, O, O, CH_3, CH_3, CH_3, CH_3	$(n\text{-}C_4H_9)_4NF$	THF, 80°, 24 hr	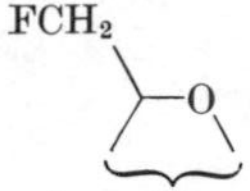 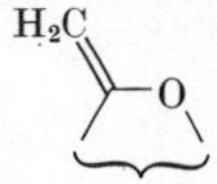 FCH_2, O, **A** (29), H_2C, O, **B** (21)	127
		$(n\text{-}C_4H_9)_4NF$	CH_3CN, reflux, 10 hr	**A** (69), **B** (31)	127
		$(n\text{-}C_4H_9)_4NF$	DMF, 50–60°, 3.5 d	**A** (53), **B** (17)	127

		$(n\text{-}C_4H_9)_4NF$	Diglycol, 100°, 24 hr	No reaction	127
	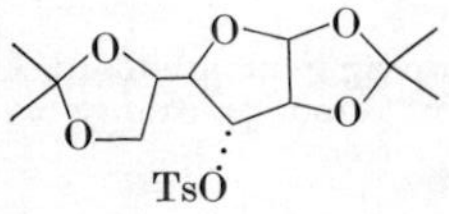 TsO	$(n\text{-}C_4H_9)_4NF$	CH_3CN, reflux, 3 5 d	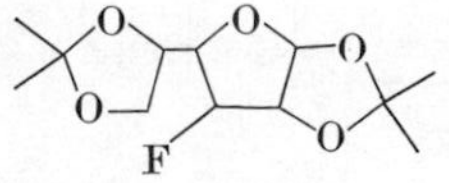 F (71)	127
	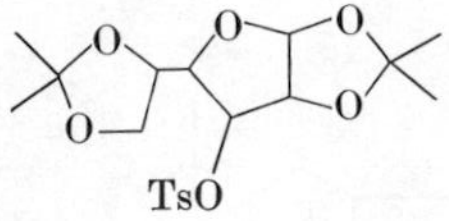 TsO	$(n\text{-}C_4H_9)_4NF$	CH_3CN, 60–65°, 48 hr	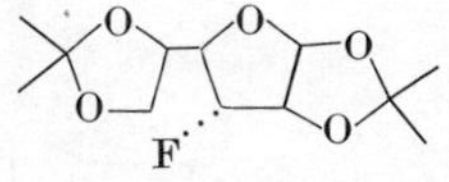 F (not isolated), 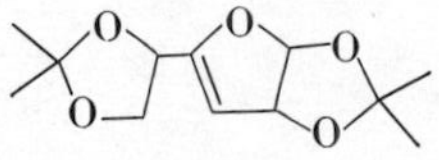 (not isolated)	338
C_{20}	α-Bromo-ε-phthalimidocaproanilide	KF	Diglycol, 130°, 2.5 hr, 25 mm	α-Fluoro-ε-phthalimidocaproanilide (60)	337
C_{21}	21-Iodoprogesterone[b]	AgF	CH_3CN, 30–40°	21-Fluoroprogesterone (63)	339
	19-Tosyl-Δ^4-10α-androsten-17β-ol-3-one acetate	LiF	DMF, reflux, 5 hr	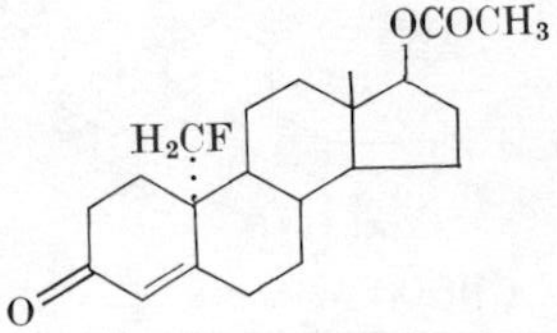 $OCOCH_3$, H_2CF, O (—)	153a

Note: References 301–386 are on pp. 404–406.

[a] A stoichiometric excess of fluoride was normally used.
[b] This older example is included to permit comparison with newer fluorinations.

TABLE XII. MONOFLUOROSUBSTITUTION OF ORGANIC HALIDES OR SULFONIC ESTERS (*Continued*)

	Halide or Ester	Metal Fluoride[a]	Condition(s)	Product(s) and Yield(s) (%)	Ref.
C_{21} (*contd.*)	21-Iodo-4,9(11)-pregnadiene-3,20-dione[b]	AgF	CH_3CN/H_2O, 3 hr, 30–40°	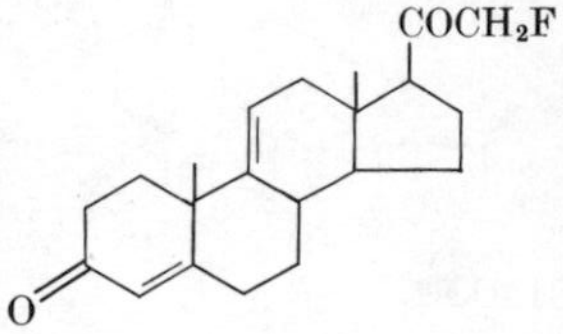 $COCH_2F$ O (—)	307
	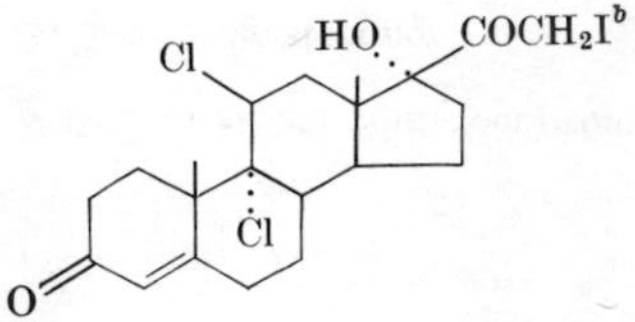 Cl HO $COCH_2I$[b] Cl O	AgF	CH_3CN/H_2O, 30–40°, 4 hr	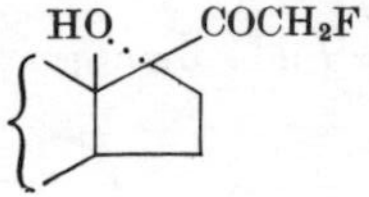 HO $COCH_2F$ (—)	308
	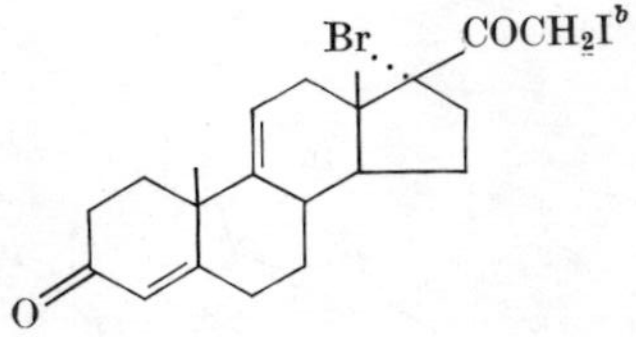 Br $COCH_2I$[b] O	AgF	CH_3CN/H_2O, 30–40°, 4 hr	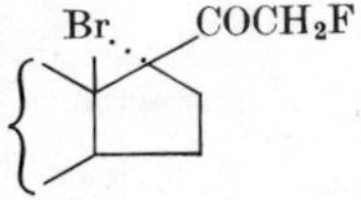 Br $COCH_2F$ (—)	308
C_{22}	6α-Fluoro-21-iodo-17α-methyl-4,9(11)-pregnadiene-3,20-dione[b]	AgF	CH_3CN/H_2O, 30–40°, 4 hr	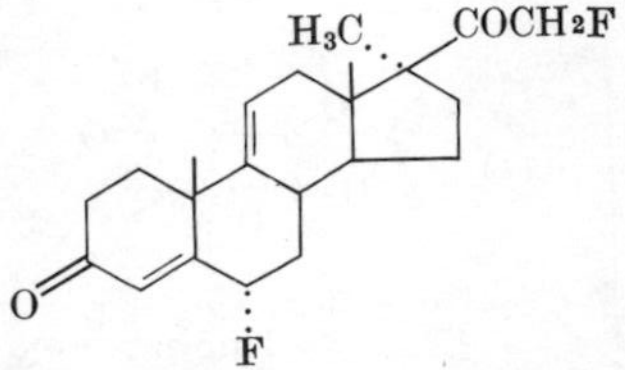 H_3C $COCH_2F$ O F (—)	309

	Substrate	Fluoride	Conditions	Product(s) and Yield(s) (%)	Refs.
C_{44}	2,3-Di-O-benzyl-4,6-di-O-methylsulfonyl-α-D-glucopyranosyl 2,3-Di-O-benzyl-4,6-di-O-methylsulfonyl-α-D-glucopyranoside	$(n\text{-}C_4H_9)_4NF$	CH_3CN, reflux, 65 min	$[\text{FCH}_2, \text{MsO}\cdots, \text{BzyO}, \text{OBzy pyranosyl}]_2\cdots O$ (53)	340a
	$[\text{YCH}_2, \text{X}, \text{BzyO}, \text{OBzy pyranosyl}]_2\cdots O$	$(n\text{-}C_4H_9)_4NF$	CH_3CN	$[\text{YCH}_2, \text{X}, \text{BzyO}, \text{OBzy pyranosyl}]_2\cdots O$	340b
	A: X = MsO···, Y = $(C_6H_5)_3CO$		**A**, 6 days	from **A**: X = F—, Y = $(C_6H_5)_3CO$ (21)	
	B: X = MsO···, Y = MsO		**B**, 5 days	from **B**: X = F—, Y = F (62)	
	C: X = MsO—, Y = $(C_6H_5)_3CO$		**C**, 18 hr	from **C**: X = F···, Y = $(C_6H_5)_3CO$ (low)	
	D: X = MsO—, Y = MsO		**D**, 4 days	from **D**: X = F···, Y = F (low; converted to diacetate for isolation)	

Note: References 301–386 are on pp. 404–406.

[a] A stoichiometric excess of fluoride was normally used.
[b] This older example is included to permit comparison with newer fluorinations.

TABLE XIII. Difluorotriphenylphosphorane or Trifluorodiphenylphosphorane for Replacement of Alcoholic Hydroxyl by Fluorine

$$ROH + (C_6H_5)_3PF_2 \xrightarrow[8\text{ hr}]{150^\circ} RF + (C_6H_5)_3PO + HF$$

$$ROH + (C_6H_5)_2PF_3 \xrightarrow[8\text{ hr}]{150^\circ} RF + (C_6H_5)_2POF + HF$$

	Alcohol	Fluorinating Reagent, $R = C_6H_5$	Ratio F Reagent/ Alcohol	Product and Yield (%)	Ref.
C_3	3-Chloropropanol	R_3PF_2	2.00	$Cl(CH_2)_3F$ (64)	112
		R_2PF_3	2.2	,, (64)	113a
C_4	1,4-Butanediol	R_2PF_3	0.35	Tetrahydrofuran (91)	113a
C_5	1-Pentanol	R_3PF_2	1.85	n-$C_5H_{11}F$ (60)	112
		R_2PF_3	2.2	,, (62)	113
	2-Pentanol	R_3PF_2	1.87	n-$C_3H_7CHFCH_3$ (52)	112
		R_2PF_3	2.2	,, (54)	113a
	Isoamyl alcohol	R_3PF_2	2.00	$(CH_3)_2CHCH_2CH_2F$ (53)	112
		R_3PF_2	0.32	$[(CH_3)_2CHCH_2CH_2]_2O$ (64)	112
	Neopentyl alcohol	R_3PF_2	1.90	$(CH_3)_2CH{=}CHCH_3$ (48)	112
C_6	Cyclohexanol	R_3PF_2	0.88	Cyclohexene (66)	112
	1,6-Hexanediol	R_3PF_2	2.81	1,6-Difluorohexane (12)	112
C_7	Benzyl alcohol	R_3PF_2	2.00	$C_6H_5CH_2F$ (33)	112
		R_2PF_3	2.2	,, (32)	113a
C_8	1-Octanol[a]	R_3PF_2	2.02	n-$C_8H_{17}F$ (78)	112
		R_2PF_3	2.2	,, (76)	113a
	2-Phenylethanol	R_3PF_2	2.05	$C_6H_5CH_2CH_2F$ (73)	112
		R_2PF_3	>2.00	,, (52)	113a

Note: References 301–386 are on pp. 404–406.

[a] A reaction temperature of 170° was used.

TABLE XIV. Alkyl Fluorides from Alkyl Trimethylsilyl Ethers and Phenyl- and Ethyl-tetrafluorophosphorane[a]

$$ROSiCH_3 + C_6H_5PF_4(+C_2H_5PF_4) \rightarrow RF + (CH_3)_3SiF + C_6H_5POF_2$$

C Atoms in R	R Group in $ROSi(CH_3)_2$	Reagent[b]	Condition(s)[c]	Product(s) and Yield(s) (%)	Ref.
C_1	Methyl	$C_6H_5PF_4 + C_2H_5PF_4$	—	CH_3F (100)	130
C_2	Ethyl	$C_6H_5PF_4 + C_2H_5PF_4$	—	C_2H_5F (100)	130
	2-Chloroethyl	$C_6H_5PF_4$	Ice-cooled	$Cl(CH_2)_2F$ (95)	129
	2-Bromoethyl	$C_6H_5PF_4$	Ice-cooled	$Br(CH_2)_2F$ (10), $Br(CH_2)_2Br$ (40)	129
C_3	Isopropyl	$C_6H_5PF_4 + C_2H_5PF_4$	—	$(CH_3)_2CHF$ (100)	130
	3-Chloropropyl	$C_6H_5PF_4$	Ice-cooled	$Cl(CH_2)_3F$ (95), $CH_2{=}CHCH_2Cl$ (5)	129
C_4	*t*-Butyl	$C_6H_5PF_4$	Ice-cooled	*t*-C_4H_9F (90)	129
		$C_6H_5PF_4 + C_2H_5PF_4$	—	,, (100)	130
	4-Chlorobutyl	$C_6H_5PF_4$	Ice-cooled	$Cl(CH_2)_4F$ (75), $CH_2{=}CHCH_2CH_2Cl$ (25)	129
C_5	1-Pentyl	$C_6H_5PF_4$	Ice-cooled	1-Fluoropentane (50) + 1-pentene (50)	129
	2-Pentyl	$C_6H_5PF_4$	Ice-cooled	2-Fluoropentane (60), pentenes (40)	129
		$C_6H_5PF_4$	—	2-Fluoropentane (100)	130
	3-Pentyl	$C_6H_5PF_4$	Ice-cooled	3-Fluoropentane (60), 2-pentene (40)	129
	Cyclopentyl	$C_6H_5PF_4$	Ice-cooled	Fluorocyclopentane (40), cyclopentene (60)	129

Note: References 301–386 are on pp. 404–406.

[a] A thorough study of this reaction with extensive examples was published too late for inclusion in the table.[129b] Direction reaction of phenyltetrafluorophosphorane with 1- and 2-pentanol has also been reported.[113b]

[b] $C_6H_5PF_4 + C_2H_5PF_4$ means a 1:1 molar ratio of the phosphoranes.

[c] The reaction time is that required to add the ether dropwise and slowly to the reagent. Unless otherwise specified, no solvent was used and the reaction was run at room temperature.

TABLE XIV. ALKYL FLUORIDES FROM ALKYL TRIMETHYLSILYL ETHERS AND PHENYL- AND ETHYL-TETRAFLUOROPHOSPHORANE (*Continued*)

$$ROSiCH_3 + C_6H_5PF_4(+ C_2H_5PF_4) \rightarrow RF + (CH_3)_3SiF + C_6H_5POF_2{}^a$$

C Atoms in R	R Group in $ROSi(CH_3)_2$	Reagent[b]	Condition(s)[c]	Product(s) and Yield(s) (%)	Ref.
C_6	Cyclohexyl	$C_6H_5PF_4$	Ice-cooled	Fluorocyclohexane (30), cyclohexene (70)	129
C_7	Cyclohexylmethyl	$C_6H_5PF_4 + C_2H_5PF_4$	—	Cyclohexylfluoromethane (100)	129
C_8	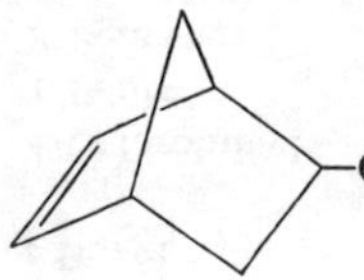 [bicyclic structure]—CH_2—	$C_6H_5PF_4 + C_2H_5PF_4$	—	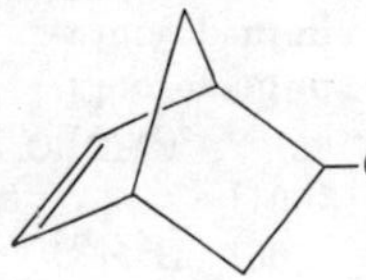 [bicyclic structure]—CH_2F (100)	130
C_{27}	Cholesterol	$C_6H_5PF_4$	Ice-cooled	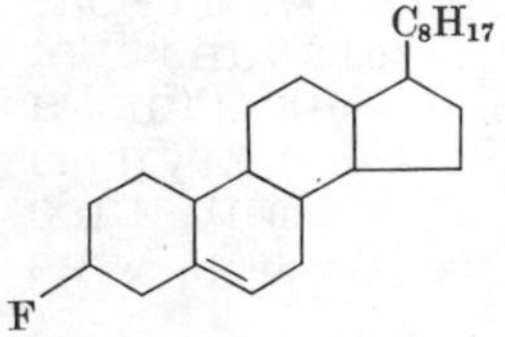 [steroid structure with F and C_8H_{17}]	129

Note: References 301–386 are on pp. 404–406.

[a] A thorough study of this reaction with extensive examples was published too late for inclusion in the table.[129b] Direction reaction of phenyltetrafluorophosphorane with 1- and 2-pentanol has also been reported.[113b]

[b] $C_6H_5PF_4 + C_2H_5PF_4$ means a 1:1 molar ratio of the phosphoranes.

[c] The reaction time is that required to add the ether dropwise and slowly to the reagent. Unless otherwise specified, no solvent was used and the reaction was run at room temperature.

TABLE XV. SUBSTITUTION REACTIONS OF HYDROGEN FLUORIDE ON CARBOHYDRATE ESTERS AND STEROID ALCOHOLS

	Reactant	Condition(s)	Product(s) and Yield(s) (%)	Refs.
C_{10}	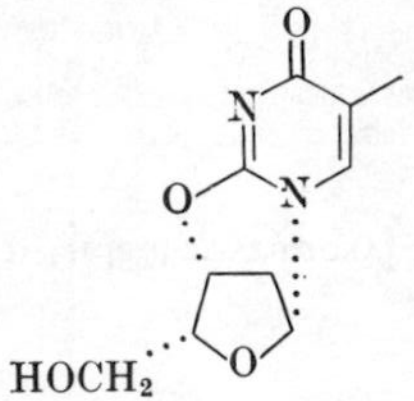 O, N, N, O, HOCH₂, O	AlF_3, 0.1% HF, dioxane, 150–170°, 4–5 hr	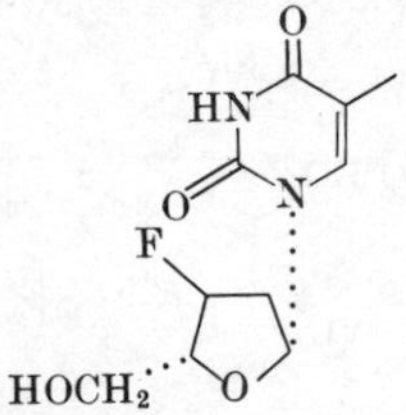 O, HN, O, N, F, HOCH₂, O **A** (34)	341a
		KHF_2, diglycol, 190°, 90 min	**A** (14)	341a
	1,2,5-Tri-O-acetyl-β-D-glucoso-3-6-lactone	HF, −78°, warm to room t, 20 min	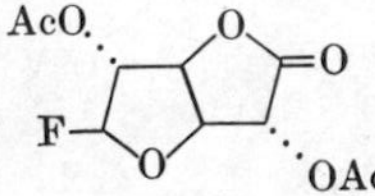 AcO, O, O, F, O, OAc (–)	341b
C_{12}	3,4,6-Tri-O-acetyl-2-deoxy-2-fluoro-β-D-mannopyranosyl fluoride	HF, −10 to 0°, 15 min room t, 10 min	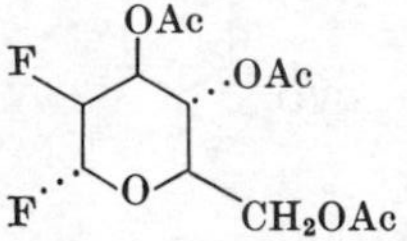 OAc, F, OAc, F, O, CH_2OAc (77) (Isomerization only)	176
C_{14}	1,2,4,6-Tetra-O-acetyl-3-deoxy-3-fluoro-α and β-D-glucopyranoses	HF, −10°, 20 min; room t, 10 min	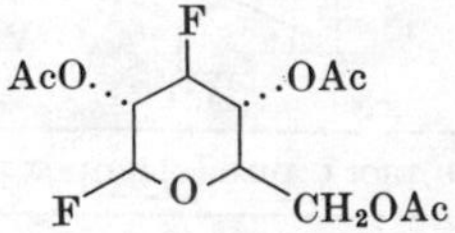 F, AcO, OAc, F, O, CH_2OAc (–)	335a

Note: References 301–386 are on pp. 404–406.

TABLE XV. Substitution Reactions of Hydrogen Fluoride on Carbohydrate Esters and Steroid Alcohols (*Continued*)

Reactant	Condition(s)	Product(s) and Yield(s) (%)	Refs.
1,2,3,6-Tetra-O-acetyl-4-deoxy-4-fluoro-β-D-galactopyranose	HF, −10°, 20 min	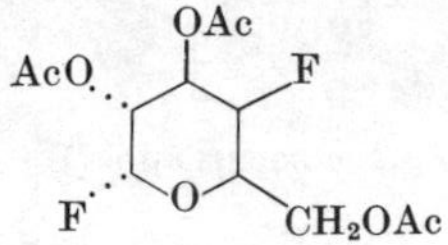 AcO, OAc, F, F, O, CH_2OAc (68)	335b
D-Glucose pentaacetate	HF, −78°, warm to room t, 20 min	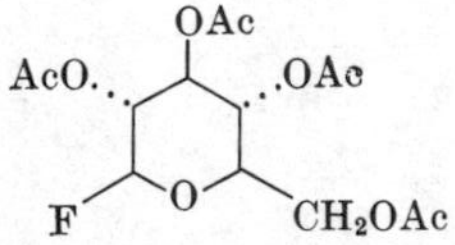 AcO, OAc, OAc, F, O, CH_2OAc (74)	149
D-Mannopyranosyl pentaacetate	HF, −78°, warm to room t, 20 min	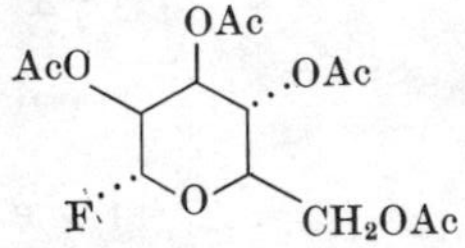 AcO, OAc, OAc, F, O, CH_2OAc (—)	149
D-Galactopyranosyl pentaacetate	HF, −78°, warm to room t, 20 min	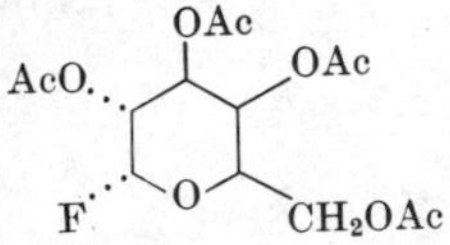 AcO, OAc, OAc, F, O, CH_2OAc (—)	149
D-Allose pentaacetate	HF, −78°, warm to room t, 20 min	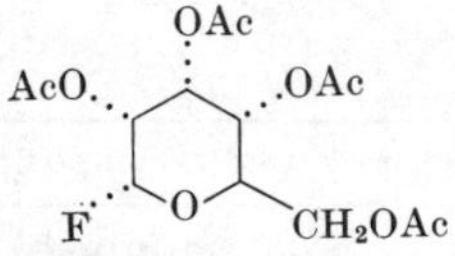 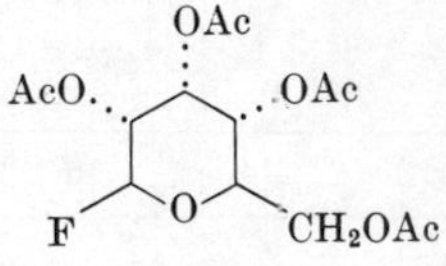 AcO, OAc, OAc, F, O, CH_2OAc , AcO, OAc, OAc, F, O, CH_2OAc Nearly equimolar mixtures (91)	149

C_{14} (*contd.*)	2-O-Acetyl-3,6-anhydro-5-O-benzoyl-β-L-furanose	HF, −78°, warm to room t, 20 min	AcO, F, O, O, OBz (70) 2-O-Acetyl-3,6-anhydro-5-O-benzoyl-β-L-furanosyl fluoride (—)	341b
C_{22}	HO, CH_2OAc, X, O X = OH, X = H	70% HF by weight in pyridine	CH_2OAc, X, F, O X = OH (—), X = H (—)	177
C_{21}	$COCH_3$, HO	HF, CH_2Cl_2/THF 6°, 18 hr	$COCH_3$, F (12)	170
C_{22}	HO, $COCH_2OMs$, CH_3, O, F	KHF_2, DMF, 110°, 15 hr	$COCH_2F$, CH_3 (—)	342

Note: References 301–386 are on pp. 404–406.

TABLE XV. SUBSTITUTION REACTIONS OF HYDROGEN FLUORIDE ON CARBOHYDRATE ESTERS AND STEROID ALCOHOLS (*Continued*)

	Reactant	Condition(s)	Product(s) and Yield(s) (%)	Refs.
C_{27}	BzO, O, =O, BzO–, O, OBz	HF, −78°, warm to room t, 20 min	BzO, O, =O, F–, O, OBz (46)	341b
	HO, OBz, BzO···, –CH_2OBz, O	HF, 15 min, −10°	HO, OBz, F–, –CH_2OBz, O **A**, BzO, OH, F–, –CH_2OBz, O **B**, Mixture of **A** + **B** (17) HO, OBz, F···, –CH_2OBz, O **C** (44)	175
		HF, 20 hr, room t	Mixture of **A** + **B** (45), **C** (15), BzO, OBz, F–, –CH_2OBz, O (5)	175
C_{28}	BzO, OBz, CH_3O–, –CH_2OBz, O	HF, 24 hr, room t	Mixture of **A** + **B** (45), **C** (12)	175

C_{29}	BzO, OBz, AcO—(furanose O)—CH_2OBz	HF, 1 hr, room t	BzO, OBz, F—(furanose O)—CH_2OBz (51), BzO, OBz, F—(furanose O)—CH_2OAc (17) (Note AcO transfer)	175
	AcO, OBz, BzO—(furanose O)—CH_2OBz	HF, 10 min, room t	AcO, OBz, F—(furanose O)—CH_2OBz (93)	175
C_{33}	Tetra-O-benzoyl-β-D-ribofuranose	HF, 10 min, room t[a]	BzO, OBz, F—(furanose O)—CH_2OBz (70)	175
C_{34}	1,3,4,6-Tetra-O-benzoyl-2-deoxy-β-D-arabinohexapyranose	HF, 10 min, room t	OBz, OBz, F—(pyranose O)—CH_2OBz (95)	174

Note: References 301–386 are on pp. 404–406.

[a] When the reaction was run for 150 hr, only 9% of the substitution product was obtained. Products from cleavage of 2- and 3-benzoyl groups and from inversion of configuration at the 2 position predominated.

TABLE XVI. Reactions of Hydrogen Fluoride with Oxiranes

$$\text{—C(—O—)C— + HF} \longrightarrow \text{—C(OH)—C(F)—}$$

	Oxirane	Condition(s)[a]	Product(s) and Yield(s) (%)	Refs.
C_4	$CH_3CH—CHCH_3$ (epoxide)	HF, $CHCl_3$/THF (5:1), 0°, 3 hr	No fluorohydrin isolated; polymer (25), 2-butanone (7), starting material (61)	178
C_5	$CH_3CH—C(CN)CH_3$ (epoxide)	HF, $CHCl_3$/THF (5:1), 0°, 3 hr	No fluorohydrin isolated; starting material (74), 3-cyano-2-butanone (—), 2-butanone (—)	178
	Cyclopentene oxirane	$(i\text{-}C_3H_7)_2NH\cdot 3\ HF$, 110°, 18 hr	2-fluorocyclopentanol (structure: ···OH, —F) (65)	163b
	$C_2H_5(F)C—CHCH_3$ (epoxide), *cis* + *trans*	HF, trace BF_3, ether room t	$C_2H_5CF_2CHOHCH_3$ (>90)	186
	1,1-Dimethyl-2-cyanooxirane	HF, BF_3·ether, ether room t	$(C_3)_2CFCHOHCN$ (75)	164
C_6	1-Ethyl-2-cyanooxirane	HF, BF_3·ether, ether room t	$C_2H_3CHFCHOHCN$ (82)	164
	1,1,2-Trimethyl-2-cyanooxirane	HF, BF_3·ether, ether room t, 18 hr	$(CH_3)_2CF(OH)(CN)CH_3$ (81)	164
	1-Cyano-1-cyclopentene oxide	HF, BF_3·ether, ether, room t, 48 hr	2-Fluorocyclopentanone cyanohydrin (50)	164
	cis-1-Ethyl-2-methyl-2-cyanooxirane	HF BF_3·ether ether room t, 4 hr	$C_2H_5CHFC(OH)(CN)CH_3$ (48)	164

	Reactant	Conditions	Product(s) (Yield, %)	Refs.
	trans-1-Ethyl-2-methyl-2-cyanooxirane	HF, BF_3·ether, ether, room t, 4 hr	$C_2H_5CHFC(OH)(CN)CH_3$ (48)	164
	Cyclohexene oxide	HF, $CHCl_3$/THF (5:1) 0°, 3 hr	No fluorohydrin isolated; polymer (60), cyclohexanone (19), starting material (10)	178
		$(i\text{-}C_3H_7)_2NH$·3 HF, 120°, 18 hr	*trans*-2-Fluorocyclohexanol (75)	163b
	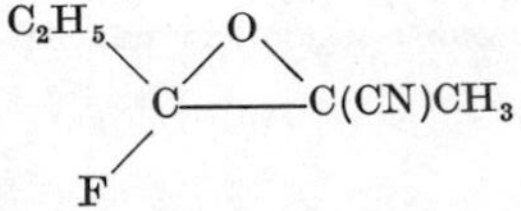 C_2H_5, O, C—C(CN)CH_3, F; *cis* + *trans*	1:1 HF:BF_3, ether, room t	$C_2H_5CF_2C(OH)(CN)CH_3$ (9)	186
	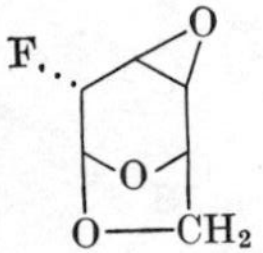 F, O, O, O—CH_2	HF, KF·HF, glycol, reflux, 2 hr	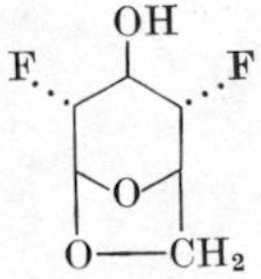 OH, F, F, O, O—CH_2 (—)	343
C_7	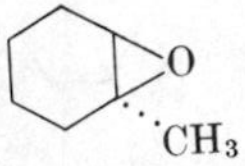 O, CH_3	HF, $CHCl_3$/THF (5:1) 0°, 3 hr	No fluorohydrin isolated; 2-methylcyclohexanone (55), polymer (10), starting material (2)	178
	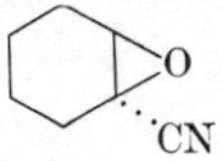 O, CN	HF, $CHCl_3$/THF (5:1), 0°, 3 hr	No fluorohydrin isolated; 2-cyanocyclohexanone (63), polymer (20)	178
		HF, BF_3·ether, ether, room t, 48 hr	2-Fluorocyclohexanone cyanohydrin (43)	164
	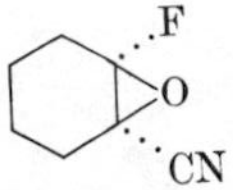 F, O, CN	6:1 HF:BF_3, ether, room t, 18 hr	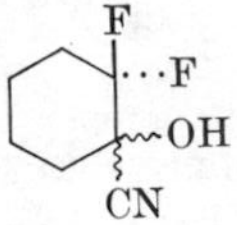 F, F, OH, CN (83)	186

Note: References 301–386 are on pp. 404–406

[a] Solvent ratios are in terms of volume.

TABLE XVI. Reactions of Hydrogen Fluoride with Oxiranes (*Continued*)

	Oxirane	Condition(s)[a]	Product(s) and Yield(s) (%)	Refs.
C_7 (*contd.*)	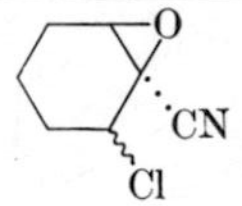	HF, BF_3·ether, ether	2-Chloro-6-fluorocyclohexanone cyanohydrin (75)	173
	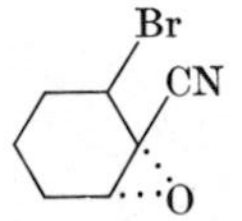	HF, BF_3·ether, ether, 6 hr	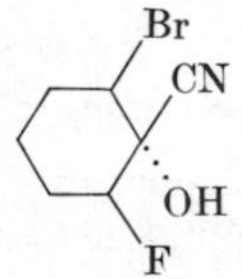 (—)	187
		HF, BF_3·ether, ether	" (75)	173
	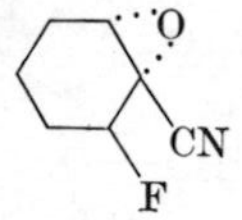	HF, BF_3·ether, ether, 7 hr	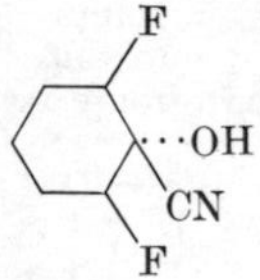 (—)	187
	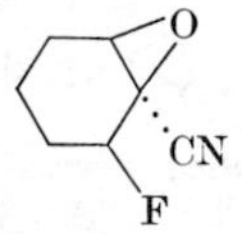	HF, BF_3·ether, ether, 3 hr	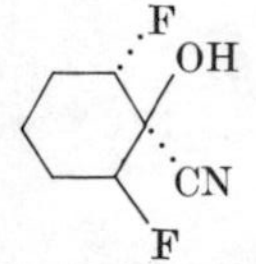 (56)	187
C_8	Phenyloxirane	$(i\text{-}C_3H_7)_2NH \cdot 3\,HF$ 105°, 24 hr	$C_6H_5CHFCH_2OH$ (53), $C_6H_5CHOHCH_2F$ (17)	163
	1-Cyanocyclohept-1-ene oxide	HF, BF_3·ether, ether, room t, 7 hr	2-Fluorocycloheptanone cyanohydrin (38)	164
	1-Methyl-2-cyanocyclohex-1-ene oxide	HF, BF_3·ether, ether, room t, 18 hr	2-Fluoro-2-methylcyclohexanone cyanohydrin (50)	164
	1-Pentyl-2-cyanooxirane	HF, BF_3·ether, ether, room t	$n\text{-}C_5H_{11}CHFCHOHCN$ (53)	164
	3-Methyl-2-cyanocyclohex-1-ene oxide	HF, BF_3·ether, ether, 48 hr	*trans*-2-Fluoro-6-methylcyclohexanone cyanohydrin (35)	173

	Reactant	Conditions[a]	Product(s) (% Yield)	Refs.
C_8 (contd.)	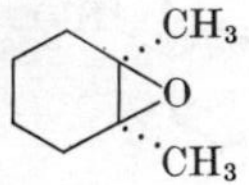 CH_3, O, CH_3	$CHCl_3$/THF (5:1), 0°, 3 hr	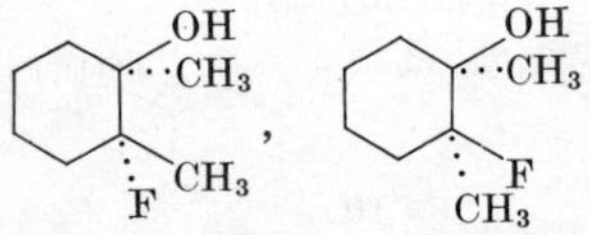 OH, CH_3, F, CH_3 , OH, CH_3, F, CH_3 (Total, 65)	178
	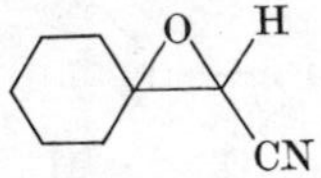 O, H, CN	HF, $CHCl_3$/THF (5:1), 0°, 3 hr	Probably 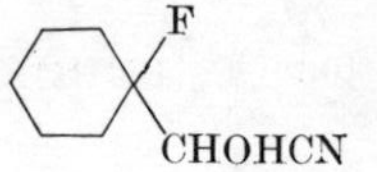 F, CHOHCN (30)	178
		HF, BF_3·ether, room t, 1 hr	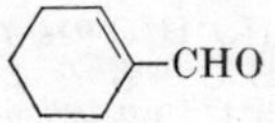 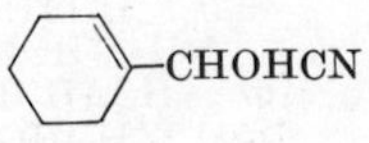 CHO (5), CHOHCN (15)	344
	1,2-Epoxy-3,6-endo-methylene-4-methylcyclohexane	HF, $CHCl_3$/THF (5:1), 0°, 3 hr	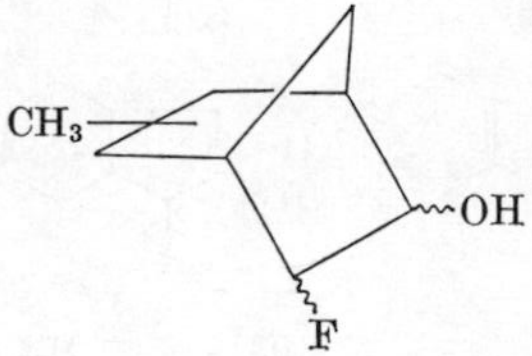 CH_3, OH, F (49)	178
C_9	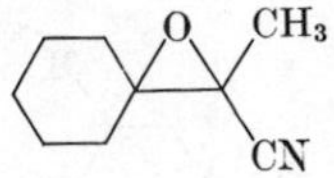 O, CH_3, CN	HF, BF_3·ether, ether, room t, 36 hr	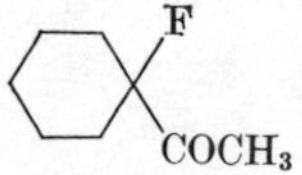 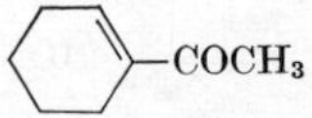 F, $COCH_3$ (55), $COCH_3$ (25)	344

Note: References 301–386 are on pp. 404–406.

[a] Solvent ratios are in terms of volume.

TABLE XVI. REACTIONS OF HYDROGEN FLUORIDE WITH OXIRANES (*Continued*)

	Oxirane	Condition(s)[a]	Product(s) and Yield(s) (%)	Refs.
C_9 (*contd.*)	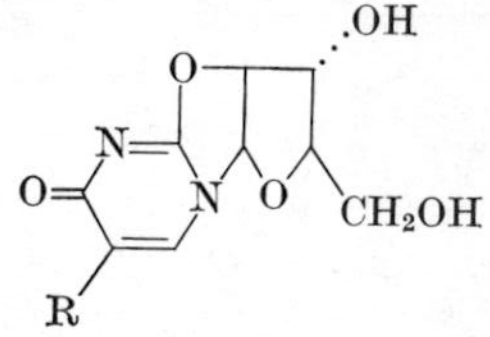 [structure: anhydro nucleoside, O=, N=, N, O, .OH, CH_2OH, R]	HF, dioxane, 95–100°, 16 hr	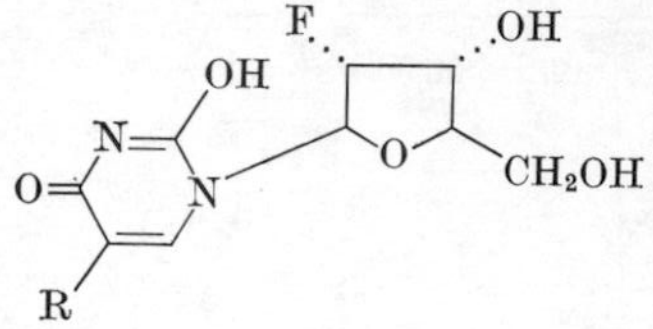 [structure: F, .OH, OH, N=, O=, N, O, CH_2OH, R] R = H (41–46); R = F (20)	345
	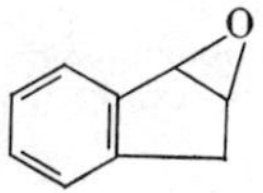 [structure: indene oxide]	$(i\text{-}C_3H_7)_2NH\cdot 3\,HF$, 78°, 5 hr	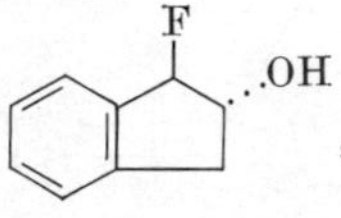 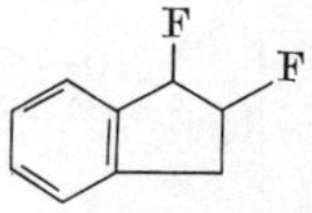 [structure: F, .OH indane] (38), [structure: F, F indane] (21)	163b
	trans-1-Methyl-2-phenyloxirane	$(i\text{-}C_3H_7)_2NH\cdot 3\,HF$, 132°, 23 hr	*erythro*-$C_6H_5CHFCHOHCH_3$ (74), *erythro*-$C_6H_5CHOHCHFCH_3$ (9)	163
	cis-1-Methyl-2-phenyloxirane	$(i\text{-}C_3H_7)_2NH\cdot 3\,HF$, 134°, 23 hr	*threo*-$C_6H_5CHFCHOHCH_3$ (9), *threo*-$C_6H_5CHOHCHFCH_3$ (76)	163
	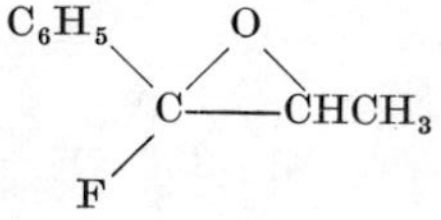 [structure: C_6H_5, O, C—$CHCH_3$, F] *cis* + *trans*	HF, trace BF_3, ether, room t	$C_6H_5CF_2CHOHCH_3$ (>90)	186
C_{10}	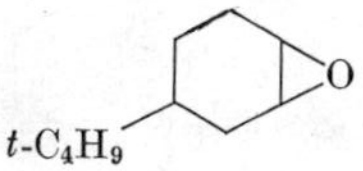 [structure: $t\text{-}C_4H_9$ cyclohexene oxide, O]	HF, $CHCl_3$/THF (5: 1), 0°, 3 hr	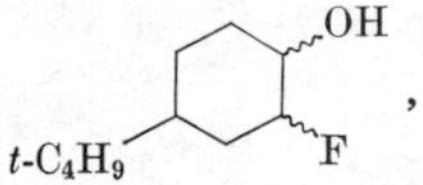 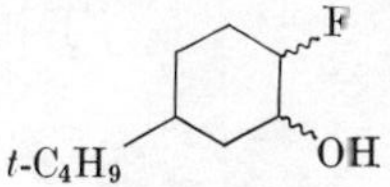 [structure: $t\text{-}C_4H_9$, OH, F], [structure: $t\text{-}C_4H_9$, F, OH] (Total, 50)	178

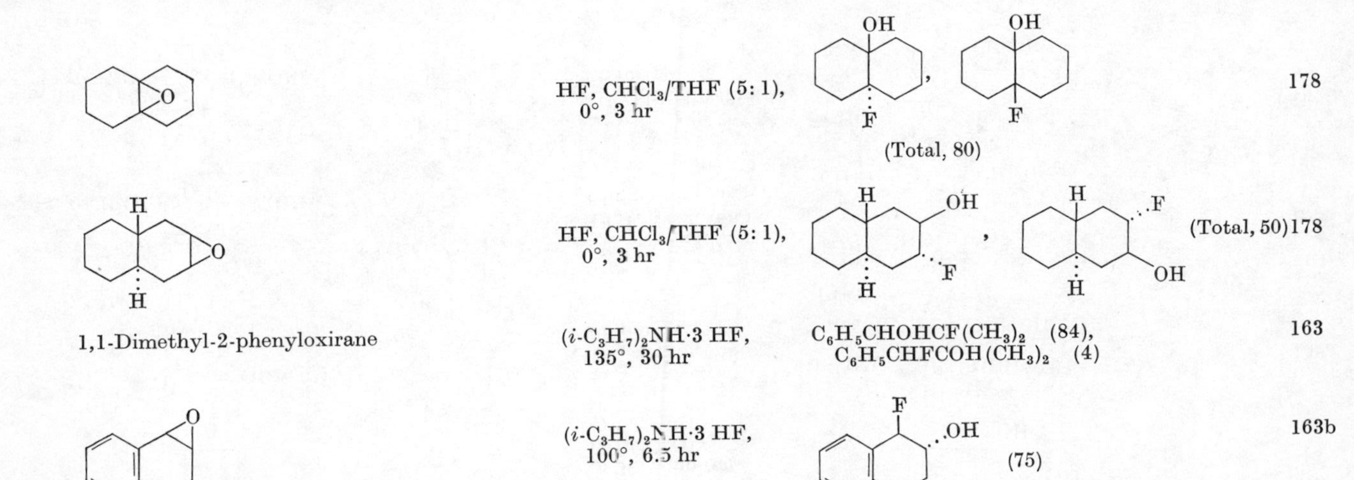

Reactant	Conditions	Product(s) and Yield(s) (%)	Refs.
(structure)	HF, $CHCl_3$/THF (5:1), 0°, 3 hr	(structures) (Total, 80)	178
(structure)	HF, $CHCl_3$/THF (5:1), 0°, 3 hr	(structures) (Total, 50)	178
1,1-Dimethyl-2-phenyloxirane	$(i\text{-}C_3H_7)_2NH \cdot 3\,HF$, 135°, 30 hr	$C_6H_5CHOHCF(CH_3)_2$ (84), $C_6H_5CHFCOH(CH_3)_2$ (4)	163
(structure)	$(i\text{-}C_3H_7)_2NH \cdot 3\,HF$, 100°, 6.5 hr	(structure) (75)	163b
(structure)	$(i\text{-}C_3H_7)_2NH \cdot 3\,HF$, 100°, 9 hr	(structure) (80)	163b
(structure)	HBF_4, BF_3·ether, ether, room t, 48 hr	(structure) (50)	346

Note: References 301–386 are on pp. 404–406.

[a] Solvent ratios are in terms of volume.

TABLE XVI. Reactions of Hydrogen Fluoride with Oxiranes (*Continued*)

	Oxirane	Condition(s)[a]	Product(s) and Yield(s) (%)	Refs.
C_{10} (*contd.*)	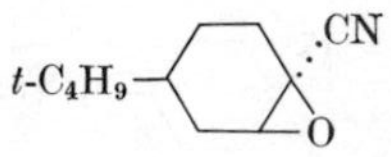 t-C_4H_9, CN, O (structure)	HBF_4, BF_3·ether, ether, room t, 48 hr	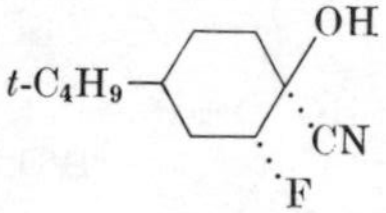 t-C_4H_9, OH, CN, F (structure) (30)	346
	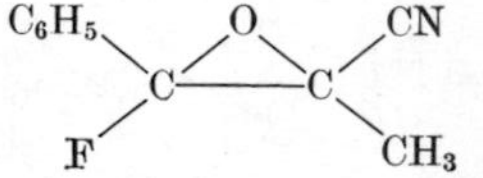 C_6H_5, F, C, O, C, CN, CH_3 (structure) *cis* + *trans*	1:1 HF:BF_3, ether room t	$C_6H_5CF_2C(OH)(CN)CH_3$ (>90)	186
C_{12}	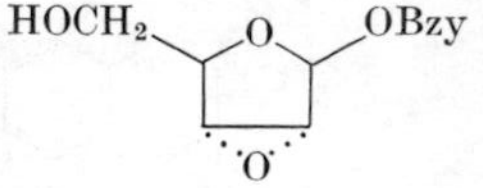 $HOCH_2$, O, OBzy, O (structure)	KHF_2, glycol, reflux, or KF, molten CH_3CONH_2	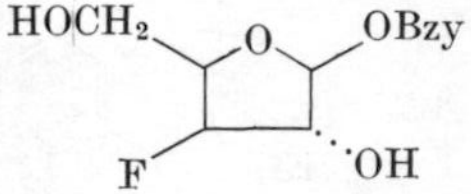 $HOCH_2$, O, OBzy, F, OH (structure)	203
C_{13}	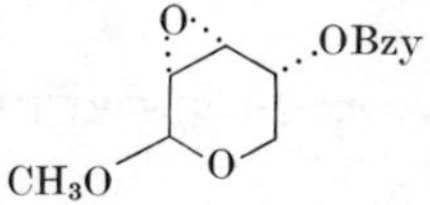 O, OBzy, CH_3O, O (structure)	KHF_2, glycol, reflux, 40 min	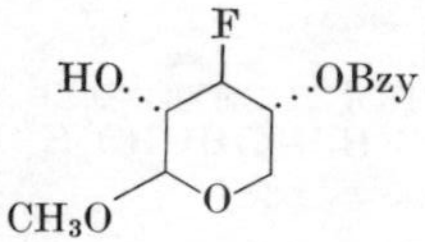 F, HO, OBzy, CH_3O, O (structure) (66)	162a
	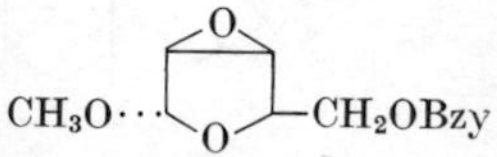 O, CH_3O, O, CH_2OBzy (structure)	KHF_2, glycol, reflux, 1 hr	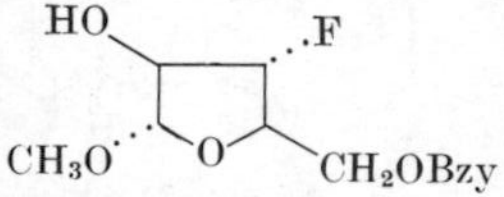 HO, F, CH_3O, O, CH_2OBzy (structure) (30)	162a
	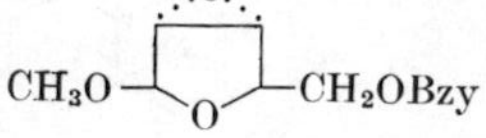 O, CH_3O, O, CH_2OBzy (structure)	KHF_2, glycol, reflux, 1.5 hr	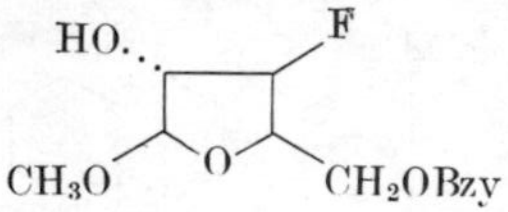 HO, F, CH_3O, O, CH_2OBzy (structure) (—)	162b

	Reactant	Conditions	Product(s) (Yield, %)	Refs.
	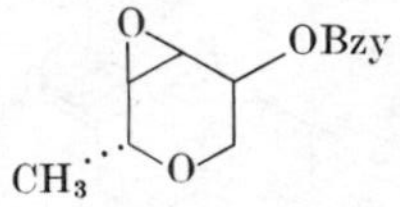	HF, dioxane, 120°, 24 hr	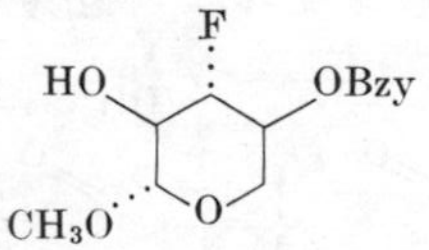 (−)	348
	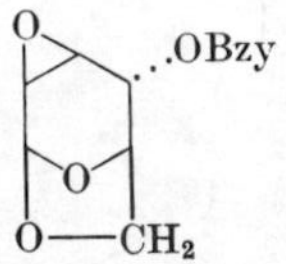	KF·HF, glycol, reflux, 2 hr, CO_2 atm	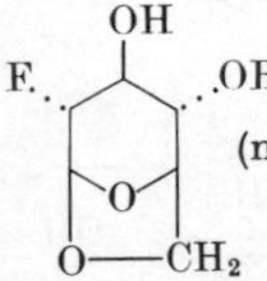 (major), 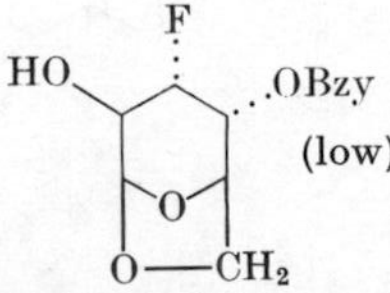 (low)	347
C_{14}	*trans*-1,2-Diphenyloxirane	$(i\text{-}C_3H_7)_2NH \cdot 3\ HF$, 115°, 24 hr	*erythro*-$C_6H_5CHFCHOHC_6H_5$ (62)	163
	cis-1,2-Diphenyloxirane	$(i\text{-}C_3H_7)_2NH \cdot 3\ HF$, 105°, 48 hr	*threo*-$C_6H_5CHFCHOHC_6H_5$ (75)	163
	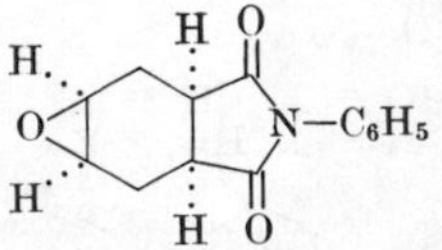	HF, $CHCl_3$/THF (5:1), 0°, 3 hr	No reaction	178
C_{15}	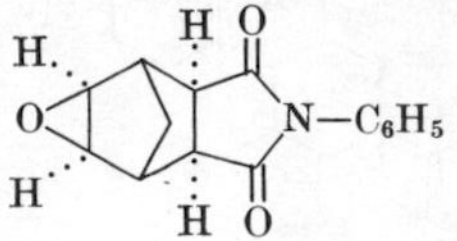	HF, $CHCl_3$/THF (5:1), 0°, 3 hr	No fluorohydrin formed; most starting material recovered.	178
C_{18}	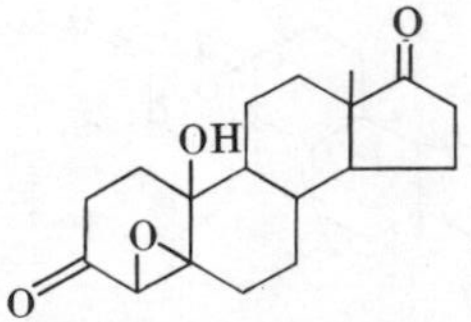	HF, $CHCl_3/C_2H_5OH$ (9:1), 0°	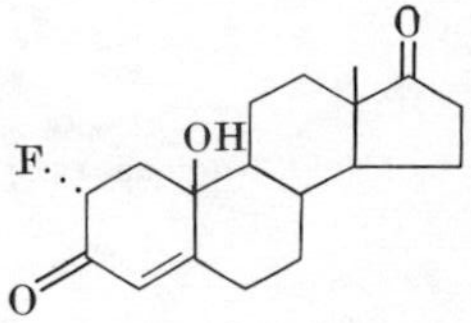 (−)[b]	179

Note: References 301–386 are on pp. 404–406.

[a] Solvent ratios are in terms of volume.

[b] Experimental details and analyses of products were not given. By-products may be present.

TABLE XVI. REACTIONS OF HYDROGEN FLUORIDE WITH OXIRANES (*Continued*)

	Oxirane	Condition(s)[a]	Product(s) and Yield(s) (%)	Refs.
C_{19}	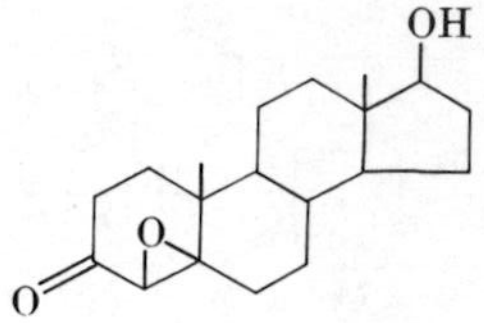 OH, O (structure)	HF, THF or DMF, 23°	No reaction	179
	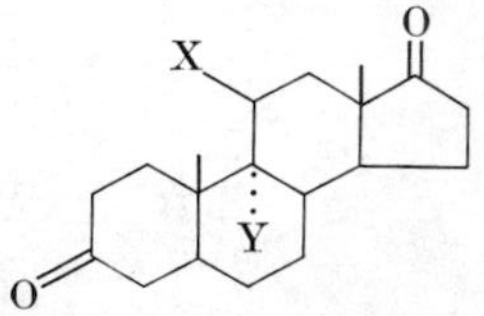 X = OH, Y = H X = H, Y = OH	70% HF by weight in pyridine	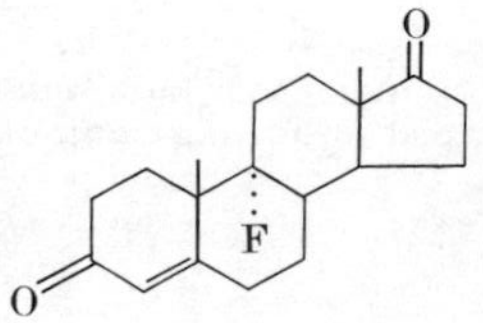 F (structure) (—)	177
	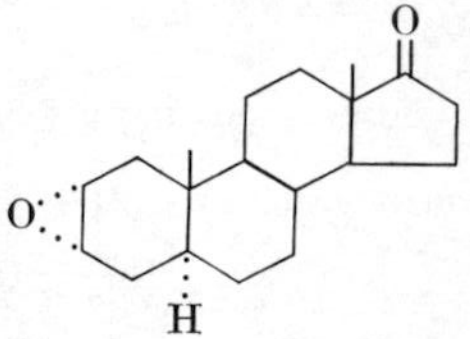 O, H (structure)	HF, $CHCl_3$/THF, −10°, 2.5 hr, standing at room t, 1½ hr	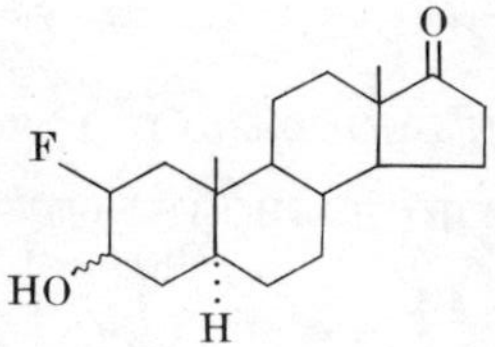 F, HO, H (structure) (—)[b]	349
	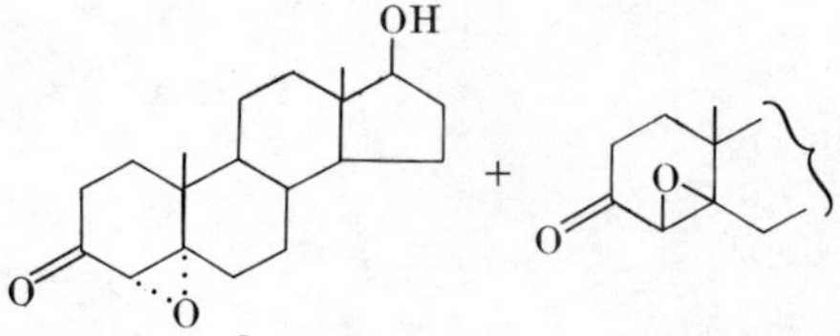 A + B A/B = 1:5	HF, $CHCl_3$ (ethanol-free), room t, 1 hr	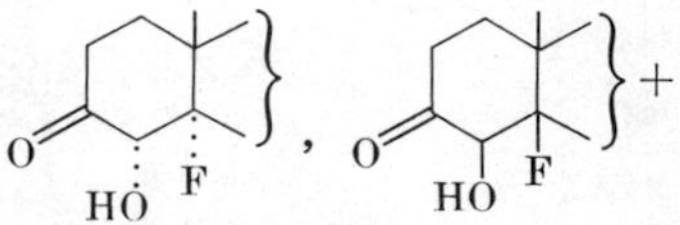 A , B′ + A′/B′ = 1:4 (Total, —)	179

	Reactant	Conditions[a]	Product (Yield, %)	Refs.
	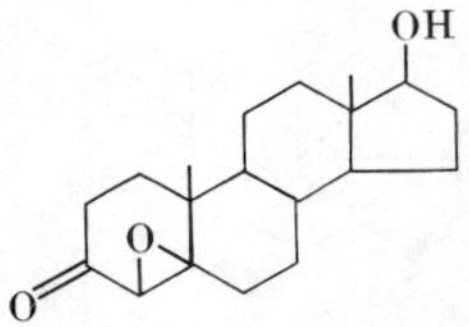	HF, $CHCl_3$/ethanol (9:1), 3 min	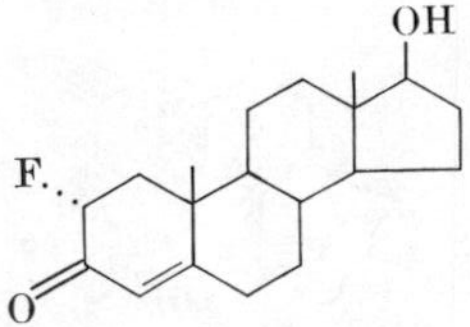 (—)	179
C_{20}	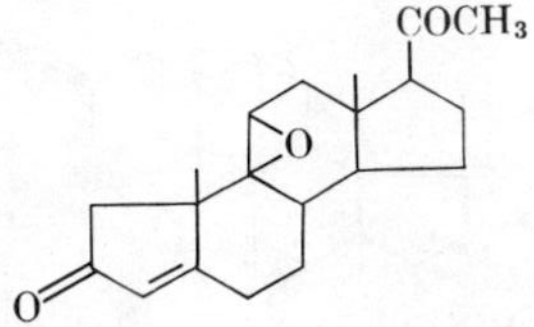	HF, 3:1 $CHCl_3$/THF, 4°, 5 hr	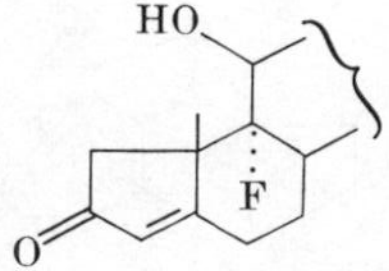 (72)	350
	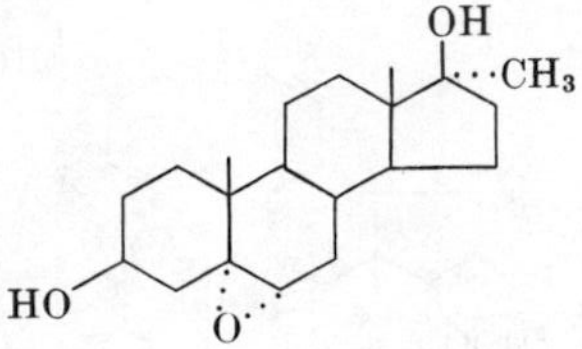	48% aq HF	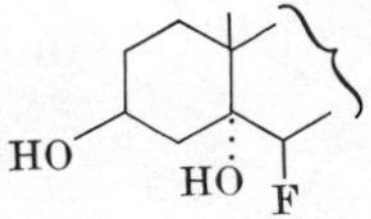 (—)	181
C_{21}	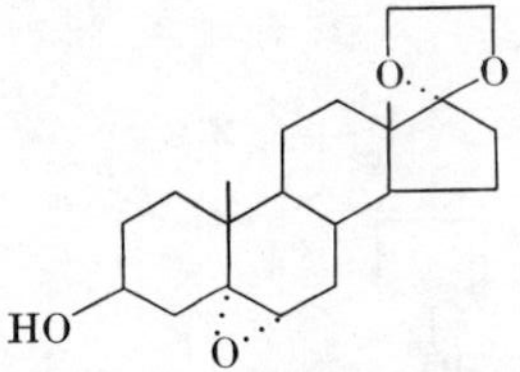	HF, $CHCl_3$	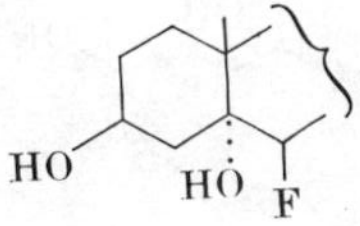 (—)	351

Note: References 301–386 are on pp. 404–406.

[a] Solvent ratios are in terms of volume.

[b] The crude 3-hydroxy compound was oxidized and then isolated as the 3-keto steroid.

TABLE XVI. Reactions of Hydrogen Fluoride with Oxiranes (*Continued*)

	Oxirane	Condition(s)[a]	Product(s) and Yield(s) (%)	Refs.
C_{21} (*contd.*)	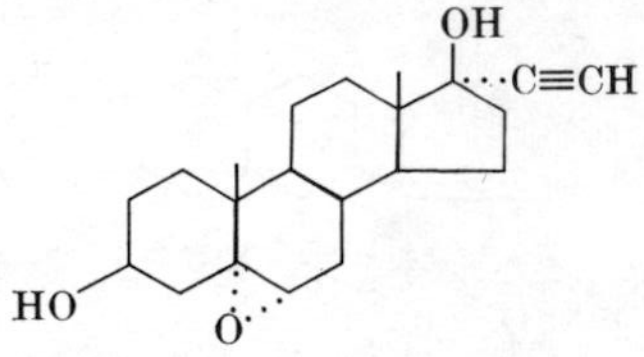 OH, C≡CH, HO, O	48% aq HF 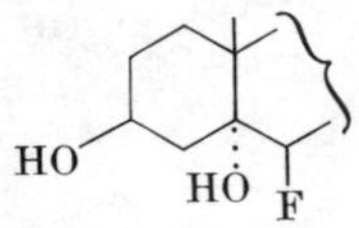	HO, HO, F (—)	181
	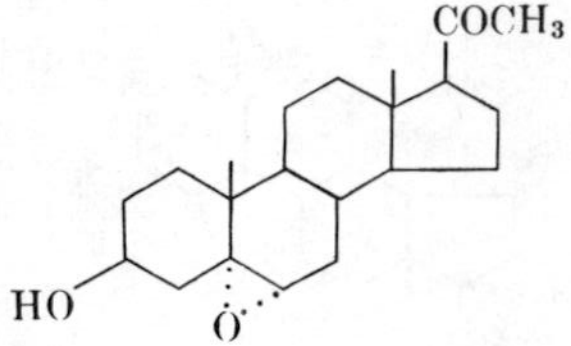 $COCH_3$, HO, O	48% aq HF 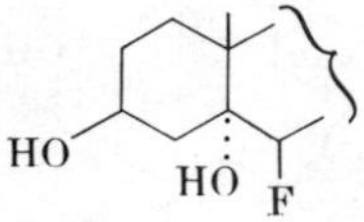	HO, HO, F (—)	181
	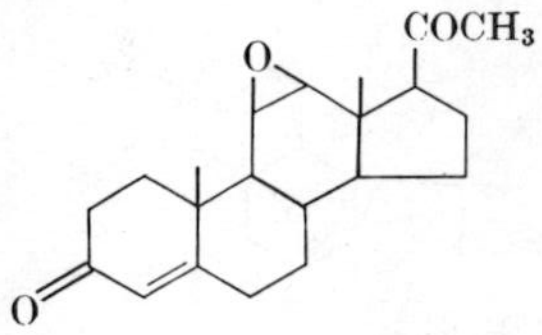 $COCH_3$, O, O	HF, $CHCl_3$ (5% ethanol), 0°, 1.5 hr 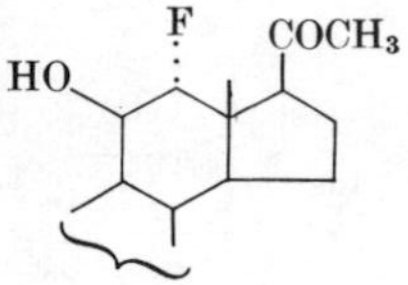	F, $COCH_3$, HO (—)	347
	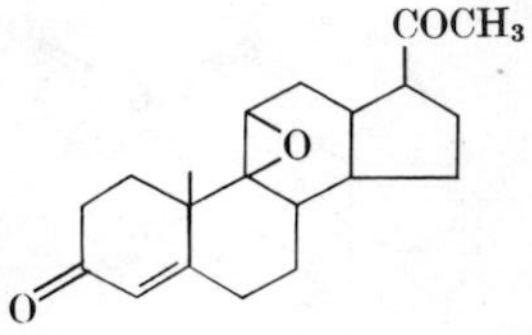 $COCH_3$, O, O	HF, $CHCl_3$ (5% ethanol), 0°, 1–2 hr[d]	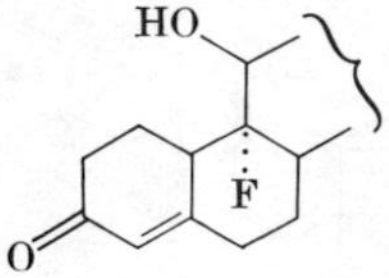 HO, F, O (60–65)[d]	353

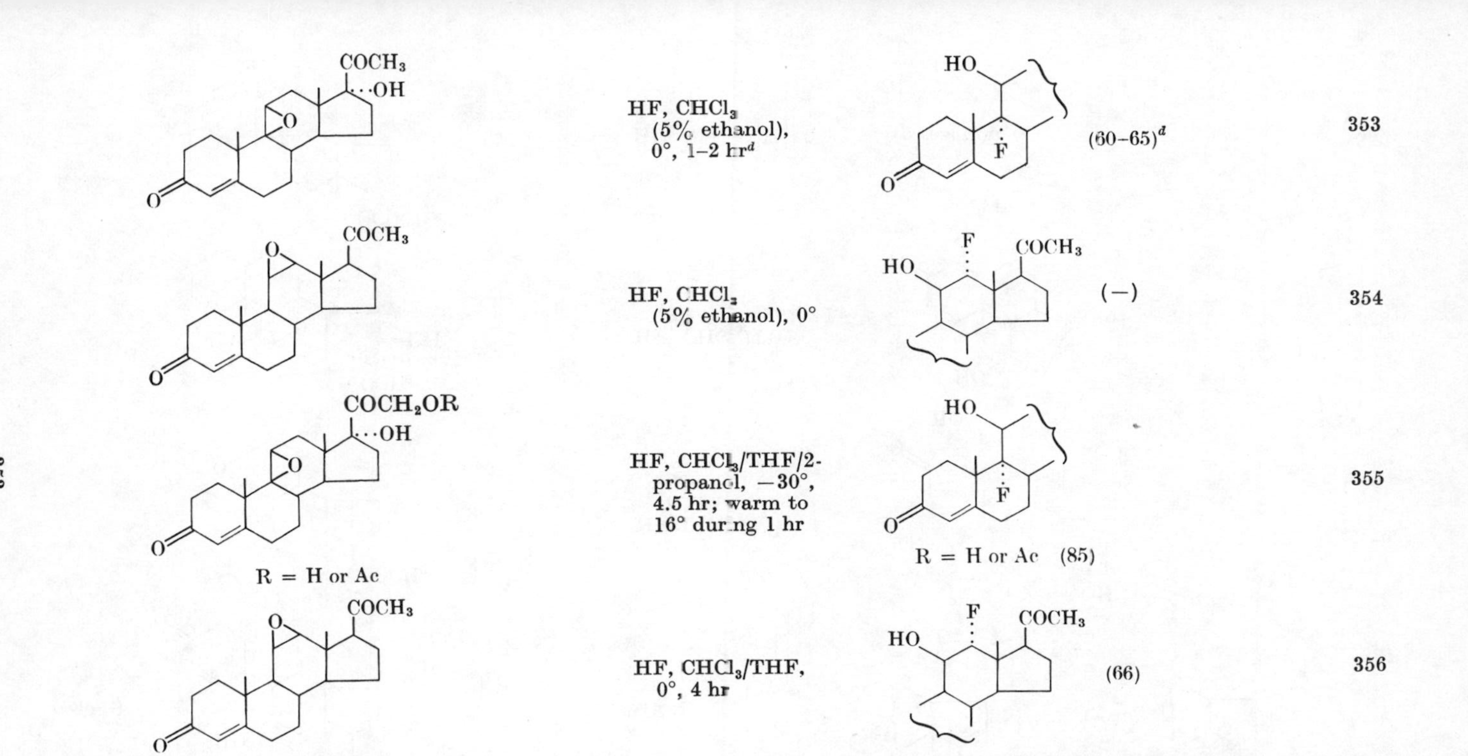

Conditions	Product (Yield, %)	Refs.
HF, $CHCl_3$ (5% ethanol), 0°, 1–2 hr[d]	(60–65)[d]	353
HF, $CHCl_3$ (5% ethanol), 0°	(—)	354
HF, $CHCl_3$/THF/2-propanol, −30°, 4.5 hr; warm to 16° during 1 hr	R = H or Ac (85)	355
HF, $CHCl_3$/THF, 0°, 4 hr	(66)	356

Note: References 301–386 are on pp. 404–406.

[a] Solvent ratios are in terms of volume.

[d] Neither yields nor experimental conditions were given in the communication on the work. Yields and experimental details are inferred from a later paper.[216]

TABLE XVI. Reactions of Hydrogen Fluoride with Oxiranes (*Continued*)

	Oxirane	Condition(s)[a]	Product(s) and Yield(s) (%)	Refs.
C_{21} (*contd.*)	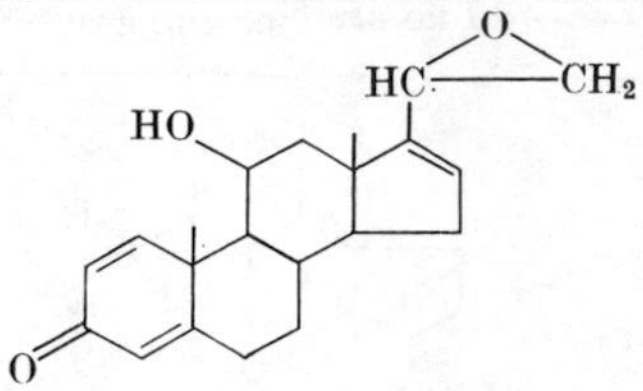 HO, HC—O—CH_2, O	HF	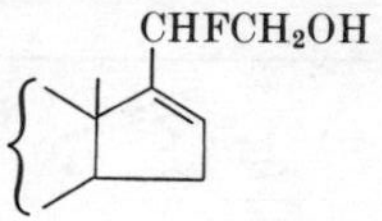 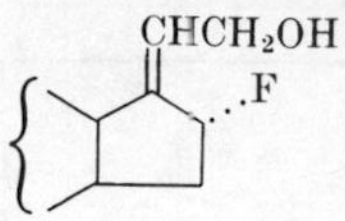 $CHFCH_2OH$ (–), $CHCH_2OH$, F (–)	357
	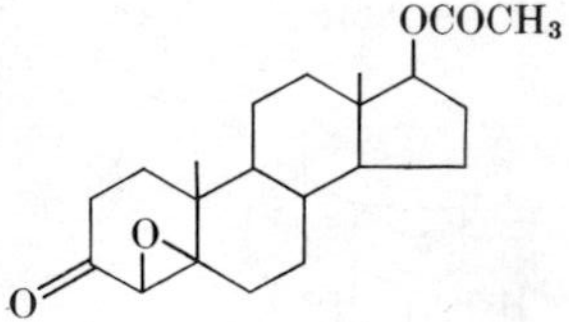 $OCOCH_3$, O, O	HF, $CHCl_3$ (ethanol-free), room t, 1 hr	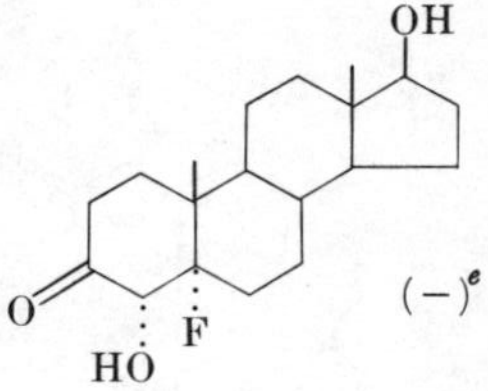 OH, O, HO, F (–)[e]	179
C_{22}	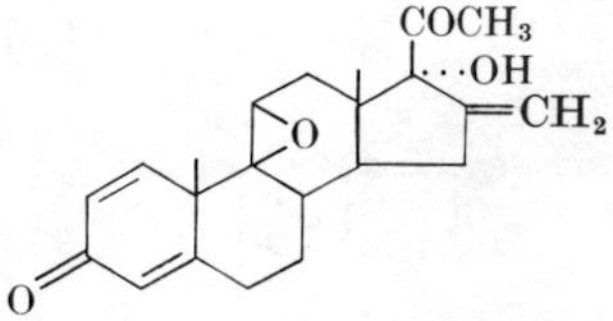 $COCH_3$, OH, CH_2, O, O	HF, $CHCl_3$/THF (3.3: 1), –25°, 2 hr	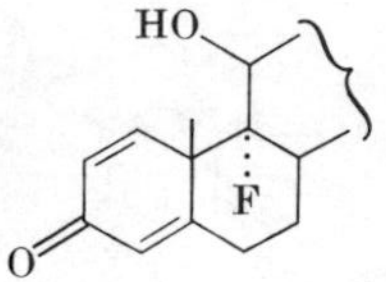 HO, F, O (51)	358
	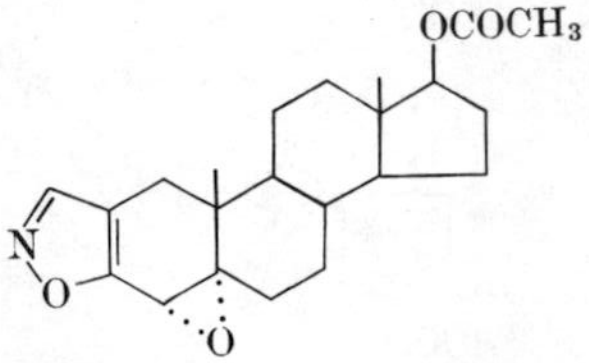 $OCOCH_3$, N, O, O	HF, ether-C_6H_6, 0°	No reaction	359

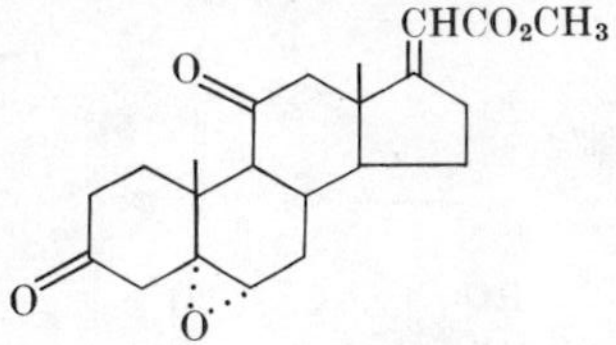 $CHCO_2CH_3$	48% aq HF	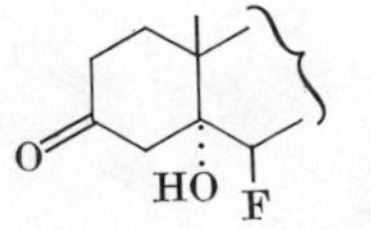 HO F (−)	181
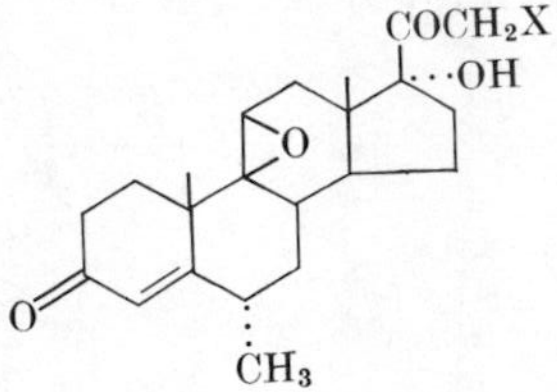 $COCH_2X$ ···OH O CH_3	7% $HClO_4$ in 48% aq HF, CH_2Cl_2, 6 hr	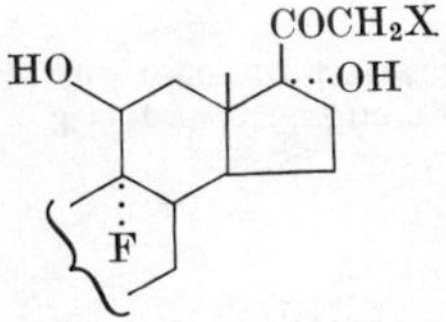 HO $COCH_2X$ ···OH F X = H (−), X = F (−)	165
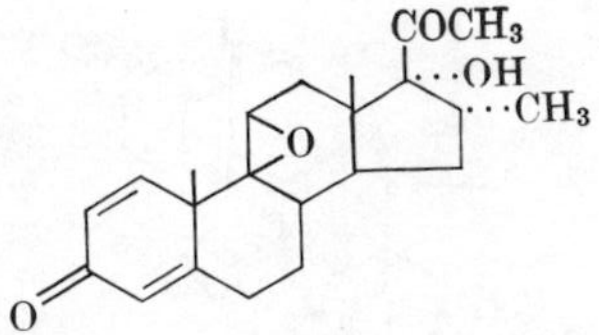 $COCH_3$ ···OH ···CH_3 O	HF, DMF, −5 to 0°, 15 hr	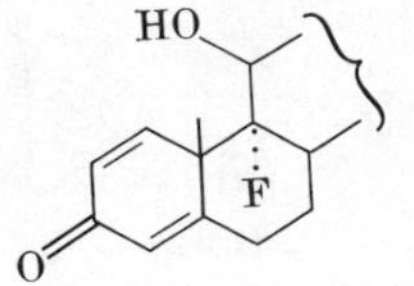 HO F (−)	360
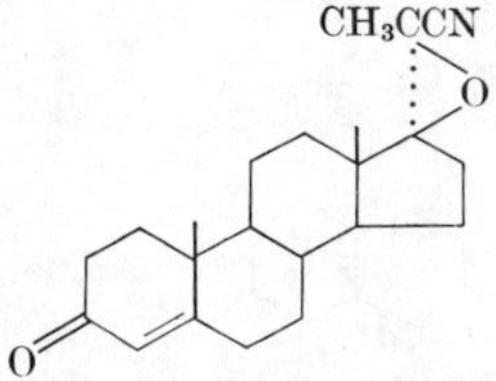 CH_3CCN O	48% aq HF, room t, 1 hr	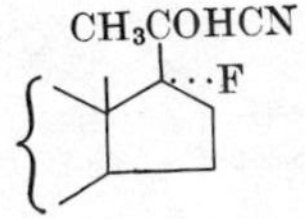 CH_3COHCN ···F (71)	361

Note: References 301–386 are on pp. 404–406.

[a] Solvent ratios are in terms of volume.

[e] The initially formed β-fluoro derivative epimerizes.

TABLE XVI. REACTIONS OF HYDROGEN FLUORIDE WITH OXIRANES (*Continued*)

	Oxirane	Condition(s)[a]	Product(s) and Yield(s) (%)	Refs.
C_{22} (*contd.*)	COCH$_3$, ···OH, =CH$_2$, O, O	HF, CHCl$_3$/THF (3.3:1), −25°, 2 hr	HO, F, O (52)	358
C_{23}	COCH$_3$, O, O, AcO, H	HF, THF/HF (1:1), room t, 5 hr	COCH$_3$, ···OH, F (~2), COCH$_3$, ···F, ···OH (5–10), H, CH$_3$, ···COCH$_3$, OH, F (5–10), 3 isomeric olefins (—) and recovered starting material	184
	COCH$_2$OAc, ···OH, O, O, O	HF:CHCl$_3$:THF/4:4:7, room t, 3 d	O, OH, F (33)[f]	362

Reactant	Conditions	Product (Yield, %)	Refs.
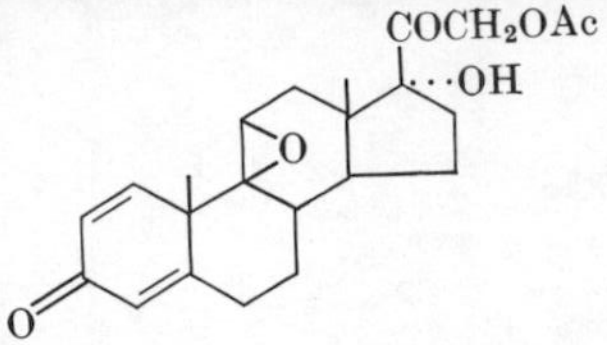 $COCH_2OAc$, ···OH, O	HF, $THF/CHCl_3$, 0°, 3.5 hr	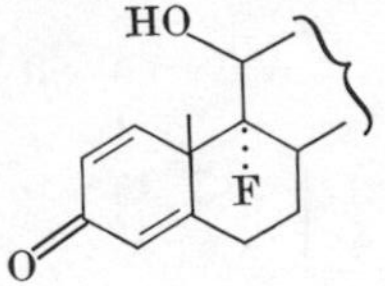 HO, F (75)	180
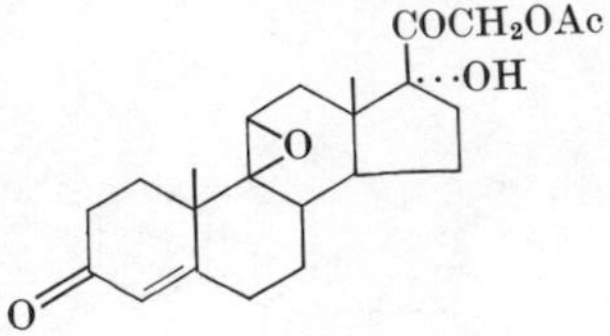 $COCH_2OAc$, ···OH, O	HF, $THF/CHCl_3$, −30°, 4 hr	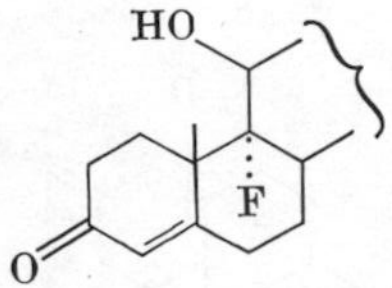 HO, F (74)	180
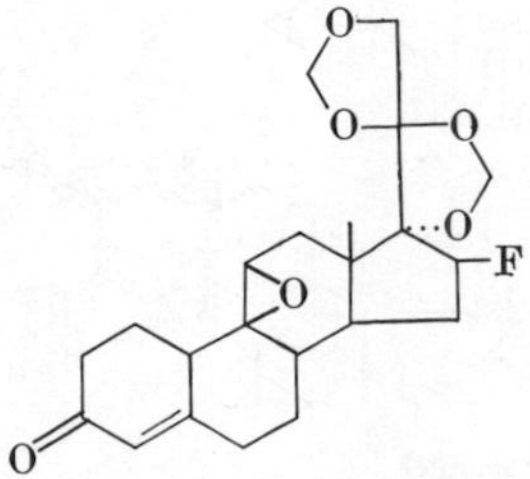 O, O, O, ···O, F, O	HF, $CHCl_3$, 0°	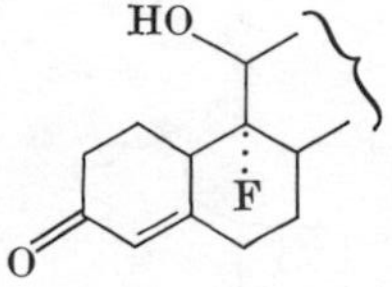 HO, F (—)	363
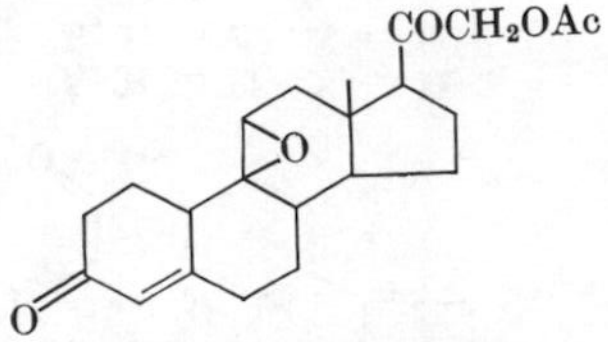 $COCH_2OAc$, O	HF, $CHCl_3$ (5% ethanol), 0°, 1–2 hr[d]	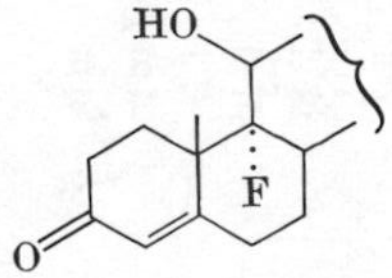 HO, F (60–65)[d]	353

Note: References 301–386 are on pp. 404–406.

[a] Solvent ratios are in terms of volume.

[d] Neither yields nor experimental conditions were given in the communication of the work. Yields and experimental details are inferred from a later paper.[216]

[f] The fluorohydrin was not isolated but, instead, was converted to 6-F-Δ^6-cortisone in an overall yield of 33%.

TABLE XVI. REACTIONS OF HYDROGEN FLUORIDE WITH OXIRANES (*Continued*)

	Oxirane	Condition(s)[a]	Product(s) and Yield(s) (%)	Refs.
C_{23} (*contd.*) 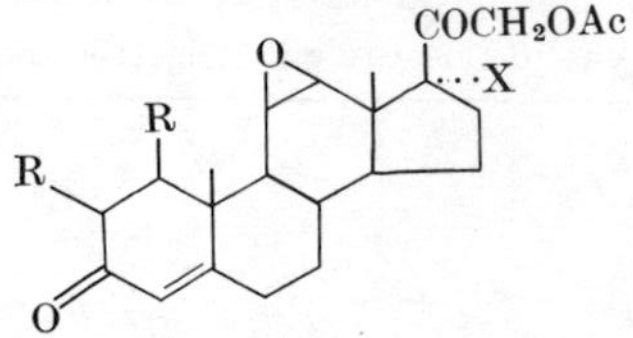	$COCH_2OAc$, X, R, R, O, O **A,** R = H, X = H **B,** R = Δ^1; X = OH	HF, $CHCl_3$/THF/2-propanol, −30°, 4.5 hr; warm to 16° during 1 hr	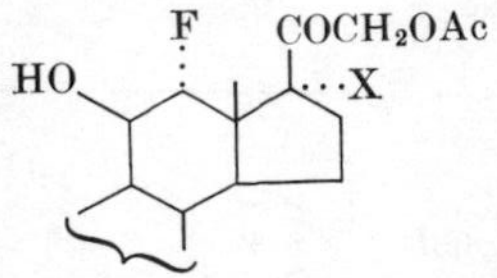 F, $COCH_2OAc$, HO, X **A′** R = H, X = H (—) **B′** R = Δ^1; X = OH (—)	355
	$COCH_2OAc$, O, O	HF, in unspecified organic solvent	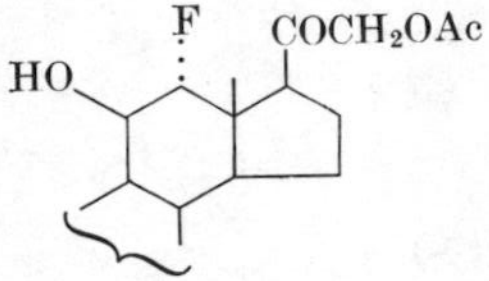 F, $COCH_2OAc$, HO (−)	364
	$COCH_2OAc$, OH, O, O	HF, $CHCl_3$, 0°, 1.5 hr	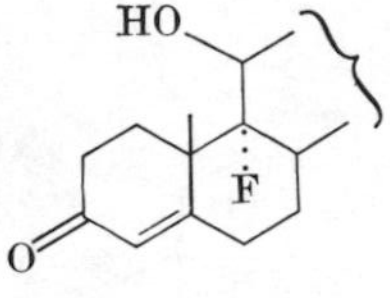 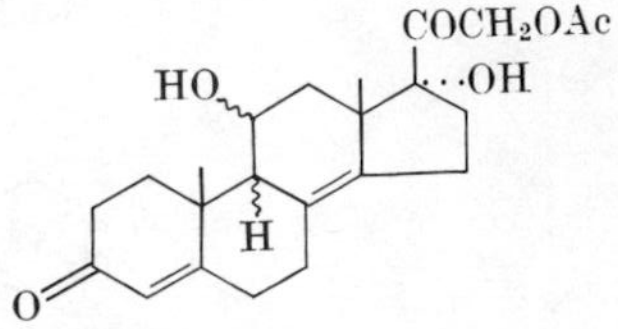 HO, F, O (51), **C** $COCH_2OAc$, HO, OH, H, O (—)	216
		HF, $CHCl_3$ (5% ethanol), 0°, 1.5 hr	**C** (60–65)	216

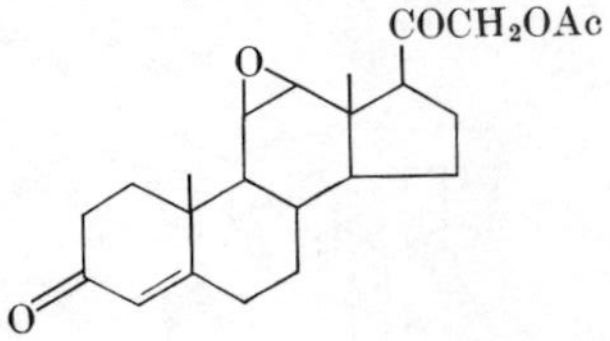

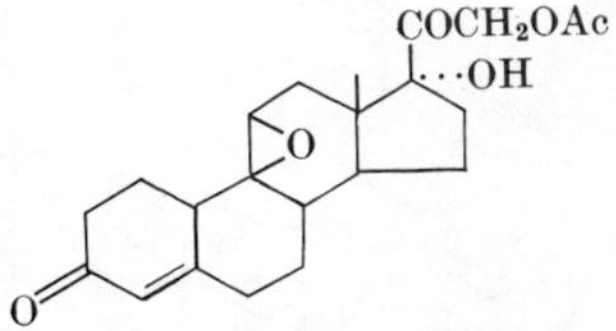

	Reactant	Conditions	Product(s) and Yield(s) (%)	Refs.
C_{24}	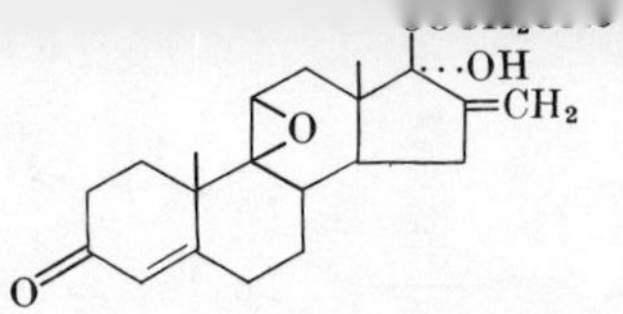 OH, $=CH_2$, O, O	HF, $CHCl_3$/THF (3.3:1), −25°, 2 hr	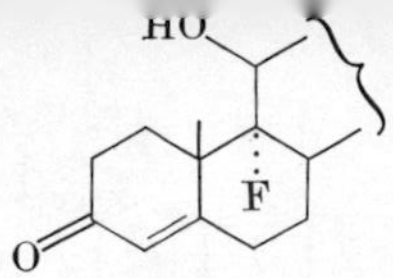 HO, F, O (45)	358
	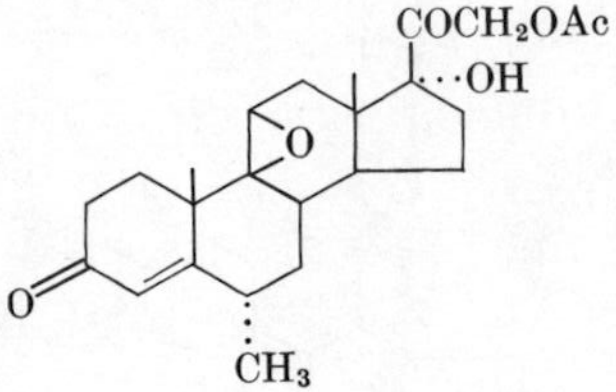 $COCH_2OAc$, OH, O, O, CH_3	HF, THF/CH_2Cl_2, −60° (—), −30° (2 hr); −20° (—)	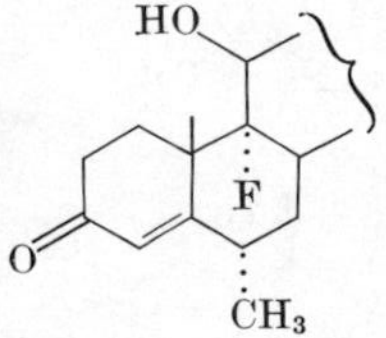 HO, F, O, CH_3 (75)	165
		48% aq HF, CH_2Cl_2, shake, 20 hr	" (38)	165
	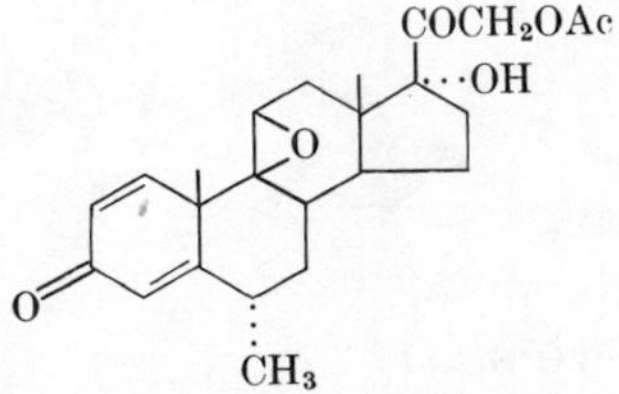 $COCH_2OAc$, OH, O, O, CH_3	48% aq HF, CH_2Cl_2, 20 hr	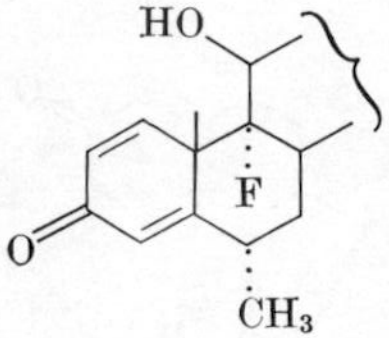 HO, F, O, CH_3 (58)	165
	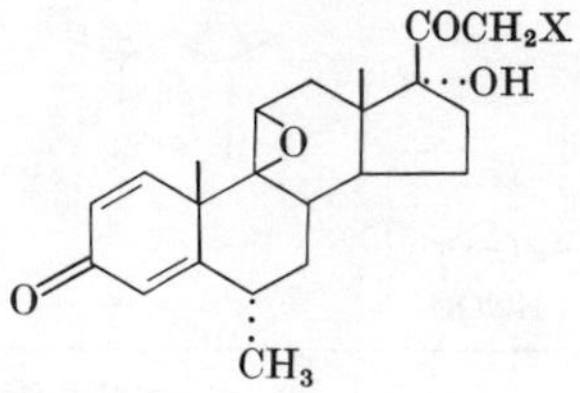 $COCH_2X$, OH, O, O, CH_3; X = H, X = F	7% $HClO_4$ in 48% aq HF, CH_2Cl_2, 6 hr	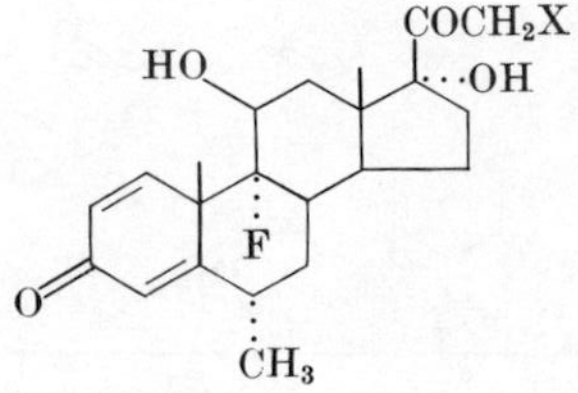 HO, $COCH_2X$, OH, F, O, CH_3; X = H (—), X = F (—)	165

Note: References 301–386 are on pp. 404–406.

[a] Solvent ratios are in terms of volume.

TABLE XVI. Reactions of Hydrogen Fluoride with Oxiranes (*Continued*)

	Oxirane	Condition(s)[a]	Product(s) and Yield(s) (%)	Refs.
C_{24} (*contd.*)	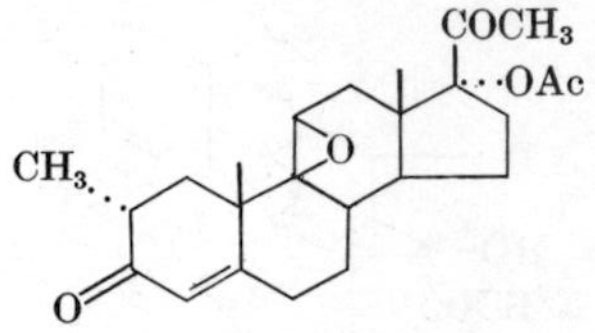 CH₃, COCH₃, OAc, O (structure)	5:1 HF, CH_2Cl_2/THF, mix at −78°, 4°, 60 hr 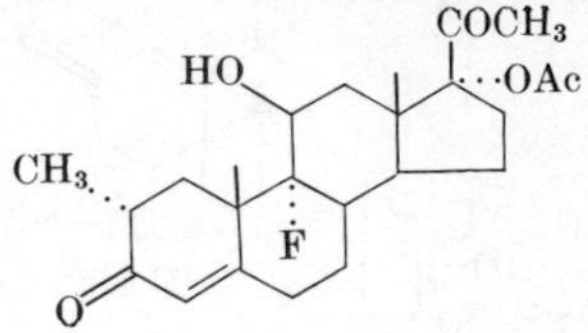	HO, COCH₃, OAc, CH₃, F, O (structure) (45)	365
	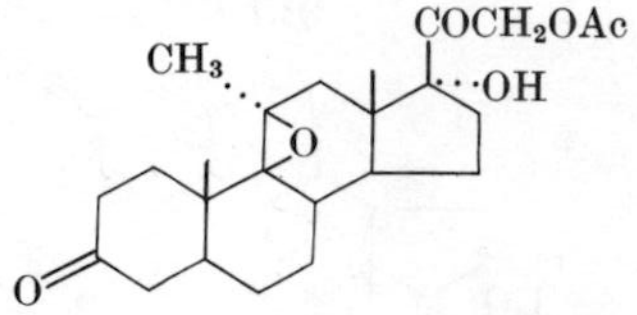 CH₃, COCH₂OAc, OH, O (structure)	48% aq HF, CH_2Cl_2, room t, stir, 5 hr 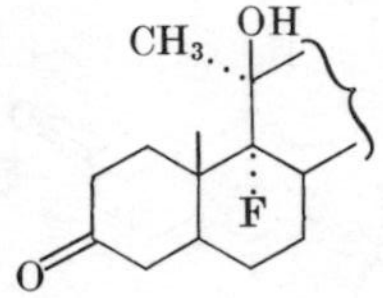	CH₃, OH, F, O (structure) (–)	366
	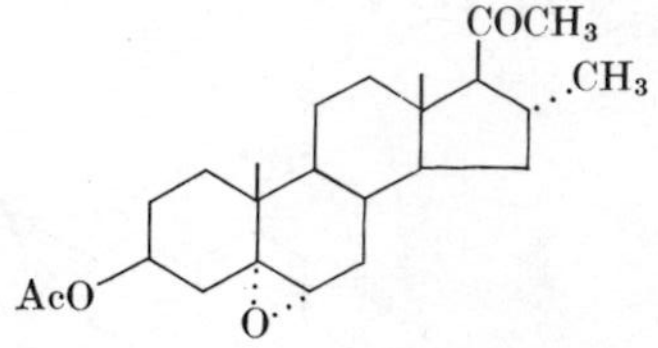 COCH₃, CH₃, AcO, O (structure)	HF, BF_3·ether, 1:1 C_6H_6-ether, room t, 24 hr 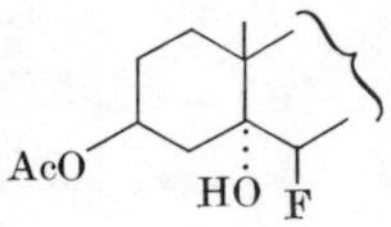	AcO, HO, F (structure) (71)	205
	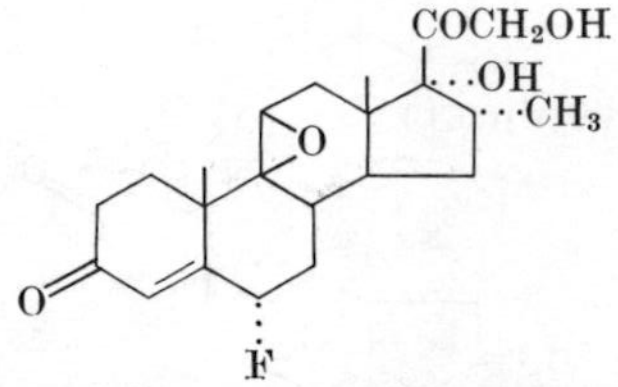 COCH₂OH, OH, CH₃, O, F (structure)	HF, CH_2Cl_2 (11% THF) 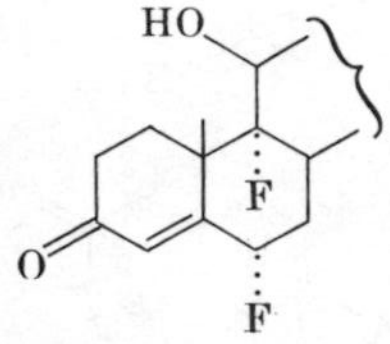	HO, F, F, O (structure) (45)	205

	Substrate	Conditions	Product(s) and Yield(s) (%)	Refs.
	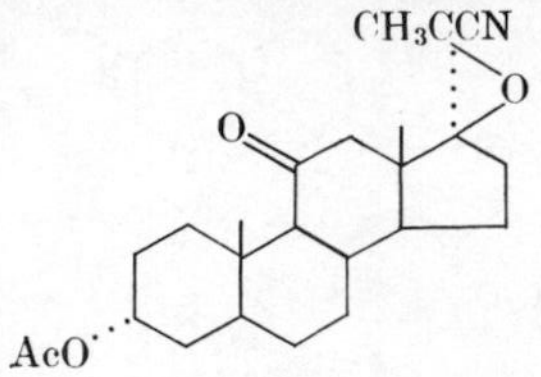 AcO, O, CH_3CCN, O	HF, $CHCl_3$/THF, 0°, 7.5 hr	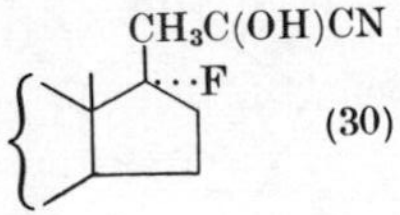 $CH_3C(OH)CN$, F (30)	367
C_{27}	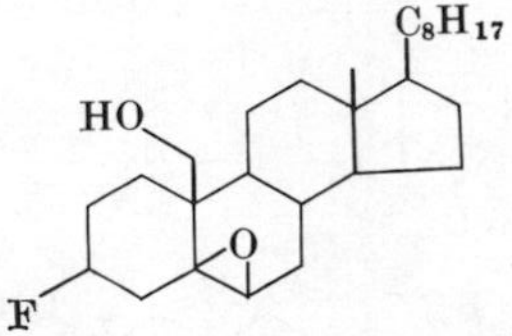 C_8H_{17}, HO, O, F	HF, $CHCl_3$, −50°, then 1 hr at 0°	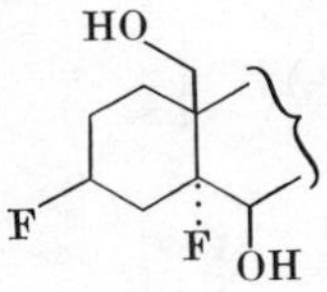 HO, F, F, OH (70)	211
	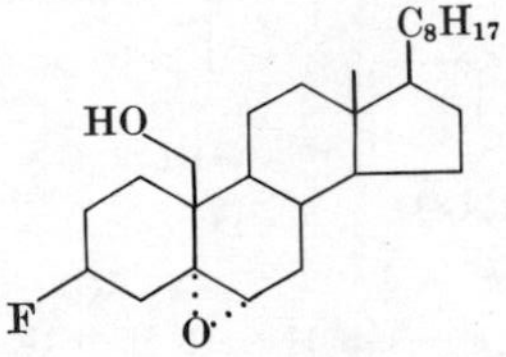 C_8H_{17}, HO, F, O	HF, $CHCl_3$, −20°, 1 hr	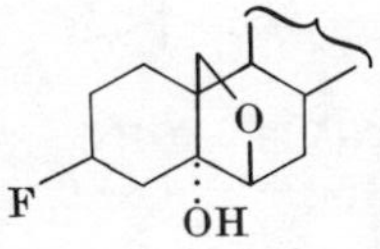 O, F, OH (67)	211
	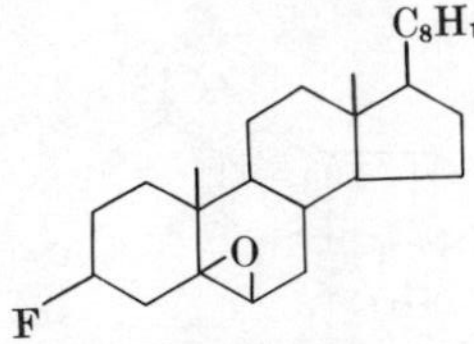 C_8H_{17}, O, F	HF, $CHCl_3$, −20°	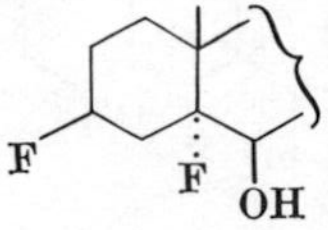 F, F, OH (88)	211

Note: References 301–386 are on pp. 404–406.

[a] Solvent ratios are in terms of volume.

TABLE XVII. REACTION OF BORON TRIFLUORIDE ETHERATE WITH OXIRANES

	Oxirane	Condition(s)[a]	Product(s) and Yield(s) (%)	Refs.
C_{18}	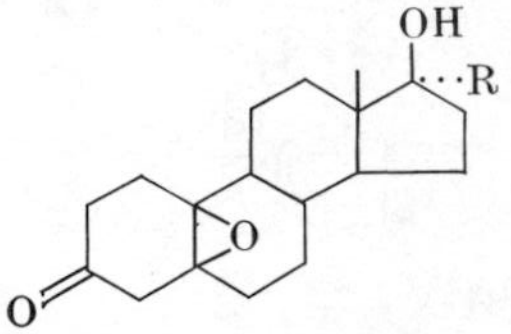 OH, R, O, O; R = H, R = HC≡C—	2:1 ether/C_6H_6, room t, 3 hr	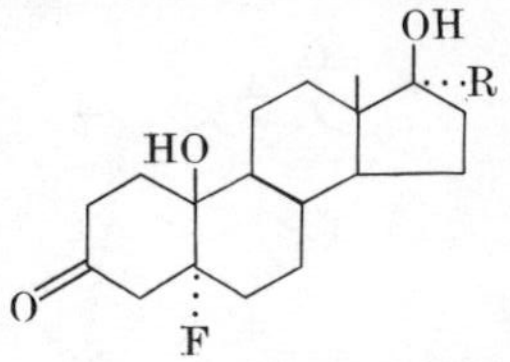 OH, R, HO, O, F; R = H (88), R = HC≡C— (78)	197
C_{22}	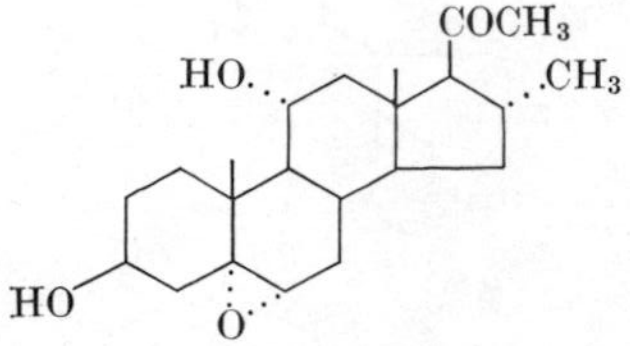 $COCH_3$, HO, CH_3, HO, O	1:1 ether/C_6H_6, room t, 3 hr	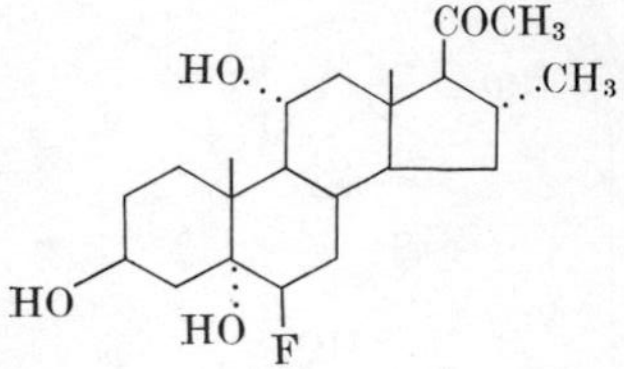 $COCH_3$, HO, CH_3, HO, HO, F (75)[b]	191
C_{23}	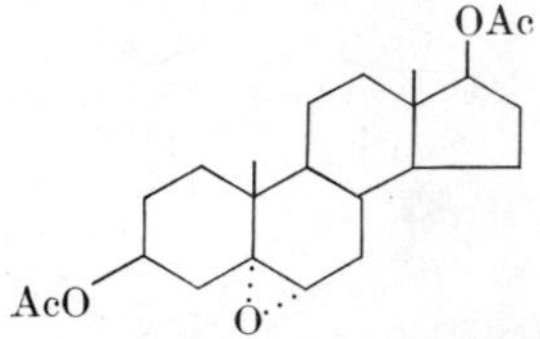 OAc, AcO, O	1:1 ether/C_6H_6, room t, 3 hr	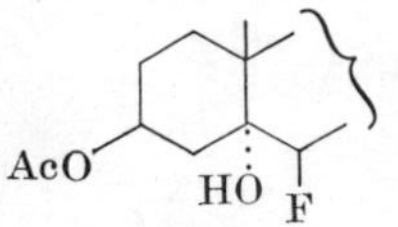 AcO, HO, F (64)	189a
	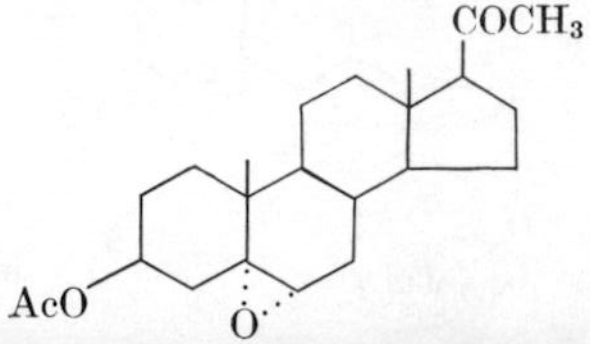 $COCH_3$, AcO, O	1:1 ether/C_6H_6, room t, 6 hr	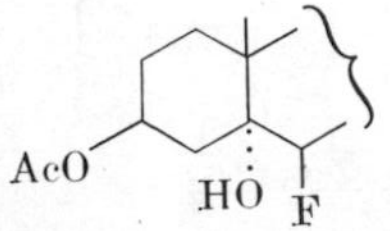 AcO, HO, F (56)	189a

	Reactant	Conditions	Product(s) and Yield(s) (%)	Refs.
	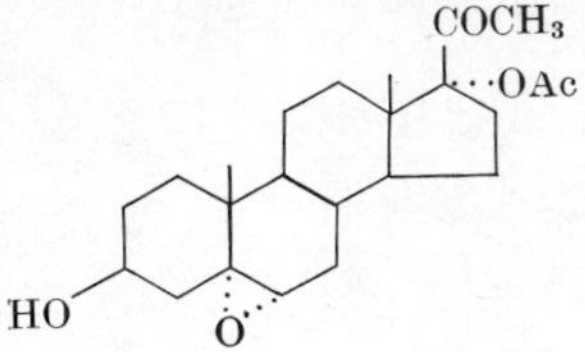 HO, epoxide; COCH3, OAc	1:1 ether/C_6H_6, room t, 12 hr	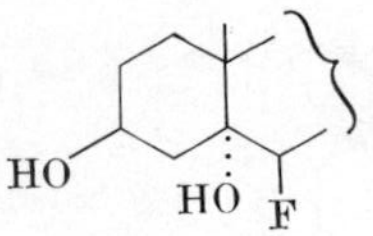 HO, HO, F (48)	189a
C_{25}	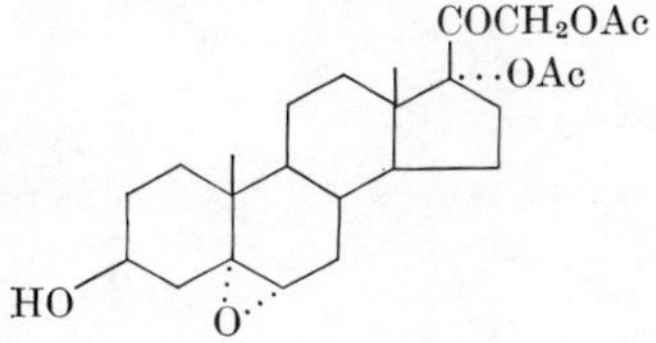 HO, epoxide; $COCH_2OAc$, OAc	1:1 ether/C_6H_6, room t, 12 hr	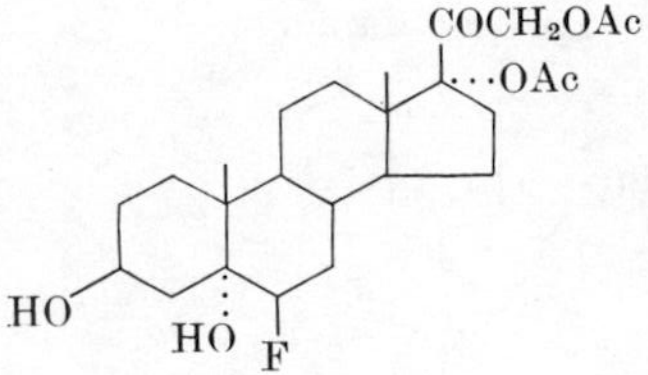 HO, HO, F; $COCH_2OAc$, OAc (36)	189b, 190
C_{27}	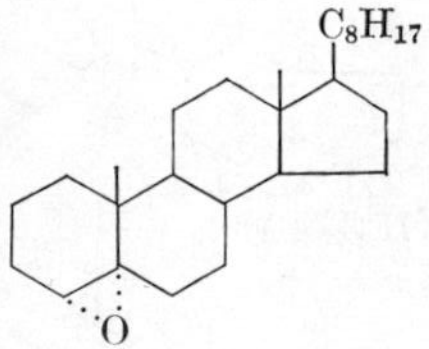 epoxide; C_8H_{17}	C_6H_6, 20°, 2 min	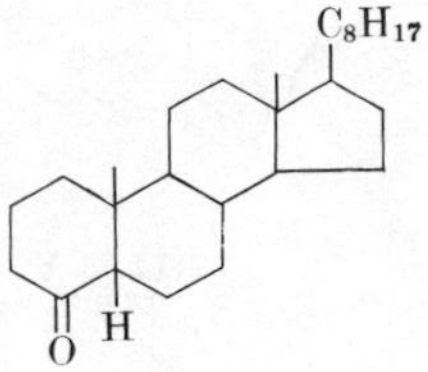 O, H; C_8H_{17} **A** (50)	188a

Note: References 301–386 are on pp. 404–406.

[a] Unless otherwise indicated, boron trifluoride etherate was used as the source of fluoride.
[b] The patent appears ambiguous on yield.

TABLE XVII. REACTION OF BORON TRIFLUORIDE ETHERATE WITH OXIRANES (*Continued*)

Oxirane	Condition(s)[a]	Product(s) and Yield(s) (%)	Refs.
C_{27} (*contd.*) [structure: C_8H_{17}]	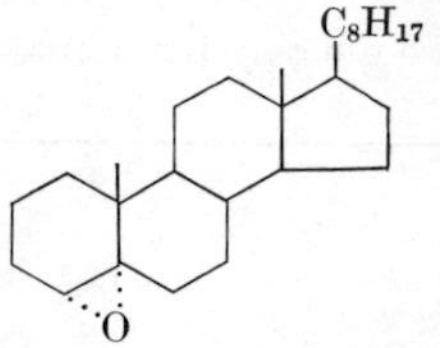 C_6H_6, 20°, 45 sec[c]	**A** (76), [structure: C_8H_{17}, OH] (11)	195
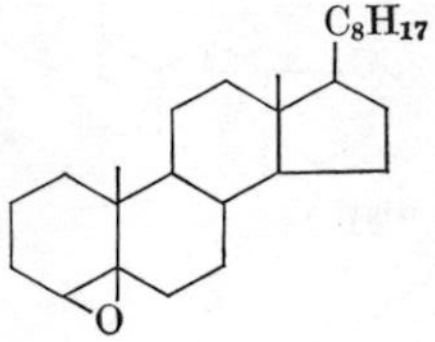 [structure: C_8H_{17}, O]	C_6H_6, 20°, 2 min[c]	**A** (41), [structure: HO, F] (3), [structure: HO] (11), [structure] (8), [structure: CHO] (7), Mixture with $\Delta^{4,6}$; [structure: C_8H_{17}, H, HO] (23), [structure: C_8H_{17}, H, HO] (3)	195

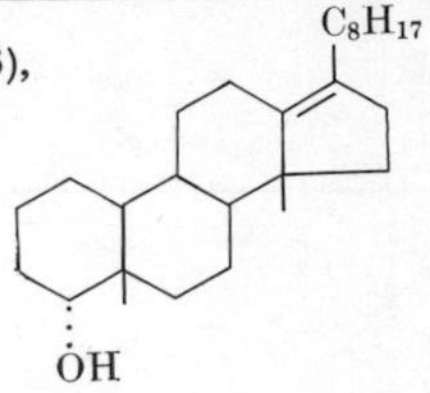

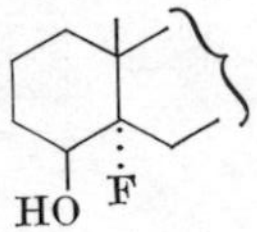

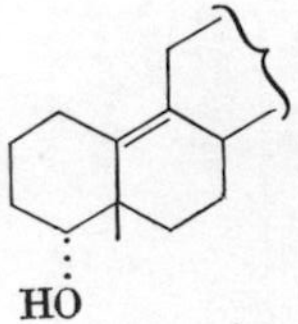

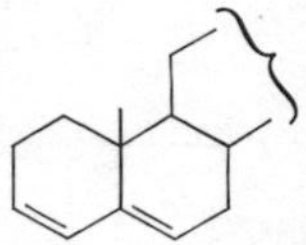

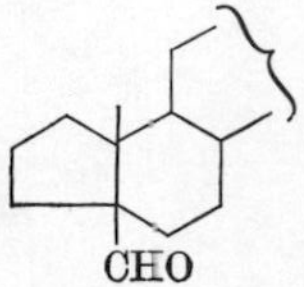

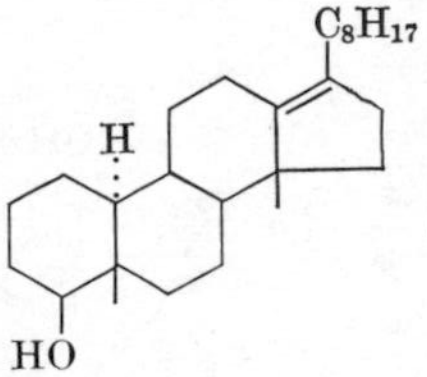

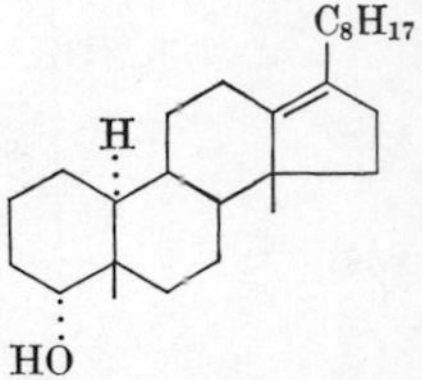

Reactant	Conditions	Product(s) and Yield(s) (%)	Refs.
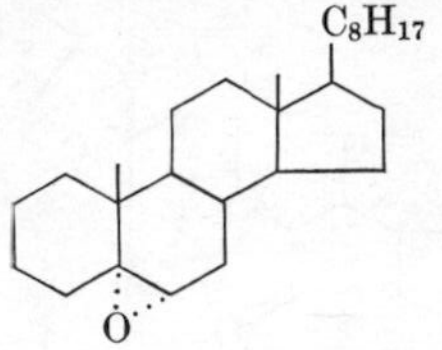 C_8H_{17} (epoxide, O)	C_6H_6, 2 min	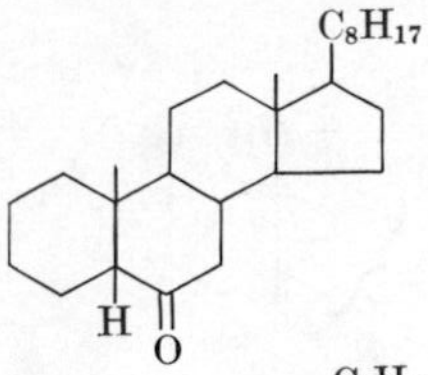 C_8H_{17}, H, O (30)	188a
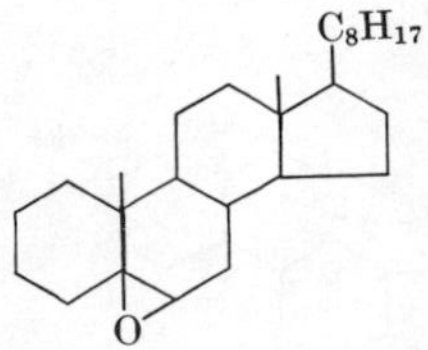 C_8H_{17} (epoxide, O)	C_6H_6, 0°, 4 min	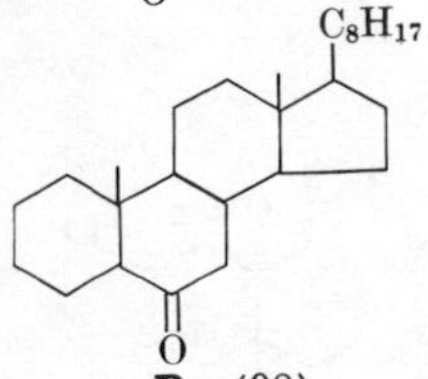 C_8H_{17}, O **B** (98)	188a
	C_6H_6, 20°, 2 min[c]	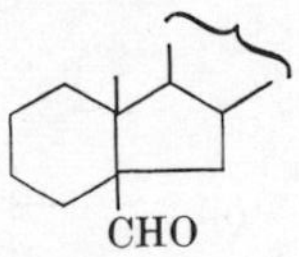 **B** (20), 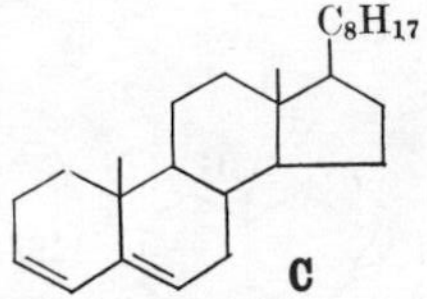 C_8H_{17} **C** (15), (Possible mixture with $\Delta^{4,6}$) 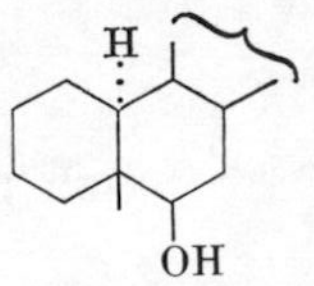 H, OH (46), CHO (18)	195

Note: References 301–386 are on pp. 404–406.

[a] Unless otherwise indicated, boron trifluoride etherate was used as the source of fluoride.

[c] The fluorohydrin obtained from this epoxide by HF addition rearranged to a similar mixture of products when treated with BF_3.ether under similar conditions.

TABLE XVII. REACTION OF BORON TRIFLUORIDE ETHERATE WITH OXIRANES (*Continued*)

	Oxirane	Condition(s)[a]	Product(s) and Yield(s) (%)	Refs.
C_{27} (*contd.*)	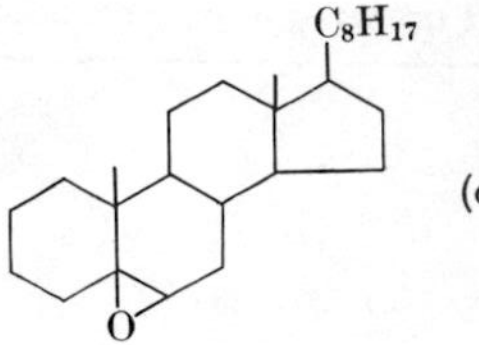 C_8H_{17}, O (*contd.*)	Ether, 20°, 2.5 hr	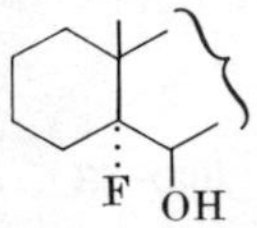 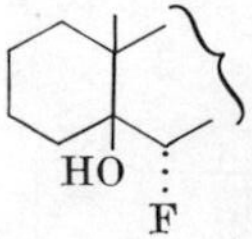 **C** (22), 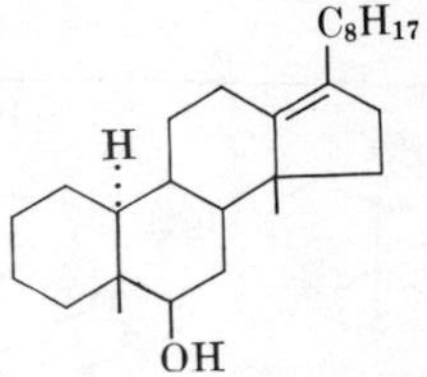 H, C_8H_{17}, OH (9), F, OH (57), HO, F (6)	195
	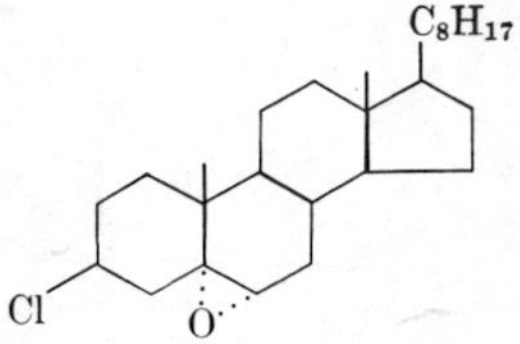 C_8H_{17}, Cl, O	C_6H_6, 20°, 18 min	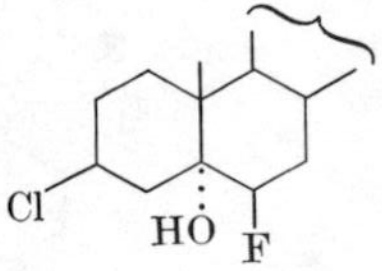 Cl, HO, F (48)	188b
	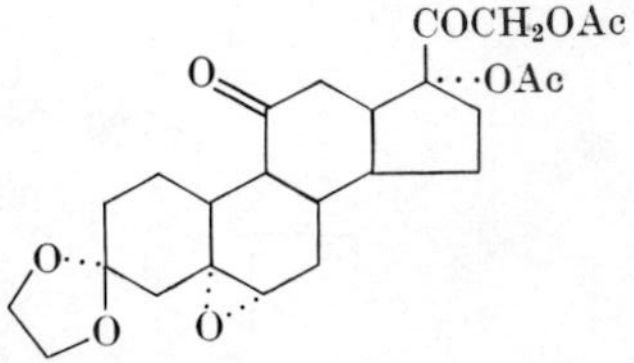 $COCH_2OAc$, OAc, O, O, O, O	1:1 ether/C_6H_6, room t	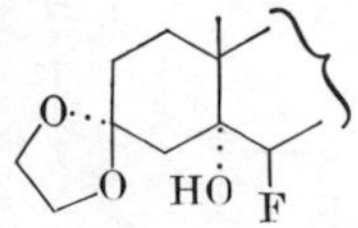 O, O, HO, F (–)	189b

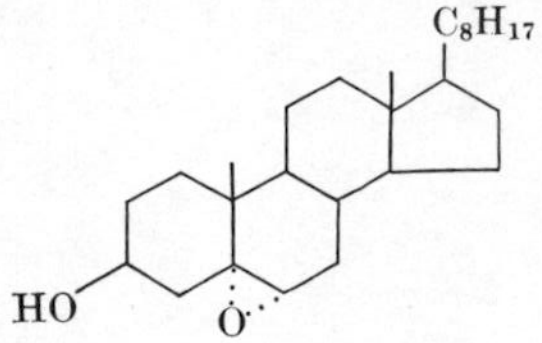

HO, O, C_8H_{17}	C_6H_6, room t, 7 min	HO, H, H, H, OH, C_8H_{17} (10), HO, H, H, H, OH, C_8H_{17} (17), HO, H, O (11), HO, OH (7), HO, HO, F, HO, H, O (4.5)	196
		D (2) Additional isomeric products were obtained by acetylation of a polar fraction.	
	1:1 ether/C_6H_6, room t, 3 hr	**D** (65)	189a

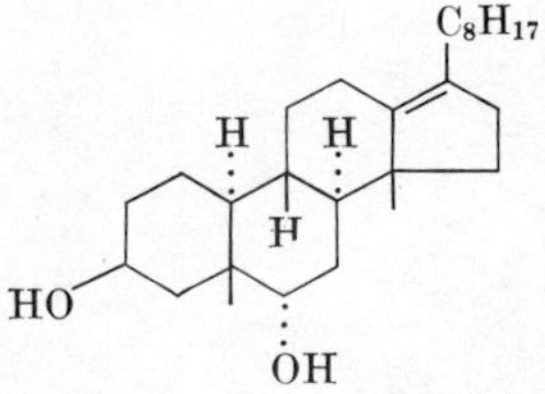

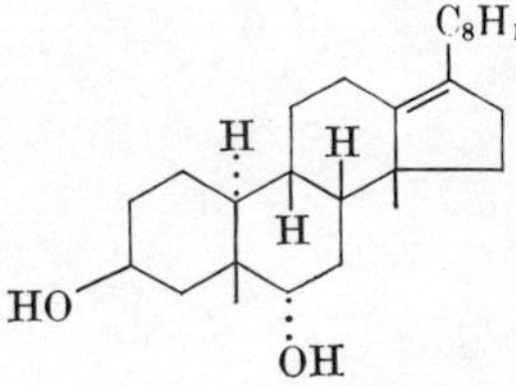

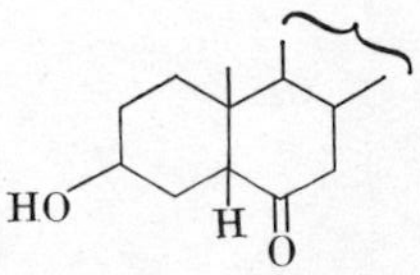

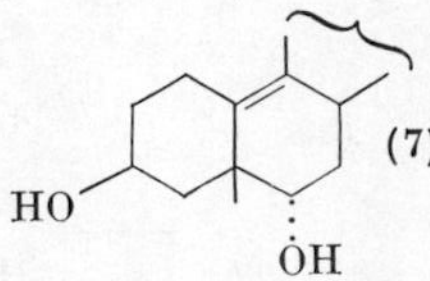

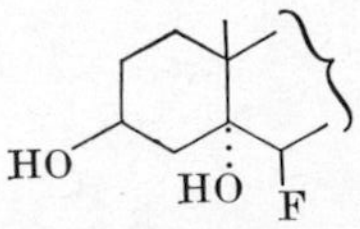

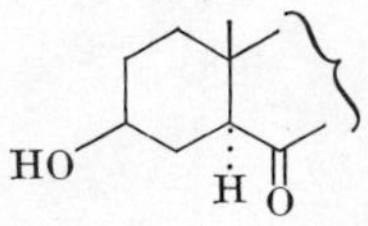

Note: References 301–386 are on pp. 404–406.

[a] Unless otherwise indicated, boron trifluoride etherate was used as the source of fluoride.

TABLE XVII. REACTION OF BORON TRIFLUORIDE ETHERATE WITH OXIRANES (*Continued*)

Oxirane	Condition(s)[a]	Product(s) and Yield(s) (%)	Refs.
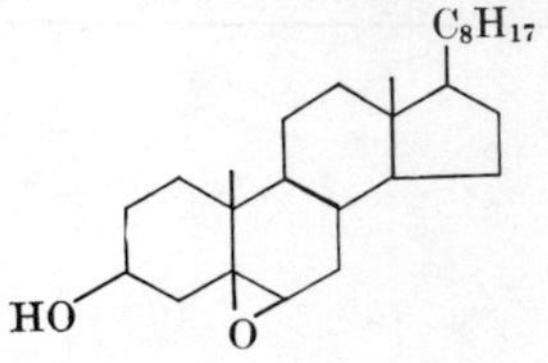 C_{27} (*contd.*) C_8H_{17} HO O	C_6H_6, room t, 7 min	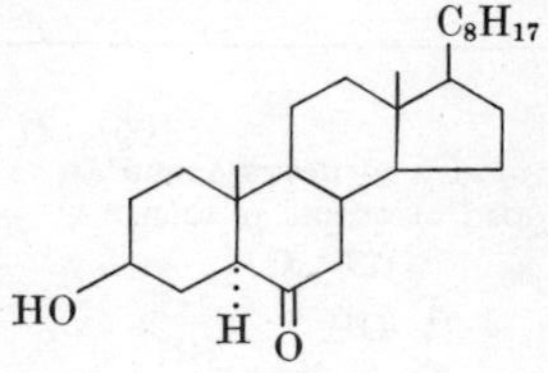 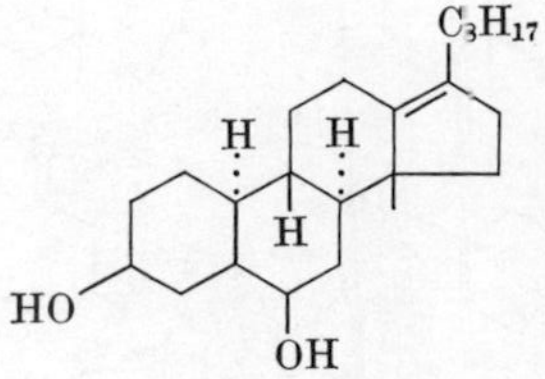 C_8H_{17} HO H O (20), C_3H_{17} H H H HO OH (40)	196b
C_{28} C_8H_{17} CH_3O O	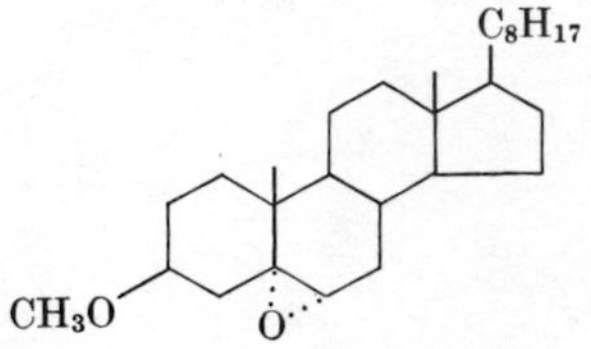 C_6H_6, room t, 7min.	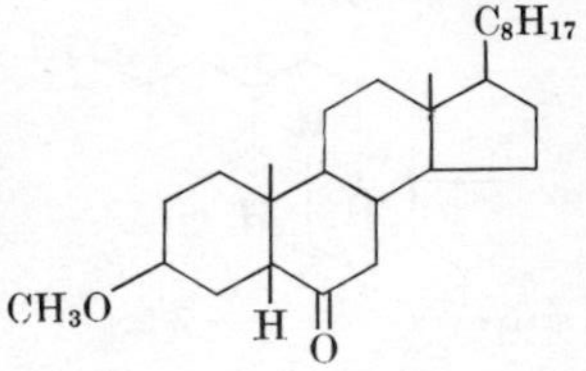 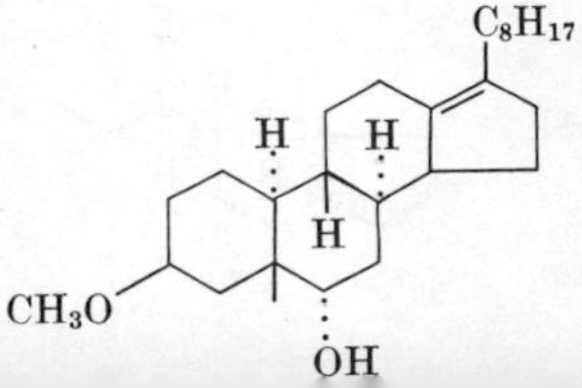 C_8H_{17} CH_3O H O (27), C_8H_{17} H H H CH_3O OH (9)	196b

			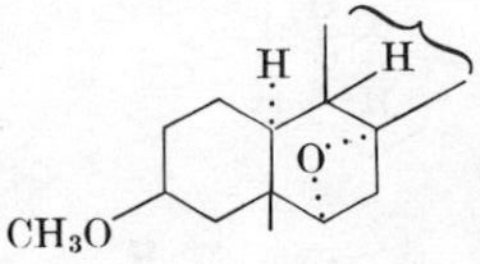 CH_3O ... OH (18), CH_3O ... H, H, O (7)	
	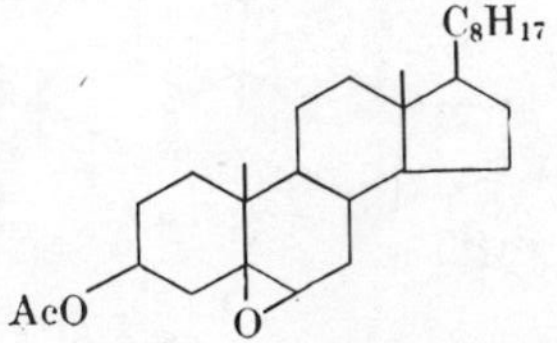	C_6H_6, 30 min, t unreported	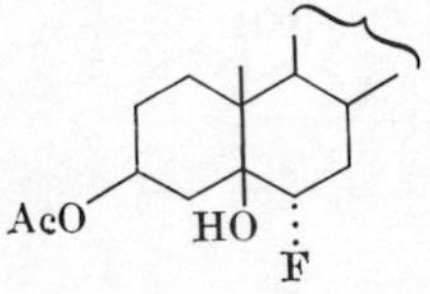 AcO ... HO, F (76)	188b
C_{29}	AcO ... O ... C_8H_{17}			
	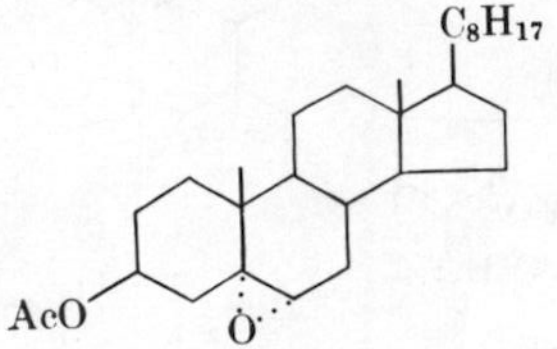 AcO ... O ... C_8H_{17}	Boron trifluoride etherate free of HBF_4, C_6H_6, 20°, 45 sec, max steroid concn	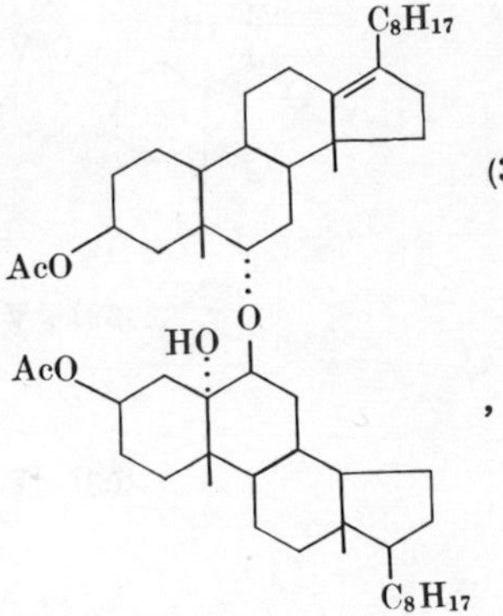 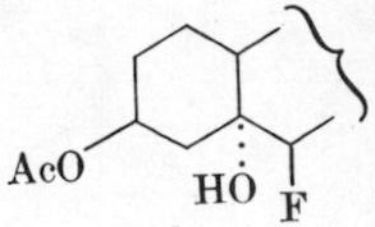 AcO ... C_8H_{17} ... O ... HO, AcO ... C_8H_{17} (32), AcO ... HO, F **A** (16)	193

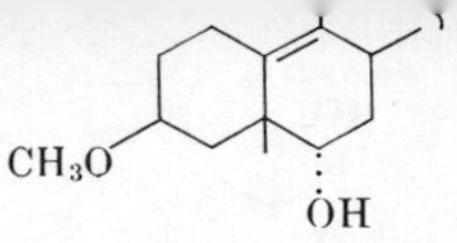

Note: References 301–386 are on pp. 404–406.

[a] Unless otherwise indicated, boron trifluoride etherate was used as the source of fluoride.

TABLE XVII. REACTION OF BORON TRIFLUORIDE ETHERATE WITH OXIRANES (*Continued*)

	Oxirane	Condition(s)[a]	Product(s) and Yield(s) (%)	Refs.
C_{29} (*contd.*)	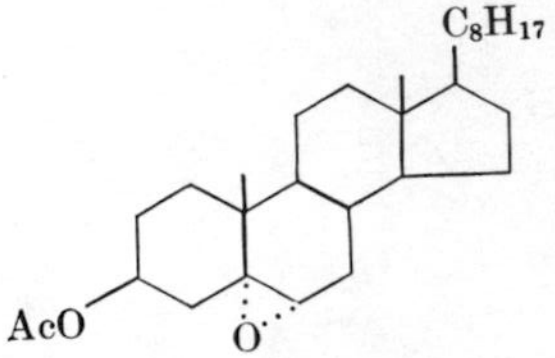 C_8H_{17} ; AcO ; O (*contd.*)	1:1 ether/C_6H_6, room t, 3 hr	**A** (80)	189a
		C_6H_6, 20°, 5 min	**A** (62)	188b
	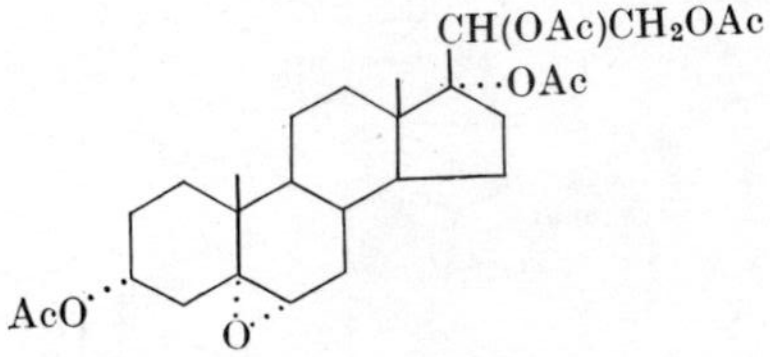 $CH(OAc)CH_2OAc$; OAc ; AcO ; O	1:1 ether/C_6H_6, room t, 3 hr	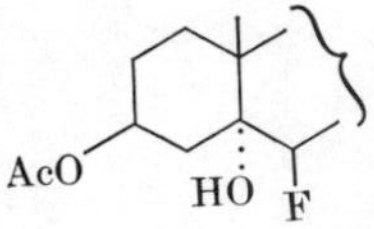 AcO ; HO ; F (—)	192
	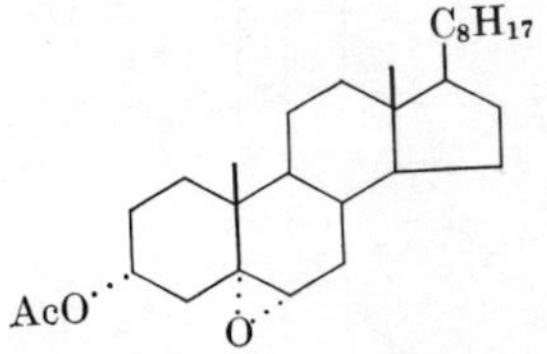 C_8H_{17} ; AcO ; O	C_6H_6, 20°, 25 sec	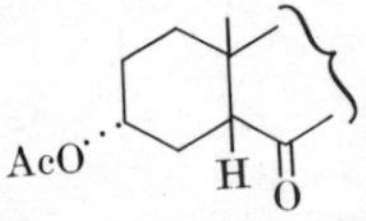 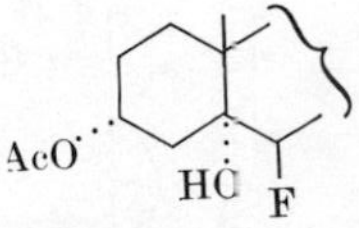 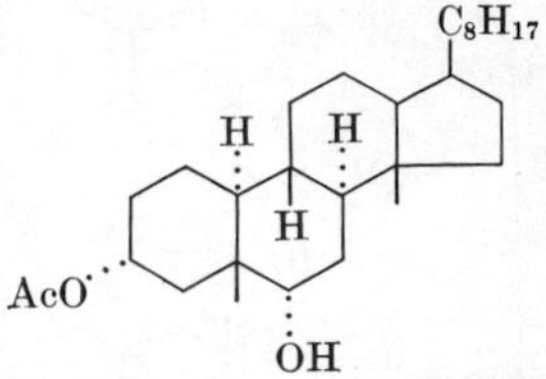 AcO ; H ; O (17), AcO ; HO ; F (8), **B** C_8H_{17} ; H ; H ; H ; AcO ; OH (37),	194

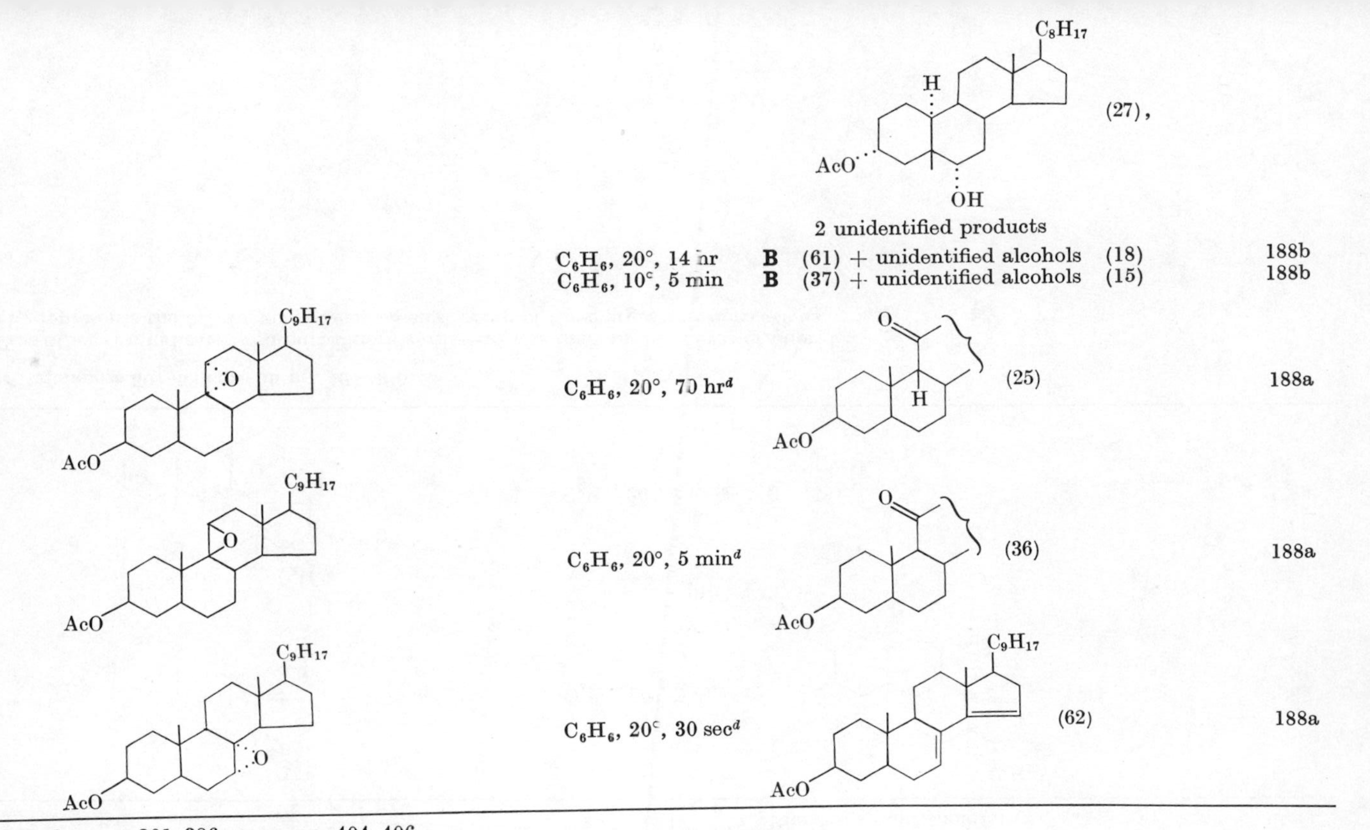

		(27), 2 unidentified products	
	C_6H_6, 20°, 14 hr	**B** (61) + unidentified alcohols (18)	188b
	C_6H_6, 10°, 5 min	**B** (37) + unidentified alcohols (15)	188b
C_{30}	C_6H_6, 20°, 70 hr[d]	(25)	188a
	C_6H_6, 20°, 5 min[d]	(36)	188a
	C_6H_6, 20°, 30 sec[d]	(62)	188a

Note: References 301–386 are on pp. 404–406.

[a] Unless otherwise indicated, boron trifluoride etherate was used as the source of fluoride.

[d] C_9H_{17} represents the 5,7-dimethyl-2-hept-3-enyl group of these ergosterol derivatives.

TABLE XVII. REACTION OF BORON TRIFLUORIDE ETHERATE WITH OXIRANES (*Continued*)

	Oxirane	Condition(s)[a]	Product(s) and Yield(s) (%)	Refs.
C_{30} (*contd.*)	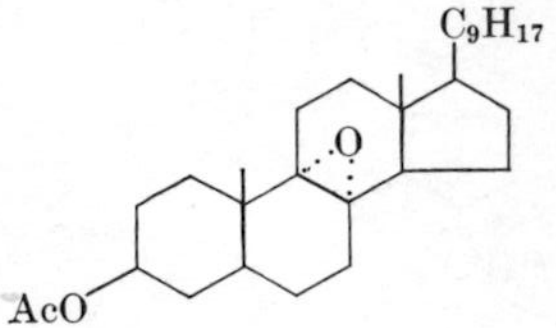 C_9H_{17} AcO	C_6H_6, 20°, 2 min[d]	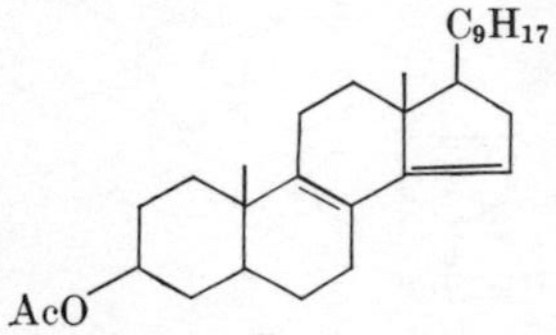 C_9H_{17} AcO **C** (—)	188a
	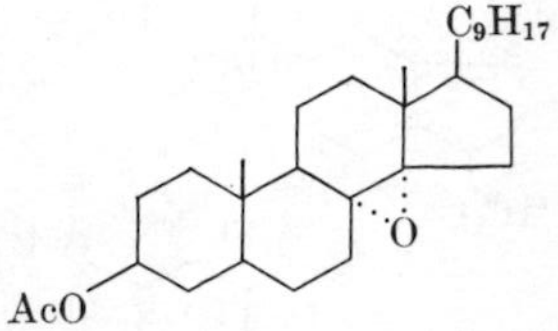 C_9H_{17} AcO	C_6H_6, 20°, 2 min[d]	**C** (88)	188a

Note: References 301–386 are on pp. 404–406.

[a] Unless otherwise indicated, boron trifluoride etherate was used as the source of fluoride.
[d] C_9H_{17} represents the 5,7-dimethyl-2-hept-3-enyl group of these ergosterol derivatives.

TABLE XVIII. Monofluorinations by Perchloryl Fluoride

	Reactant	Base Used To Form Anion, Reaction Condition(s)	Product(s) and Yield(s) (%)	Refs.
C_1	$HCF(NO_2)_2$	HMPT, 3–5°, 1.5 hr	$CF_2(NO_2)_2$ (50)	235a
	$HC(NO_2)_3$	HMPT, 20°, 2 hr	$FC(NO_2)_3$ (92)	235a
		DMF, 20°, 2 hr	" (38)	235a
		Glyme 20°, 45 hr	" (8)	235a
C_2	$Na^+[C(NO_2)_2CN]^-$	$CH_3OCH_2CH_2OH$, 54 hr	$FC(NO_2)_2C({=}NH)OCH_2CH_2OCH_3$ (74)	235b
	$K^+[C(NO_2)_2CH_3]^-$	79% aq CH_3OH, 25–40°, 7 hr	$FC(NO_2)_2CH_3$ (54)	233
	$HC(NO_2)_2CH_3$	HMPT, 20°, 14 hr	$FC(NO_2)_2CH_3$ (61)	235a
		DMF, 20°, 4 hr	" (2)	235a
	$K^+[C(NO_2)_2CH_2OH]^-$	Glyme, 20°, 13 hr	$FC(NO_2)_2CH_2OH$ (20)	235b
		DMF, <15°	" (25)	233
	$NH_4^+[C(NO_2)_2CONH_2]^-$	Glyme, 20°, 2 hr	$FC(NO_2)_2CONH_2$ (—)	235b
	$HC(NO_2)_2CH_2OK$	DMF, 15°	$FC(NO_2)_2CH_2OH$ (25)	368
C_3	$K^+[C(NO_2)_2C_2H_5]^-$	79% aq CH_3OH, 35–40°, 3.5 hr	$FC(NO_2)_2C_2H_5$ (59)	112
	$K^+[C(NO_2)_2CH_2OCH_3]^-$	$(CH_3)_2CO$, 20°, 15 min	$FC(NO_2)_2CH_2OCH_3$ (64)	235b
C_4	$C_2H_5NH_3^+[C(NO_2)_2CONH_2]^-$	Glyme, 20°, 8 hr	$FC(NO_2)_2CONH_2$ (14)	235b
	$Na^+[C(NO_2)_2CH_2CH{=}CH_2]^-$	70% aq CH_3OH, reflux 40 min, 20°, 2 hr	$FC(NO_2)_2CH_2CH{=}CH_2$ (50)	369
C_5	$K^+[C(NO_2)(CN)CO_2C_2H_5]^-$	CH_3CN, 5–10°, 2–3 hr	$FC(NO_2)(CN)CO_2C_2H_5$ (25)	20
	$K^+[C(NO_2)_2CH_2CH_2NHCOCH_3]^-$	Glyme, 1.5 hr	$FC(NO_2)_2CH_2CH_2NHCOCH_3$ (93)	235b
	$K^+[C(NO_2)_2CH_2CH_2COCH_3]^-$	THF, 20°, 3 hr	$FC(NO_2)_2CH_2CH_2COCH_3$ (65)	235b
	$K^+[CNO_2(CO_2CH_3)_2]^-$	DMF	$(CH_3O_2C)_2CFNO_2$ (70)	235b
	$K^+[C(NO_2)_2CH_2CH_2CO_2CH_3]^-$	CH_3OH, 37°	$FC(NO_2)_2CH_2CH_2CO_2CH_3$ (93)	112
	$(CH_3)_2CHCO_2CH_3$	$(C_6H_5)_3CNa$, ether, 0°, 20 min	$(CH_3)_2CFCOC(CH_3)_2CO_2CH_3$ (30)	329
	$(O_2N)_2CH(CH_2)_3CH(NO_2)_2$	HMPT, 3–5°, 2 hr	$FC(NO_2)_2(CH_2)_3C(NO_2)_2F$ (88)	235a
	$Na^+[C(NO_2)_2CH_2OCH_2\overset{O}{CH—CH_2}]^-$	70% aq CH_3OH, 20°, 4 hr	$FC(NO_2)_2CH_2OCH_2\overset{O}{CH—CH_2}$ (53) $FC(NO_2)_2CH_2OCH_3$ (12)	369
C_6	$N{\equiv}CCH(CH_3)CO_2C_2H_5$	$NaOC_2H_5$, C_2H_5OH, −10 to −15°	$HN{=}C(OC_2H_5)CF(CH_3)CO_2C_2H_5$ (75)	228
		K, DMF, 0–15°	$N{\equiv}CCF(CH_3)CO_2C_2H_5$ (40)	228

Note: References 301–386 are on pp. 404–406.

TABLE XVIII. MONOFLUORINATIONS BY PERCHLORYL FLUORIDE (*Continued*)

	Reactant	Base Used To Form Anion, Reaction Condition(s)	Product(s) and Yield(s) (%)	Refs.
C_6 (*contd.*)	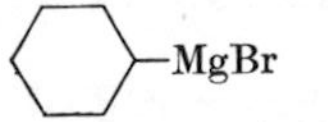 —MgBr	THF/ether/HMPT (1:1:1), −78°, inverse addition to ClO_3F	$C_6H_{11}F$ (68), C_6H_{12} (25)	232
	C_6H_5Li	THF/ether (1:1), −78°	C_6H_5F (42), C_6H_6 (12)	232
	C_6H_5MgBr	Ether, −78°	C_6H_5F (36), C_6H_6 (50)	232
C_7	Diethyl malonate	See Table VII, p. 227	See Table VII, p. 227	225
	$Na^+[C(NO_2)(CO_2C_2H_5)_2]^-$	CH_3CN, 5–10°	$FC(NO_2)(CO_2C_2H_5)_2$ (90)	20
		DMF, −15°	" (70)	234
	$N{\equiv}CCH(C_2H_5)CO_2C_2H_5$	Na, C_2H_5OH, 10–15°	$HN{=}C(OC_2H_5)CF(C_2H_5)CO_2C_2H_5$ (80)	228
		Na, $C_6H_5CH_3$, 10–15°	$N{\equiv}CCF(C_2H_5)CO_2C_2H_5$ (80)	228
	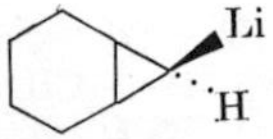 Li, H	THF/pet. ether (2:1), inverse addition to ClO_3F	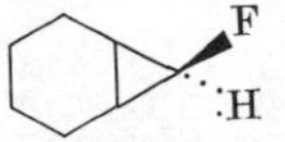 F, H (83) Norcarane (17)	232
	$C_6H_5CH_2Li$	THF/ether (1:1), −78°, inverse addition to ClO_3F	$C_6H_5CH_2F$ (50), bibenzyl (7)	232
	$C_6H_5CH_2MgBr$	HMPT/THF (1:1), −78°, inverse addition to ClO_3F	$C_6H_5CH_2F$, **A** (39), toluene, **B** (23), bibenzyl, **C** (27)	232
		HMPT/THF (1:1), −78°	**A** (10), **B** (20), **C** (65)	232
	$K^+[C_6H_5C(NO_2)_2]^-$	CH_3OH, 25°, 1½ hr	$C_6H_5CF(NO_2)_2$ (95)	233
	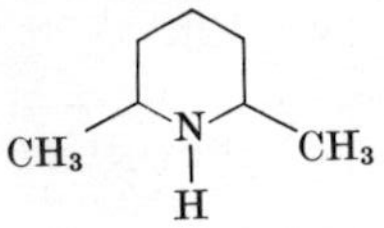 CH_3, N, H, CH_3	2% NaOH	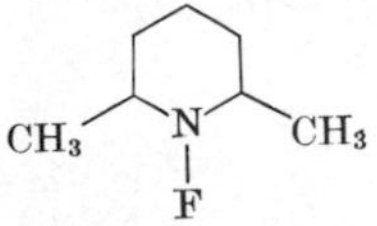 CH_3, N, F, CH_3 (−) (easily decomposed oil; not purified)	241

	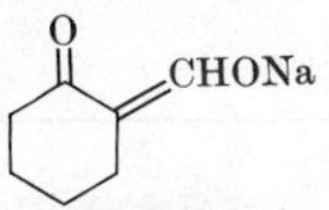 [ring structure: O, =CHONa]	C_2H_5OH (eliminates formyl)	2-Fluorocyclohexanone (40)	236
C_8	$K^+[C(NO_2)_2C_6H_5]^-$	CH_3OH, 25°	$FC(NO_2)_2C_6H_5$ (95)	368
	$HC(CH_3)(CO_2C_2H_5)_2$	Na, $C_6H_5CH_3$, 10–15°	$FC(CH_3)(CO_2C_2H_5)_2$ (84)	229
	$N{\equiv}CCH(C_3H_7\text{-}n)CO_2C_2H_5$	Na, C_2H_5OH, 10–15°	$HN{=}C(OC_2H_5)CF(C_3H_7\text{-}n)CO_2C_2H_5$ (75)	228
		Na, $C_6H_5CH_3$, 10–15°	$N{\equiv}CCF(C_3H_7\text{-}n)CO_2C_2H_5$ (48)	228
	$N{\equiv}CCH(C_3H_7\text{-}i)CO_2C_2H_5$	Na, C_2H_5OH, 10–15°	$HN{=}C(OC_2H_5)CF(C_3H_7\text{-}i)CO_2C_2H_5$ (72)	228
		Na, $C_6H_5CH_3$, 10–15°	$N{\equiv}CCF(C_3H_7\text{-}i)CO_2C_2H_5$ (80)	228
	Bromocyclooctatetraene	n-C_4H_9Li, ether, −75°	Fluorocyclooctatetraene (10)	370
	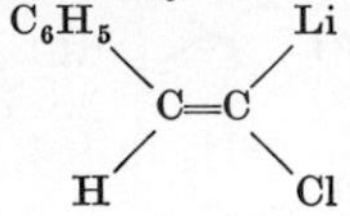 [structure: C_6H_5, Li; C=C; H, Cl]	Ether, −100°, 15 min	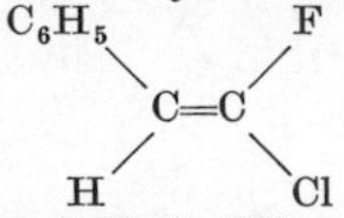 [structure: C_6H_5, F; C=C; H, Cl] (45), $C_6H_5C{\equiv}CCl$ (18), *trans*-$C_6H_5CH{=}CHCl$ (20)	232
	$C_6H_5COCH_2Li$	THF/$C_6H_5CH_3$ (2:1)	$C_6H_5COCH_2F$ (44), $C_6H_5COCH_3$ (32)	232
	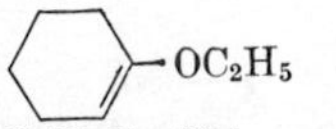 [ring structure: OC_2H_5]	Pyridine, 0°	2-Fluorocyclohexanone (—)	238
C_9	$C_6H_5C{\equiv}CLi$	Ether/pentane (1:1), −78°	$(C_6H_5CH{=}CH)_2$ (32)	240
	$N{\equiv}CCH(C_4H_9\text{-}n)CO_2C_2H_5$	Na, C_2H_5OH, 10–15°	$HN{=}C(OC_2H_5)CF(C_4H_9\text{-}n)CO_2C_2H_5$ (77)	228
		Na, $C_6H_5CH_3$, 10–15°	$N{\equiv}CCF(C_4H_9\text{-}n)CO_2C_2H_5$ (47)	228
	$N{\equiv}CCH(C_4H_9\text{-}i)CO_2C_2H_5$	Na, C_2H_5OH, 10–15°	$HN{=}C(OC_2H_5)CF(C_4H_9\text{-}n\text{-})CO_2C_2H_5$ (80)	228
		Na, $C_6H_5CH_3$, 10–15°	$N{\equiv}CCF(C_4H_9\text{-}i)CO_2C_2H_5$ (78)	228
	$N{\equiv}CCH(C_4H_9\text{-}sec)CO_2C_2H_5$	Na, C_2H_5OH, 10–15°	$HN{=}C(OC_2H_5)CF(C_4H_9\text{-}sec)CO_2C_2H_5$ (80)	228
		Na, $C_6H_5CH_3$, 10–15°	$N{\equiv}CCF(C_4H_9\text{-}sec)CO_2C_2H_5$ (46)	228
	$HC(CH_3)_2COC(CH_3)_2CO_2CH_3$	$(C_6H_5)_3CNa$, ether	$FC(CH_3)_2COC(CH_3)_2CO_2CH_3$ (30)	329
	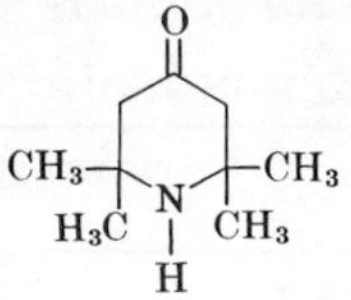 [ring structure: O; CH_3, H_3C, N–H, CH_3, CH_3]	Ether	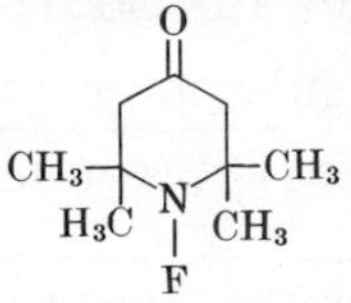 [ring structure: O; CH_3, H_3C, N–F, CH_3, CH_3] (80)	241
		2% NaOH	" (50)	

Note: References 301–386 are on pp. 404–406.

TABLE XVIII. MONOFLUORINATIONS BY PERCHLORYL FLUORIDE (*Continued*)

	Reactant	Base Used To Form Anion, Reaction Condition(s)	Product(s) and Yield(s) (%)	Refs.
C_9 (*contd.*)	$C_6H_5CONHCH_2C(NO_2)_2H$	CH_3ONa, CH_3OH, −20°, 1.5 hr	$HN{=}C(OC_2H_5)CH_2C(NO_2)_2F$ (82)	235b
C_{10}	$N{\equiv}CCH(CH_2CH_2CO_2C_2H_5)CO_2C_2H_5$	Na, C_2H_5OH, 10–15°	$C_2H_5OC(NH)CF(CH_2CH_2CO_2C_2H_5)$-$CO_2C_2H_5$ (90)	228
		Na, $C_6H_5CH_3$, 10–15°	$N{\equiv}CCF(CH_2CH_2CO_2C_2H_5)CO_2C_2H_5$ (71)	228
	β-Lithionaphthalene	THF/ether (1:1), −78°	β-$C_{10}H_7F$ (45), $C_{10}H_8$ (30)	232
	Lithioferrocene	Ether/THF (1:2), −70°	Fluoroferrocene (10)	247
C_{11}	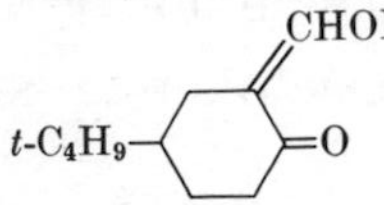 CHOH, t-C_4H_9, O	Na, C_2H_5OH, 0° (eliminates formyl)	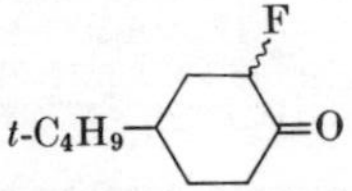 F, t-C_4H_9, O; *trans* (9), *cis* (49)	236a
	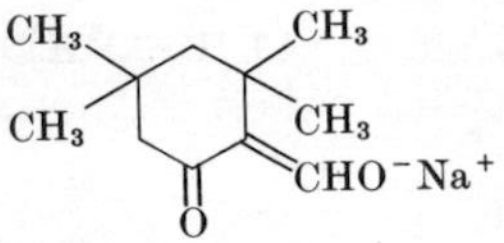 CH_3, CH_3, CH_3, CH_3, O, CHO^-Na^+	C_2H_5OH, 0–5°, 30 min; then KOAc, 100° (eliminates formyl)	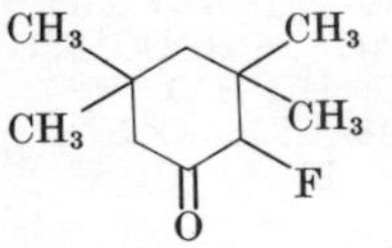 CH_3, CH_3, CH_3, CH_3, O, F (58)	237
	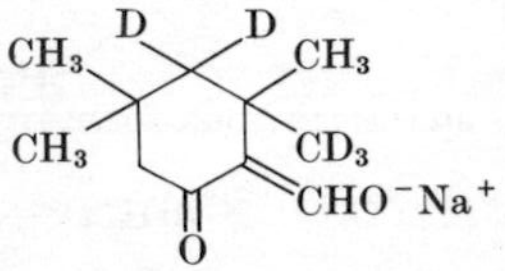 D, D, CH_3, CH_3, CH_3, CD_3, O, CHO^-Na^+	C_2H_5OH, 0–5°, 30 min; then KOAc, 100° (eliminates formyl)	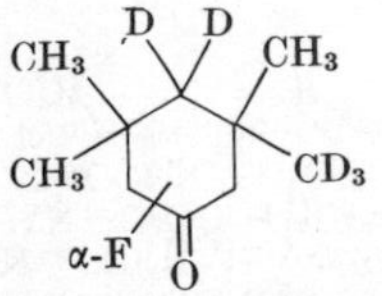 D, D, CH_3, CH_3, CH_3, CD_3, α-F, O (30); 4 isomers	237
C_{12}	$HC(C_5H_{11}$-$n)(CO_2C_2H_5)_2$	Na, $C_6H_5CH_3$, 10–15°	$FC(C_5H_{11}$-$n)(CO_2C_2H_5)_2$ (75)	229
	$N{\equiv}CCH(CH_2C_6H_5)CO_2C_2H_5$	Na, C_2H_5OH, 10–15°	$HN{=}C(OC_2H_5)CF(CH_2C_6H_5)CO_2C_2H_5$ (75)	228
		Na, $C_6H_5CH_3$, 10–15°	$N{\equiv}CCF(CH_2C_6H_5)CO_2C_2H_5$ (49)	228

	$n\text{-}C_{12}H_{25}Li$	THF/pet. ether (2:1), −78°, inverse addition to $FClO_3$	$n\text{-}C_{12}H_{25}F$ (39), $n\text{-}C_{12}H_{26}$ (20)	233
	2,2′-Dilithiobiphenyl	THF/$C_6H_5CH_3$ (2:1), inverse addition to $FClO_3$	2,2′-Difluorobiphenyl (94)	232
C_{13}	$HC(C_6H_{13}\text{-}n)(CO_2C_2H_5)_2$	Na, $C_6H_5CH_3$, 10–15°	$FC(C_6H_{13}\text{-}n)(CO_2C_2H_5)_2$ (70)	229
	$(CH_3O_2C)_2C(NHCOCH_3)CH_2CH(CO_2CH_3)_2$	CH_3ONa, CH_3OH	$(CH_3O_2C)_2C(NHCOCH_3)$, $CH_2CF(CO_2CH_3)_2$ (80)	371
C_{14}	(3,4-methylenedioxyphenyl)–$CH_2CH(CO_2C_2H_5)_2$	—	(3,4-methylenedioxyphenyl)–$CH_2CF(CO_2C_2H_5)_2$ (−)	372
	$HC(C_7H_{15}\text{-}n)(CO_2C_2H_5)_2$	Na, $C_6H_5CH_3$, 10–15°	$FC(C_7H_{15}\text{-}n)(CO_2C_2H_5)_2$ (64)	229
	$p\text{-}CH_3OC_6H_4CH(CO_2C_2H_5)_2$[a]	NaH, DMF/C_2H_5OH, 0–10°	$p\text{-}CH_3OC_6H_4CF(CO_2C_2H_5)_2$ (85) (86, 91)[a]	230, 231
C_{15}	$HC(C_8H_{17}\text{-}n)(CO_2C_2H_5)_2$	Na, $C_6H_5CH_3$, 10–15°	$FC(C_8H_{17}\text{-}n)(CO_2C_2H_5)_2$ (64)	229
	$p\text{-}C_2H_5OC_6H_4CH(CO_2C_2H_5)_2$[a]	NaH, DMF/C_2H_5OH, 0–10°	$p\text{-}C_2H_5OC_6H_4CF(CO_2C_2H_5)_2$ (92) (70, 70)[a]	230, 231
	Diethyl (1-naphthyl)malonate[a]	NaH, DMF/C_2H_5OH, 0–10°	$(1\text{-}C_{10}H_7)CF(CO_2C_2H_5)_2$ (77) (77, 85)[a]	230, 231
C_{16}	$HC(C_9H_{19}\text{-}n)(CO_2C_2H_5)_2$	Na, $C_6H_5CH_3$, 10–15°	$FC(C_9H_{19}\text{-}n)(CO_2C_2H_5)_2$ (57)	229
	CH_3, CH_3, C_6H_5, CH_3, O, $=CH_2O^-Na^+$	C_2H_5OH, 0–5°, 30 min; then KOAc, 100° (eliminates formyl)	CH_3, CH_3, C_6H_5, CH_3, O, F: A, —F (69); B, ···F (5); CH_3, CH_3, C_6H_5, CH_3, F, F, O (3)	237
	C_6H_5, CH_3, CH_3, CH_3, $=CHO^-Na^+$	C_2H_5OH, 0–5°, 1 hr; then KOAc, 100° (eliminates formyl)	C_6H_5, CH_3, CH_3, CH_3, O, F: C, —F (78); D, ···F (9); C_6H_5, CH_3, CH_3, CH_3, F, F (2)	237

Note: References 301–386 are on pp. 404–406.

TABLE XVIII. MONOFLUORINATIONS BY PERCHLORYL FLUORIDE (*Continued*)

	Reactant	Base Used To Form Anion, Reaction Condition(s)	Product(s) and Yield(s) (%)	Refs.
C_{17}	$HC(C_{10}H_{21}\text{-}n)(CO_2C_2H_5)_2$	Na, $C_6H_5CH_3$, 10–15°	$FC(C_{10}H_{21}\text{-}n)(CO_2C_2H_5)_2$ (71)	229
	$p\text{-}(i\text{-}C_4H_9)C_6H_4CH(CO_2C_2H_5)_2$[b]	NaH, DMF/C_2H_5OH, 0–10°	$p\text{-}(i\text{-}C_4H_9)C_6H_4CF(CO_2C_2H_5)_2$ (70) (86)[b]	230, 231
C_{19}	$HC(C_{12}H_{25}\text{-}n)(CO_2C_2H_5)_2$	Na, $C_6H_5CH_3$, 10–15°	$FC(C_{12}H_{25}\text{-}n)(CO_2C_2H_5)_2$ (87)	229
	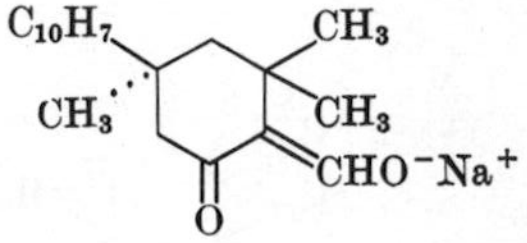 [structure: $C_{10}H_7$, CH_3, CH_3, CH_3, O, CHO^-Na^+] $C_{10}H_7$ = 1-napthyl	C_2H_5OH, 0–5°, 30 min; then KOAc, 100° (eliminates formyl)	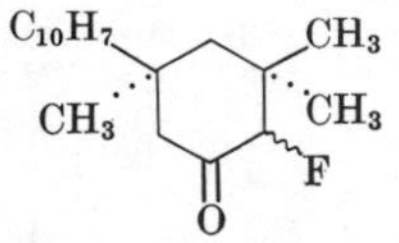 [structure: $C_{10}H_7$, CH_3, CH_3, CH_3, F, O] A, —F (73) B, ···F (5)	237
C_{21}	$HC(C_{14}H_{29}\text{-}n)(CO_2C_2H_5)_2$	Na, $C_6H_5CH_3$, 10–15°	$FC(C_{14}H_{29}\text{-}n)(CO_2C_2H_5)_2$ (89)	229
C_{22}	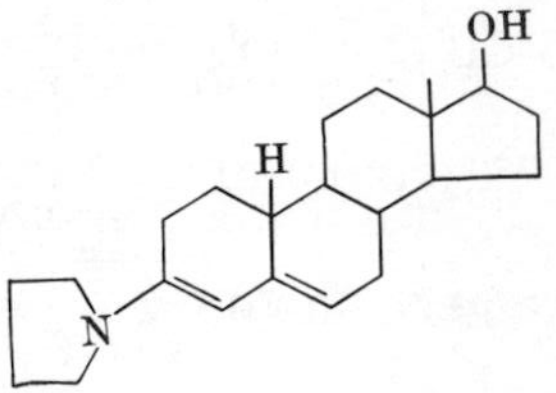 [structure: OH, H, N]	90% methanol −25°, 10 min	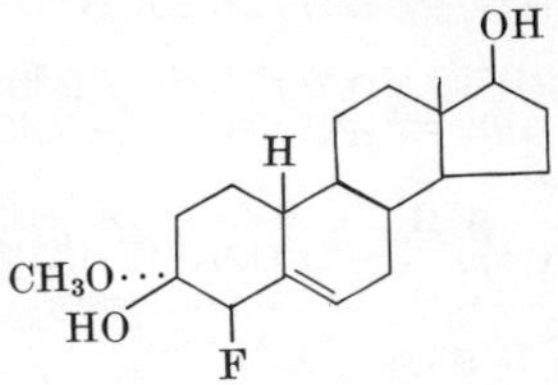 [structure: OH, H, CH_3O···, HO, F] (55)	373
C_{23}	Androsta-5,14,16-triene-3β,17-diol diacetate	THF/H_2O, room t, 50 min	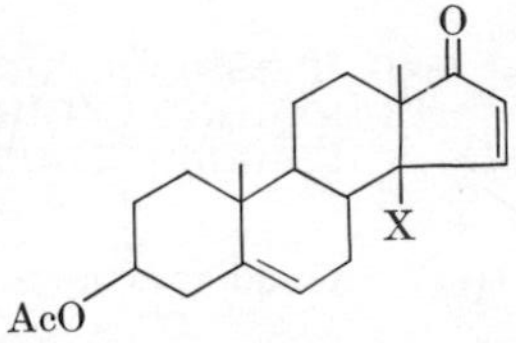 [structure: O, X, AcO] X = F (49), X = OH (8)	374

	Reactant	Conditions	Product(s) and Yield(s) (%)	Refs.
	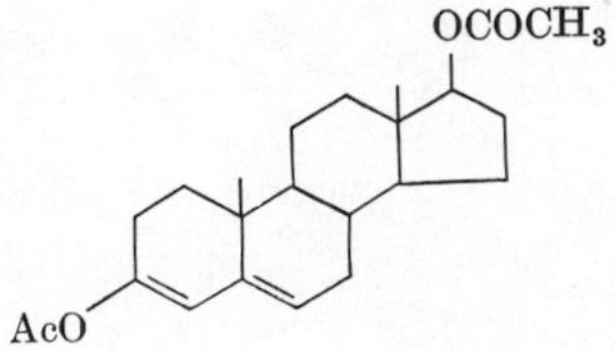 AcO, $OCOCH_3$ (structure)	95% aq dioxane, 22°, 4 hr (diluted to 60% during reaction)	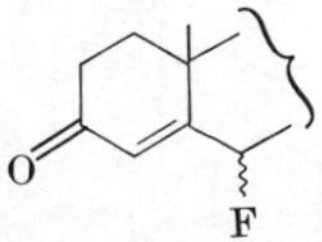 O, F (structure) —F(30), ···F(12), also three nonfluorinated steroids (18)	239
	$HC(C_{16}H_{31}\text{-}n)(CO_2C_2H_5)_2$	Na, $C_6H_5CH_3$, 10–15°	$FC(C_{16}H_{31}\text{-}n)(CO_2C_2H_5)_2$ (89)	229
C_{24}	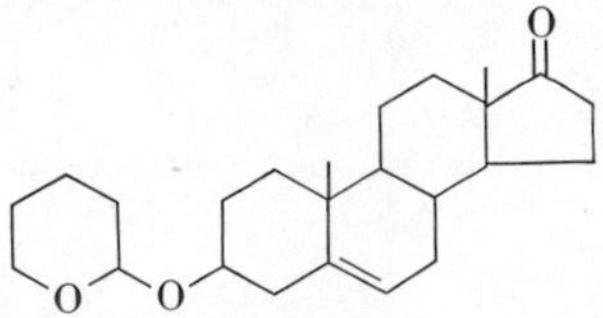 O (structure)	$LiN(C_3H_7\text{-}i)_2$, $C_6H_5CH_3$/THF (2:1)	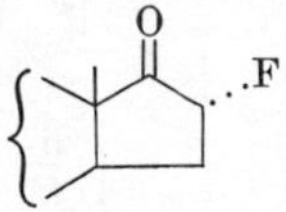 O, F (structure) (71)	232
C_{25}	$HC(C_{18}H_{37}\text{-}n)(CO_2C_2H_5)_2$	Na, $C_6H_5CH_3$, 10–15°	$FC(C_{18}H_{37}\text{-}i)(CO_2C_2H_5)_2$ (89)	232
C_{27}	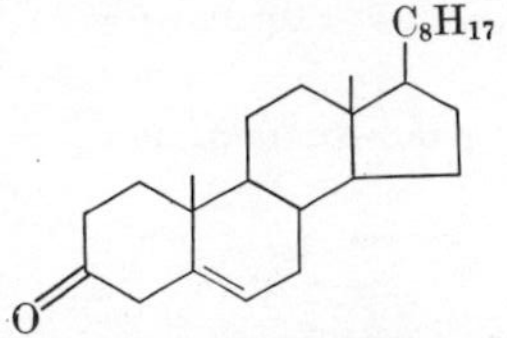 C_8H_{17}, O (structure)	$LiN(C_3H_7\text{-}i)_2$, $C_6H_5CH_3$, −78°	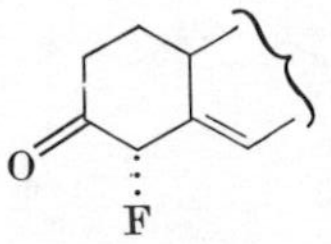 O, F (structure) (45)	232

Note: References 301–386 are on pp. 404–406.

[a] This reaction was also carried out with the mono- and di-potassium salts formed by saponification of the ester groups.
[b] This reaction was also carried out with the dipotassium salt obtained by the saponification of the ester groups.

TABLE XIX. Fluorination with Nitrosyl Fluoride

	Reactant	Condition(s)[a]	Product(s) and Yield(s) (%)	Refs.
C_5	2-Methylbut-2-ene	0°, CCl_4	$[(CH_3)_2CFCH(CH_3)N(\rightarrow O)]_2$ (12, major), $(CH_3)_2CFCOCH_3$ (—)	259b
C_{19}	3-Fluoro-2-androsten-17-one	25°, 16 hr, CH_2Cl_2	(—)[b]	375
C_{20}	Estr-4-en-17β-ol acetate	0°, CH_2Cl_2	(14)[c], (23)	259b

C_{21}	Androst-4-en-17β-ol acetate	0°, CH_2Cl_2	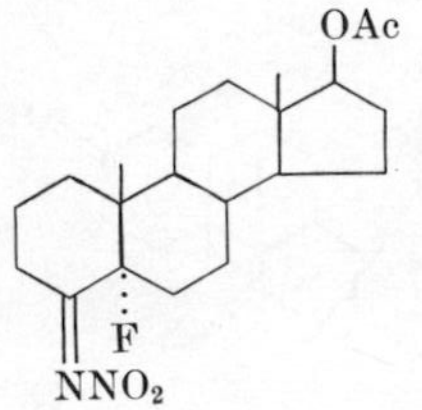 OAc, F, NNO_2 (67)	259b
	Pregna-4-16-diene-3,20-dione	0°, CH_2Cl_2	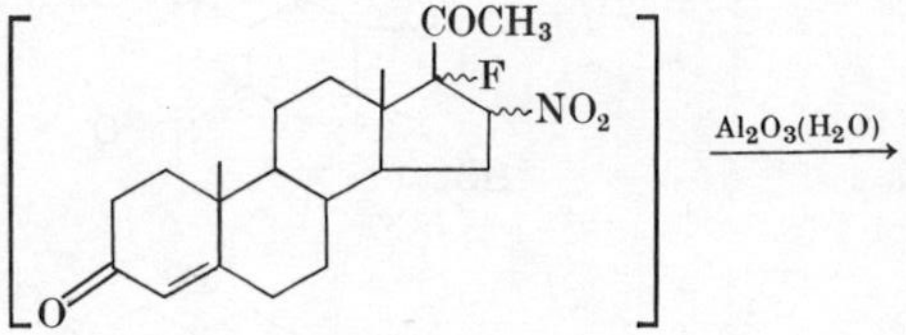 [$COCH_3$, F, NO_2, O] $\xrightarrow{Al_2O_3(H_2O)}$ $COCH_3$, NO_2 (16)[d]	259b

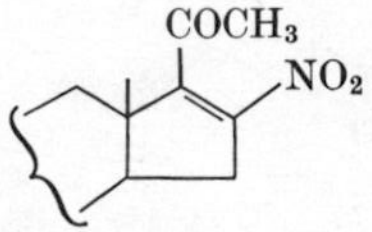

Note: References 301–386 are on pp. 404–406.

[a] Unless indicated otherwise, nitrosyl fluoride was added to reaction mixture as a slow stream.

[b] This product was isolated and purified as the corresponding 2-ketone by chromatography. Several additional examples of 2,keto-3,3-difluoro steroids were reported to have been prepared by this method.[375] Also 17,17-difluoro-16-keto (or 16-nitrimino) steroids have been prepared in a similar way.[376]

[c] This product was isolated as 5α-fluoro-4-ketone by chromatography.

[d] Only nitroolefin was isolated from chromatography workup.

TABLE XIX. FLUORINATION WITH NITROSYL FLUORIDE (*Continued*)

	Reactant	Condition(s)[a]	Product(s) and Yield(s) (%)	Refs.
C_{21} (*contd.*)	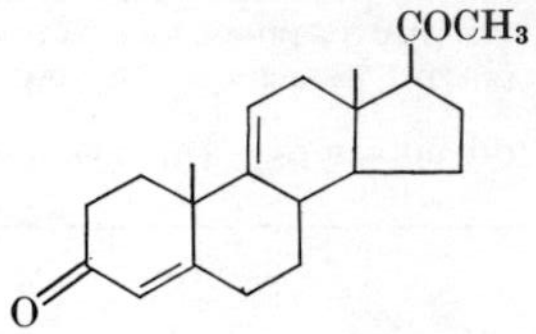 COCH$_3$	3°, 10 d, CH_2Cl_2	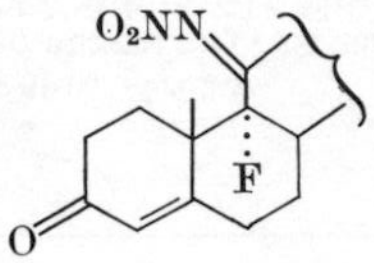 O_2NN, F, O (10)	263
	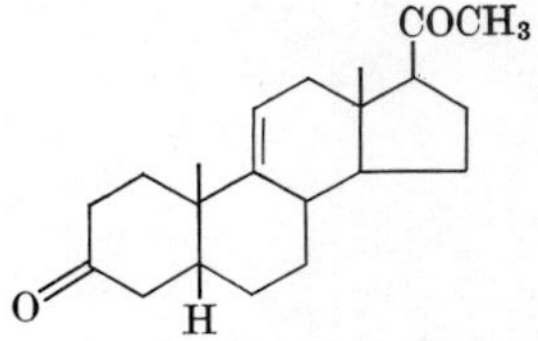 COCH$_3$, H, O	3°, 10 d, CH_2Cl_2	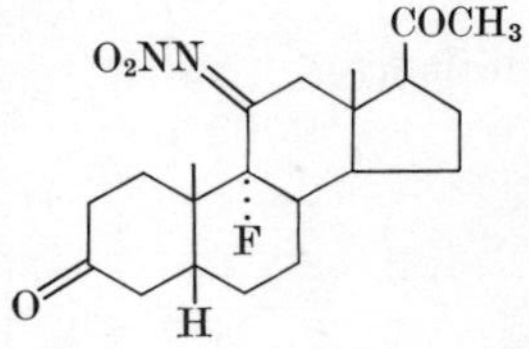 O_2NN, COCH$_3$, F, H, O (–)	263
	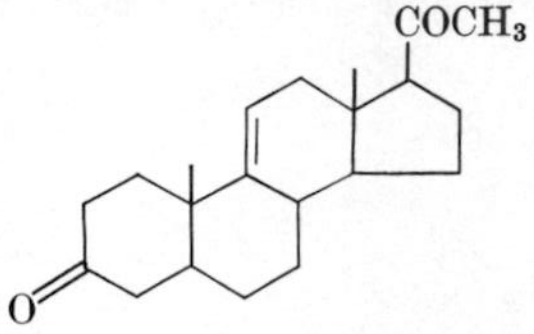 COCH$_3$, O	3°, 10 d, CH_2Cl_2	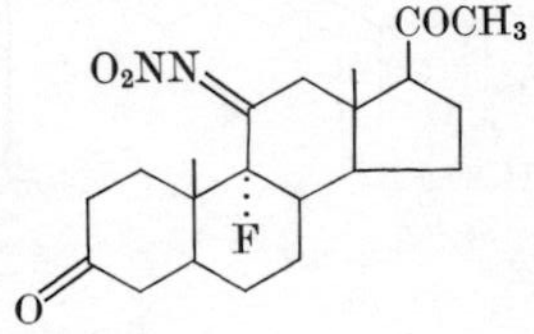 O_2NN, COCH$_3$, F, O (15)	263
C_{22}	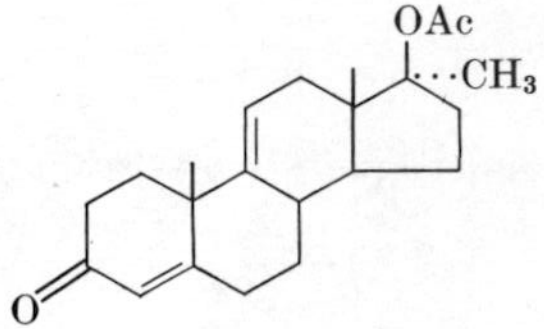 OAc, CH$_3$, O	3°, 10 d, CH_2Cl_2	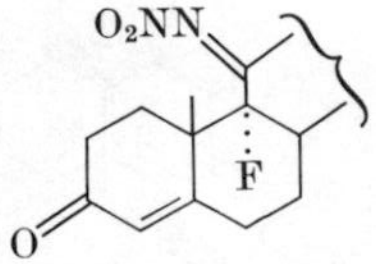 O_2NN, F, O (23)	263

	Estr-5-ene-3β,17β-diol diacetate	0°, CH_2Cl_2 or CCl_4	(20)[e],	259b, 377
			(23)	
C_{23}	Androst-5-ene-3β,17β-diol diacetate	0°, CH_2Cl_2 or CCl_4	(52)	259b

Note: References 301–386 are on pp. 404–406.

[a] Unless indicated otherwise, nitrosyl fluoride was added to reaction mixture as a slow stream.
[e] On standing, the dimer reverted to monomer and tautomerized to the 5α-fluoro-6-oxime.

TABLE XIX. Fluorination with Nitrosyl Fluoride (*Continued*)

	Reactant	Condition(s)[a]	Product(s) and Yield(s) (%)	Refs.
C_{23} (*contd.*)	3β-Hydroxypregna-5,16-dien-20-one acetate	0°, CCl_4 or CH_2Cl_2	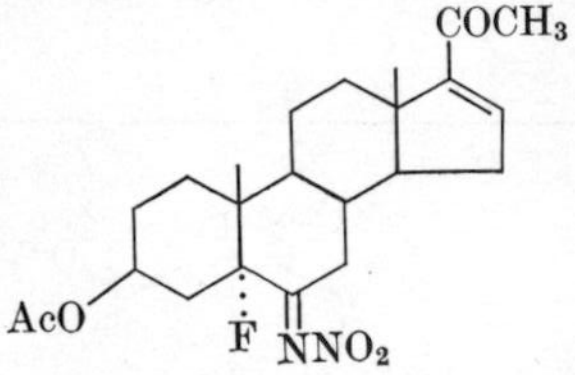 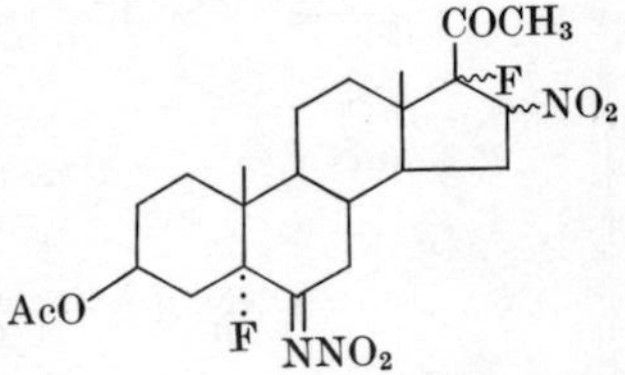 [structure: $COCH_3$, AcO, F, NNO_2] (—)[f], [structure: $COCH_3$, F, NO_2, AcO, F, NNO_2] (—)[g] $\xrightarrow{Al_2O_3(H_2O)}$ 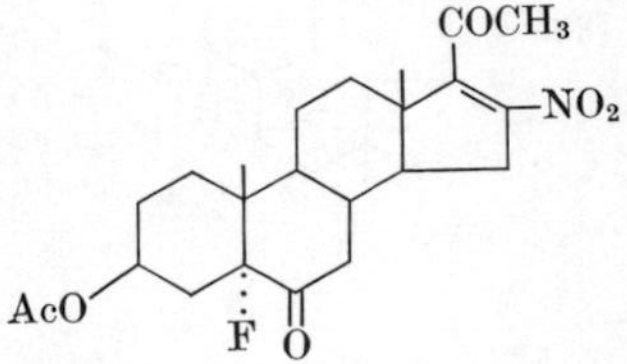 [structure: $COCH_3$, NO_2, AcO, F, O] (9)	259b
	3β-Hydroxy-5α-pregn-9(11)-en-20-one acetate	3°, 10 d, CH_2Cl_2	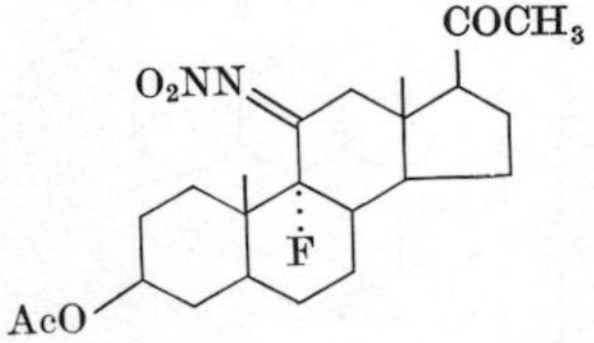 [structure: $COCH_3$, O_2NN, F, AcO] (45)	263

	Reactant	Conditions[a]	Product(s) and Yield(s) (%)		Refs.
	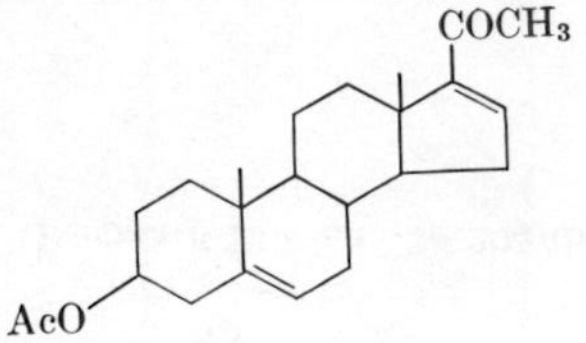 COCH$_3$ / AcO (structure)	0°, 3 hr, CH_2Cl_2	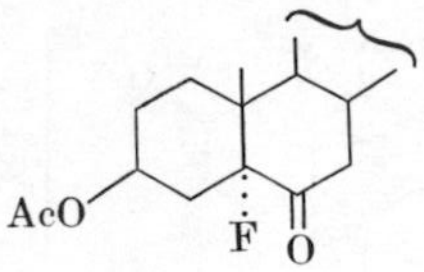 AcO, F, O (structure)	(55)[h]	378
		0°, 5 hr, CH_2Cl_2/ glyme, $NOBF_4$	"	(66)[h]	378
C_{24}	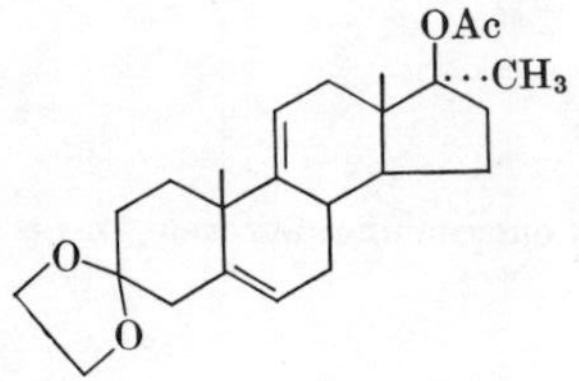 OAc, CH$_3$ (structure)	3°, 10 d, CH_2Cl_2	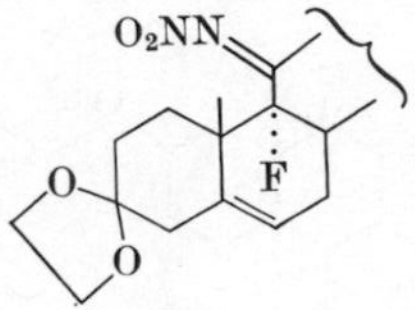 O_2NN, F (structure)	(70)	263
C_{25}	Pregn-4-ene-3β,20β-diol diacetate	0°, CCl_4	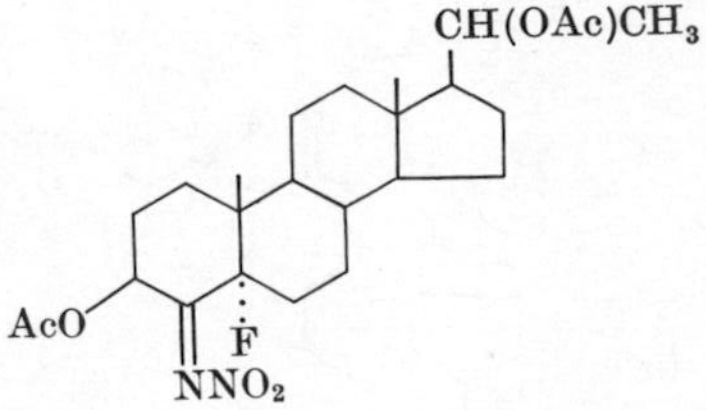 CH(OAc)CH$_3$, AcO, F, NNO$_2$ (structure)	(—)	259b

Note: References 301–386 are on pp. 404–406.

[a] Unless indicated otherwise, nitrosyl fluoride was added to reaction mixture as a slow stream.
[f] The product was isolated as the corresponding 5α-fluoro-6-ketone (≤1%) by chromotography on neutral alumina.
[g] This product not isolated but converted to 5α-fluoro-6-keto-16-nitroolefin in the workup.
[h] The products was isolated by chromatography on neutral alumina.

TABLE XIX. FLUORINATION WITH NITROSYL FLUORIDE (*Continued*)

	Reactant	Condition(s)[a]	Product(s) and Yield(s) (%)	Refs.
C_{25} (*contd.*)	17α,20;20,21-Bismethylenedeoxy-3β-hydroxypregn-5-en-11-one 3-acetate	0°, CCl_4	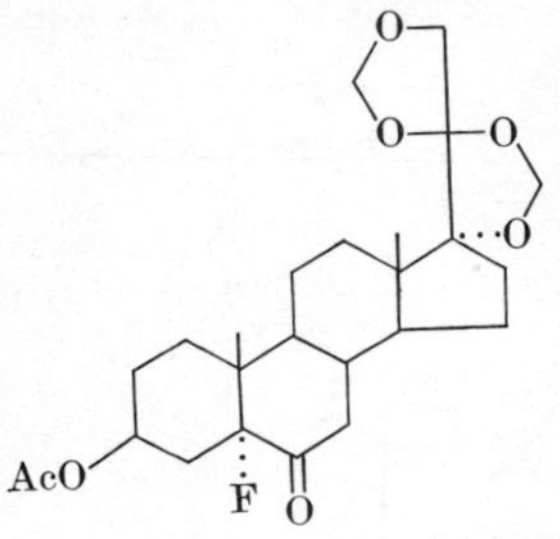 (50) AcO, F, O, O, O, O, O	379
	17-α-Acetoxypregnenolone acetate	CCl_4, 0°	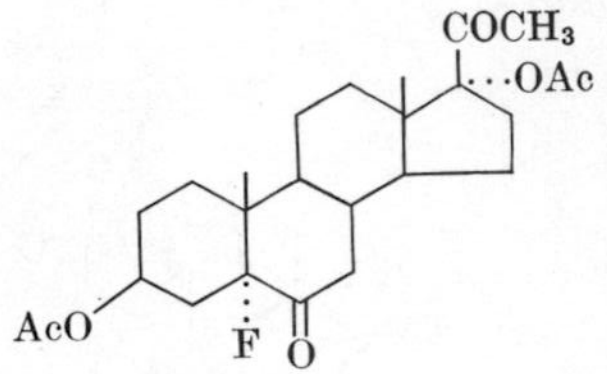 COCH₃ / OAc, AcO, F, O (76)[i,j]	260
	Pregna-5,16-diene-3β,20α-diol diacetate	—	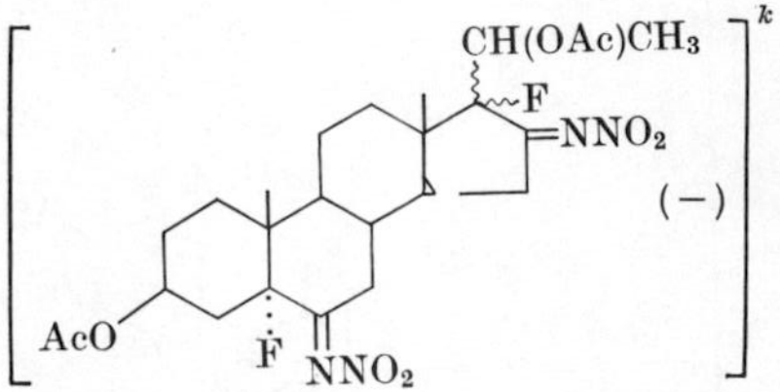 [CH(OAc)CH₃, F, =NNO₂, AcO, F, NNO₂ (−)][k] $\xrightarrow{Al_3O_3(H_2O)}$	259b

Reactant	Conditions	Product(s) and Yield(s) (%)	Refs.
		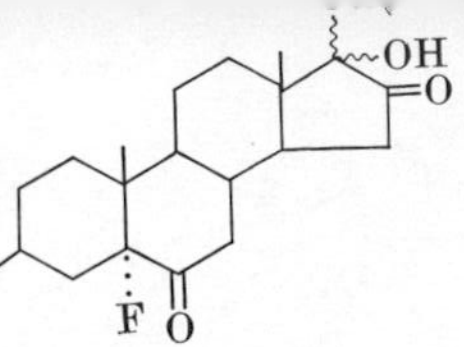 (28)	
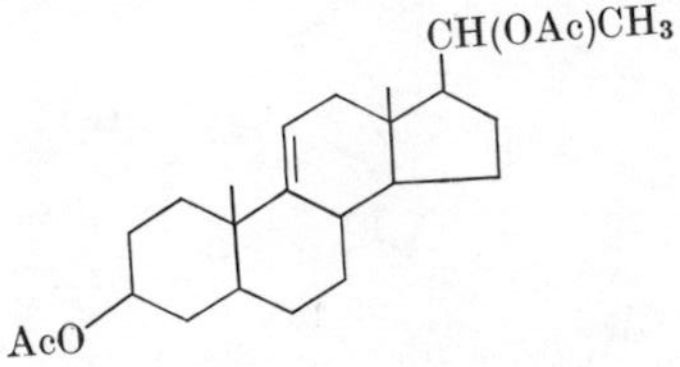 AcO-steroid, CH(OAc)CH$_3$	3°, 10 d, CH_2Cl_2	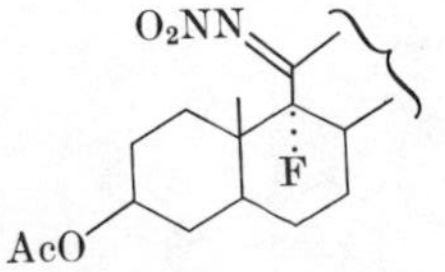 O_2NN=, F, AcO (69)	263
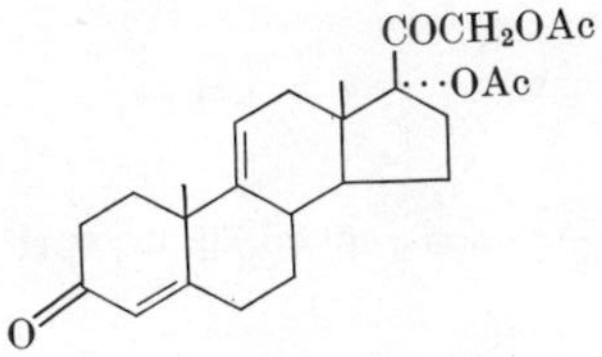 COCH$_2$OAc, ···OAc, O=	3°, 10 d, CH_2Cl_2	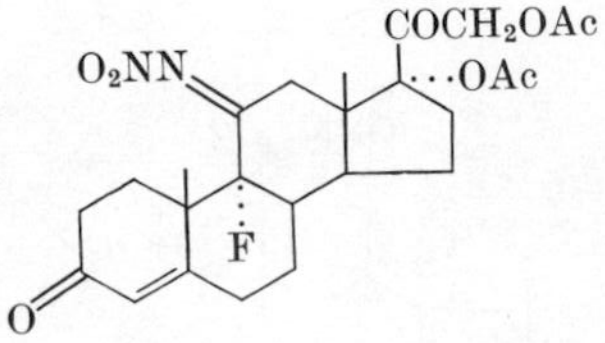 O_2NN=, COCH$_2$OAc, ···OAc, F (10–15)	263
3β,17α-Diacetoxypregn-5-en-20-one	0°, CH_2Cl_2 or CCl_4	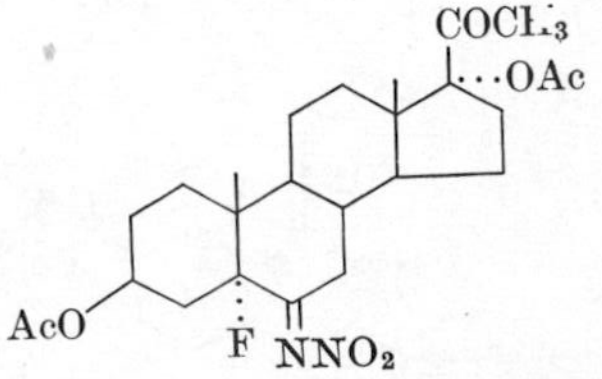 COCH$_3$, ···OAc, AcO, F, NNO$_2$ (68)	259b

Note: References 301–386 are on pp. 404–406.

[a] Unless indicated otherwise, nitrosyl fluoride was added to reaction mixture as a slow stream.

[i] The 5α-fluoro-6-nitrimine intermediate was not isolated.

[j] A series of 5α-fluoro-6-keto steroids of the pregnane and androstane families was also prepared by this method without isolation of the intermediate 5-fluoro-6-nitrimines.[259b,380]

[k] This material was neither isolated nor characterized.

TABLE XIX. Fluorination with Nitrosyl Fluoride (*Continued*)

	Reactant	Condition(s)[a]	Product(s) and Yield(s) (%)	Refs.
C_{25} (*contd.*)	Pregn-5-ene-3β,20-diol diacetate	0°, CH_2Cl_2 or CCl_4	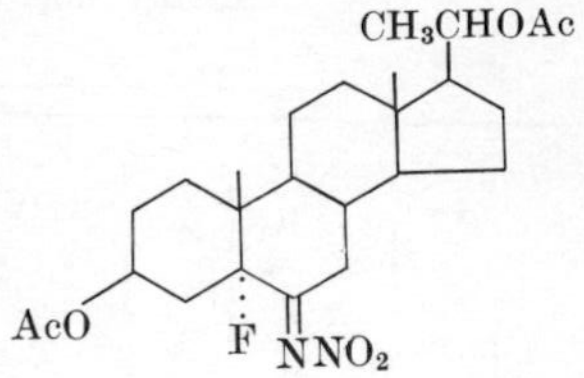 CH_3CHOAc; AcO; F; NNO_2 (70)	259b
	16α-Methylpregn-5-ene-3β,20β-diol diacetate	0°, CH_2Cl_2 or CCl_4	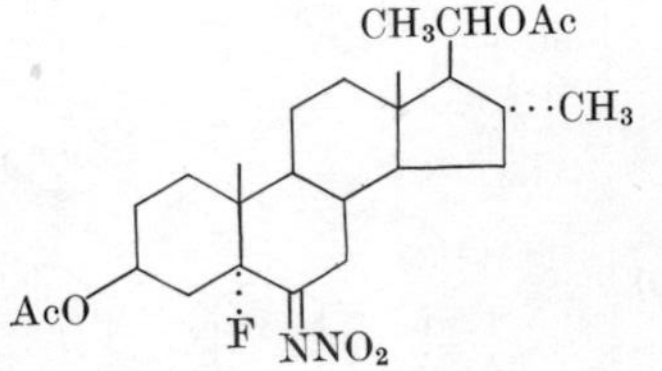 CH_3CHOAc; CH_3; AcO; F; NNO_2 (67)	259b, 380
C_{26}	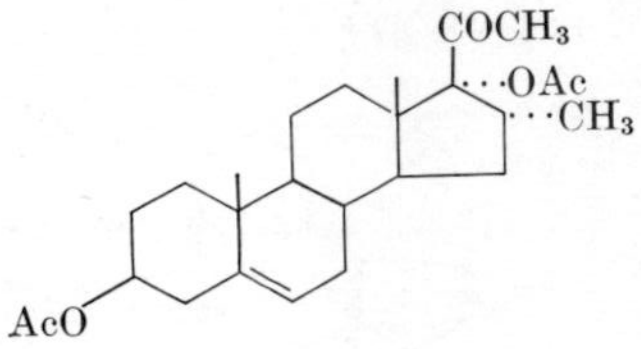 $COCH_3$; OAc; CH_3; AcO	27°, 6–8 hr, CH_2Cl_2	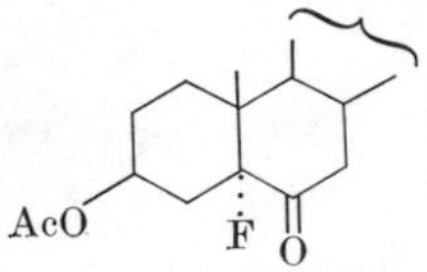 AcO; F; O (69)[h]	381
C_{27}	Cholest-2-ene	20–25°, CH_2Cl_2	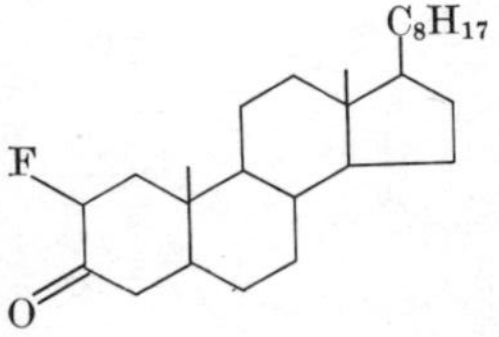 C_8H_{17}; F; O (8),	259b

		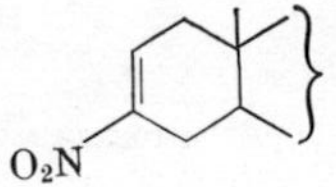 O_2N (15)	
Cholesteryl chloride	CH_2Cl_2 or CCl_4	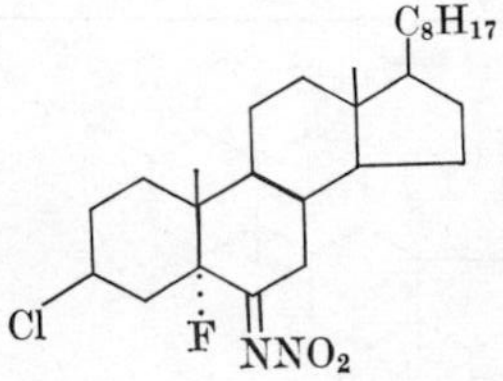 C_8H_{17}, Cl, F, NNO_2 (69)	259b
Cholesterol	CH_2Cl_2 or CCl_4	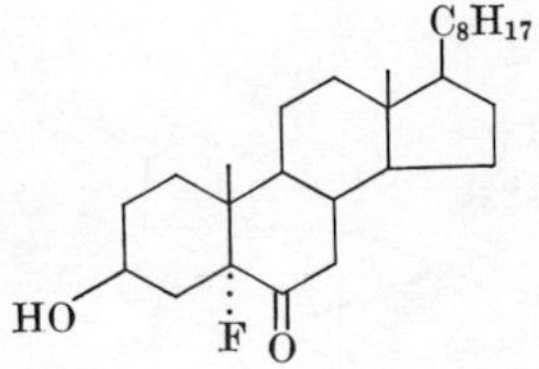 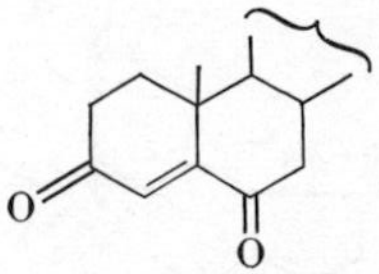 C_8H_{17}, HO, F, O (–)[h],	259b
		O, O (–)[l]	

Note: References 301–386 are on pp. 404–406.

[a] Unless indicated otherwise, nitrosyl fluoride was added to reaction mixture as a slow stream.

[h] The products were isolated by chromatography on neutral alumina.

TABLE XIX. Fluorination with Nitrosyl Fluoride (*Continued*)

	Reactant	Condition(s)[a]	Product(s) and Yield(s) (%)	Refs.
C_{27} (*contd.*)	Cholest-4-en-3-one	Autogenous pressure, 25°	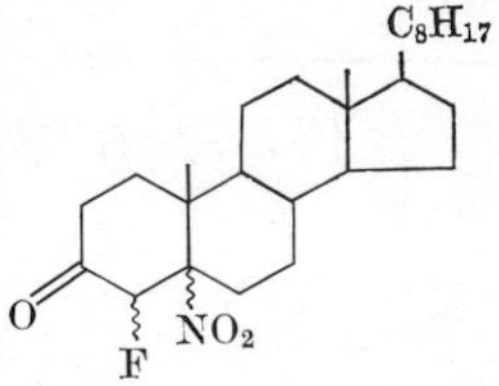 C_8H_{17} O F NO_2 (4)	259b
C_{28}	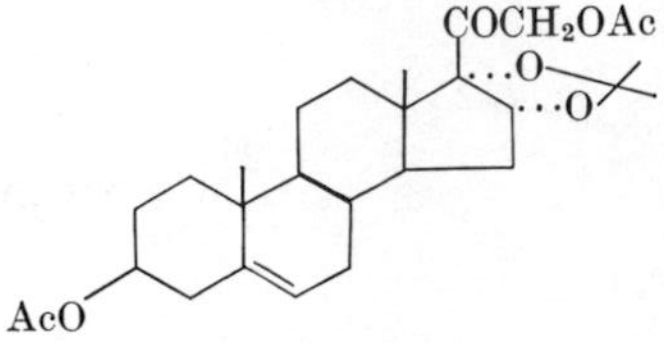 $COCH_2OAc$ O O AcO	0°, 3 hr, CH_2Cl_2, NaF	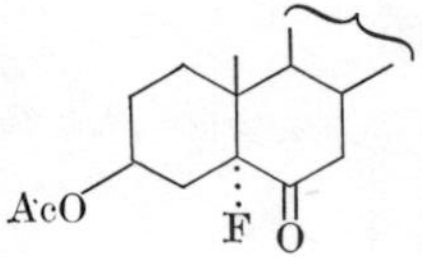 AcO F O (33)[h]	382
C_{29}	Cholesteryl acetate	CH_2Cl_2 or CCl_4	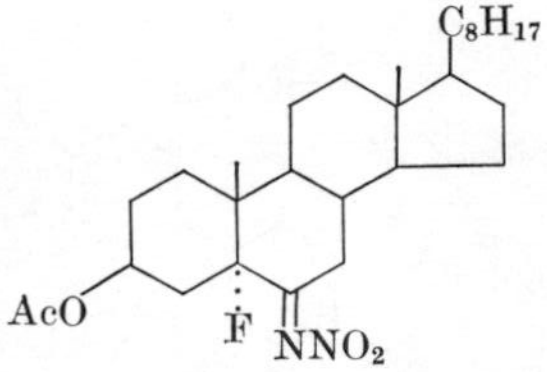 C_8H_{17} AcO F NNO_2 (72)	259

Note: References 301–386 are on pp. 404–406.

[a] Unless indicated otherwise, nitrosyl fluoride was added to reaction mixture as a slow stream.

[h] The products were isolated by chromatography on neutral alumina.

	Reactant	Condition(s)[a]	Product(s) and Yield(s) (%)	Refs.
C_2	Ethylene	$h\nu$, gas phase, 282 torr, N_2 diluent	$CF_3OCH_2CH_2F$ (100)	267b
	Acetic acid	$CFCl_3$, −78°, $h\nu$	CH_2FCO_2H (68)	383a
	Ethylamine	HF, −78°, $h\nu$	$FCH_2CH_2NH_2$ (—)[b]	383a
C_3	Cyclopropane	$h\nu$, gas phase, N_2 diluent	$CF_3OCH_2CH_2CH_2F$ (<5)	267b
	D-Alanine	HF, −78°, $h\nu$, 1.5 hr	$FCH_2CH(NH_2)CO_2H$ (41)	383
C_4	Isobutyric acid	−78°, $CFCl_3$, $h\nu$	$(CH_3)_2CFCO_2H$ (31), $FCH_2CH(CH_3)CO_2H$ (39)	277, 383a
	L-Azetidine-2-carboxylic acid	−78°, HF, $h\nu$	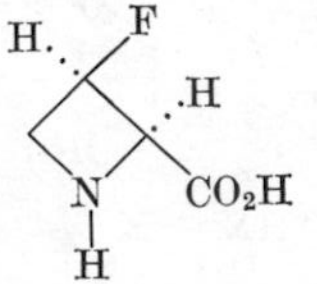 3-fluoroazetidine-2-carboxylic acid [structure: H, F; H; N–H; CO_2H] (53)	277, 383a
	Uracil	−78°, $CH_3OH/CFCl_3$, followed by $(C_2H_5)_3N$[c]	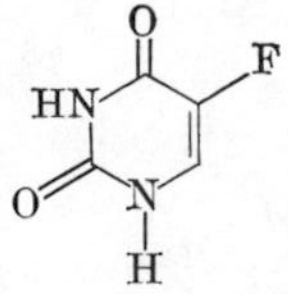 5-fluorouracil [structure: HN, O, F, N–H] **A** (84)	281
		$CFCl_3/H_2O/CF_3CO_2H$, −78° to room t	**A** (85), fluorohydrin (—)[d]	283, 384
	Cytosine	$CH_3OH/CFCl_3$, −78°, 5 min	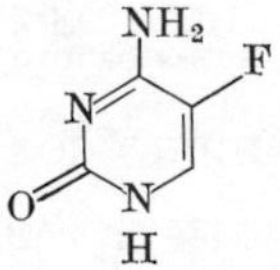 5-fluorocytosine [structure: NH_2, F, N, O, N–H] (85)[c]	282
C_5	1-Methyluracil	$CH_3OH/CFCl_3$, −78°	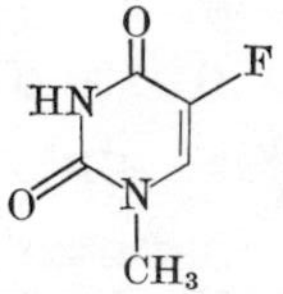 5-fluoro-1-methyluracil [structure: HN, O, F, N–CH_3] (84)[c]	281

[a] The hypofluorite is CF_3OF unless otherwise specified.
[b] Di- and tri-fluorosubstitution products are also reported for ethylamine and other amines.
[c] The product was isolated from workup with 10% triethylamine in 50% aq methanol.

TABLE XX. FLUORINATION WITH PERFLUOROALKYL HYPOFLUORITES (*Continued*)

	Reactant	Condition(s)[a]	Product(s) and Yield(s) (%)	Refs.
C_5 (*contd.*)	Piperidine	HF, −78°, *hν*, 2.5 hr	$FC_5H_9N \cdot HF$ (—)	383
C_6	Benzene	Addition during 1 hr at −78° in $CFCl_3$, *hν*	C_6H_5F (65), $C_6H_5OCF_3$ (10)	277, 383
		Addition during 1 hr at −78° in $CFCl_3$, no *hν*	C_6H_5F (17), complex mixture	277
		hν, gas phase	C_6H_5F (5), $C_6H_5OCF_3$ (5)	267b
	Cyclohexane	−78°, $CFCl_3$, *hν*[e] (also with C_2F_5OF)	Cyclohexyl fluoride (44)	277, 383a
	L-Isoleucine	−78°, HF, 1 hr, *hν*	$FCH_2CH_2CH(CH_3)CH(NH_2)CO_2H$[f] (39)	277, 383a
	ε-Aminocaprolactam	HF, −78°, *hν*, 5.5 hr	Fluoro-ε-aminocaprolactam (—)[g]	383a
C_7	Toluene	−78°, $CFCl_3$, *hν*	2-$FC_6H_4CH_3$ (34), $C_6H_5CH_2F$ (25)	277, 383a
	Anisole	−78°, $CFCl_3$, *hν*	2-$FC_6H_4OCH_3$ (38) 3- and 4-$FC_6H_4OCH_3$ (4)	277, 383a
	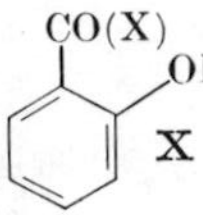 CO(X), OH; X = OH	2.3 $CHCl_3/CFCl_3$, CaO, 0°	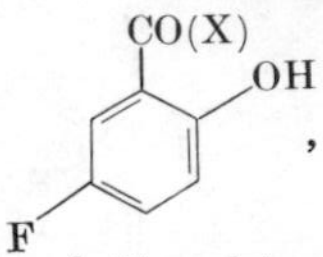 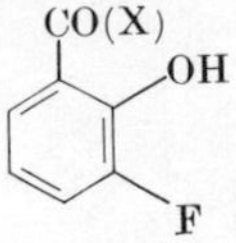 CO(X), OH, F; A (4 parts) X = OH, CO(X), OH, F; B (1 part) X = OH (Total, 70)	276a, 278b
	″ X = NH_2	″	A, X = NH_2 B, X = NH_2 (—)	278b
C_8	2,5-Dimethylphenol	$CFCl_3$, CaO, −75°	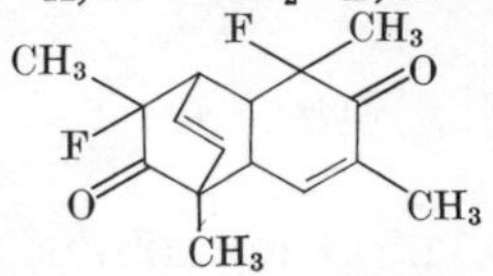 CH_3, F, O, F, CH_3, O, CH_3, CH_3 (−)	276a, 278
	Benzofuran	$CFCl_3$,CaO, −75°	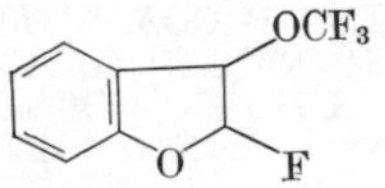 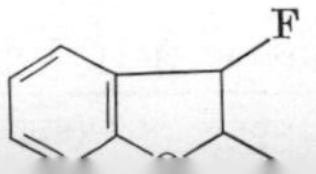 OCF_3, O, F (−), F (−)	276a, 278b

C_9	Cytidine	$CH_3OH/CFCl_3$, −78°, 5 min	H_2N–(pyrimidinone)–N–(tetrahydrofuranyl)–CH_2OH (55)[e]	282
	3,4-Di-O-acetyl-D-arabinal	−70°, $CFCl_3$	(3,4-di-O-acetyl-2-fluoro-pyranose with X) X = F (—), X = OCF_3 (—)	385a
	3,4-Di-O-acetyl-D-xylal	−70°, $CFCl_3$	(3,4-di-O-acetyl-pyranose with X, Y) X: F····, Y: CF_3O··· (5); X: F····, Y: F···· (5); X: F—, Y: CF_3O— (26)[h]; X: F—, Y: F— (42)[h]	385
C_{10}	Naphthalene	2 moles CF_3OF, $CFCl_3$, CaO, −78°	(tetrahydronaphthalene with F, OCH_3 / F, OCH_3), (tetrahydronaphthalene with F, OCH_3 and F, F),	278b

Note: References 301–386 are on pp. 404–406.

[a] The hypofluorite is CF_3OF unless otherwise specified.
[b] Di- and tri-fluorosubstitution products are also reported for ethylamine and other amines.
[c] The product was isolated from workup with 10% triethylamine in 50% aq methanol.
[d] An unstable intermediate is obtained that converts to 5-fluorouracil. The 85% yield of 5-fluorouracil includes the decomposition of this intermediate.
[e] No reaction without irradiation.
[f] The product was isolated as *trans*-3-methyl-L-proline by hydrolysis.
[g] The position of fluorine substitution was not determined; di- and tri-fluoro products were also formed. Polycaprolactam was also fluorinated.
[h] Structure given in second publication did not agree with that in preliminary communication and appeared to be in error.

TABLE XX. FLUORINATION WITH PERFLUOROALKYL HYPOFLUORITES (*Continued*)

	Reactant	Condition(s)[a]	Product(s) and Yield(s) (%)	Refs.
C_{10} (*contd.*)	Naphthalene (*contd.*)	2 moles CF_3OF, $CFCl_3$ CaO, −78°	1,2,3,4-Tetrafluoro-1,2,3,4-tetrahydronaphthalene (−)	
	1-Aminoadamantane	−78°, HF, *hν*	1-Amino-4-fluoroadamantane (27), 1-amino-3-fluoroadamantane (25)	277 383a
	3′,5′-Di-O-acetyl-2′-deoxyuridine	—	5-Fluoro-uracil N-nucleoside with OH and CH_2OH (55)[e]	281
C_{12}	3,4,6-Tri-O-acetyl-D-galactal	−70°, $CFCl_3$	2-F, 3,4-OAc, 6-CH_2OAc pyranose with X at C-1; X = F (37–55); X = OCF_3 (39–40)	386, 385a

	Reactant	Conditions[a]	Product(s) and Yield(s) (%)	Refs.
C_{12} (*contd*).	3,4,6-Tri-O-acetyl-D-glucal	−78°, $CFCl_3$	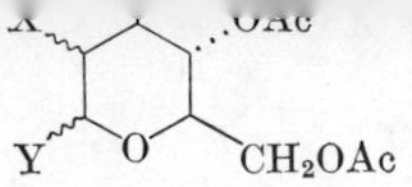 (pyranose ring with X, OAc, Y, CH_2OAc) X = F· · ·, Y = CF_3O· · · (26); X = F· · ·, Y = F· · · (34); X = F—, Y = CF_3O— (6); X = F—, Y = F— (8)	280
	N-Acetyl-α-naphthylamine	1:1 $CHCl_3$, $CFCl_3$, CaO, −78°	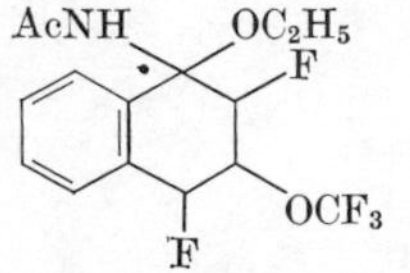 (AcNH, OC_2H_5, F, OCF_3, F substituted tetrahydronaphthalene) (−)	276a, 278b
	N-Acetyl-β-naphthylamine	$CHCl_3$/$CFCl_3$, CaO, −75°	1-$FC_{10}H_6$NHAc-2 (>50)	276a, 278b
C_{14}	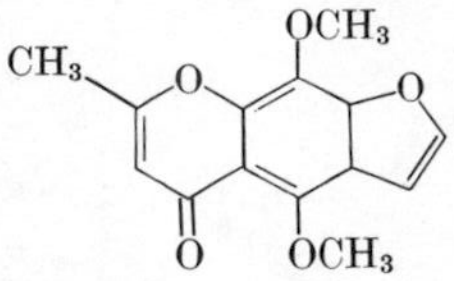 (CH_3, OCH_3, OCH_3 furochromone) (khellin)	$CFCl_3$, −75°	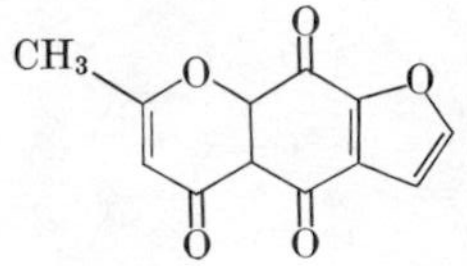 (CH_3 furochromone quinone) (−)	276a
	cis-Stilbene	2:3 $CHCl_3$/$CFCl_3$, CaO, −78°	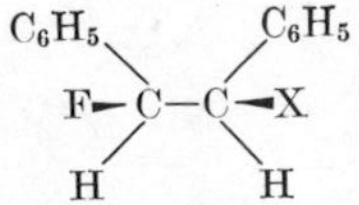 (C_6H_5, C_6H_5, F, X, H, H) Mixture of X = F (—), OCF_3 (—)	245, 278b
	trans-Stilbene	2:3 $CHCl_3$/$CFCl_3$, CaO, −78°	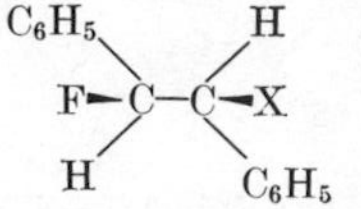 (C_6H_5, H, F, X, H, C_6H_5) Mixture of X = F (—), OCF_3 (—)	245, 278b

Note: References 301–386 are on pp. 404–406.

[a] The hypofluorite is CF_3OF unless otherwise specified.
[e] No reaction without irradiation.

TABLE XX. Fluorination with Perfluoroalkyl Hypofluorites (*Continued*)

	Reactant	Condition(s)[a]	Product(s) and Yield(s) (%)	Refs.
	$C_6H_5C{\equiv}CC_6H_5$	[i]	$C_6H_5CF_2CF(OCF_3)C_6H_5$ (—)	245, 279
	$(p\text{-}ClC_6H_4)_2C{=}CCl_2$	$CFCl_3$, −20°, 90 min	$(p\text{-}ClC_6H_5)_2(CF_3O)CCFCl_2$ (—)	278b
	4-Nitro-4′-acetamidodiphenyl sulfone	CF_3CO_2H, −12°, *hν* 2.5 hr	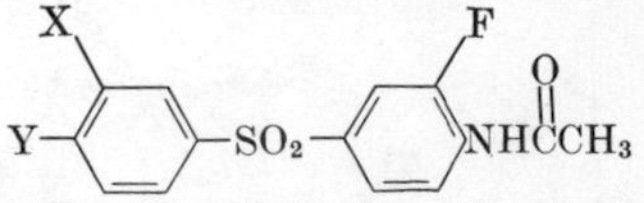 **A**, Y = NO_2, X = H (—)	383a
C_{15}	2′,3′,5′-Tri-O-acetyluridine	$CHCl_3/CFCl_3$, −78°	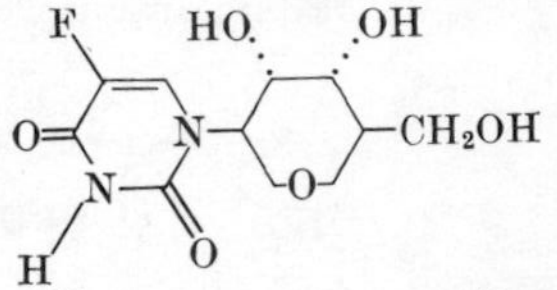 (80)	281
C_{16}	4,4′-Bis(acetamino)diphenyl sulfone	CF_3CO_2H, −12°, *hν*, 2.5 hr	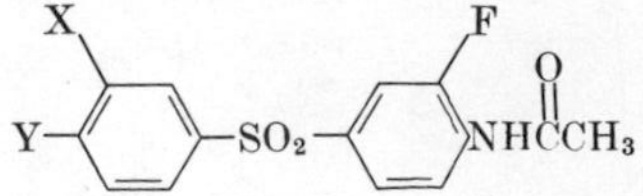 **A**, Y = $CH_3\overset{O}{\overset{\|}{C}}NH$, X = H, X = F (1:1 mixture) (—)	383
C_{17}	4-Acetamido-1-(2,3,5-tri-O-acetyl-β-D-arabinofuranosyl)-2-pyrimidinone	$CHCl_3$, −78°	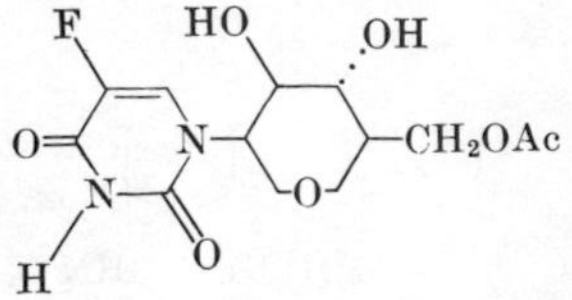 (83)[c]	282
C_{17}	Grieseofulvin (X = Y = H)	$CFCl_3$, CaO, −78° (also 1:1 $CHCl_3/CFCl_3$, CaO, −78°)	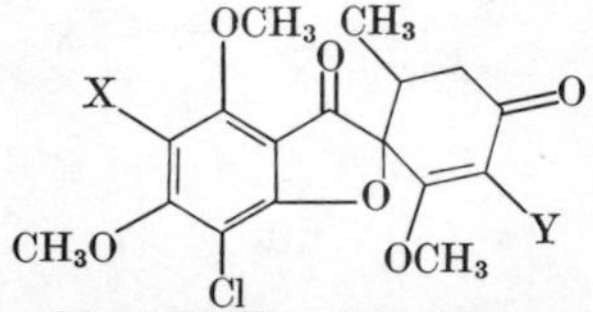 X = F, Y = H (67) X = H, Y = F (15)	276b, 278b

	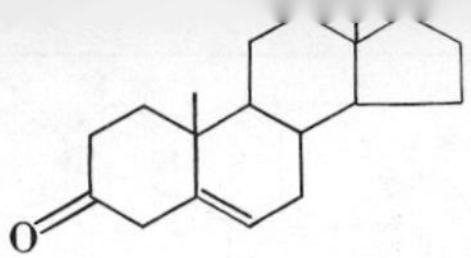 (structure)	$CFCl_3$, CaO, −78°	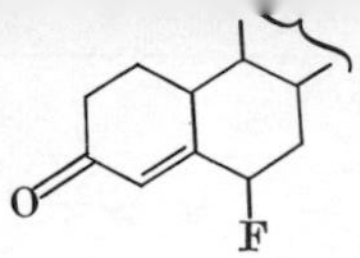 (structure) (--)	278b
C_{19}	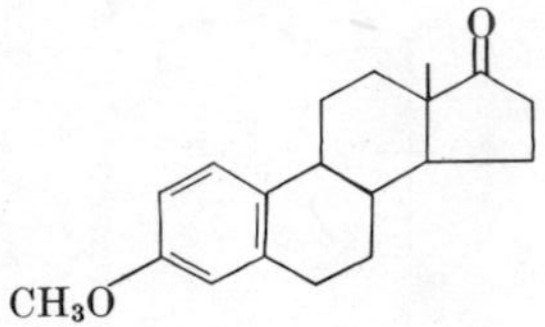 (structure) CH_3O	$CFCl_3$	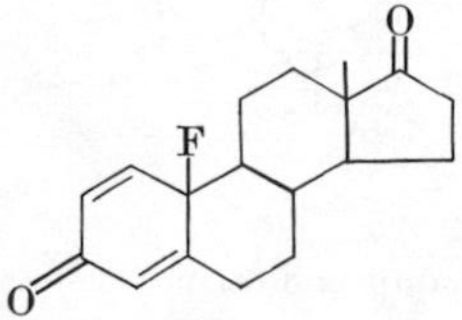 (structure) (—) **B**	276a
C_{20}	Estrone acetate	$CFCl_3$, −75°	**B** (—)	276a
	2-Methyl-1-(*p*-methylsulfinylbenzyl)-indene-3-acetic acid	$CFCl_3$, −15°, $h\nu$, 15 min	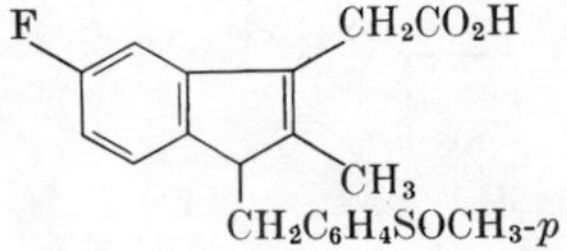 F, CH_2CO_2H, CH_3, $CH_2C_6H_4SOCH_3$-*p* (structure)	383*a*
C_{21}	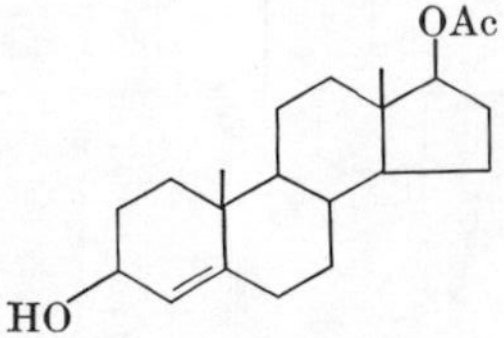 (structure) OAc, HO	$CFCl_3$, −75°	Complex mixture	279

Note: References 301–386 are on pp. 404–406.

[a] The hypofluorite is CF_3OF unless otherwise specified.
[c] The product was isolated from workup with 10% triethylamine in 50% aq methanol.
[i] No conditions were stated; probably CF_3OF was used at −75° in $CFCl_3$.
[j] This product was obtained from an alkaline workup.

TABLE XX. FLUORINATION WITH PERFLUOROALKYL HYPOFLUORITES (*Continued*)

	Reactant	Condition(s)[a]	Product(s) and Yield(s) (%)	Refs.
	Testosterone acetate	$CFCl_3$, CaO, −78°	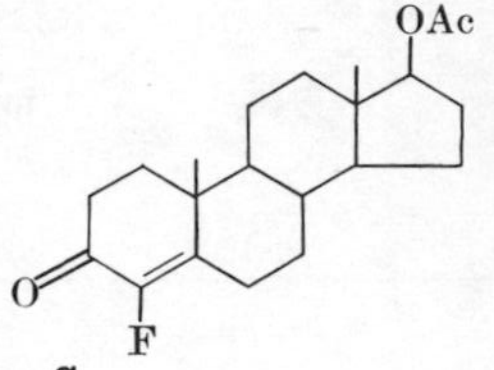 OAc; O; F (—)[j]; **C**	278b, 279
C_{22}	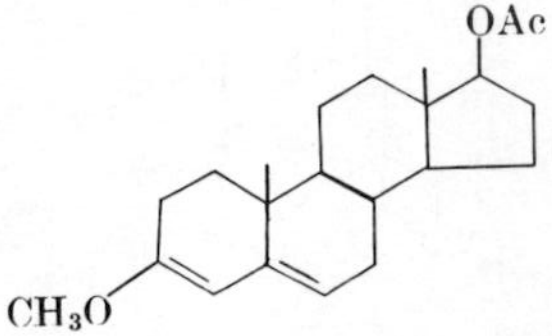 OAc; CH_3O	$CFCl_3$, −75°	**C** 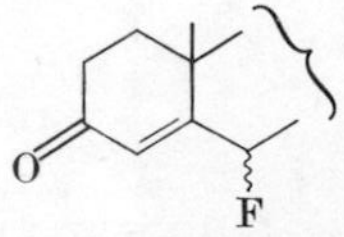 (—), O; F (−)	278
C_{23}	Pregnenolone acetate	$CFCl_3$, CaO, −75°	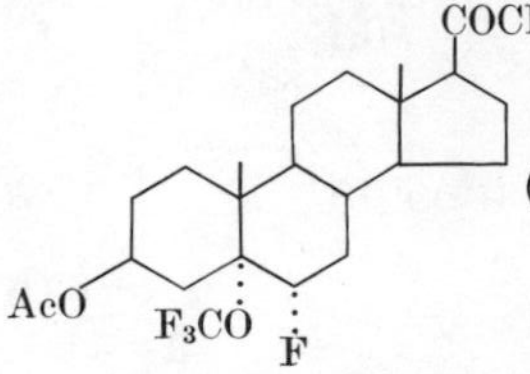 $COCH_3$; AcO; F_3CO; F (Modest yield), rearranged fluorinated products	279
	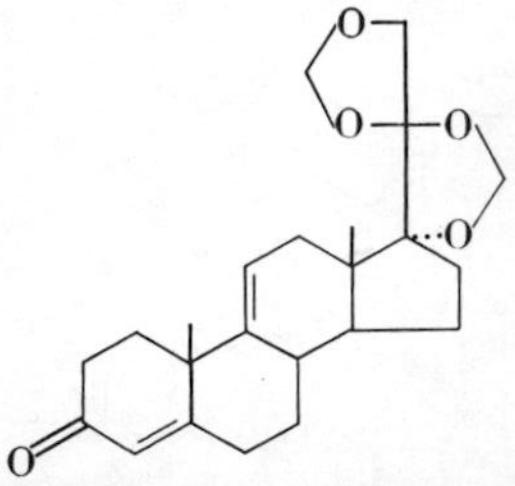 O; O; O; O; O	1:5 $CHCl_3/CFCl_3$, CaO, −78°	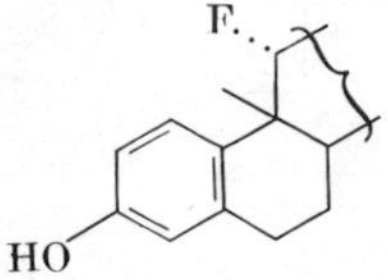 F; HO (Major product)	278b 279

	Reactant	Conditions[a]	Product(s) and Yield(s) (%)	Refs.
C_{23} (*contd.*)	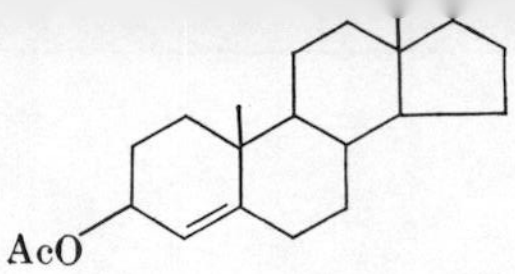 AcO	$CFCl_3$, CaO, −78°, 2 hr	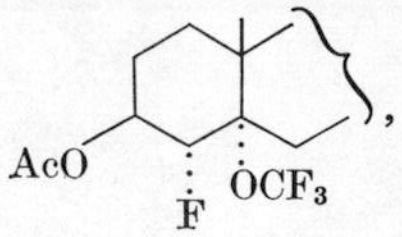 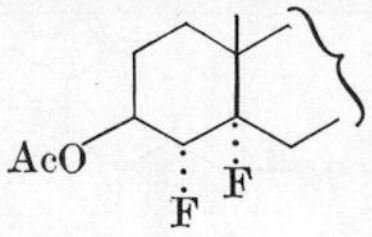 AcO, F, OCF_3 (major), AcO, F, F (minor)	278b, 279
	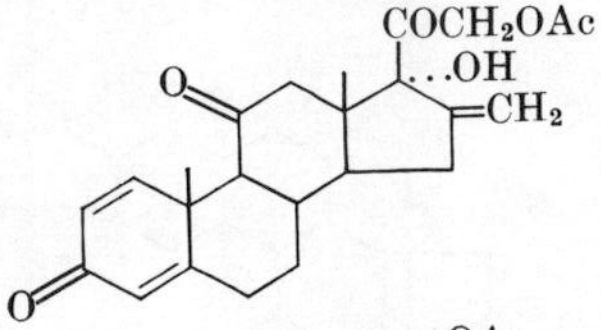 $COCH_2OAc$, ...OH, $=CH_2$, O, O	$CFCl_3$, −75°	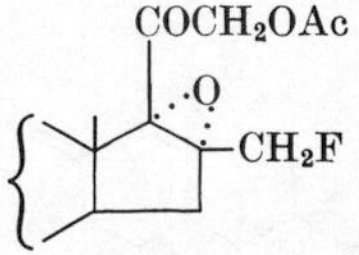 $COCH_2OAc$, ..O, CH_2F (−)	279
	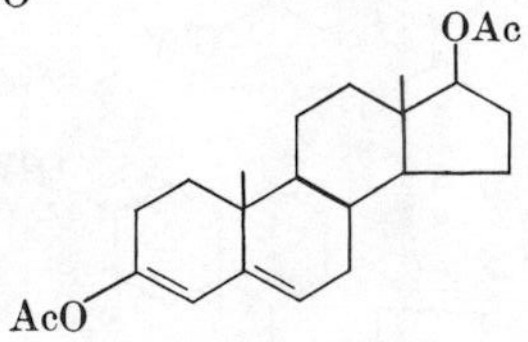 OAc, AcO	$CFCl_3$, CaO −75°	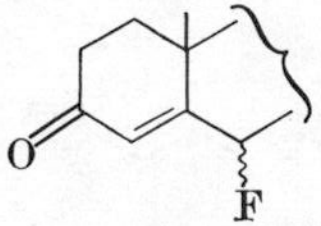 O, F (−), Mixture of 6F— and 6F···; 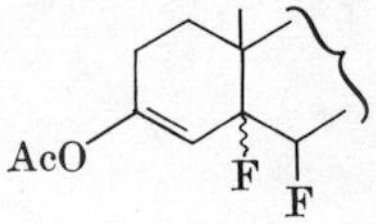 AcO, F, F (−)	278
C_{24}	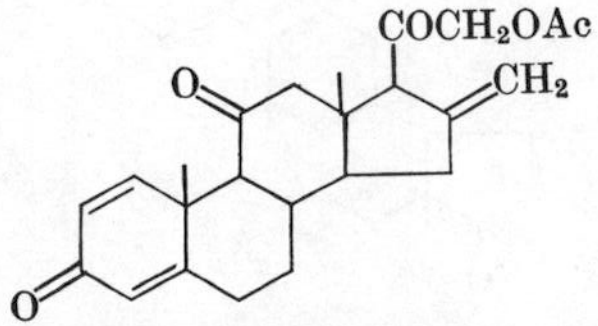 $COCH_2OAc$, CH_2, O, O	$CFCl_3$, CaO, −78°	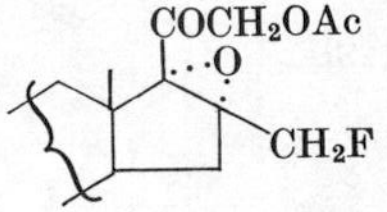 $COCH_2OAc$, ..O, CH_2F (—)	278b

Note: References 301–386 are on pp. 404–406.

[a] The hypofluorite is CF_3OF unless otherwise specified.

TABLE XX. FLUORINATION WITH PERFLUOROALKYL HYPOFLUORITES (*Continued*)

Reactant	Condition(s)[a]	Product(s) and Yield(s) (%)	Refs.
C_{25} (OAc; N-pyrrolidinyl)	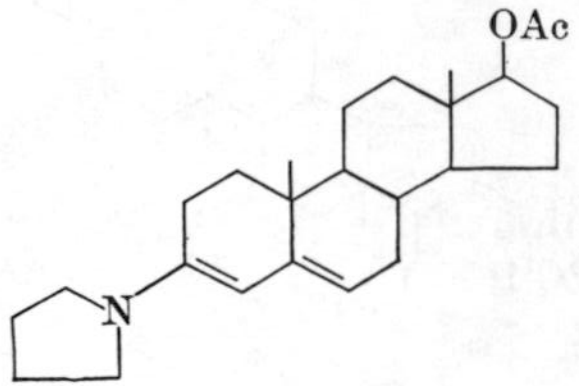 $CFCl_3$, CaO, −75°	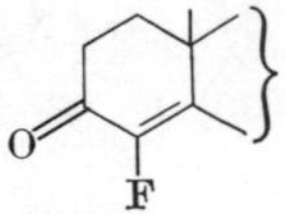 O; F (35)	278
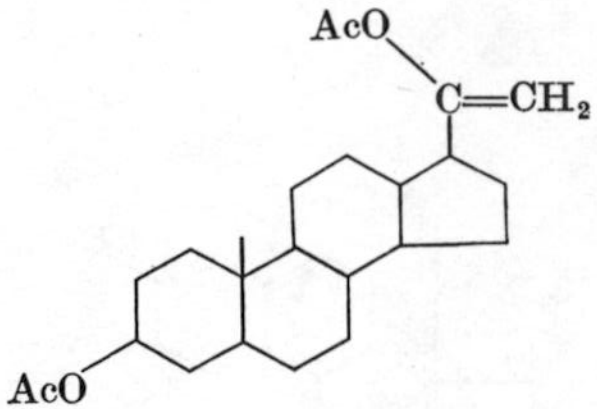 (AcO; $C{=}CH_2$; AcO)	$CFCl_3$, CaO, −78°	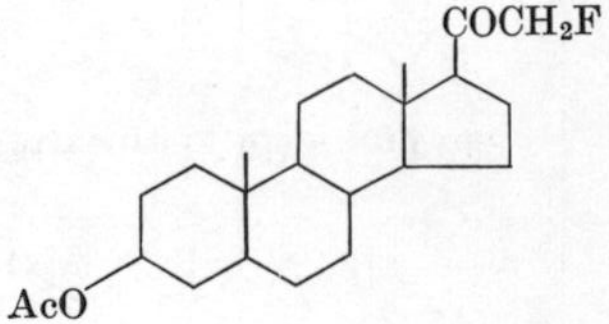 $COCH_2F$; AcO (−)[j]	278b
C_{27} (AcO; AcO; AcO; H)	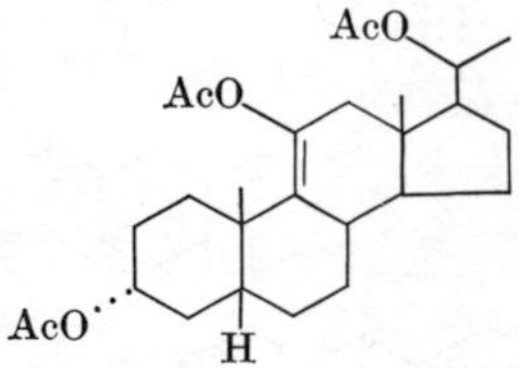 2:1 $CFCl_3/CH_2Cl_2$, CaO, −78°, 2 hr	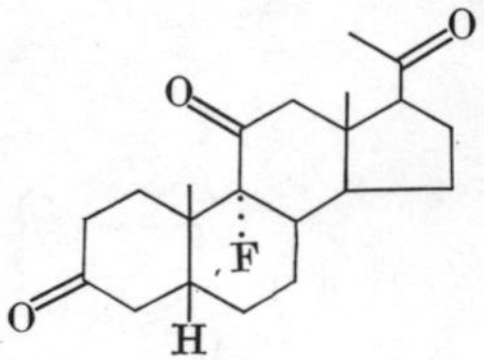 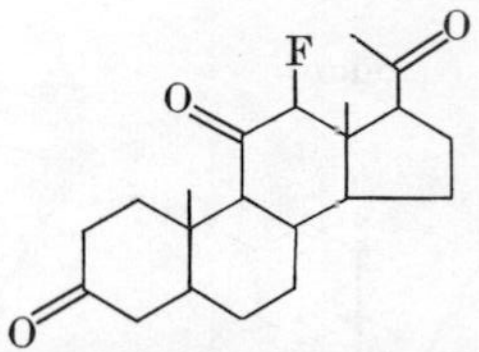 O[j]; O; F; O; H (Good),	278
		O; F; O; O (−)[j]	

Steroid	Reactant	Conditions	Product(s) and Yield(s) (%)	Refs.
	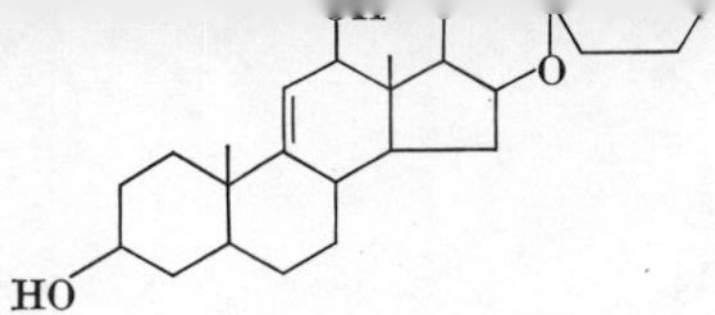 HO (steroid, spiroketal)	$CFCl_3$, −75°	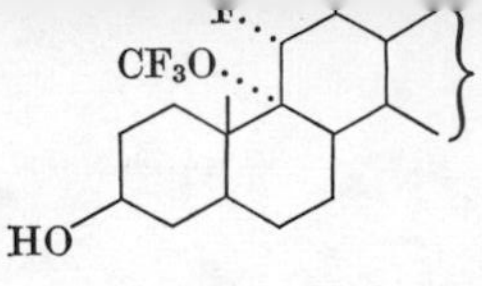 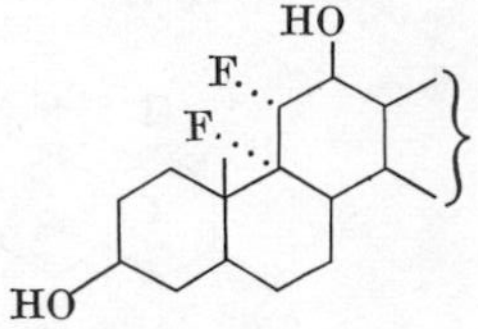 CF_3O, F, HO (major), HO, F, F, HO (minor)	279
C_{28}	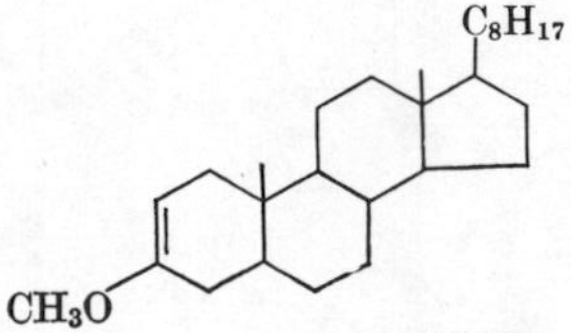 C_8H_{17}, CH_3O	$CFCl_3$, CaO, −75°	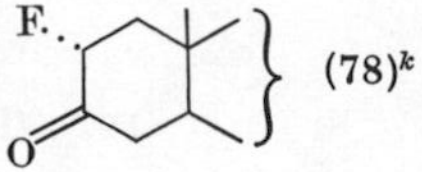 F, O (78)[k]	278
C_{29}	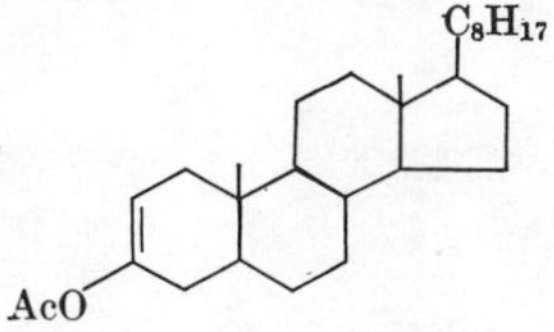 C_8H_{17}, AcO	$CFCl_3$, CaO, −75°[l]	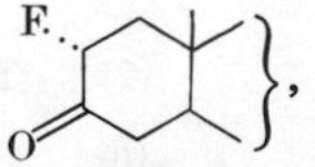 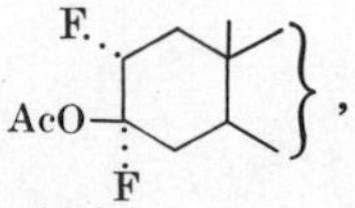 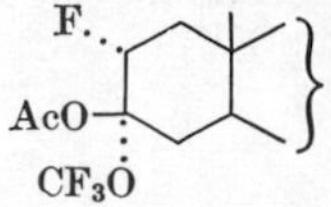 F, O (—), F, AcO, F (—), F, AcO, CF_3O (—)	278

Note: References 301–386 are on pp. 404–406.

[a] The hypofluorite is CF_3OF unless otherwise specified.

[j] This product was obtained from the alkaline workup.

[k] An unstable nonketonic adduct also formed and decomposed to 2α-fluorocholestanone.

[l] $CFCl_3$ is the preferred solvent, but CCl_4, $CHCl_3$, or CH_2Cl_2 can be added to improve solubility. It is also stated that acetone, ether, methanol, or tetrahydrofuran can be used with caution.

TABLE XX. FLUORINATION WITH PERFLUOROALKYL HYPOFLUORITES (*Continued*)

Reactant	Condition(s)[a]	Product(s) and Yield(s) (%)	Refs.
C_{29} (*contd.*)	$(CF_3)_2CFOF$, $-75°$, $CFCl_3$ (CaO, 1 hr)	**D** 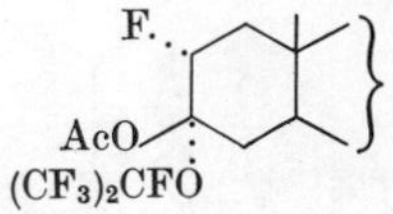 (—), F, AcO, $(CF_3)_2CFO$ (—)[k]	278
	$(CF_3)_3COF$, (2 mol equiv), $-75°$, $CFCl_3/CH_2Cl_2$	**D** (40)[m]	273
	$(CF_3)_2(C_2F_5)COF$, (4 mol equiv), $-75°$, $CFCl_3/CH_2Cl_2$	**D** (45)[l]	273
	SF_5OF (1 mol equiv), $-75°$, $CFCl_3/CH_2Cl_2$	**D** (60)	273
	$CF_2(OF)_2$ (0.5 mol equiv), $-75°$, $CFCl_3/CH_2Cl_2$	**D** (65)[m]	273
	CF_3OF (1 mol equiv), $-75°$, $CFCl_3/CH_2Cl_2$, CaO	**D** (62)	273
C_{30} 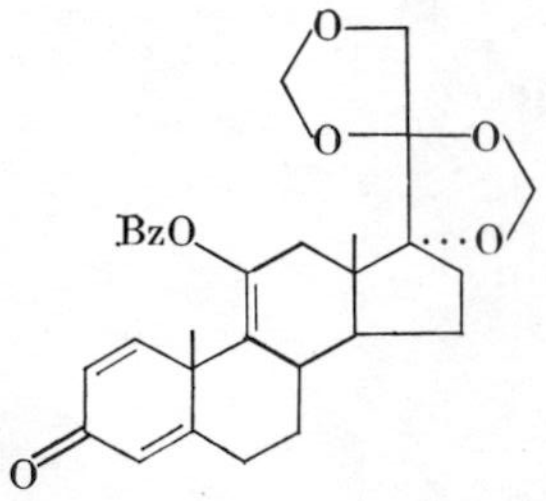 BzO	CF_3OF, 1 mol equiv, $-78°$, $CFCl_3/CH_2Cl_2$, CaO	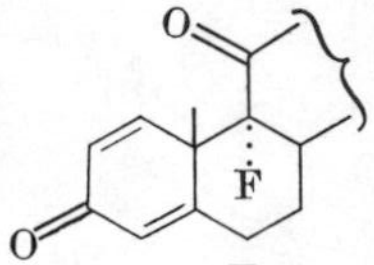 O, O, F **E** (52)[j]	273
	$(CF_3)_2(C_2F_5)COF$ (large excess), $-78°$, $CFCl_3/CH_2Cl_2$	**E** (Trace)[j]	273
	SF_5OF (1 mol equiv), $-78°$, $CFCl_3/CH_2Cl_2$	**E** (49)[j]	273

	Substrate	Conditions	Product(s) and Yield(s) (%)	Refs.
	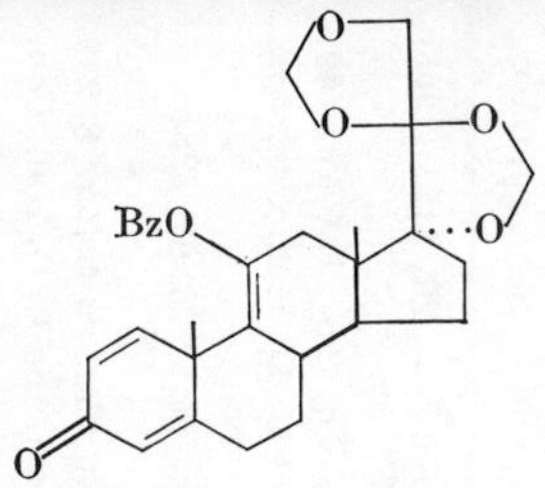 BzO ... O O O O (steroid structure)	$CF_2(OF)_2$ (0.5 mol equiv), $-78°$, $CFCl_3/CH_2Cl_2$	**E** (51)[j]	273
C_{31}	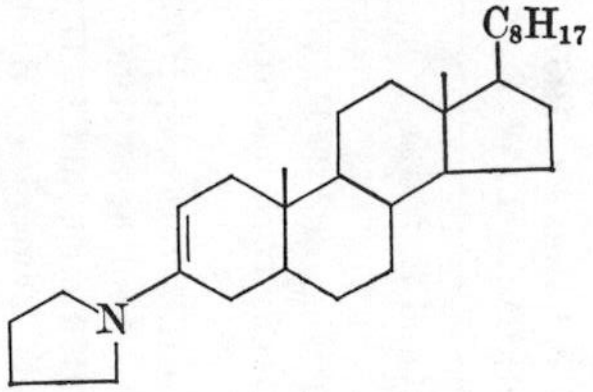 C_8H_{17}, N (steroid structure)	$CFCl_3$, CaO, $-75°$	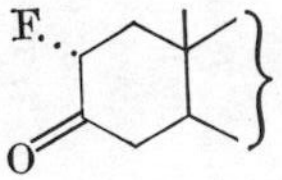 F, O (structure) (−)	278

Note: References 301–386 are on pp. 404–406.

[a] The hypofluorite is CF_3OF unless otherwise specified.

[e] No reaction without irradiation.

[j] This product was obtained from an alkaline workup.

[l] $CFCl_3$ is the preferred solvent, but CCl_4, $CHCl_3$, or CH_2Cl_2 can be added to improve solubility. It is also stated that acetone, ether, methanol, or tetrahydrofuran can be used with caution.

[m] The addition product was observed but not fully characterized.

REFERENCES TO TABLES X–XX

[301] C. L. Thomas, U.S. Pat. 2,673,884 (1954) [*C.A.*, **49**, 4007 (1955)].

[302] H. Schmidt and H. Meinert, *Angew. Chem.*, **72**, 493 (1960).

[303] (a) M. Hudlicky, *J. Fluorine Chem.*, **2** (1972/73); (b) P. W. Kent and J. E. G. Barnett, *J. Chem. Soc.*, Suppl. II, **1964**, 6196.

[304] K. R. Wood, P. W. Kent, and D. Fisher, *J. Chem. Soc.*, *C*, **1966**, 912.

[305] P. W. Kent and M. R. Freeman, *J. Chem. Soc.*, *C*, **1966**, 910.

[306] A. Bowers, U.S. Pat. 3,139,446 (1964) [*C.A.*, **61**, 10744 (1964)].

[307] H. Reimann and D. H. Gould, U.S. Pat. 3,009,928 (1961) [*C.A.*, **56**, 6049c (1962)].

[308] H. Reimann and D. H. Gould, U.S. Pat. 3,009,929 (1961) [*C.A.*, **56**, 11672e (1962)].

[309] H. Reimann and D. H. Gould, U.S. Pat. 3,009,932 (1961) [*C.A.*, **56**, 10245c (1962)].

[310] H. Laurent, K. Prezewowsky, H. Hofmeister, and R. Wiechert, Ger. Pat. 2,107,315 (1972) [*C.A.*, **77**, 126934r (1972)].

[311] A. Cohen and E. D. Bergmann, *Tetrahedron*, **22**, 3545 (1966).

[312] Z. Arnold, *Collect. Czech. Chem. Commun.*, **28**, 2047 (1963).

[313] G. Kowollik, K. Gaertner, and P. Langen, *Tetrahedron Lett.*, **1969**, 3863.

[314] B. Cavalleri and A. Cometti, *Farm. Ed. Sci.*, **25**, 565 (1970) [*C.A.*, **73**, 130745q (1970)].

[315] S. Landa, J. Burkhard, and J. Vais, *Z. Chem.*, **7**, 388 (1967).

[316] R. L. Autrey and P. W. Scullard, *J. Amer. Chem. Soc.*, **90**, 4924 (1968).

[317] (a) D. E. Ayer, U.S. Pat. 3,105,078 (1963) [*C.A.*, **60**, 427b (1964)]; (b) *J. Med. Chem.*, **6**, 608 (1963).

[318] (a) L. H. Knox, U.S. Pat. 3,257,426 (1966) [*C.A.*, **65**, 13796d (1966)]; (b) U.S. Pat. 3,184,484 (1965) [*C.A.*, **63**, 11676b (1965)].

[319] L. H. Knox, U.S. Pat. 3,417,111 (1968) [*C.A.*, **70**, 68628m (1969)].

[320] D. E. Ayer, U.S. Pat. 3,169,133 (1965) [*C.A.*, **62**, 13211b (1965)].

[321] (a) J.-L. Borgna and M. Mousseron-Canet, *Bull. Soc. Chim. Fr.*, **1970**, 2210; (b) **1970**, 2218.

[322] L. H. Knox, U.S. Pat. 3,210,388 (1965) [*C.A.*, **64**, 786b (1966)].

[323] M. Mousseron and J.-L. Borgna, Fr. Pat. 1,602,905 (1971) [*C.A.*, **75**, 141055g (1971)].

[324] D. E. Ayer, U.S. Pat. 3,118,894 (1964) [*C.A.*, **60**, 12072d (1964)].

[325] P. Crabbé, H. Carpio, and E. Velarde, *Chem. Commun.*, **1971**, 1028.

[326] M. E. Wolff and W. Ho, *J. Org. Chem.*, **32**, 1839 (1967).

[327] J.-C. Brial and M. Mousseron-Canet, *Bull. Soc. Chim. Fr.*, **1969**, 1758.

[328] C. Wilante and L. Deffet, Belg. Pat. 738,964 (1970) [*C.A.*, **73**, 109261u (1970)].

[329] W. J. Gensler, Q. A. Ahmed, and M. V. Leeding, *J. Org. Chem.*, **33**, 4279 (1968).

[330] K. Wallenfels and F. Witzler, *Tetrahedron*, **23**, 1359 (1967).

[331] E. Elkik, *Bull. Soc. Chim. Fr.*, **1964**, 2254.

[332] J.-J. Delpuech and C. Beguin, *Bull. Soc. Chim. Fr.*, **1967**, 791.

[333] G. Schröder, G. Kirsch, J. F. M. Oth, R. Huisgen, W. E. Konz, and U. Schnegg, *Chem. Ber.*, **104**, 2405 (1971).

[334] B. Cavalleri, E. Bellasio, G. G. Gallo, and E. Testa, *Farm., Ed. Sci.*, **23**, 1127 (1968) [*C.A.*, **71**, 12684e (1969)].

[335] (a) L. D. Hall, R. N. Johnson, A. B. Foster, and J. H. Westwood, *Can. J. Chem.*, **49**, 236 (1971). (b) A. B. Foster, J. H. Westwood, B. Donaldson, and L. D. Hall, *Carbohyd. Res.*, **25**, 228 (1972).

[336] N. B. Chapman, R. M. Scrowston, and R. Westwood, *J. Chem. Soc.*, *C*, **1967**, 528.

[337] E. D. Bergmann and A. Cohen, *Isr. J. Chem.*, **5**, 15 (1967).

[338] (a) J. S. Brimacombe, A. B. Foster, R. Hems, and L. D. Hall, *Carbohyd. Res.*, **8**, 249 (1968); (b) J. S. Brimacombe, A. B. Foster, R. Hems, J. H. Westwood, and L. D. Hall, *Can. J. Chem.*, **48**, 3946 (1970).

[339] P. Tannhauser, R. J. Pratt, and E. V. Jensen, *J. Amer. Chem. Soc.*, **78**, 2658 (1956).

[340] (a) L. Hough, A. K. Palmer, and A. C. Richardson, *J. Chem. Soc.*, *Perkin I*, **1972**, 2513. (b) L. Hough, A. K. Palmer, and A. C. Richardson, *J. Chem. Soc.*, *Perkin I*, **1973**, 784.

[341] (a) G. Etzold, R. Hintsche, G. Kowollik, and P. Langen, *Tetrahedron*, **27**, 2463 (1971);

(b) L. D. Hall and P. R. Steiner, *Can. J. Chem.*, **48,** 2439 (1970).

[342] H. Laurent, K. H. Kolb, and R. Wiechert, Ger. Offen. 2,010,458 (1971) [*C.A.*, **75,** 141062 g (1971)].

[343] J. Pacák, J. Podešva, and M. Černý, *Chem. Ind.* (London), **1970,** 929.

[344] G. Stork, W. S. Worrall, and J. J. Pappas, *J. Amer. Chem. Soc.*, **82,** 4315 (1960).

[345] J. F. Codington, I. L. Doerr, and J. J. Fox, *J. Org. Chem.*, **29,** 558, 564 (1964).

[346] J. Cantacuzène and R. Jantzen, *Tetrahedron,* **26,** 2429 (1970).

[347] J. Pacák, Z. Točík, and M. Černý, *Chem. Commun.*, **1969,** 77.

[348] N. F. Taylor, R. F. Childs, and R. V. Brunt, *Chem. Ind.* (London), **1964,** 928.

[349] R. E. Counsell and P. D. Klimstra, U.S. Pat. 2,980,710 (1961) [*C.A.*, **55,** 24838e (1961)].

[350] F. L. Weisenborn, U.S. Pat. 2,950,289 (1960) [*C.A.*, **56,** 6057d (1962)].

[351] J. S. Mihina, U.S. Pat. 2,992,241 (1961) [*C.A.*, **55,** 26040i (1961)].

[352] J. Fried and J. E. Herz, U.S. Pat. 2,963,492 (1955) [*C.A.*, **56,** 6045i (1962)].

[353] J. Fried, J. E. Herz, E. F. Sabo, A. Borman, F. M. Singer, and P. Numerof, *J. Amer. Chem. Soc.*, **77,** 1068 (1955).

[354] J. E. Herz, J. Fried, and E. F. Sabo, *J. Amer. Chem. Soc.*, **78,** 2017 (1956).

[355] R. F. Hirschmann and R. Miller, Ger. Pat. 1,035,133 (1958) [*C.A.*, **55,** 3657f (1961)].

[356] P. A. Diassi, J. Fried, R. M. Palmere, and E. F. Sabo, *J. Amer. Chem. Soc.*, **83,** 4249 (1961).

[357] B. J. Magerlein, R. D. Birkenmeyer, and F. Kagan, *J. Amer. Chem. Soc.*, **82,** 1252 (1960).

[358] F. von Werder, H. J. Mannhardt, K. H. Bork, H. Metz, K. Brückner, and K. Irmscher, *Arzneim.-Forsch.*, **18,** 7 (1968) [*C.A.*, **69,** 19386b (1968)].

[359] H. C. Neumann, G. O. Potts, W. T. Ryan, and F. W. Stonner, *J. Med. Chem.*, **13,** 948 (1970).

[360] G. Muller and R. Bardoneschi, Fr. Pat. 1,224,139 (1960) [*C.A.*, **56,** 6043 (1962)].

[361] R. Deghenghi and R. Gaudry, *Can. J. Chem.*, **39,** 1553 (1961).

[362] K. Brückner, B. Hampel, and U. Johnsen, *Chem. Ber.*, **94,** 1225 (1961).

[363] R. G. Berg and G. D. Laubach, U.S. Pat. 2,980,670 (1961) [*C.A.*, **55,** 18,816c (1961)].

[364] D. Taub, R. D. Hoffsommer, and N. L. Wendler, *J. Amer. Chem. Soc.*, **78,** 2912 (1956).

[365] J. A. Hogg, F. H. Lincoln, and W. P. Schneider, U.S. Pat. 2,992,244 (1961) [*C.A.*, **55,** 26037a (1961)].

[366] A. Wettstein and G. Anner, Ger. Pat. 1,078,572 (1971) [*C.A.*, **55,** 22382d (1961)].

[367] H. L. Herzog, M. J. Gentles, H. M. Marshall, and E. B. Hershberg, *J. Amer. Chem. Soc.*, **82,** 3691 (1960).

[368] M. J. Kamlet, U.S. Pat. 3,624,129 (1971) [*C.A.*, **76,** 33936g (1972)].

[369] E. F. Witucki, G. L. Rowley, M. Warner, and M. B. Frankel, *J. Org. Chem.*, **37,** 152 (1972).

[370] D. E. Gwynn, G. M. Whitesides, and J. D. Roberts, *J. Amer. Chem. Soc.*, **87,** 2862 (1965).

[371] V. Tolman and K. Veres, *Coll. Czech. Chem. Commun.*, **32,** 4460 (1967).

[372] B. Cavalleri, E. Bellasio, and E. Testa, *Farm., Ed. Sci.*, **24,** 402 (1969) [*C.A.*, **70,** 114773e (1969)].

[373] M. Neeman, Y. Osawa, and T. Mukai, *J. Chem. Soc., Perkin I,* **1972,** 2297.

[374] J. Pataki and G. B. Siade, *J. Org. Chem.*, **37,** 2127 (1972).

[375] W. C. Ripka, U.S. Pat. 3,629,301 (1971) [*C.A.*, **76,** 24604v (1972)].

[376] G. A. Boswell, Jr., and W. C. Ripka, personal communication.

[377] (a) G. A. Boswell, Jr., A. L. Johnson, and J. P. McDevitt, *Angew. Chem., Int. Ed. Engl.*, **10,** 140 (1971); (b) *J. Org Chem.*, **36,** 575 (1971).

[378] Neth. Pat. 7113713 (1972), assigned to E. I. du Pont de Nemours and Co., (≡ W. C. Ripka, U.S. Pat. 3,718,673).

[379] G. A. Boswell, Jr., and W. C. Ripka, U.S. Pat. 3,641,005 (1972) [*C.A.*, **76,** 141213c (1972)].

[380] G. A. Boswell, Jr and W. C. Ripka, U.S. Pat. 3,546,259 (1970) [*C.A.*, **76,** 86012j (1972)].

[381] Neth. Pat. 7105386 (1971), assigned to E. I. du Pont de Nemours and Co.

[382] Neth. Pat. 7200329 (1972), assigned to E. I. du Pont de Nemours and Co.

[383] (a) J. Kollonitsch, Ger. Pat. 2,136,008 (1972) [*C.A.*, **76** 154488w (1972)]. (b) J. Kollonitsch, L. Barash, F. M. Kahan, and H. Kropp, *Nature* **243,** 346 (1973)

[384] R. H. Hesse and D. H. R. Barton, Ger. Pat. 2,149,504 (1972) [*C.A.*, **77,** 126674f (1972)].

[385] (a) R. A. Dwek, P. W. Kent, P. T. Kirby, and A. S. Harrison, *Tetrahedron Lett.*, **1070,** 2987; (b) C. G. Butchard and P. W. Kent, *Tetrahedron*, **27,** 3457 (1971).

[386] J. Adamson and D. M. Marcus, *Carbohyd. Res.*, **13,** 314 (1970).

AUTHOR INDEX, VOLUMES 1–21

CHAPTER AND TOPIC INDEX, VOLUMES 1–21

Many chapters contain brief discussions of reactions and comparisons of alternative synthetic methods which are related to the reaction that is the subject of the chapter. These related reactions and alternative methods are not usually listed in this index.

In this index the volume number is in BOLDFACE, the chapter number in ordinary type.

SUBJECT INDEX, VOLUMES 1–21

Since the table of contents provides a quite complete index, only those items not readily found from the contents pages are listed here.

Numbers in BOLDFACE type refer to experimental procedures.